THE INDUSTRIES OF SCOTLAND

W.K. Ritchie

21. IV. 94

THE INDUSTRIES OF SCOTLAND

Their Rise, Progress and Present Condition

A Reprint of the classic work by

DAVID BREMNER

with a new introduction by
John Butt and Ian L. Donnachie

REPRINTS OF ECONOMIC CLASSICS

AUGUSTUS M. KELLEY, PUBLISHERS
NEW YORK 1969

Published in the United States of America by
Augustus M. Kelley, Publishers, New York
Library of Congress No. 69–11242

This work first appeared as a series of articles
in *The Scotsman* in 1868
First published in book form by Adam & Charles Black in 1869
© 1969 new introduction John Butt, Ian L. Donnachie

Printed in Great Britain by
Redwood Press Limited Trowbridge Wiltshire

INTRODUCTION
TO THE 1969 EDITION

On Monday, 27 January 1868, *The Scotsman* carried the following advertisement on its front page :

> Industries of Scotland—A Series of Articles on this subject, tracing in a Systematic and Comprehensive form the Rise and Progress of the more Important Industrial Occupations carried on in Scotland, describing the more interesting Processes and Appliances, and giving accurate details as to the Condition, Earnings and Education of the Working Classes has been commenced in the Weekly Scotsman and will be continued from Week to Week.

The articles duly appeared each Saturday throughout 1868. None of them was signed, this being regarded as a model of correctness. Their author, David Bremner, evidently retained copyright for he published them in 1869 through the firm of Adam and Charles Black of Edinburgh. About the author little is known save that he is listed in the Edinburgh & Leith Directory 1869–70 (p 22) as sub-editor of 3 Sciennes Hill Place and does not appear after this date. There is no mention of him in the history of *The Scotsman* which is particularly sad, since he pioneered industrial reporting in Scotland and by so doing contributed to the rise of the industrial correspondent.

By comparing his articles with his book, it is clear that Bremner made a number of changes, especially in order. Many articles were linked to form chapters ; numerous additions had been made ; careful revision of the subject-matter had occurred. Thus, Bremner's book marked a return to the tradition of Sir John Sinclair away from the dilettantism of the travellers' accounts in which the previous century had been so prolific.

This explains why his book is so valuable a source for students of Industrial History. And if Business History can be merely defined as the study of enterprises and their *milieu*, he can be claimed as an early pioneer. His purposes, apart from literary and historical, were to allay the agitation—which he thought was misplaced—after the Paris exhibition of 1867 about the state of British industry and to reduce contemporary shrieks of confusion to sensible debate by providing the facts as he saw them in Scotland. His own work, he hoped, would stimulate others to examine the 'political economies' of England and Ireland.

As a commentator on the contemporary industrial scene, Bremner possessed certain clear advantages. First, he was prepared to leave his desk and to see for himself. Obviously able to use the *cachet* provided by the name of his newspaper, he selected the firms of special interest to him and generally secured access to their premises. It was this capacity to impress entrepreneurs and managers which was a prerequisite for the case-studies with which this volume is liberally endowed. Unlike later industrial correspondents such as the writers for *The Mercantile Age*, Bremner did not merely provide an uncritical commercial 'puff' for the businesses which he visited. He delved extensively into earlier printed accounts of industrial conditions, making great use of the *Statistical Accounts*, parish histories, trade journals, blue books and local history society transactions. Not content with this he dabbled in manuscript sources, being one of the earliest to use the records of the Board of Trustees. He enjoyed the advantage of writing at the beginning of a great capital goods boom, which may have stimulated his optimism, after the trading uncertainties of the American Civil War and before the sharp collapse and bank failures of the late 1870s.

Periodicals like the *Penny Magazine* in the 1840s had produced supplements on manufacturing processes, and there were several encyclopaedic volumes in the 1850s, notably Professor Andrew Ure's *Dictionary of Manufactures*, embracing the principal technical developments available to entrepreneurs, but Bremner was unique in his regional orientation and in his regard for the past. A variety of industries, some well-established and others—for example, floor-cloth, india-rubber and oil—recent introductions, were discussed. Nowadays, criticisms are easy to make. Glaswegians will complain, as did the reviewer for *Engineering* (9 July 1869) that Bremner did not do justice to 'the extent and variety of the chemical and engineering industries of Glasgow'. Several volumes would still be needed to make suitable amends, although the British Association handbooks of 1876 and 1901 made gallant attempts. Also, there is little on the institutional framework of Scottish industry and businesses—the odd reference to the importance of particular banks but nothing of real value on the commercial structure. From time to time, Bremner mentions business amalgamation, for instance in his chapter on railways, because of its recent news-worthiness, but on the general question of the concentration of capital, he is silent. This is surprising because some of the firms he visited were prime examples of the process, and Scottish law encouraged the company form in business and, indirectly, interlocking directorships. He also occasionally concerns himself with the elaborate delineation of capital formation within specific sectors and sometimes illustrates through his case-studies his knowledge of the financial structure. On the subject of profits in business, he is diplomatically mute. For these and other reasons he will not entirely satisfy the severe scrutiny and rigorous requirements of modern economic historians—but how many of their precursors would they leave uncriticized?

Bremner's account of the Scottish coal-mining industry evidently owed much to the blue books, especially to the *Royal Commission on the Employment of Children in Mines* (1843) and to the *Reports of the Mines Inspectors*; indeed, he singles out Tremenheere's reports and quotes them. He begins by indicating the extent of dependence of the Scottish economy on coal and goes on to provide an accurate and careful account of coal-winning methods and conditions of labour and safety in mines. Bremner comments at length, in humanitarian fashion, on living conditions in colliery villages—housing, water supply and wages. He attacks the truck system, details the scale of evasion of existing legislation, points out the need for new by describing the scale of extortion and its effects. He describes the progress of unionisation without attacking it, a rare event indeed. For every point he makes, he finds an illustration, and there is an excellent description of his trip to the pit bottom at the Emily pit, Arniston (Midlothian), at that time the deepest pit in the East of Scotland. His regard for statistical detail is to some extent neutralised by the fact that he was writing before the gigantic expansion of the Scottish coal industry which occurred in the thirty years following his book. Whereas Scottish output of coal was just over 12 million tons in 1866, by 1908 it had reached 39 million tons. By 1900 Lanarkshire alone produced more coal than the whole Scottish industry of Bremner's day. If his humanitarianism occasionally cloys through an overdose of condescension (cf p 10: 'Before proceeding to visit the pit, we acted on good counsel, and donned a capacious suit of pilot cloth, which, though of most uncouth cut, proved to be quite an aristocratic costume when brought into contrast with the habiliments of the dusky fellows below') he still somehow conveys a sensible impressionistic commentary on the way miners lived, worked, dressed, wasted their substance, and died.

In his treatment of the iron industry Bremner makes a somewhat artificial division between iron-smelting and iron working. In his assessment of the growth of the Scottish iron industry there is little explanation of the fluctuations in its fortunes and virtually no analysis of the market structure and its effects. But his narrative is painstaking, accurate and punctuated with numerous statistics relating for instance to the number of furnaces, the levels of employment, output, prices, and exports. He also recognised that by 1869 the competitive advantages of the Scottish pig-iron industry were beginning to slip away ; he points to the dangerous extent of competition from Cleveland pig and Cumberland ore. Like some latter-day imaginative tourist to Palermo or Naples who might survey the wonders of Etna or Vesuvius, Bremner from a steeple of the parish church in Coatbridge viewed at nightfall the startling spectacle of the flames from fifty blast furnaces. Yet his romanticism did not obscure his capacity to make a careful and more mundane description of the Gartsherrie ironworks of William Baird & Company. The historian of technology will find much of significance in Bremner's account of the Carron Ironworks, casting and moulding at Falkirk and of malleable iron production at any one of the 400 puddling furnaces which existed at the time of his account.

In his account of the shipbuilding industry, Bremner gave the best history of the development of steam navigation by Scottish engineers as yet in print. Historians interested in this subject who neglect this account do so at their peril, since most other sources uncritically accept the claims of Henry Bell or William Symington without giving due credit to Patrick Miller of Dalswinton and James Taylor of Cumnock who were also important contributors to the economic exploitation of the marine engine. Although he examines other shipbuilding regions of Scotland, Bremner rightly concentrated on Clydeside, thoroughly examining the basis for the growth of this industry there. But like his comments on coal, his account inevitably suffers because he could not have foreseen the rapid growth made by the industry after 1870.

The historian of transport will find considerable detail in Bremner's chapter on the railways of Scotland. Not only had he first-hand knowledge of the structure of this industry which through amalgamations altered shortly before he compiled his account, but also he consulted a wide range of trade journals and newspapers. The consideration of the railway mania of 1845 and the later rationalisation of the industry is perhaps too superficial even if heavily larded with statistics, but at least it provides a responsible starting point for those concerned with the history of Scottish railway companies. The Callander & Oban Railway was the only important line then under construction, although the Highland Railway was being extended. Gone were the wild days of financial speculation in railway shares ; sober, steady working of capital equipment to secure a maximum return on investment was the standard business response. He makes reference to the railway workshops but says little about the private builders. Writing for an Edinburgh newspaper he might be forgiven for spending more space on coach-making, a long-established local industry slowly adapting from the age of the brougham to the age of the horse omnibus.

.Another mainly Edinburgh industry was the manufacture of plate and jewellery. This employed a labour force of about 2000 in 1868 on pay scales which seem surprisingly low, as Bremner points out, for the level of skill required. In a miscellany relating to metal processing and light engineering, Bremner indicates a broad range of industries of growing significance to the British economy—machine tools, machinery of many sorts, copper and brass founding, lead and type-founding. Many of these products reflected the growing sophistication of urban markets and services. For instance, copper and brass founding was linked with the expansion of industries like shipbuilding, brewing and distilling and with the growth of services like gas and water supply and sanitation.

Since we still unfortunately lack a definitive history of the Scottish woollen industry, Bremner's lengthy and detailed account remains the best printed source. His analysis of the structure of the industry in the eighteenth century owes much to Patrick Lindesay's *The Interest of Scotland Considered* (1733) and David Loch's *Essays on the Trade, Commerce, Manufactures and Fisheries of Scotland* (1778), and he follows their view that the Scottish industry, despite its widespread geographical distribution, was faced with an overwhelmingly superior English competitor. Historians of technology and all those concerned with the diffusion of invention will learn much from his account of the mechanisation of the industry, especially from his detailed studies of the Border towns of Galashiels, Hawick, Selkirk, Innerleithen, Jedburgh, Dumfries and Langholm. Apart from describing these centres of tweed manufacture, Bremner considers carpet manufacture, hosiery, blankets and heavy woollens in other industrial centres as far apart as Stirling and Inverness.

Considering its journalistic origins, his chapter on linen and jute is a model of scholarship. It is true that he perhaps misses the significance of the rôle of flax imports on the geographical location of the linen industry, but, leaning heavily on the second edition of Alexander Warden's classic account of the industry (1868), he painstakingly examines all the areas in which production was important. The concentration of the coarse linen industry in Angus, Fife and Perthshire has been commonly noticed by historians, but Bremner had travelled widely and was able to indicate the diversity of production within this region from personal experience : coarse cloth in Kirkcaldy, fine linen and damasks in Dunfermline, coarse linen and jute bagging in Dundee.

A number of inaccuracies mar his account of the Scottish cotton industry. For instance, although David Dale was involved in financing cotton mills at New Lanark, Catrine, Ayr, Blantyre and Stanley, he never owned Rothesay mills. It was another Glasgow banker, Robert Carrick, who was the key figure in this enterprise. However, it should be said in mitigation that Dale in 1785, before New Lanark was in full production, bid £2000 for Rothesay mills, but this was not acceptable to Carrick and his partners. Possibly, Bremner was embarrassed by the wealth of secondary sources, and this may explain why he spends more words on British experience in general than one might expect. Yet his account of the growth of the cotton industry in the West of Scotland is substantially correct. The fine linen trade, the extensive trade in cotton yarn, the weaving of mixtures and the regional attempt to specialise in imitations of and substitutes for East Indian cloths together spawned the new cotton-spinning firms. Thus, calicoes, bengals, zebras and other exotic products became part of the normal weaving out-put of the handlooms of Paisley and Glasgow. It was the demand for fine yarns that brought the two primary cotton areas of Britain—Lancashire and the West of Scotland—together, and the importation of English techniques ushered in the factory age. Highland migration, Irish immigration and the high family-unit size of both groups ensured an adequate supply of labour when Scots were reluctant to serve. Bremner gives a reasonably full account of unionisation in the cotton industry, of attempts at minimum wage agitation and of violence among Glasgow cotton-spinners in the first four decades of the nineteenth century. His most detailed treatment he reserves for the 1860s, when Scotland shared the lean years of the cotton famine caused by interruptions to cotton supply during the American Civil War.

Anachronistic survivals also interested Bremner. A significant remnant of the domestic system was the embroidering of muslins, which, according to the census of 1861, employed over 7200 women, mainly in the West of Scotland. Merchant firms operating from warehouses, usually in Glasgow, put out work over great distances, as their forebears had done for over a century. This passing industrial organisation had suffered several shocks, and in Bremner's time, financial crises in 1857, 1866 and 1879 produced a heavy mortality among the merchant houses concerned in it. This unfortunate industry should be compared with the experience of another former craft, the manufacture of fishing nets. Bremner noticed the beginnings of mechanisation which proceeded apace after 1880 because of the rapid growth of the white-fish industry.

Other textiles, which were to grow in significance after Bremner's day, the manufacture of paper, wallpaper and floor-cloth, he merely notices. The oldest, paper-making, was long established around Edinburgh, and in 1868, nearly 10,000 workers produced 30,000 tons of paper per annum from eighty machines. The two most important manufacturers, Alexander Cowan & Sons of Penicuik and Alexander Pirie & Sons of

Aberdeen, he visited. Urban development and the uneven but definite growth of real incomes encouraged firms in the larger centres of population to cater for changes in consumer demand and domestic fashion. Although the real significance of income movements after 1870 awaits detailed examination—and it may well be that real wages rose but not fast enough, for the good of the economy as a whole—Bremner notices significant pointers to the trend. For instance, he tells us that Wylie & Lochhead included wallpaper manufacture and home decoration in their house-furnishing business from 1861 ; they were later to branch out as undertakers in competition with the 'co-ops'. In 1847 Michael Nairn developed floor-cloth manufacture in Kirkcaldy, and in the twenty years to Bremner's articles it is not surprising that this new industry showed a remarkable rate and range of innovation, showing constant attention to the solution of production problems. By 1878, 1300 people were employed by this industry in Kirkcaldy alone.

Although there is no specific chapter on the Scottish chemical industry, and Bremner did not visit St Rollox, Scotland's greatest heavy chemical plant, there are several devoted to ancillary activities. Bremner's historical method, indeed, can be best seen in his description of calico-printing and turkey-red dyeing. He gives a history, followed by an account of technical changes, and then many details of prevailing capital equipment, employment, wages, and output. He exemplifies, finally, by reference to William Stirling & Sons of Glasgow, the most important Vale of Leven firm. An antique industry was leather, but undergoing transformation. Bremner noted growing concentration within the industry which was most marked in 1867 in the hide market, Glasgow, Edinburgh and Aberdeen being dominant. It is a pity that he did not visit Maybole (Ayrshire) which by 1886 was to have five large factories producing 200,000 pairs of shoes per annum. The day of the small country tannery and shoemaker was virtually over. In contrast, a very new industry, on Bremner's own doorstep, was india-rubber which rapidly developed in Edinburgh. Similarly, the Lothians and Fife saw the rapid expansion of the mineral oil industry, particularly after 1864, when the dominant patent covering James Young's process of dry distillation expired.

It is no reflection on Bremner's work as a whole to declare that his chapter on glass is perhaps the poorest in the book. Those interested in the history of Dumbarton will find no account of what was once the largest glassworks in Scotland. He is much better describing the history and development of the quarrying and earthenware industries. About 10,000 were employed in the manufacture of earthenware, bricks and tiles in 1868, but the scale of activity was markedly different within these branches. Half this labour force was employed in fourteen potteries, the most extensive of which was Bell & Company of Glasgow. The other half was employed in 122 brick and tile works, widely distributed throughout Scotland but particularly important in Lanarkshire and the Lothians. The best account of granite quarrying and polishing, slate working and pavement cutting, associated mainly with Aberdeenshire, Argyll, Angus and Caithness, but often locally prominent elsewhere, is provided here.

Several essays are devoted to the food industries and food processing. The most significant are those concerned with brewing and distilling. Most of the evidence which Bremner used, perhaps inevitably, comes from Edinburgh, but he also details the operation of the Excise, the number of producers at different times, exports and foreign markets. Distilling had been shrouded in romantic mystery, and until Bremner's more prosaic account, most historians, if they mentioned the industry at all, saw it in terms of the heroic Highlander with his sma' still 'tweeking' the nose of the Excise in some secluded glen. Alfred Barnard, a later commentator, was more fortunate than Bremner, because there was a great expansion of the whisky industry after 1870; there was, therefore, more of moment to see, for instance the evolution of the Distillers' Company. Yet Bremner did notice a significant increase in the scale of operations by individual distillers, and his statistics, with which he was, as usual, very generous, show an increase in average output of 250 per cent between 1824 and 1864.

From these comments on Bremner's methodology and subject matter, it should be obvious that this work is still an indispensable source for the modern social and economic historian. Recent bibliographical guides may be consulted in *Studies in Scottish Business History* (edited by P. L. Payne) and in J. Butt, *The Industrial Archaeology of Scotland*. But there is no substitute for Bremner's classic study. As one reviewer put it so pleasantly: 'Mr. Bremner does not profess to treat his subjects with anything approaching to scientific detail, but simply as things which should be "generally known".'

University of Strathclyde, 1968

John Butt
Ian L. Donnachie

THE

INDUSTRIES OF SCOTLAND

THEIR

RISE, PROGRESS, AND PRESENT CONDITION.

BY

DAVID BREMNER.

EDINBURGH:
ADAM AND CHARLES BLACK.
MDCCCLXIX.

PREFACE.

A word of preliminary explanation of the object and scope of this book may not be superfluous, the more so that it takes a line which I am not aware has been taken in any previously published work. I have sought to outline the history of such branches of Scotch industry as merit notice by their extent or other peculiarity, to track out their modest beginnings, and to follow their subsequent development. This is what Bacon called mechanical history, or the history of industrial arts; and I venture to say that the history of Scotch industry is peculiarly rich in that profitable knowledge which Bacon held to belong to such investigations. The main object of the book, however, is to describe the actual state of the chief branches of industry in Scotland; and I thought I should best accomplish this by restricting myself to a plain narrative of judiciously chosen facts. The reader must expect to meet with few general reflections: it is hoped that many of these will be suggested without prompting on the part of the author.

Another consideration has been kept in view. Since the Paris Exhibition, which revealed the surprising progress made of late years by our foreign competitors in the industrial arts, there has been much lively discussion on technical education. The discussion would be much more profitable were the disputants more correctly informed of the actual state of, and progress recently made in, the industries of Great Britain.

I trust that this book may be of some little service in imparting such information, so far as Scotland is concerned, and that others may do for England and Ireland what I have attempted to do for Scotland.

It is proper to mention that the substance of the following pages originally appeared in a series of articles printed during last year in the weekly issue of the *Scotsman* newspaper. The articles were most favourably received as fair and accurate accounts of the branches of trade to which they related, and it is in accordance with a generally expressed desire that they are now reprinted in a more permanent form. The text has been subjected to careful revision; numerous additions have been made; and where it was considered essential, the latest statistics have been given.

D. B.

EDINBURGH, *March* 29, 1869.

CONTENTS.

INDUSTRIES OF SCOTLAND.

COAL MINING.

THE EARLY HISTORY OF COAL—OBJECTIONS TO ITS BEING USED AS FUEL—FIRST
ATTEMPTS AT COAL MINING—SLAVERY IN THE MINES—THE SCOTCH COAL-
FIELDS—VISIT TO A COLLIERY—DESCENT INTO A PIT—THE MINERS AT WORK
—PERILS OF THE PITS—SOCIAL CONDITION OF THE MINERS—THEIR EARN-
INGS, STRIKES, AND UNIONS—THE HOUSES IN WHICH THEY LIVE—THE
MEANS PROVIDED FOR EDUCATING THEIR CHILDREN.

WHEN it is considered how much the manufacturing interests of
the country and many of the comforts of life depend upon coal, it
becomes easy to understand the anxiety evinced by political econo-
mists as to the results that would probably follow the exhaustion of
the supply of that material. From coal we derive the force which
turns the mill, propels the steamboat, draws the railway train, and
performs a thousand other offices tending to economise time, lessen
labour, and increase and multiply our enjoyments; and even a temporary stoppage of the supply would be one of the greatest calamities
that could befal us. It is within a comparatively recent period in
the history of the country, however, that coal has risen into importance. Its existence and combustible qualities were known in very
early times; but beyond being regarded as a curiosity, no attention
seems to have been paid to it; and up till about six centuries ago,
no attempt was made to use it as fuel.

The earliest documents in which it is mentioned are "The Saxon
Chronicle of Peterborough," written in the year 852, and Bishop
Pudsey's "Boldon Book," dated 1180. Newcastle coal is first
alluded to in a charter granted to the inhabitants of the town in
1234, conferring the right to dig the mineral. The first mention of
coal in Scotland is found in a charter granted in 1291 to the Abbot

and Convent of Dunfermline, giving them the privilege of digging
coal in the lands of Pittencrieff; but the first workers of the mineral
are supposed to have been the monks of Newbattle Abbey. A vein
of coal which crops out on the banks of the Esk was worked by the
latter, not as a mine, but in the fashion of a quarry. Though the
monks appreciated the value of coal thus early, it does not appear to
have found favour with the people generally until several centuries
afterwards. Wood and peat were the materials commonly used as
fuel, and in the houses of the wealthier classes charcoal was burned.
It was only when wood began to get scarce, and, as a consequence,
went up in price, that attention was turned to the "black stones;"
but such was the prejudice against them on account of the disagreeable smoke they gave out, that those who were disposed to give
them a fair trial met with opposition on all hands.

In the beginning of the fourteenth century, the London brewers
and smiths, finding the high price of wood pressing hardly upon their
returns, resolved to make some experiments with coal; but immediately an outcry was raised against them by persons living near the
breweries and forges, the King was petitioned, and a law was passed
prohibiting the burning of coal within the city. Those who tried
it, however, found the new fuel to be so much superior to wood that
they persisted in its use. But so determined were the Government
to suppress what was regarded as an intolerable nuisance, that a law
was passed making the burning of coal in London a capital offence;
and it is recorded that one man at least was executed under that
law. As a contrast to these facts, it may be mentioned that the
London of the present day consumes annually between six and seven
million tons of the once despised and rejected mineral.

It would appear that the ladies were most bitterly opposed to the
use of coal for domestic purposes. They considered the smoke to
be ruinous to their complexions, and would not attend parties at
houses in which the objectionable fuel was used. Some persons
went the length of refusing to eat food of any kind that had been
cooked on a coal fire.

In the account of Scotland given by Eneas Sylvius, who visited
the country in the fourteenth century, it is stated that the poor
people who begged at the church-doors received for alms "pieces of
stone, with which they went away quite contented." "This species
of stone," he adds, "whether with sulphur or whatever inflammable
substance it may be impregnated, they burn in place of wood." A
description of Scotland, written in the beginning of the sixteenth
century, says, "There are black stones also digged out of the ground,

which are very good for firing; and such is their intolerable heat, that they resolve and melt iron, and therefore are very profitable for smiths and such artificers as deal with other metals."

The popular prejudice against coal, and the want of appliances for digging it out of the earth, combined to prevent its coming into general use as a substitute for wood and turf until about the close of the sixteenth century, when it was recorded that "the use of coal beginneth to grow from the forge into the kitchen and halle." In the early part of last century, coal was suddenly raised into importance by the invention of the steam-engine; and since then it has been one of the most valuable agents in spreading civilisation, and in promoting the welfare of mankind.

The history of coal mining, like that of most other industrial pursuits, is chiefly a record of experiments, disappointments, and ultimate successes—a steady contest with difficulties, and a gradual improvement of appliances to overcome these. The first miners of coal would find the work easy enough, as they doubtless confined their attention to the out-croppings in river-banks and valleys. It would not be until coal began to grow in popular favour, and the superficial supplies became exhausted, that real difficulties would be encountered.

The first step in the direction of mining was the driving into the coal-seams of tunnels—called "ingaen e'es" (ingoing eyes) by the miners. Only a small extent of the seams could be worked in that way, however, while the tunnels were rendered dangerous by the accumulation of foul air, as well as by the want of mechanical skill in the workers to protect themselves from the masses of superincumbent strata which were constantly falling. Where the seams dipped downwards, water accumulated, and no little labour was expended in baling out the workings, or in the formation of draining levels where these were practicable. The remains of some of the levels in the earliest known collieries show them to have been of vast extent, and their construction, with the appliances used by the pioneers of mining, must have been a most arduous undertaking. A number of those ancient levels are still in operation in Fife and the Lothians.

After the mode of working the coal by means of shafts descending to the seams was adopted, various contrivances for raising the coal and keeping the pits clear of water came into use. In some cases, both coal and water were drawn up by a winch worked by men; in others, horse-gins were employed for hoisting the coal, and chain-and-bucket engines for the water; while in a few instances the elevating power was derived from common waterwheels. A steam-engine was first employed to work a coal-pit in England in 1680, and in Scotland in 1762. Few of those early pits were carried beyond a depth of twenty or thirty fathoms; but even at that depth the difficulty of working them was enormous. All the collieries are now worked by steam-power, and recently that agent has been applied to machinery for excavating the coal.

As the depth of the pits was increased by the miners seeking out and working lower seams of coal, what was considered to be an almost insurmountable difficulty arose. There were no means of ventilating the mines, and the accumulations of gas became so troublesome as to cause operations to be suspended altogether in many cases, after immense cost had been incurred in sinking shafts to the lower seams. The want of ventilation, while attended by great danger to the workmen, threatened ruin to the coal proprietors, and the prospect was anything but cheering as to the future of the coalfields. But emergencies of this kind have rarely occurred without bringing to the front some person fitted to cope with them, and so in this case a genius was not wanting. A working smith, employed at one of the Durham collieries, having observed that the fire in his forge caused a strong current of air to rush in, bethought him that he would rid the mines of foul air if he could succeed in causing a draught in them by means of a fire. The first experiments were conducted in this way: A cylindrical stove, about three feet long and two feet in diameter, was filled with burning coal and lowered halfway down one of the shafts of a mine. Immediately there was a rush of air up that shaft and down the other, and the result was considered to be highly satisfactory. As the works advanced to some distance from the bottom of the shaft, however, it was found that the gas again accumulated, and it became necessary to adopt some contrivance for drawing off the gas and injecting fresh air into the mine. A large furnace was constructed at the mouth of the shaft, and wooden pipes leading to the furnace were laid through the workings. The furnace drew its sole supply of air from those pipes, and that, of course, caused a rush of fresh air down the shaft and to all the points to which the pipes extended. This plan, adopted in 1760, was considered to be most effective, and again the works were pushed forward. Like other inventions which had been regarded as perfect in their time, however, the pipe system of ventilation came to be looked upon as being not quite so efficient after all, and other plans were proposed.

The lot of the early miners and coal-bearers in Scotland was rendered hard enough by their having to work in the face of many dangers and difficulties to the removal of which science had not then been applied ; but their condition was made more wretched by a system of bondage or serfdom which prevailed. On entering a coal mine, the workers became bound to labour therein during their whole lifetime ; and in the case of sale, or alienation of the ground on which a colliery was situated, the right to their services passed to the purchaser without any special grant or agreement. The sons of the collier could not follow any occupation save that of their father, and could labour only in the mine to which they were held to be attached by birth. Tramps and vagabonds, who were not sufficiently wicked to deserve hanging, and on whom prison accommodation would only be wasted, were sometimes consigned by the Lords of Justiciary to life-long service in the collieries and salteries. Every man thus disposed of had riveted on his neck a collar, on which was engraved the name of the person to whom he was gifted, together with the date. The collar was intended as a check upon deserters ; and constables were highly rewarded when they brought back a fugitive.

Though serfdom had a considerable time previously become extinct, so far as all other classes of workers were concerned, colliers and salters were not liberated until towards the close of last century, and the custom of celebrating the anniversary of their emancipation has not yet died out. The Act which set them free was passed on the 23d May 1775, and was entitled "An Act for altering, explaining, and amending several Acts of Parliament of Scotland, respecting colliers, coal-bearers, and salters, &c." The preamble and headings of the Act will show its purport. These were as follow :—

"Whereas many colliers, coal-bearers, and salters in Scotland are in a state of slavery or bondage, bound to the collieries and salt-works, where they work for life, and are sold with the mines : be it enacted that—

"1. No person shall be bound to work in them in any way different from common labourers.

"2. It shall be lawful for the owners and lessees of collieries and salt-works to take apprentices for the legal term in Scotland.

"3. All persons under a given age, now employed in them, to be free after a given day.

"4. Others of a given age not to be free till they have instructed an apprentice."

Up till the year 1843, children of tender years and women were employed to do underground work in the coal mines of Scotland, as well as in those of England. An inquiry into the condition of children employed in factories revealed the existence of a system of mismanagement and mercenary cruelty which excited considerable surprise and indignation, and a law was passed to put an end to the evil. Attention was then drawn to the condition of children in other employments, and Lord Ashley procured the appointment of commissioners for inquiring into the employment of children generally. In investigating the state of matters existing in mines and collieries, the commissioners found that, while the case of the children was extremely bad, that of the women similarly employed was no less pitiable. The report presented to Parliament by the commissioners excited a thrill of horror all over the country, and led to the speedy passing of a measure—brought into the House of Commons by Lord Ashley, now Earl of Shaftesbury, on 7th June, 1842—prohibiting the employment of boys under the age of ten years, limiting the period of apprenticeship, and putting a stop to the employment of women.

From the report of the commissioners, it would appear that the condition of the women and children employed in the collieries in the east of Scotland was as bad as existed anywhere. Many children five or six years of age were employed. In the west of Scotland, the youngest children in the pits were eight years old ; but in some of the English pits, infants of four years were to be found. In the east of Scotland pits, women were generally employed, but in the west they were rarely met with. Before winding apparatus came into use, the labour assigned to the women and children was to carry the coal on their backs from the place where it was excavated to the pit-mouth. The journey along the pit-bottom was bad enough; but the ascent of the wet and slimy wood stairs leading up the shaft was extremely difficult and perilous, and accidents were of daily occurrence. The weight of coal carried on each journey by some of the women was, according to reliable evidence, four and a-half cwts. After the application of machinery to draw up the coal, the women and children were solely occupied in dragging the coal from the place where the miners were at work to the bottom of the shaft. In pits where women were not employed, the coal was drawn on sledges by ponies. About the beginning of the present century rails were introduced into the pits, and the coal was drawn in "hurleys," or wheeled boxes, to which boys and girls were yoked by a rude kind of harness, known as the "girdle and chain." This mode of harnessing is not yet extinct, but it is used only to a limited extent, and the drawers are stout lads more fitted for the work than were the puny children and wretched girls who were formerly employed.

Regarding the places in which those poor creatures had to work, the report stated that, "in the east of Scotland, where the side roads do not exceed from twenty-two to twenty-eight inches in height, the working places are sometimes 100 and 200 yards distant from the main road ; so that females have to crawl backwards and forwards with their small carts in seams in many cases not exceeding twenty-two to twenty-eight inches in height. The whole of these places, it appears, are in a most deplorable state as to ventilation. The evidence of their sufferings, as given by the young people and the old colliers themselves, is absolutely hideous." On the main roads of some pits, the coal was carried on the backs of girls and women ; and in one of the pits a sub-commissioner found a girl, only six years old, carrying half-a-hundredweight of coal, and making fourteen journeys a-day, each journey being equal to ascending to the top of St Paul's Cathedral.

The evidence as to the moral degradation of the women was shocking in the extreme ; and on all sides the necessity for abolishing the employment of females in the pits was forcibly urged. One old Scotchwoman, Isabel Hogg, said to the commissioners,—" You must just tell the Queen Victoria that we are quiet, loyal subjects ; women-people here don't mind work, but they object to horse-work ; and that she would have the blessing of all the Scotch coal-women if she would get them out of the pits and send them to other labour." Not only was the work degrading and severe, and carried on under circumstances the most adverse to personal comfort, but the hours of labour were long and irregular. In the latter respect, the collieries of the east of Scotland were again pointed to as shameful examples. In them the labour was often continued on alternate days, at least fifteen and even eighteen hours out of the twenty-four. One girl, seventeen years of age, said,—" I have repeatedly wrought the twenty-four hours ; and after two hours of rest and my pease-soup, have returned to the pit and worked other twelve hours." The labour was generally uninterrupted by any regular time set apart for rest or refreshment ; what food was taken in the pit being eaten while the work went on.

In a number of Scotch pits women and children had never been employed ; and before the passing of Lord Ashley's Bill, some of the coal proprietors, into whose pits such labour had been introduced, had given orders for its exclusion. The change was not altogether satisfactory to those affected by it, as one of its results was a serious reduction in the family earnings. By the introduction of rails for the hurleys to run upon, and other improvements, the men and boys were enabled to earn more money ; and the ultimate result was a very marked amendment in the moral and social condition of the mining communities.

Since 1843 there has been additional legislation regarding mines, embracing—Restrictions as to the employment of females and boys; wages, and their mode of payment; the appointment of inspectors, with large powers and onerous duties; rules for the regulation of coal mines, collieries, and ironstone mines; penalties for the non-observance of the Acts, &c.

The following " rules and regulations of the great seam pit-bottom at Newbattle Colliery," in force at the beginning of the present century, throws some light on the manners of the miners and female coal-bearers, and their mode of working at that period. We copy the document as it stands on the official books of the colliery :—

" 1st, It is agried amongest the men that all Desputs and controvries a rising in the pit Botom shall be Decided by 2 men who shall be chosen as commites, whos Determination shall be finiel and binding on all parties.

" 2nd, It is agried that every Birer shall keep her own Border or Lair. Whoever shall inchroch on ther nebhour property, so as rise any desturbance, the commities shall be sent for, & the man or woman that is fownd in the wrong shall be fined of 1s. for every transgison of this kind not to be forgivin.

" 3d, Be it liquis agried that every man shall have his own fair and regular turn of tubs riding; and if any man or woman shall take ther nebhour turn by force or frawd or strength against ther nebhour, will the person that took ther los the tub sent up, it not being ther own fair turn.

" 4th, But as the coal is so varible in its nature that sume may have coals in the morning, others not till afternoon, them that has them in morning must set them away for to serve the saile; but when ther nebhour who was behind in the morning & gets his coal through the day he must get up his turns that he was behind.

" 5th, As it is a prevaling custom amang Birers to curse and swear, and call others vile and scandles reproachfull names without a caus, the person so offending shall be find of 1s. starling for every offence of this kind not to be forgiven.

" 6th, And if it can be proven that the pit botom man dos not pay due attention to these reglations, through fear of sume and through favor to others, he shall be find of — starling; and he is not keep the gen [gin horse] stabled upon any account.

" 7th, It is agried that if any collier or Birer shall Break any of the above reglations, and rise a desturbance to that degrie of passion that the Lift ther hand and strik ther nebhour with ther hand, or foot, or stick, or ston, or coal, or any other thing that can hurt or egure one another, the person so offending shall pay 5s. of a fine not to be forgiven ; and, lastly, all those fines to be lifted from the coal greve by the commities on that day the offence is commited, and to be keept of the offending person on ther pay day."

The carboniferous system of Scotland has received considerable attention from geologists, and its nature and extent have been frequently described. Though fragmentary strata of coal occur in the Western Islands and at one or two other points, the great coal-fields occupy a well-defined position, extending across the country in the line of the valleys of the Forth and Clyde ; and their superficial area is calculated to be about seventeen hundred and fifty square miles, or one-seventeenth part of the surface of Scotland. The uppermost of the coal strata is found at Fisherrow, and between it and the Old Red Sandstone, which forms the floor of the coal formation, there are three hundred and thirty-seven alternations of strata, having a thickness in the aggregate of five thousand feet. In the thickest part there are sixty-two seams of coal, counting the double seams as one, and about one-half of these are workable. The depth of strata at Musselburgh is, however, exceptional ; and the average depth is estimated to be about three thousand feet, of which the coal seams occupy one hundred and twenty-six feet. The thickest bed of coal in the Lothians field is thirteen feet ; but at Johnstone, in Renfrewshire, there is a seam one hundred feet in thickness. This latter owes its extraordinary bulk to the overlapping of the coal strata during some great convulsion in the locality. The most important of the coal-fields is the Clydesdale, on which one-half of the entire number of collieries in Scotland are situated. Thirteen counties lie over or touch upon the coal-fields, and of these Lanarkshire has by far the largest share of the store. Judging from the number of collieries possessed by each, Ayrshire, Fifeshire, and Stirlingshire come next in order. In nearly all the counties, more or less valuable beds of ironstone, shale, and limestone are intermixed with the coal. The Scotch cannel or parrot coals are very valuable on account of the high proportion of gas and oil which they yield. The Boghead variety gives one hundred and twenty gallons of crude burning oil, or fifteen thousand cubic feet of gas per ton ; and the brown Methil ninety gallons of oil, or ten thousand cubic feet of gas per ton. The cannel coal found at Wemyss, Fifeshire, is carved into various articles of a useful and ornamental character—such as picture-frames, inkstands, brooches, &c.

The deepest coal pit in Scotland is at Nitshill in Renfrewshire, and the most extensive individual colliery, while at the same time the deepest, is Mr Dixon's Shawfield pit at Govan. The deepest in the eastern district is the Emily pit at Arniston, belonging to Mr Christie, who is one of the most extensive coalmasters in Scotland. It is one hundred and sixty fathoms in depth—fifteen less than the Nitshill pit.

As Mr Christie's is a well-appointed colliery, and one which displays the two modes of working coal, an account of a visit paid to it may be interesting. The colliery is situated near the line of the North British Railway, about a mile north from Gorebridge Station, and has three working shafts, the deepest of the three descending to what is known as the "splint" seam, at a depth of one hundred and twenty-five fathoms, and to the "parrot" seam, thirty-five fathoms farther down. The rise and dip shafts are about seven hundred and thirty yards apart, but the workings with which they communicate open into each other. After the accident at the Hartley pit a few years ago, it was made compulsory to have two shafts for each colliery; but the Arniston colliery, and many others in Scotland, were long before that time furnished with two outlets.

Before proceeding to visit the pit, we acted upon good counsel, and donned a capacious suit of pilot cloth, which, though of most uncouth cut, proved to be quite an aristocratic costume when brought into contrast with the habiliments of the dusky fellows below. Under the guidance of one of the managers, we first inspected the above-ground fittings of the Emily pit. These consist of a large engine-room, containing the winding engines. The drums of these engines, on which the rope is wound, are ten feet in diameter, and fitted with a powerful break, which ensures the greatest safety and nicety in raising and lowering the cages in the shaft. The rope to which the cages are attached is one and a-half inch in diameter, and composed of wire. It passes over a pair of immense pulleys fixed about thirty-six feet above the pit mouth, and is thence led on to the drums of the winding engines in the engine-room. This apparatus is the most important connected with a colliery, and its management requires extreme care. Attached to the winding-drums is an index, which shows the exact position or progress of the cage in the shaft, and by observing the index the engineman can stop the cage within an inch of any desired point; and he is able to deposit it on the pit bottom so gently, that those who occupy it are unconscious of its having come to a stop, and that, too, after it has passed through the shaft at a rate of something like twenty-five miles an hour. Close by the winding-engine room is an apartment containing the pumping-engine —a ponderous piece of mechanism erected over the compartment of the shaft which contains the pumps. This engine is 400-horse power nominally. The cylinder is eighty inches in diameter, and the piston has a 12-feet stroke. The cylinder is placed in an inverted position over the pit, and the piston-rods, of which there are two, are directly connected with the pump-rods. There are five columns of pumps, the one discharging into the other, the internal diameter of which increases from twelve inches at the pit bottom to seventeen inches at the top. Though the pumps discharge thirty-nine thousand gallons of water per hour from the bottom of the pit, they have to be kept going almost incessantly in order to keep the pit clear. With reference to the pumps, a curious fact, illustrating the extent of one of the difficulties with which miners have to contend, may be mentioned. When the pit is working at full power, thirty tons of material are put out per hour ; while the quantity of water that has to be raised in the same time would weigh one hundred and seventy-four tons. In addition to the winding and pumping engines, there are a donkey-engine for feeding the boilers, and a steam-crane for hoisting out the pumps when repairs are necessary. The crane is fitted with a wire rope capable of bearing a strain of forty tons. The steam for the engines is generated in six immense boilers. Immediately adjoining are extensive workshops for engineers, smiths, and carpenters, a large number of whom are employed in keeping the working gear of the colliery in order.

The inspection of the machinery and workshops having been concluded, we ascended to the elevated bank or platform which surrounds the mouth of the shaft ; and, while waiting the arrival of lamps to light us through the pit, had an opportunity of seeing how the coal was brought out. The shape of the shaft is an oblong square, with the sides bulged out a little. It measures fifteen feet one way and nine the other, and is divided into three equal compartments, in two of which the cages are worked, while the pumps are enclosed in the third. The cages are simply composed of an iron framework floored with wood, and having a sheet-iron roof of semi-circular form. Each cage is sufficiently large to admit of two hurleys or " tubs " being brought up at a time ; and the winding-gear is so adjusted that while one cage is ascending the other is descending. The cages travel from bottom to top of the shaft in thirty seconds when laden with coal, but when the freight is a living one the speed is considerably reduced. As we stood and watched the cages emerge alternately, slimy and dripping as if they came from the depths of some subterranean lake, our intention to descend into the dark abyss threatened to evaporate. But before a resolution to defer the venture was formed, our guide appeared with a lighted lamp in each hand, and, with a reassuring smile, invited us to step into the cage. The invitation was accepted, but not without a certain feeling of dread, as the "situation" brought vividly to. mind the recollection of many a catastrophe which had befallen persons making a journey similar to that on which we had now entered. Men who work in or about coal mines may make light of the perils which surround them, but few persons descend a shaft for the first time without experiencing a very keen sense of danger.

When we had entered the cage, and had received a few words of instruction as to holding on and keeping steady, the word "right" was passed. The first motion of the cage was upwards for a few inches, to relieve the self-acting stoppers on which the cage rested at the mouth of the shaft. Then the engine was reversed, and we were off. A feeling of giddiness was experienced as the cage glided down, steadily, swiftly, and almost noiselessly ; but that wore off ere half the distance to the pit bottom had been accomplished. The daylight did not accompany us far, and the black and oozy walls of the shaft absorbed so much of the light of the lamps that we were left in a dismal gloom. Suddenly something rushed past and excited a current of air which nearly extinguished our feeble illuminators. It was the ascending cage; we had now got half way down, and were some distance under the sea-level. As we sped downwards, the walls of the shaft became very wet, and big drops of water pattered upon the iron canopy overhead, while showers of spray entered the cage on all sides. Gradually the fall of water increased. The drops had grown to streams, and the spray to little jets, when we became conscious of a slackening of the speed of the cage. Simultaneously with this, our ears caught a confused sound of voices, and in another moment we had alighted.

The first objects that met our eyes were a number of men engaged in various ways about a train of "tubs" which had just been brought forward from the workings. The besmudged countenances of the men, seen imperfectly by the light of the lamps which they carried on their foreheads, well accorded with the surrounding blackness and gloom. The fellows were cheerful withal, and set about their work with a will—laughing, "chaffing," and singing, in defiance of the depressing influences around them. A number of horses are employed in the pit to draw the "tubs" on the main roads, and are lodged in a stable near the bottom of the shaft.

The animals do not seem to suffer any bad effect from confinement in the pit, being as sleek and well-conditioned as those of their kind that are privileged to roam in green pastures and bask in the sunshine. The roadways are arched over with brick for some distance, but the roof beyond consists of rock. The main roads have been excavated to a height sufficient to allow the horses to pass; but the branch roads are no higher than the thickness of the coal seam, which is about three feet. The seam is the most valuable in the field, as it contains the "parrot" or gas coal. The latter is found in a layer, varying from eight to nine inches in thickness, enclosed between two layers of good household coal, each of which is about a foot thick. Though we were now nearly a thousand feet beneath the surface of the earth, and more than half that depth below the level of the sea, the air was fresh and the temperature summer-like, and we were assured that all through the twenty miles of roads and passages in the pit it was the same. We did not advance into the workings here, as a better opportunity for seeing the miners at work would be afforded in the "Kailblades" seam, in order to reach which the Emily shaft had to be ascended, and a descent made by another a few hundred yards distant ; for though, as already stated, there is underground communication between the shafts, the passage from the one seam to the other may be made more readily and comfortably by the "overland route."

Leaping through the rushing shower of big water-drops, which came from the shaft, we were once more in the cage ; and signals having been duly exchanged by the man at the pit-bottom and the engineman, we began to ascend—slowly, according to pre-arrangement, in order that a view might be obtained of the pumps and the entrances into the various seams which have been opened in the pit. When the cage had ascended above the denser portion of the shaft drippings, a hasty peep upward was ventured upon. A speck of light no bigger than might be covered by one's hand was all that was visible, even the huge cable which supported the cage was lost to view in the distance. After a few brief pauses, for the purposes above stated, we rapidly glided into daylight. Inspired with confidence by what had been already accomplished, the descent of the other shaft was made without any very decided apprehension of danger.

This time we had to go down about ninety fathoms only, and into a region almost entirely free from water. There were no horses in the seam, the drawing, or rather "putting," of the coal being done by boys or lads. Passing the group of men employed at the pit bottom we advanced into the workings, preceded by our guide, who endeavoured to divert attention from the difficulties of the path by explaining the formation of the coal strata, which glistened on one side of our path—a roughly-built stone wall forming the other. "So toilsome was the road to trace," however, that neither geological nor statistical gossip served to make us unconscious of its disagreeableness and terrors. Our path—a main roadway, be it recollected—was about four feet in width, and barely so much in height. The bottom of it was laid with a line of rails, and the space between the rails was wet and muddy. Overhead, ugly rents yawned, and fragments of rock protruded in a most threatening way. In order to protect the head from knocks and the feet from stumbling, a sharp look-out had to be kept above and below. Progress was frequently interrupted by the passing of coal-laden hurleys, which were pushed along the rails by lads carrying lights on their foreheads; and an occasional pause was made to take advantage of some gaps in the roof, which permitted us to obtain some rest by standing erect. Roads branched off to right and left at intervals, and the openings of certain of them were provided with doors, to shut which after passing is an imperative rule of the pit. The purpose of these doors is to guide the air-current on its way through the workings, and neglect in attending to them would destroy the ventilation of the mine. At certain points, the air-current could be heard sighing along the galleries, or whistling through the chinks of the doors; and so strong was it at times, that great care was required to keep the lamps alight. Knowing that by this time muscles unaccustomed to such difficult pedestrianism would be wearied and sore, our guide hailed a passing "putter," into whose "tub" we were right glad to take a seat, and complete the remainder of the journey by rail. On and on we whirled through the terrible gloom, assured that we had not far to go, but without seeing any sign to indicate that the desired goal—the "face" at which the miners were at work—was near. By-and-by a confused noise began to break on the ear, and a look ahead revealed a number of lights flickering and moving about as mysteriously as wills-o'-the-wisp. Human voices pitched to the lowest notes could then be distinguished amid a chorus of dull thuds; and a few yards further on our carriage was brought to a stand: we had reached the "face." There, in a series of recesses branching off the road to the left, were the miners, who, in going to and from their work, have to traverse the path we had just passed over.

Entering one of the recesses, technically known as a "room," we had a closer view of the miner and his mode of working. The dimensions of the room would be about twelve feet wide by twenty long, and the height from floor to ceiling was exactly three feet. The miner, after cutting a deep niche along the lower part of the seam, commenced to cut two perpendicular slits about six feet apart. After he had reached a certain depth, the coal began to crack, and in a moment or two the mass, detached by its own weight, fell and broke up into fragments with a noise resembling the breaking of a wave on a pebbly beach. The coal in this seam is soft, and neither gunpowder nor wedges are required, as in some cases, to bring it down. The work, nevertheless, is very hard and irksome—though we were told it was mere child's play when compared with the labour of excavating the "low seams," the depth of which is only from twenty-two to twenty-four inches. In a three-feet seam, the miner can kneel while working; but in thin seams, he has to lie at length on his side, and in some cases water pours down on him continuously. As the coal is broken away from the face, it is shovelled aside, and committed to the care of the " putter," who fills it into his " tub" and wheels it along to the pit bottom. This is very severe toil for boys, but those engaged in it were stout and healthy, and appeared to be nowise discontented with their lot.

While seated on the floor of this room, we were favoured with an explanation of the two systems of working coal, both of which are followed in the Arniston colliery. The systems are respectively designated " stoop-and-room" and " long wall." It is a matter of great importance that the miner should be able to extract as much as possible of the coal in the various seams; and to enable him to do that, various plans have been proposed and tried; but of these only the two we have named have come into favour. The " stoop-and-room" system, which was followed in the part of the pit in which we were seated, consists in driving passages or " rooms" through the coal, leaving " stoops" or pillars of coal between, of sufficient strength to support the roof. The rooms are from twelve to twenty feet wide, and the pillars or " stoops" ten to twenty yards square. The pillars are allowed to remain until the limit of the seam is reached, when the miners turn back and work away the pillars, using wood props to prevent the roof from falling. This is the most precarious part of the miner's work, and requires the exercise of great skill and care to prevent accidents. After a certain proportion of the pillars have been removed, the wood props are taken out, and the superincumbent strata allowed to settle down.

The operation of removing the props has to be performed with great caution, and is intrusted only to picked men. The noise made by the beds of rock as they break down, is described as being peculiar and terrific. Not more than one-tenth of the coal is lost by this method, whereas by the plan pursued in early times only one-half was got out. The "long wall" system is considered by mining engineers to be the most advantageous, as it admits of the coal being worked out thoroughly at once. According to this plan, the miners work along a continuous face of the seam, cutting out the coal completely, and allowing the roof to settle as they advance, care being taken to preserve roads by throwing up parallel lines of stone and waste, and using wood props occasionally. As the roof collapses, it is blasted down in the roads, to keep them sufficiently high for the loaded tubs to pass. This mode of working is not so perilous as it appears to inexperienced persons ; for the roof does not fall at once. It subsides gradually, and if the miners advance at a steady rate, they may calculate on being from fifty to eighty yards in front of the place where the roof comes into contact with the floor. Both systems are liable to considerable modification, but the above is a rough statement of their chief features.

Another dismal and spine-racking journey brought us to the district of the mine which is worked on the "long wall" system. The strata in the pit, it may be mentioned here, lie at an angle of about 20° to the horizon, and the miners work upwards on the slope. The "long wall" workings have been carried forward to a considerable distance from the main road, and, in order to reach the face, we had to go through one of the narrow roads kept open amid the fallen rocks by means of the protection walls already referred to. As the immense weight of the stone overhead had crushed the walls considerably, and had thrown them down in some places, the original dimensions of the tunnel were much reduced, and the average height and width were less than three feet. It was necessary at some parts to travel on "all fours," a mode of progression rendered very disagreeable by a thick layer of finely pulverised coal which covered the floor of the passage. There are a number of similar roads in the pit, and it is through them that the coal is brought from the "face" to the main roads. The "putter" fills the coal into a box mounted like a sledge on iron-shod slides ; and as the slope of the road is more than sufficient to cause the sledge to descend of its own accord, he has to seize it by the front, and, walking backward, guide it through the tunnel, and prevent it from travelling too rapidly. After transferring the coal to a hurley, he has to get into harness, and drag the sledge up to the "face" again. There are no rails in these narrow roads, and their absence makes the work severe and hazardous. At intervals in the pit are several self-acting inclines. These are laid with double lines of rails, and by means of a drum and tackle, the "tubs" are let down and pulled up with very little labour—the full ones in their descent causing the empty ones to ascend.

We next visited the air furnace, an immense brick structure, communicating with a shaft extending to the surface of the earth. A huge fire is kept almost constantly burning in the furnace, which causes a strong rush of air from the workings. Before the air, entering by the working shaft, can reach the furnace, it has to traverse every part of the mine, and that accounts for the pureness of the atmosphere in even the remotest nooks. The gas given off by the coal is diluted and rendered harmless by the current of fresh air; and were it not for the particles of coal which fly off at every stroke of the pick, the atmosphere in which the miners work would be as pure as that breathed by the most favourably-situated workmen above ground.

The miners enter the pit between five and six o'clock in the morning; but before they do so, an inspection of the workings is made by the "viewer," in order to ascertain the state of the ventilation. They remain in the pit until two in the afternoon, the eight hours' spell of work being relieved by a brief interval for breakfast. As the pit under notice is free from fire-damp, naked lights are used. These consist of small tin lamps. The flame is fed with tallow instead of oil, as the former gives off less smoke than the latter. Very little trouble is required to keep the lamps in trim, and though the light appears dim to unaccustomed eyes, the miners find it sufficient for their purpose.

When the miners stop work, another class of men enter the pit these are the "reddsmen" and "brushers," whose duty it is to examine and repair the roads, remove any stones that may have fallen, and see that the roof is secure throughout the workings. About 500 men and boys are employed in and about the Arniston and other pits leased by Mr Christie. In addition to the pits visited, he has six others in operation. Having gleaned the information and experience above recorded, and satisfied to the utmost a feeling of curiosity as to the nature of the miner's occupation, we set out for the pit bottom, and in due time emerged into the sunshine.

The miner's avocation is a very perilous one. From the moment he sets his foot in the cage to descend to his work, he is in constant danger of a violent death, or of injury that may render life a burden

B

to him. The winding gear may give way, and dash his body into fragments on the pit bottom; or, after he arrives safe at the "face," a mass of rock may descend from the treacherous roof, and crush out his life. He is in danger of being suffocated by foul air, and of being scorched to death by the ignition of the fearful fire-damp. These and other risks he has to encounter daily; and when he is deposited safe at bank after his toil is over, one may fancy that, if he has any feeling at all, it will be something akin to that of the soldier who at the close of a battle finds his head upon his shoulders and his limbs unfractured. The mining statistics for 1866 show that in the collieries of England, Wales, and Scotland, no fewer than 1484 lives were in that year lost by accident. The total number of miners employed was 320,663, so that one person was killed out of every 216 employed. The year was an unusually fatal one, however, the explosions at the Oaks and Talk o' the Hill collieries—the one involving a loss of 361 lives, and the other of 91 lives—having occurred during its course. In 1865 the number of lives lost was 984, or 500 fewer than in 1866. In proportion to the quantity of coal raised in Scotland, the loss of life is considerably less than in England or Wales. The total quantity of coal raised in Britain in 1866 was 100,728,881 tons; and as the number of lives lost was 1484, we find that one life was sacrificed for every 67,877 tons of coal. In Scotland, 12,034,638 tons were raised, and 77 lives lost, so that for each person killed, 161,252 tons were got.

For the purposes of the Mines Inspection Act, Scotland is divided into two districts. The eastern district includes the Lothians, Fifeshire, Clackmannanshire, Kinross-shire, part of Perthshire, the eastern division of Stirlingshire, and the upper division of Lanarkshire. The western district embraces the lower division of Lanarkshire, the western division of Stirlingshire, and the counties of Ayr, Dumbarton, Renfrew, Dumfries, and Argyle. The proportion of lives lost to the quantity of coal raised in the year 1866 in Scotland was :—

Districts.	No. of Miners.	Tons of Coal Raised.	Lives Lost.	Tons of Coal per Life Lost.	No. of Collieries.
Eastern district,	21,200	6,100,000	32	190,625	254
Western district,	20,046	5,934,638	45	131,880	218
Totals, ...	41,246	12,034,638	77	161,252	472

The difference of death-rate in favour of Scotch mines is owing chiefly to their comparative freedom from fire-damp. Of the deaths in Scotland, 7 only (or 9 per cent.) are attributed to the explosion of that dangerous gas; while in England about 45 per cent. of the deaths in 1867 were caused by it, and in the preceding year about 18 per cent. The 77 deaths in Scotch mines are thus classified :— By explosions, 7; falls of coal and roof, 38; in shafts, 21; miscellaneous, 7; above ground, 4. The ages of the persons killed ranged from thirteen to seventy years; and though the cause of death in most cases was such that no care or foresight could avert it, yet it is evident that in several instances death was the result of neglecting the most ordinary precautions for ensuring safety. Previous to the passing of the Mines and Collieries Act, which came into operation in 1843, and made it illegal to intrust the winding machinery to any person under fifteen years of age, it was no unusual thing to find the engines in charge of mere children—boys of twelve, eleven, and even nine years, and many lives were lost in consequence. During the inquiry which was instituted before the passing of the Act referred to, the chief constable of Oldham stated that he was not aware of a single case in which children were not employed as engineers. He mentioned an instance in which four boys were killed in consequence of the neglect of an engineer nine years of age, who, while the engine was winding up his companions, was attracted from his post by a mouse on the hearth. By the statute of 1861, at every colliery there must be established certain general rules to be observed by the owner and agent, and also special rules for the conduct and guidance of the person acting in the management of such colliery, and of all persons employed in and about the same, as under the particular circumstances of such colliery may appear best calculated to prevent dangerous accidents. These general and special rules, and improved machinery, have gone far to lessen the fatality of the mines, but the perils of those who work in them must always be great.

The miner holds an humble position in the industrial ranks. His occupation does not require much skill, nor has it any tendency to incite him to intellectual pursuits. Where his own interests are not directly concerned, he rarely intrudes; and the great body of society beyond the coal fields would become almost unconscious of his existence, had it not an occasional reminder in the records of the terrible disasters which sometimes overtake him. His intercourse with the rest of mankind is limited, and often the circle of his intimate acquaintance does not extend beyond the little community attached to the pit in which he works. His occupation is peculiar, and quite distinct from that of any other class of workmen, one of its effects being the creation in him of a desire for exciting amusements ; and the means taken to gratify that desire have something to do with the low position he occupies in the social scale. He long had an unhappy notoriety—of which he has not yet got completely rid— for drinking, poaching, and other irregularities ; and his neighbours of other occupations were prone to regard him as a rough sort of fellow. Even when he lived in a town, he failed somehow to get absorbed into the great industrial body. Within late years, however, a change has been coming over him, and his old manners and habits are yielding to the influence of education. Still, there appears to be a want of sympathy between him and the mason, the carpenter, the tailor, the shoemaker, and other tradesmen, and they rarely associate. Were his social position to be regulated by the amount of his earnings, the miner would stand above a large proportion of the working classes ; but he appears to be indifferent to rank, provided he is allowed to enjoy life according to his own notions. It is but ninety years since he was a slave—or, strictly speaking, only sixty-eight years, because the Act of 1775 was hampered with restrictions which prevented him from obtaining full freedom, and his emancipation was not completed till 1799, when a new Act was passed, and it was not until a considerable time after he was set free that he began to raise himself in the social scale.' Indeed, the work of reformation can scarcely be said to have begun until the passing of 'Lord Ashley's Bill in 1843 for the abolition of female labour in the pits. It is not to be wondered, then, that traces of old habits, superstitions, and prejudices, are still discernible, especially among the aged people.

The rapid development of the coal and iron trades in the west of Scotland led to an immense influx of Irish labourers between 1830 and 1850 ; and as they were generally very ignorant, they retarded for a time the general progress of improvement. The liberality, however, of the employers, in establishing schools at every colliery, is daily effecting a change ; and with the advent of another generation the traces of degradation will probably disappear, and there is evidence to lead to a hope that the miner will come to occupy a much improved position in society. In the Lothians, where the relations between master and servant have been little disturbed by strikes or fluctuations in trade, the miners are superior in every respect to the same class in Lanarkshire and the West of Scotland generally ; and the same may be said of the Fife men. This arises chiefly from the fact that, while the eastern miners are almost without exception Scotsmen, whose forefathers for several generations have followed the same avocation in the same locality, a great proportion of those in the west are Irishmen, mostly of a very rough type.

As a rule, the sons of miners follow the occupation of their fathers, and begin to work when they reach twelve years of age—by which time they are now fairly proficient in reading, writing, and arithmetic. After they commence to work, however, they are encouraged to make further progress in education, and for that purpose evening classes are taught at most of the schools. The period of apprenticeship is four years, and the father and sons generally work in partnership. The daughters of the miners find employment on the farms, or at the brickworks and factories in the vicinity of their homes, and an increasing proportion of them go into domestic service. The sons live under the parental roof until they reach eighteen or twenty years of age, when they take wives and begin housekeeping for themselves. Intermarriage with members of other classes was formerly a thing almost unknown, but now such marriages are not infrequent in certain districts. Ignorance of domestic economy, and a want of care for domestic comfort on the part of their wives, have been the means of keeping back many well-disposed men among the miners ; but now that women who have had some experience of domestic service are to be had for wives, a better state of things is beginning to prevail.

The wages of miners, which are paid according to piecework, vary considerably in different districts, and are liable to considerable fluctuation. In some cases the quantity of coal a man may put out in a day is limited by mutual consent, or in accordance with a rule of the Union ; in others, the working hours are limited, each man being allowed to put out as much as he can in the stated time ; and again, there are collieries at which there is no limitation as to time or quantity. Exactly a century ago the wages paid to the men, all serfs, who worked at Newbattle Colliery, were as follow :—Grieve, 7s. a-week ; oversman, 10s. ; banksman, 6s. 7½d. ; bottom-man, 6s. 7½d. ; miners, from 7s. to 8s. 4d. The miners used candles in those days, and these were supplied without charge. The average wage at the same colliery is at present 4s. 6d. a-day. What a miner gets for his labour cannot always be stated in shillings and pence, however, as he sometimes enjoys special advantages in the way of a free house, cheap education for his children, and the like. Thus, the average wage at Dalkeith Colliery is 3s. 6d. a-day ; but the miners are provided with good houses rent-free, and have in addition

other privileges. In the east of Scotland wages have not fluctuated
so much as in other quarters. An understanding seems to prevail
between the men and their employers, which allows the work to go
on steadily, no matter what the state of the coal market may be.
In 1851 the average wage of miners in Scotland, according to a
statement published by a Glasgow firm, was 2s. 6d. a-day; in 1854,
it was 5s. A gradual fall then took place; and in 1858 the average
was 3s.; below which sum it has not fallen, the figures for the
six succeeding years being respectively 3s. 6d., 4s., 4s. 6d., 5s. 6d.,
and 4s. 9d. From these sums about 3d. a-day falls to be deducted for
light, sharpening tools, &c. The wages are paid fortnightly, and that
period embraces ten working days in some districts, eleven in others.

When a boy of twelve years enters a coal pit, he is attached to his
father or some other man, and becomes what is known technically as
a "quarter-man." The miner with whom he works is entitled to
put out one-fourth more coal than if he worked without assistance;
and from the price received for the extra quantity he pays the boy,
whose duty it is to fill the coal into the "tubs" and convey it to the
pit bottom. At fourteen, the boy becomes a "half-man;" at sixteen,
a "three-quarter-man;" and at eighteen, he assumes the title of
miner, performs a man's work, and draws a man's pay. When the
boy ranks as a "quarter-man," he usually receives 1s. a-day; when
a "half-man," 2s.; and when a "three-quarter-man," 3s. These rates
are, however, subject to variation according to the amount of wages
received by the men. From this it will be seen that, when the
miner's family includes two or three sons able to go into the pits, the
total earnings must amount to a considerable sum.

As a class, the miners could afford to live pretty comfortably, but
the great body of them have yet to acquire provident habits. They
are to a large extent victims of the pass-book system, and are rarely out
of debt to the provision-dealer; while many of them draw their wages
in advance, thereby incurring considerable loss, in the shape of a
heavy percentage which is charged by some of the employers. It
will scarcely be credited, but it is a fact, that many coalmasters in
Lanarkshire take most unmerciful advantage of the improvident
habits of the colliers, by charging 5 per cent. for all money advanced
between pay-days. Thus—if it be advanced the day after the pay,
5 per cent., or 125 per cent. per annum, is charged; if advanced
only two days before it is due, the same 5 per cent., or 900
per cent. per annum, is charged. This is a crying shame. If the
masters who charge such ruinous rates of interest (some of them
under pretence of discouraging the practice of lifting money in

advance), were to establish savings' banks among their men, they would soon enable them to save as much as would carry them from one pay-day to the next. It is needless to say that the practice has a demoralising tendency. Though few miners are to be found among depositors in savings' banks, numerous friendly and benefit societies exist among them. The advantages of such societies seem to be fully appreciated, though the treasurers have not always been faithful.

Co-operative stores have been opened at several places ; but, except in a few cases, the success of these is yet doubtful. A number of miners in the west of Scotland are connected with the Free Colliers' and Free Gardeners' Lodges. In some parts, so-called yearly or half-yearly benefit societies are got up by small tradesmen, who collect fees from the members on every pay-day, undertaking, of course, to give a certain amount of sick or funeral money when such is needed. On the average, a very small sum is required to meet such contingencies. Instead, however, of dividing the balance of the funds among the members, as in ordinary cases, the promoter of the society compels them to take goods from him to the amount of their respective dividends—reserving to himself, of course, a certain sum in consideration of his trouble and risk. When the miner and his wife, excited by the jingle of money in their pockets at the fort-night's end, drop into the shops of the promoters of such societies, there is generally not much difficulty in prevailing upon them to join ; but the wary keep aloof. At a number of collieries and iron-works in the west, stores are kept by the proprietors, from which the men are required—or at least expected, which is pretty much the same thing—to purchase all their provisions, &c. It has been found difficult to legislate for the suppression of such stores, and the matter is at present left to be dealt with according to the ideas of fair-play and liberty which exist among the owners of them. The Truck Act was aimed at their extermination, but the spirit of the Act is evaded while its letter is complied with. In the year 1860 a committee of the Social Science Association on Trades' Societies and Strikes, received the following statement on the truck system from the representative of the Scottish Miners' Association :—

" Of the mining counties in Scotland, there is in Clackmannan no truck shop, in Mid and East Lothians one, in Fifeshire three, in Stirlingshire a few, in Renfrewshire they are beginning to be established, in Linlithgowshire there is a truck shop to nearly every colliery. The truck shops provide all articles of subsistence and clothing, with one exception, drink. The Truck Act renders it necessary for the masters to have a separate pay-office. But this office they take care shall be close to the truck shop ; sometimes it is separated only by a partition. They pay the men at the long interval of a fortnight, or even of a month, and in the meantime allow them, upon application, subsistence-money from day to day, or even on the half-day. This subsistence-money the miner is practically compelled by penalties to carry to the truck shop ; for, if not, the subsistence allowance is stopped, and he must wait for his pay till the end of the fortnight or month ; or he is shifted to a less favourable part of the mine, or he is altogether dismissed. Dismissal has, indeed, become more common under the new system of employ-ment, which has substituted for a contract of fourteen days a contract terminable at a day's notice. To such an extent is truck carried, that even if the truck shop has not in store the articles required, the miner is not supplied with cash, which he might lay out where he would, but with tokens which certain shopkeepers in the town will recognise, and on receiving them supply articles to the extent of the value of the tokens. These tokens, however, have afterwards to be returned by the shopkeepers to the truck shop to be exchanged into cash, and the rate of exchange is a deduction of 3s. in every 20s., in favour of the truck shop. This loss, amounting to 15 per cent., the shopkeeper has, of course, taken care has already fallen upon the miner."

In the main, the above statement still holds good ; but we learn that some employers have shown that, in establishing stores at their works, they have no selfish motive, since they supply a quality of goods superior to what can be obtained for the same money at the village shops in their neighbourhood, while the workmen are free to choose where or how they shall spend their money.

As already stated, the relations between the east of Scotland miners and their employers have been little disturbed by disputes as to work or wages. In the west the case has been different—strikes being of frequent occurrence ; and it is remarkable that all the strikes have been for wages, none of them to get rid of any griev-ance in respect of bad management or bad ventilation. In 1832 the Lanarkshire miners were out on strike for four months ; and, instead of getting the advance they demanded, had to return to work for less wages than they had when they went out. Three years after-wards the first Union of the Scotch miners was established. In 1837 a reduction was made from 5s. to 4s. in the daily wage of the miners, and a general strike took place. Four months of idleness and privation were spent without result, the Union was dissolved, and the men returned to work at the terms offered. Wages gradually declined until, in 1842, they reached so low as 2s. 6d., and even 1s. 8d. a-day. The Union was then resuscitated, and the men went out on strike. An increased demand for coal and iron sprang up in the meantime, and prices advanced to such an extent, that the masters were enabled to grant to the miners an advance of from 1s. to 2s. 6d. a-day. During the strike, affairs in Lanarkshire wore such a threatening appearance that a military force was held in readiness at Coatbridge. After this success the Union was again allowed to decline, and in the course of two years wages had undergone a considerable reduction. The Union lever was once more applied, with apparent success, for at the end of three months the masters yielded ; but it was evident that the men could not have held out much longer, for they had worked only a few weeks at the advanced rate, when the masters, who found that the prices they were receiving would not enable them to pay their men so much, intimated that a reduction was to be made to the previous rate, and to that reduction the men submitted quietly. In 1847 there was a great strike, in which the men, after standing out fourteen weeks for 5s. a-day for the " darg," or minimum quantity of coal put out daily, accepted the masters' terms of 3s. a-day, and in a few weeks afterwards submitted to a reduction to 2s. a-day. In 1852 the Scottish Miners' Association was formed, " for the protection of miners' rights and privileges, by providing funds for the support of members out of work." The Association is composed of local societies, each holding its own money, and remitting only what may be required to cover the necessary expenses of the general Association. The entry-money is sixpence, and each member has to contribute one penny a-week for the purposes of the society. There is a central board, consisting of three persons, who summon a conference of delegates from district societies when any matter of general interest comes up for consideration. A year after the Association was established another strike for an advance of wages took place. But some of the men considered that they had endured quite enough of needless privation on previous occasions, and refused to join in the contest. They were accordingly subjected to abuse, and military, yeomanry, and police were called out to protect them. Hundreds of men who had never been in a pit were found willing to take employment as colliers; and when, at the end of four months' idleness, the men gave in, they in many instances found their places occupied. The result of the turn-out was a considerable reduction in the value of labour. Any commotion in the coal trade generally tells on the iron trade, and *vice versâ;* and in the year of the strike referred to, the quantity of pig-iron made in Scotland was 80,000 tons less than in the preceding year, while the price went up considerably. " In November 1855," says Mr Tremenheere, in his official report on the mining districts, " the men forced the masters into granting an advance of from 4s. to 5s. per day. Many of the masters, since they could not afford to continue that rate, before long felt obliged to take their stand against it, and to reduce the rate again to 4s. A strike took place. It became general in the districts concerned at the beginning of March 1856; and after from 30,000 to 40,000 men had been idle for sixteen weeks, and had inflicted enormous loss upon their masters and the public, and great suffering upon themselves and their families, the event proved that they were again wrong, for they resumed work at their masters' terms." In 1860 a strike of a most disastrous nature took place in Lanarkshire—the loss to the district being estimated at L.200,000. Frequent disputes have since occurred, chiefly in the west.

For the most part miners reside in houses specially built for their accommodation in the vicinity of the collieries. In the early days of coal mining the houses were of the most wretched description; and even yet a large proportion of them are deficient in the ordinary requisites of human habitations. To convey some idea of the present condition of the houses of the mining population, we shall give some account of what came under observation in a journey through the Mid-Lothian and Lanarkshire collieries, in both of which the best and worst classes of miners' houses may be seen. In Mid-Lothian there are a number of mining villages of the old type. These were built when people's notions of personal comfort were not quite so refined as in the present day, and probably the first occupiers of them were content; but the case has come to be different. No working man is likely to be less particular than the miner as to the mere stone-and-lime comforts of his home, since it must be a poor place indeed that will not look comfortable in contrast with the damp and gloomy recesses of the mine; but even he has come to think that something is due to him in the way of providing a better lodging for himself, his wife, and little ones. For a number of years improvements have been in progress. Old houses have in some cases been patched, altered, and provided with coal-sheds, &c., and others have been removed to make way for more commodious structures. Where the opening of new pits has necessitated the erection of new houses, these have been built on an improved plan. A bit of ground is attached to all the houses; but hitherto the miners' horticultural taste and skill have been almost entirely devoted to the rearing of cabbages and leeks. Flowers are rarely met with near the old houses, and their cultivation seldom gets beyond the range of a dilapidated tea-pot on the window-sill; but some of the plots in front of the new cottages are nicely laid out. It is still possible to find not a few specimens of the old domiciles displaying all their primitive unhealthiness and ugliness. Their mean masonry, founded on the surface without any excavation, rises to a height of little over five feet, and they are roofed with tiles, though we were informed that the first covering was a combination of turf and straw. The floor is composed of native earth, and its uneven surface has by constant treading been rendered almost as compact as stone. In front of the fire-place a brick or two has been let into the floor, and the vicinity of the doorway is similarly strengthened against the wear of frequent feet. The walls have never been plastered, but successive coats of whitewash have made them air-tight, if not beautiful. The window—some of the houses have two, however—is about two feet square. There is no ceiling beneath the tiles, but the want of it has to some extent been supplied by nailing mats upon the rafters, and overlaying them with a thick coat of whiting—an arrangement which, while it may improve the appearance of the interior, certainly detracts from its healthiness, or rather increases its unhealthiness. Strictly speaking, the houses consist of only one apartment measuring twelve feet by fifteen; but by a peculiar arrangement of the furniture a small closet is formed, in which a bed is fitted up. Occasionally a house of this kind may be found occupied by a miner, his wife, and four, six, or even eight of a family. Neither ash-pit nor drain is provided, and the surroundings of the dwellings are consequently in a most insalubrious condition. Notwithstanding these unfavourable circumstances, a creditable effort at cleanliness and tidiness is sometimes observable in even the meanest of the hovels.

The houses provided by the Duke of Buccleuch for the men employed in the Dalkeith Colliery, though built a considerable time ago, have few equals. They are well constructed and commodious, and are of various sizes, to suit the requirements of different families. All have two or more apartments, and are supplied with water and water-closets. Large spaces of ground are attached to the houses, and may be used for drying clothes or as a playground for the children. No rent is charged, and the people appear to be well cared for and contented. His grace makes liberal provision for men incapacitated for work ; and widows receive fortnightly pensions, and are provided with houses, coal, &c., free of charge. Invalids are supplied with wine, beef, soup, &c., from Dalkeith House. The effect of this kindness is visible upon all engaged about the colliery. They are generally more regular in their habits than people connected with some of the other collieries in the county, and rarely leave the employment of their own accord.

The Marquis of Lothian owns two hundred and sixty miners' houses, among which are to be found some of the best of the kind in Scotland, together with some of the worst. The Newbattle Colliery, with which they are connected, is one of the oldest in the county, and has never been leased, the successive Marquises keeping the working of it in their own hands. The earlier houses of the miners were miserable thatched hovels ; but all the houses built within the past thirty or forty years are of a superior description. The present Marquis, who takes much interest in the welfare of his work-people, commenced a few years ago to work extensive reform in the houses.   Only a few cottages of the very old type remain, and the dwellings by which they are being superseded are very comfortable and commodious, some of them containing four or five apartments. The rooms, though small, are lofty and well ventilated. The walls are of brick, the floors of glazed tiles, and the roofs of slate. They are well planned, and externally have some architectural pretensions. All things considered, the houses are well furnished; and it is a noteworthy fact that, though most of the people, while living in the old houses, appeared to be careless as to the quality or condition of their furniture, they were no sooner removed into one of those new roomy domiciles than they displayed quite a contrary taste. It is true that some of the new houses appear to be tenanted by people who cannot appreciate the change, yet the foregoing remarks hold good in the majority of cases.   The new houses are supplied with water, have flower-gardens in front, and kitchen-gardens and coal-houses behind. The rents charged vary from L.1, 10s. to L.3, 18s. per annum; and, as elsewhere, the rent is deducted from the fortnightly pay of the men. The houses at the other collieries in Mid-Lothian are of a mixed kind, and in the case of many of them, as already hinted, there is urgent need for improvement. Comfortable houses have a powerful effect in elevating the tastes and habits of the working classes—a fact which should be borne in mind by all who have their welfare at heart.

In Lanarkshire a great majority of the miners' houses are of a very poor kind, and many of them have only one apartment. They are arranged either in closely-built rows or confined squares, and the people are literally huddled together in them. It is no uncommon thing to find a family of six or seven persons living together in one room measuring not more than fourteen feet square, and who yet consider that they have accommodation to spare for one or two lodgers. The Irish, it appears, are especially given to overcrowding their dwellings—against the ventilation of which, too, they carefully guard. One favourable circumstance is that their furniture does not occupy much space. The sitting accommodation rarely consists of more than two chairs or a rudely constructed "form." The fire is kept burning continually, its use during the night being to dry the "pit-clothes" of the men ; and as these are often wet, and always dirty, the vapour they give out adds considerably to the pollution of the atmosphere breathed by the crowded sleepers. The houses at some collieries are of very slim construction, and are constantly getting out of repair. Not unfrequently they are wrecked by the subsidence of ground caused by the withdrawal of the coal beneath them. For these houses a rent of from 3s. 6d. to 5s. a-month is charged. At some collieries one-half of the houses have two apartments, and these are occupied by the better class of workpeople. The usual rent of a two-roomed house is from 6s. to 9s. a-month. These remarks, of course, apply to the worst class of houses. Some of the coalmasters have done a great deal towards providing comfortable dwellings for their workmen, and more is still being done. At Overtown, near Wishaw, and at Motherwell, a large number of houses of an improved kind have recently been built ; and at Gartsherrie and Govan the workmen have good houses provided for them. But there is no blinking the fact that, in the aggregate, the dwellings of the mining population of the west of Scotland are far from what they should be. On questions of work and wages the miners are very sensitive, but they have never made any movement for having their dwellings improved.

A serious obstacle in the way of sanitary improvement in the mining villages of the west, is the want of anything like a regular or adequate supply of water suitable for domestic purposes. Many of the larger villages have had a supply brought to them, but in the smaller and more remote hamlets great hardship is endured in consequence of the scarcity of water. In some instances the younger children or girls of the family have to carry the indispensable element in pitchers from a distance of one or two miles, and then the water acquires a value which prevents anything like a free use of it for purposes of cleanliness. The eaves-droppings are collected in barrels, carefully covered and locked; but the rain supply is uncertain, and in summer especially, we believe, some poor families endure great privations. The local authorities under the Public Health (Scotland) Act 1867, have ample power to deal with this matter, and it is to be hoped that they will do something to remove so great an evil.

Abundant facilities exist for the education of the children of colliers. In connection with almost every colliery of any extent there are one or more schools provided by the coalmasters; and though the schools are only indifferently appreciated by the parents, the children attend pretty regularly. It appears, however, that at some of the large works in Lanarkshire it has been found necessary to use a little pressure in order to get the parents to take an interest in the education of their children. In several cases the father has deducted from his wages a certain sum for every child he has who ought to be at school, whether the child attends or not; and the effect of this rule is generally found to be, that the father comes to think that since he has to pay the fee the child may as well be sent to school. The fee charged at the Duke of Buccleuch's colliery school is 3d. a-week for each child. At the Marquis of Lothian's schools the charge is 1d. a-month for each branch. In no case do the fees nearly meet the expenses of the school, but the proprietors contribute whatever additional money may be required.

One or two medical officers are attached to each colliery, and all the men pay a small sum weekly for their services. The employers defray all extra charges on account of accidents.

After the coal passes out of the hands of the miners, its distribution over the country and to foreign parts gives employment to many thousands of persons, and most railway companies draw a large amount of revenue from it. At the coal depôts of the principal towns a considerable number of men are employed in removing the coal from the railway waggons or canal boats, piling it into heaps, weighing, and filling it into carts. Then a large staff of porters are required to shovel it into cellars or carry it up stairs. All these men being unskilled labourers, receive a small wage—not more, we believe, than from 14s. to 18s. a-week.

The declared annual value of the coal exported from Scotland is about half-a-million sterling. In 1866 it was L.515,805, divided over the principal ports in the following proportions :—Glasgow, L.51,493; Leith, L.79,777; Greenock, L.38,835; Grangemouth, L.50,244; Ardrossan and Troon, L.73,642; Dundee, L.16,101;

Borrowstounness, L.93,671 ; Kirkcaldy, L.53,528 ; other ports, L.58,424. The quantity of coal, cinders, and culm represented by the above would be about 1,500,000 tons. Our customers for coal are scattered all over the world. Some cargoes are sent even to San Francisco, though the freight to that port from Glasgow is 50s. a-ton, or more than seven times the cost of the coal at the port of shipment.

THE MANUFACTURE OF IRON.

THE ORIGIN AND PROGRESS OF THE MANUFACTURE OF IRON IN SCOTLAND—
STATISTICS OF THE TRADE—EFFECTS OF OVER-SPECULATION—COATBRIDGE
AND ITS FURNACES—DESCRIPTION OF THE GARTSHERRIE IRONWORKS—THE
SMELTING PROCESS—INVENTION OF THE HOT BLAST AND ITS EFFECT ON
THE TRADE.

THOUGH the existence of ironstone in the Scotch coal measures was
known many years previously, no attempt was made to turn it to
account until the year 1760, when the Carron Ironworks were
established. Only one kind of ironstone was then used—namely,
the argillaceous or "clayband;" for the more valuable carbonaceous
or "blackband" was not discovered till the beginning of the present
century. These two varieties are known as the coal measure ironstones, and are found in all the great coal fields of Britain except
those of Northumberland, Durham, and Lancashire. Though there
are nineteen kinds of iron ore known to the mineralogist, it has been
calculated that nine-tenths of the iron produced is derived from the
clayband and blackband ironstones, the relative value of which is
thus stated in a paper read before the Scottish Society of Arts by
Mr Ralph Moore, Government inspector of Mines :—" Clay ironstones
contain from thirty to fifty per cent. of metallic iron. Before being
melted, they are mixed with coal, and calcined in kilns or large heaps,
to drive off the carbonic acid gas, sulphur, and other impurities.
This description of ironstone is found in seams or bands, and in
nodules, throughout the whole of the measures, but is most plentiful
in the lower part of the section. Blackband ironstone is a carbonate
of iron, laminated with coal, generally in sufficient quantity for cal-
cination without further admixture of coal ; and leaves, when
calcined, a metallic coke containing from fifty to seventy per cent. of
metallic iron. This description of ironstone is found in seams or
bands in well-defined positions in the measures, but these are neither

persistent in position nor equable in quality. Sometimes the seam is wanting altogether, or so thin as to be unworkable ; at other times, the coaly element so predominates that its metallic value is of small amount, while not unfrequently it contains nothing but coal. A good blackband ironstone contains from two to eight per cent. of coal. When it contains more than twenty per cent. of coal, it is of little value unless mixed with clayband, which uses up the excess of coal. It is more easily melted than clayband, and requires less coal ; and the weekly produce of a furnace from blackband is fifty per cent. greater than from clayband."

Deposits of hematite, or red iron ore, have been discovered recently in Haddingtonshire, Dumfriesshire, and Kirkcudbright, and operations for utilising the ore are in progress.

From the establishment of the Carron Ironworks in 1760 till 1788, the quantity of iron produced in Scotland did not exceed 1500 tons per annum ; but during the succeeding eight years a number of new furnaces were erected in the counties of Lanark, Fife, and Ayr. In 1796 the number of furnaces was seventeen, and the quantity of iron made in that year was 18,640 tons. Thirty-three years afterwards the production had reached 29,000 tons; and from that figure the invention of the hot-blast process raised it to 75,000 tons in 1836. By that time the construction of railways had begun to open a new market for the iron-merchant ; and to supply the demand, many new ironstone pits were opened, and furnaces erected. In the ten years from 1835 to 1845, the production of iron increased about 700 per cent.— the quantity made in the latter year being 475,000 tons. Ten years later, 825,000 tons were made ; and another decade gave an addition of 339,000—the quantity manufactured in 1865 being 1,164,000 tons. Over speculation, and the consequent financial crisis in 1866, operated most prejudicially on the iron trade, and the production for that year fell to 994,000 tons. Matters gave promise of taking a turn for the better in 1867 ; but the promise was realised only to a limited extent, the revelations made in connection with several of the principal railway companies, and other causes, having had an unfavourable effect on the trade. The total quantity made in 1867 was 1,031,000 tons—an increase of 37,000 as compared with the preceding year, but 133,000 tons short of the production of 1865. The quantity shipped in each of the past ten years has not varied much. In 1867 it was 593,277 tons, of which 338,364 tons were for foreign ports, and the remainder went coastwise. Of the quantity shipped to foreign ports, France took 60,500 tons ; Germany and Holland

99,600 tons; Belgium, Denmark, Sweden, and Norway, 20,100; Russia, 9600; Spain and Portugal, 5100; Italy, 14,200; United States, 117,300; British America, 43,000; East Indies, China, Australia, and South America, 14,000. The largest quantity ever consumed in Scotland in one year was 532,000 tons, in 1865. In 1867, the quantity taken was 420,262 tons. The make of malleable iron in 1867 was 142,800 tons, being a reduction of 11,400 tons as compared with 1866.

The rapid development of the iron trade has not been peculiar to Scotland. England, Wales, and a number of Continental countries have had a similar experience, arising from the same causes—namely, the formation of railways, the substitution of iron for wood in the construction of ships, and its increased application, in the form of machinery, to the industrial arts. The number of blast furnaces in Scotland is 164, of which 108 were in blast during 1867. Each furnace produced on the average about 9546 tons. One furnace gives employment, directly and indirectly, to fully two hundred men and boys, so that the number of persons engaged in the production of pig-iron during the year could not be less than 22,000, with wages ranging from 2s. to 6s. a-day. It will thus be seen that the "damping out," or stoppage of a furnace is a serious matter for the working population in the iron districts. Were the entire number of furnaces in blast, employment would be given to upwards of 33,000 men and boys, while the annual production would exceed 1,500,000 tons. The number of ironstone miners in Scotland is about 13,000, and the largest quantity of ore put out in a year was 2,500,000 tons in 1857.

The occupation of the ironstone miner differs little from that of the coal-miner, and the two occupy nearly the same position as regards wages, &c. The ironstone seams are generally only from six to eighteen inches in thickness, so that in taking out the ore a considerable quantity of rock has to be excavated. As the miner advances, he builds up behind him as much as possible of the stone and rubbish, and sends out the ironstone and surplus material in small waggons or "hutches."

The price of pig-iron has been subject to considerable fluctuations. In 1854, the mean average price of a ton of Scotch pig-iron was 79s. 7d.; it was 54s. 4d. in 1858; and 49s. 3d. in 1861. In 1866 the market was much disturbed by the operations of certain bold speculators, who forced the top price up to 82s. 6d.—an increase of 21s. on the price at the close of the preceding year. Then came financial disasters; and the brief space of four weeks witnessed a fall of 31s. 6d. a ton. After rising and falling several times subsequently, the price at the close of 1866 was 54s. 6d. In order to enable them to overcome the effects of the crisis brought about by the speculators alluded to, the ironmasters resolved to reduce the production, and forty furnaces were "blown out." Nearly eight thousand men and boys were thus thrown idle, while at the same time the wages of the men retained were considerably reduced. In the course of 1867 a number of the furnaces which had been stopped were set on, and about one-third of the men who had been thrown idle were restored to work. The average price of pig-iron during 1867 was 53s. 6d. per ton. The lowest figure reached in the course of the year was 51s. 3d., in the month of March, and the highest 55s. 6d., in October.

From the convenient situation and facilities for transport enjoyed by Scotch ironmasters, coupled with the cheapness of labour, it might be thought that no English or other producers of the metal could undersell them; but it is, nevertheless, a fact that something like 70,000 tons of pig-iron were in 1867 imported into Scotland, from Middlesborough, in Yorkshire, where ironstone costs less than one-fourth of its value in Scotland. The iron is, however, of a much lower quality than the native Scotch, and is used for mixing with the latter for the production of certain kinds of material. A small quantity of fine pig-iron is brought from West Cumberland by some of the malleable iron makers, who use it to mix with and improve the quality of the native iron. It is expected that, when the Solway Junction Railway is opened, a large quantity of Cumberland iron ore will be brought into Scotland. In 1868 the price of malleable bars ranged from L.6, 15s. to L.7, 5s. a ton; plates, L.8, 10s.; and rails, L.6 to L.7. Cast-iron pipes were quoted at from L.4, 15s. to L.6 per ton, and railway chairs at from L.3, 12s. 6d. to L.4.

The most valuable deposits of ironstone are in Lanarkshire and Ayrshire, and in the former county two-thirds of the pig-iron made in Scotland is produced. The blast furnaces are chiefly concentrated in the vicinity of Coatbridge, Airdrie, and Wishaw, all of which towns were rapidly raised to importance by the development of the mineral treasures which lay beneath and around them. Coatbridge stands within a crescent of blast furnaces, and in the town are a large number of rolling mills, forges, and tube works, the hundred chimneys of which form quite a forest of brickwork capped with fire.

Though Coatbridge is a most interesting seat of industry, it is anything but beautiful. Dense clouds of smoke roll over it incessantly, and impart to all the buildings a peculiarly dingy aspect. A coat of black dust overlies everything, and in a few hours the visitor finds his complexion considerably deteriorated by the flakes of soot which fill the air, and settle on his face. To appreciate Coatbridge, it must be visited at night, when it presents a most extraordinary and—when seen for the first time—startling spectacle. From the steeple of the parish church, which stands on a considerable eminence, the flames of no fewer than fifty blast furnaces may be seen. In the daytime these flames are pale and unimpressive; but when night comes on, they appear to burn more fiercely, and gradually there is developed in the sky a lurid glow similar to that which hangs over a city when a great conflagration is in progress. For half-a-mile round each group of furnaces, the country is as well illumined as during full moon, and the good folks of Coatbridge have their streets lighted without tax or trouble. There is something grand in even a distant view of the furnaces; but the effect is much enhanced when they are approached to within a hundred yards or so. The flames then have a positively fascinating effect. No production of the pyrotechnist can match their wild gyrations. Their form is ever changing, and the variety of their movements is endless. Now they shoot far upward, and breaking short off, expire among the smoke; again spreading outward, they curl over the lips of the furnace, and dart through the doorways, as if determined to annihilate the bounds within which they are confined; then they sink low into the crater, and come forth with renewed strength in the shape of great tongues of fire, which sway backward and forward, as if seeking with a fierce eagerness something to devour.

The most extensive ironmasters in Scotland are Messrs Baird and Co., who own forty-two blast furnaces, employ nine thousand men and boys, and produce about three hundred thousand tons of pig-iron per annum, or one-fourth of the entire quantity made north of the Tweed. Twenty-six of their furnaces are situated in various parts of Ayrshire, and the remaining sixteen are concentrated at Gartsherrie, in the neighbourhood of Coatbridge. Gartsherrie Ironworks are the largest in Scotland, and it is stated there is only one establishment in Britain which has a greater number of furnaces. The quantity of pig-iron made is one hundred thousand tons per annum, and the number of men and boys connected with the works is three thousand two hundred. More than a thousand tons of coal are consumed every twenty-four hours; and, as showing how well chosen is the site of the works, it may be mentioned that nineteen-twentieths of the coal required is obtained within a distance of half-a-mile from the furnaces. One coal-pit is situated close to the furnaces, and has been in operation since the works were established, forty years ago.

The coal from this pit is conveyed to the furnaces by means of a self-acting incline. Most of the ironstone was at one time obtained from pits in the nighbourhood, but now it has to be brought from a distance of from two to twenty miles ; and a complete system of railways connects the pits with the works. The total length of the railways is about fifty miles, and the traffic is carried on by means of six locomotives and an immense number of trucks. The establishment is also connected with the great railway systems of the country, and possesses additional facilities for transport in a branch of the Monklands Canal, which has been carried through the centre of the works. For the canal traffic, there is a fleet of eighteen barges, of about sixty tons each ; and eight of these are screw steamers. A great proportion of the manufactured iron is sent out by the canal.

As the Gartsherrie Ironworks have a wide-spread reputation for producing iron of a superior quality, and are among the best organised manufactories in the country, a description of them may be interesting.

The furnaces, sixteen in number, stand in two rows, one on each side of the canal, and about forty yards distant from it. A constant supply of coal and ironstone can be reckoned upon, and therefore only a small stock is kept at the works. The mineral trains are worked with unfailing regularity, and their cargoes are deposited conveniently for immediate use. There is thus no superfluous shovelling about of the materials, nor is any expense incurred by piling them into heaps. The proportions of ironstone, coal, and limestone, laid down are exactly what are required in the process of smelting. Manual labour has, by a variety of ingenious appliances, been reduced to a minimum, and the amount of waste is infinitesimal. Everything is done according to a well-defined system, and nothing connected with the works is considered to be too insignificant to merit attention. No heaps of rubbish are allowed to accumulate, no scraps of iron or cinder lie about, and every nook and cranny about the vast place is as tidily kept as it can possibly be. The workmen are liberally treated, but they must do their work carefully and well. Negligence and irregularity are unfailingly punished, while merit is as certainly rewarded. All the men employed about the furnaces, even the firemen and engineers of the blast engines, are paid according to the quantity and quality of iron produced. This arrangement is found to work admirably, as each man knows that, by attending to his work, he is not only putting money into the pockets of his fellow-labourers, but also improving his own earnings.

Before the ironstone is ready for smelting it has to be calcined, which operation is performed at the pits. The object of calcining is to separate carbonic acid, water, sulphur, and other deleterious substances, which are volatile at a red heat ; and it is performed in this way :—A layer of rough coal is first laid down, and on that the ore, mixed with a certain quantity of small coal, is piled. The blackband ironstone, as it contains a large proportion of carbon, requires less coal to calcine it than the clayband. When the heap is completed, fire is applied to the windward side, and combustion goes on gradually until the desired effect is produced. When the ore cools, it is ready for the furnace ; but when the heat has been too intense, the ore is found to have run into large masses, the breaking up of which takes a considerable amount of labour.

Having been built at different periods, the Gartsherrie furnaces are of various patterns. The general shape is cylindrical, the diameter twenty-two feet, and the height sixty feet. The Nelson Monument, on Calton Hill, Edinburgh, would, were it less lofty, bear a close resemblance to one of the most recently erected furnaces. The furnace is fed from the top, and, in order to protect the " fillers," the mouth of it is surrounded by a light wall of brick, pierced with convenient openings. This brick wall is so much thinner than the main wall of the furnace on which it stands, that a gallery or footway several feet in width is left clear all round. Externally, there are four arched recesses in the base of the furnace, three of which are occupied by the " tuyeres," or pipes conveying the " blast ;" while the fourth contains a doorway by which the " slag " is drawn off, and also the opening through which the molten iron is discharged. The interior of the furnace consists of a circular cavity, seven and a half feet in diameter at the lower part or hearth. At a height of five or six feet from the bottom of the hearth, the cavity begins to increase in diameter, until, at half the height of the furnace, it measures eighteen feet across. It is then gradually contracted, and at the top the diameter is eleven feet. The materials with which the furnace is fed are roasted ore, coal, and limestone. The proportions of these vary according to their quality. In some cases, a small quantity of red-iron ore or hematite is used along with the blackband ironstone, and then the proportions of what are called a " charge " are these :—Coal, about 10 cwt.; roasted ore, 6½ cwt.; red ore, ½ cwt.; and lime, 2⅝ cwt. About sixty " charges " are thrown into the furnace in the course of twelve hours, and at six o'clock in the morning and at six at night the furnace is " tapped " and the iron run off. The chemical changes undergone by the materials introduced into the furnace are thus described :—The iron ore consists of iron, oxygen, and sand, and the object of the iron smelter is to separate the two latter substances from the former. The coal introduced has two functions to fulfil—in part it is burned so as to raise the contents of the furnace to such a high temperature that they will be enabled to act on each other ; and, at the same time, it carries away the oxygen which was originally in combination with the iron in the roasted ore. The lime plays the part of a flux, and combines with the sandy matter to form a slag. During the whole operation, hot air is being constantly forced in at the lower part of the furnace, so as to aid in the necessary combustion. The roasted iron ore being thus deprived of its oxygen by the coal, and of its sand by the lime, allows the other constituent—the iron—to trickle down through the mass of red-hot cinders to the lower part or hearth of the furnace.

In front of each furnace is a level piece of ground covered with coarse sand, in which before the " tapping " takes place a number of small furrows are formed. These communicate with larger channels leading from the opening in the furnace ; and when the iron is let out, it runs along the main channels in a glowing, bubbling stream, and distributes itself into all the hollows. The large channels are called "sows," and the small ones " pigs;" hence the term "pig-iron." Two men are employed to feed each furnace. One fills half a charge of coal into a large iron barrow, and the other half a charge of the other materials into a second barrow. The men and the barrows reach the staging communicating with the mouth of the furnace by means of a hydraulic lift. The coal is thrown in first, and the other materials immediately afterwards. The occupation of the " fillers " appears to be a somewhat dangerous one, as the flames at times shoot out upon, and almost surround them. Two men are employed at the hearth scooping out the slag and cinders with a huge spoon suspended from a crane, and from time to time stirring up the contents of the furnace.. This is very severe labour, and the faces of the men engaged in it have a half-roasted appearance. The slag is poured into iron trucks, and, when it consolidates, is wheeled away to be emptied on the waste heap—which, it may be mentioned, contains as much material as would build a copy of the Great Pyramid. The pig moulds are formed in the sand by boys, the operation being a very simple one.

Up till about forty years ago the air forced into blast-furnaces was cold, and the process of smelting was slow, and also costly, in consequence of the great quantity of coal that was required. In 1827, Mr J. B. Neilson, engineer of the Glasgow Gas-Works, conceived the idea of heating the air before injecting it into the furnace; and two years afterwards a most successful trial was given to the invention at the Clyde Ironworks. With the cold blast coke had to be used, and 8 tons 1¼ cwt. of coal converted into coke was required to reduce one ton of iron. It was found that when heated air was employed the coal might be used raw, and that 2 tons 13¼ cwt. was sufficient to smelt a ton of iron, including 8 cwt. required for heating the air. This discovery gave an extraordinary impetus to the iron trade, and the patentee and his partners are said to have realised L.300,000 by the invention. At Gartsherrie there are three immense engines for generating the blast—two for one range of furnaces, and one for the other. The engines are on the beam principle, and their united "duty" is equal to about 500 horse power. The steam cylinder of the largest is five and a-half feet in diameter, and ten feet deep, and the air cylinder is ten feet in diameter and depth. The air cylinders are simply gigantic pumps, which force the air into receivers, whence it flows at an equal pressure through the tubes of the heating oven, and into the furnace. By passing through the oven the temperature of the air is raised to 800°. It has been calculated that the quantity of air thrown into a blast-furnace in full work exceeds in weight all the solid material used in smelting.

In the vicinity of Gartsherrie there are about five hundred houses belonging to Messrs Baird & Co., and occupied by their workmen. Nearly all the houses have two apartments, and a few have a third room. A bit of garden ground is attached to each house, and all are supplied with water and gas at a cheap rate. The miners get as much coal as they require without payment—only they must dig it out for themselves; and the other workmen are charged only 3s. 6d. for a cartload. Liberal provision is made for the education of the children of the workpeople. There are three schools in direct connection with the works, each being divided into separate apartments for infants, boys, and girls. The workmen seem to appreciate highly all that has been done for their welfare, and few of them leave the place. They own one of the most successful co-operative stores in the country. It is managed by a committee of the workmen, but its prosperity is in a great measure owing to the fostering care of the employers, who, however, have no interest in the concern beyond seeing that it is properly conducted. There are seven hundred members in the society, nearly all of whom are heads of families, and the business done amounts to about L.1200 a-month. In addition to general grocery goods, wines, spirits, butcher meat, and potatoes, are sold in the store.

MANUFACTURES IN IRON.

THE following Act passed by the Scottish Parliament in 1686,
throws some light on the origin of working in cast iron in this
country :—

" His Majesty and Estates of Parliament, taking into consideration
the great advantage that the nation may have by the trade of Founding, lately brought into the kingdom by John Meikle, for casting of
balls, cannons, and other such useful instruments, do, for encouragement to him, and others in the same trade, statute and ordain, that
the same shall enjoy the benefit and priviledges of a manufacture in
all points as the other manufactures newly erected are allowed to
have by the laws and Acts of Parliament, and that for the space of
nineteen years next following the date hereof."

The Carron Ironworks may, however, claim to be the birth-
place of the Scotch iron trade in its most important form ; and
long had the reputation of being the most extensive foundry in
Britain. Though now surpassed in extent, they retain their old
reputation for producing work of a superior kind, and in several
branches of production they may be said to have no rivals. The
works were established in 1760, under a chartered company, projected by Dr Roebuck, of Sheffield, who appears to have been the
first to appreciate the value of the iron ores of Scotland. The operations of the Company have all along embraced the digging and
smelting of the ore, the manufacture of the iron into an endless
variety of articles, and the sending of these into every market of the

world. Notwithstanding the great extension of the iron trade that has taken place in recent years, the Carron Works have not been much enlarged ; and though improved appliances have been introduced, the buildings maintain pretty much their original appearance, allowance being made for the effects produced by the smoke and dust which have swept around them for a century. The establishment is situated on the bank of the Carron River, about two miles from Falkirk, and may be most conveniently reached from Grahamston railway station.

The operations carried on within the works were long kept secret from the outside world, and the rule which excluded Burns when he went to view the place is yet relaxed only in the case of persons who are not likely to use to the prejudice of the company any knowledge they may gain while witnessing what is going on. The poet gave vent to his feelings on the occasion referred to in the following lines, which he scratched on a pane of a window in the inn at Carron:—

> " We cam' na here to view your warks,
>   In hopes to be mair wise,
> But only, lest we gang to hell,
>   It may be nae surprise ;
> But when we tirled at your door,
>   Your porter dought na hear us ;
> Sae may, should we to hell's yetts come,
>   Your billy Satan sair us !"

If the history of Carron Ironworks were minutely written, it would be a record of much interest, as showing the many stages of improvement through which the manufacture of iron has passed. It must suffice here to mention the change which has taken place in the motive power at the works. The site of the establishment was chosen on account of the abundant and convenient supply of water which could be made available for driving the machinery. The blast was created, and the tilt-hammers, lathes, and other machines were driven by water applied over a large number of wheels. As the premises were extended the supply of water became inadequate, and somewhat anomalous means of overcoming this difficulty were devised. While James Watt was working out his improvements on the steam-engine, he entered into partnership with Dr Roebuck of the Carron Ironworks, and a joint-patent was taken out for a condenser. This partnership was not a fortunate one for Watt. During the time he was associated with Dr Roebuck, however, he erected a large steam-engine at Carron, and that was the anomalous contrivance alluded to. Instead of the power of the engine being applied directly to the machinery, it was merely employed to pump back into a reservoir the water that had passed over the water-wheels, and so enable it to be used again and again. The engine was fitted with four pumps, which raised to a height of thirty-six feet forty tons of water per minute. This old servant of the company has been sadly neglected. Though it has been allowed to remain in its original position, nothing has been done to prevent it falling into decay. The engine-room is crumbling into ruins, and the iron-work is black and furrowed by oxidation. As one of the earliest engines ever made, this piece of mechanism is an object of much interest to men of science; and it is to be regretted that the very little care necessary for its preservation has not been taken. The engine, which is on the atmospheric principle, has a cylinder six feet in diameter, by eight feet in depth, and the beam is about thirty feet in length. The steam was supplied by three cast-iron boilers, two of which are globular in shape, and measure fifteen feet in diameter. About thirty years ago this engine was superseded by one of improved design, applied directly to the machinery, and since then the use of water power has gradually died out. The machinery in the engineers' shops is, however, kept in motion by a powerful turbine wheel. A few years ago a splendid beam engine was added to the establishment. It was made by the company's own workmen, and supplies the blast to all the furnaces. The steam cylinder is six feet in diameter, and the piston has a stroke of ten feet. The blast cylinder is one hundred and four inches in diameter, and ten feet deep.

On approaching the works by the long irregularly built street leading in an almost direct line from Grahamston, the visitor's eye is first attracted by the flames of five blast furnaces which stand on the south side of the works. The smaller flames issuing from the chimneys of the cupola and air-furnaces next arrest attention ; and a nearer approach brings into view a whole forest of chimneys, shooting up from amid vast ranges of brick-built workshops. On getting inside the boundaries of the establishment the mere sight-seer would probably be somewhat disappointed. The great extent of the place does not become apparent until the various departments are visited in succession; nor can it be said that externally the workshops present an inviting appearance. But within those ragged-looking and smoke-begrimed structures, processes go on which illustrate some of the grandest developments of human ingenuity; and in no individual establishment, in this country at least, can such a variety of operations in the manufacture of iron be seen. As one passes through the place, the roar of furnaces, the clash of machinery, and the clatter of anvils, fall upon the ear from all sides, and combined with the irregular nature of the roadways, the immense and apparently confused piles of iron, old and new, and of finished and unfinished articles of every conceivable form, produce a most bewildering effect on persons unaccustomed to such sounds and scenes.

Persons who are privileged to visit the works are first shown the various processes in smelting iron, which are similar to those already described as being practised at Gartsherrie, and are then conducted through the other departments, commencing with the pattern-shop. The latter is a large three-story building, on the lower floor of which are a saw-mill and other machinery for preparing wood. On the middle floor the patterns are made; and the upper is filled from end to end with a vast collection of patterns of articles of all sorts, from a spittoon to the cylinders of a 200-horse power engine. As the cost of making the patterns is considerable—those here collected representing many thousands of pounds—they are carefully preserved. The patterns are chiefly made of wood, and considerable skill is required for their construction. When the patterns are completed, they pass into the hands of the moulders, who take an impression or mould of them in sand. The more simple the outline and plainer the surface of the article, the more easy it is to form the mould. For instance, nothing could be more simple than the operation of making a mould for the heater of a dressing-iron; but the moulding of a tea-kettle requires considerable skill on the part of the workman.

A few years ago the Prince of Wales visited Carron for the purpose of inspecting the works, and expressed a desire to see the process of making a mould. The moulding of a common three-legged pot was shown him, and no better illustration could be given of the moulder's work. The patterns for a pot consist of nine pieces—two for the body, three for the feet, and two for each of the ears. The body pieces have been formed by taking a completed pot, denuding it of feet and ears, and cutting it vertically into two pieces. These pieces the moulder takes, and, placing the severed edges together, lays them down on his bench with the bottom upward. He then encloses the pattern in a circular casing, which he fills up with sand. The sand is rammed down all round and over the pattern, care being taken during this process to insert the feet pieces, and also a wooden plug to form a " gate " through which to pour the metal. The moulder then turns the box over, and fills the inside of the pot with sand. The next part of the operation is to take out the pattern and leave open and entire the space it occupies. The advantage of having the casing and the pattern in sections now becomes manifest. The upper section of the casing is unfastened and taken off, when it is seen that the sand bears an impression of the bottom of the pot. The side pieces are in like manner removed, leaving the body pattern clear. The latter is carefully lifted off, one-half at a time, exposing the "core," or globular mass of sand which represents the interior of the pot. The surface of the sand is next thickly dusted with ground charcoal, and rubbed quite smooth—a process which makes the iron take a finer "skin" than it would otherwise do. The feet and ear pieces having been withdrawn, all that is now necessary is to put the casing together again, fasten it tight up, and prepare the "gate" by pulling out the plug and rounding off the edges of the hole. So compact does the sand become, that the completed mould may be moved about freely without sustaining injury. An expert hand can mould a pot of the largest size in from fifteen to twenty minutes.

After a certain number of moulds have been prepared, the workmen proceed to "cast" them. The molten metal is carried from the furnaces in huge ladles, and appears to be as fluid as water. When it is poured into the mould, gas is at once generated, which finds its way through the sand, and, issuing from the joints of the casing, becomes ignited, and burns with a beautiful purple flame. Were the gas not allowed to escape, the mould would burst, and the consequences to the workmen would be most disastrous. It is a curious fact that, while a few drops of water would ruin a mould, the boiling metal may be poured in from a height of a couple of feet without disturbing a particle of the sand. When the metal has cooled sufficiently, it is dug out of the sand and taken to the dressing shops, where roughnesses are removed. Articles cast in several pieces are then carried to the fitting shops, where they are put together. Kettles and stew-pans, which are to be tinned, are first annealed, and then passed to turners, who put a smooth and bright surface on the inside. The tinning is then done, the handles are put on, the outsides japanned, and the completed goods removed to the warehouse. Portions of many of the articles are of malleable iron, such as the handles of kettles and pans, and in making these a large number of smiths are employed.

The division of labour system is extensively applied in the works, and the result is, that the men in the various departments display extraordinary expertness. When a boy enters on his apprenticeship, he chooses, or has chosen for him, the branch of work which he is to follow, and to that he adheres. Let us suppose that a boy selects pot-moulding. After some preliminary training, he is entrusted with the making of pots of the smallest size. As he advances in years, so does the size of his pots increase; and by the time that grey hairs come, he finds his hands employed upon vessels so capacious that each might contain a dozen of those he made in his early days. This is one of the peculiarities of life at Carron; and though it looks as if designed to remind the men of the flight of time and the growth of years upon them, it is simply the result of promotion by seniority. The mould for a small pot requires nearly as much time to make as that of a large one; but there is a difference of price in favour of the latter, and these the older hands claim the privilege of making. Another peculiarity of the pot-making branch is the mode of payment. A man agrees to make a certain number of pots for half-a-crown, and he is allowed one shilling of premium on every hundred he produces. Taken altogether, the men employed in moulding make higher wages than those in the other departments, and it is no unusual thing for one of them to receive as much as L.3 for a week's work; but the general wage of the class may be set down at about 25s. a-week.

Though the reputation of Carron is now chiefly based on its production of what may be called domestic iron-work—such as stoves, grates, cooking ranges, boilers, pots, rain-pipes, &c.—at one time it was closely identified with the manufacture of cannon and shot. The now obsolete piece of ordnance known as the "carronade" was there brought to perfection, and derived its name from the works. None of those guns have been made since 1852, about which time the revolution which has taken place in the construction of implements of war commenced. Among the heaps of old iron in the works may be seen one or two condemned castings of carronades, which show the mode of manufacture. The guns were cast solid, in an upright position; and in order to ensure closeness of texture, the mould was filled up for a distance of two feet above the muzzle of the gun. This superfluous mass was cut off, and the gun bored to the required calibre.

The company possess, and work for themselves, extensive mines of coal, iron, and lime, some of which are in the immediate vicinity of the establishment. The raw material is brought in by a railway which approaches close to the furnaces; and additional facility for carrying is afforded by a canal three miles in length, extending from the centre of the works to Grangemouth. For the conveyance of goods to the east and west, sixteen canal boats are employed. Six steam-vessels are owned by the company, and chiefly occupied in carrying the produce of the foundry to London, where the company have an extensive warehouse. Attached to the works is a farm of four hundred acres ; and no fewer than five villages in the vicinity are dependencies of the company, and many of the houses have been built by them.

The company employ nearly two thousand men and boys, whose labours are never affected by fluctuations in the markets for the productions—as, when a temporary slackening of demand takes place, the company go on making stock goods ; and as a precaution against any contingency that might interfere with the supply of raw material, an immense stock is always kept on hand. These circumstances are, of course, a great advantage to the men ; and one of the results is that very few leave the service of the company, so that the great body of them are natives of the locality, whose forefathers for three or four generations had worked in the place. There are no Irish among them ; and, in the aggregate, they are an exemplary body of workmen. They have three principal benefit societies and a number of minor ones. The most important society has been in existence for many years, has accumulated a considerable amount of capital, and holds an interest in the company. It has a membership of seven hundred, mostly heads of families. A co-operative store has flourished in connection with the works for upwards of forty years. No special provision for the education of the children of the workmen was provided by the company until recently, when a large and commodious schoolroom was erected by them.

The Falkirk Iron works, which are situated in the immediate vicinity of Carron, deserve notice, both on account of their extent and the kind of goods they produce. They were started in 1819 by a company chiefly, if not entirely, composed of workmen from Carron. The beginning was not a pretentious one, but the concern prospered. In 1848 the establishment was acquired by the present proprietors, Messrs Kennard, who had been shareholders for many years in the old company. Since then it has steadily risen in importance, and is now the largest foundry in Scotland, with the exception of Carron. Nine hundred men and boys are employed ; and when ordinarily busy, upwards of 300 tons of castings are turned out per week. The buildings, which cover eight acres of ground, have during the last few years been almost entirely reconstructed, and considerable additions are being made. The most improved appliances are in use in every department. During the Crimean war, 16,000 tons of shot and shell were made at the foundry for Government ; and guns of all sizes, from 4 to 18-pounders, for use on board mercantile vessels, are manufactured in considerable numbers. With these exceptions, the productions of the firm are associated with the arts of peace, and they range from bridges of the largest size to ornamental inkstands and fancy castings of the most delicate patterns possible in cast iron. The castings for some of the principal iron bridges in India, Italy, and Spain have been made at the Falkirk Ironworks. The heaviest pieces of work recently executed were the columns for the Solway Viaduct. This bridge is supported on groups of cast-iron columns, securely fixed in the bed of the Firth, and strengthened by diagonal bracings of malleable iron. The columns were cast in ten and twenty feet lengths, which were readily bolted together, and made as strong as if the entire column were cast in one piece. A large number of fountains for the Calcutta Water Company were turned out a short time ago. These were of a neat design, and bore the arms of the company, and the maxim "Waste not, want not," in English and Hindustani. Those for the East Indies; sugar-pans for the West Indies; tubular telegraph-posts for South America; grates, pots, and pans for the million; and beautiful objects of art for homes in many lands—might be seen piled side by side in the elegant arcade and extensive shipping warehouses, testifying alike to the wide connection of the firm and to the merit of Scottish workmanship.

Messrs Kennard have devoted great attention to the production of cast-iron goods of an artistic kind; and it is stated that no establishment in Britain possesses such a valuable collection of patterns for stoves, grates, umbrella-stands, garden-seats, verandahs, iron stairs, balconies, and fancy articles—such as inkstands, card-trays, mirror-frames, statuary groups, &c. The designs of these articles are without exception beautiful, and they are being manufactured in constantly increasing variety. When any new article is to be produced, a drawing of it is first made, and from that a modeller forms a pattern in wood, wax, or plaster. From the pattern a cast is taken in tin, and from the tin copy, which is nicely chased up, the moulder makes the impression in sand from which the iron is cast. A smoother surface is thus given to the iron than would be the case were a wood pattern used. In all cases the details of the pattern are sharpened in the iron after casting, by filing. Though no model seems to be too difficult for the moulder to make in one piece, yet, as a matter of convenience, most articles of any size or complexity are made in several pieces. A specimen of work from the moulding-shop in which the ornamental castings are made was shown at the Exhibition of 1862, along with a variety of other castings, and excited a great deal of interest, as showing the capabilities of the sand-moulding process. It was a small figure of a stag browsing; and, in order to its being cast in one piece, the mould had to be made in upwards of one hundred parts, each part being simply a clod of moist sand held together by compression.

There are six other foundries in the neighbourhood of Falkirk, which give employment in the aggregate to six hundred men. Glasgow, however, is the centre of the iron trade, as indeed of nearly all other trades in Scotland, and produces a great amount of iron goods of every kind. The foundries and machine shops of Edinburgh, Leith, Dalkeith, Kirkcaldy, Dundee, and Aberdeen, also turn out a considerable quantity of machinery, &c.

Fire-grates and stoves form a large portion of the produce of several establishments, and one in the Falkirk district is devoted exclusively to making stoves after American patterns. A "ventilating fireplace," recently patented by Mr J. D. Morrison, surgeon-dentist, Edinburgh, may be noticed here, as it bids fair to revolutionise the modes of warming and ventilating hitherto in use. Mr Morrison's invention consists of a combination of the open fireplace and close stove, possessing the pleasant appearance of the former, and the active heating qualities of the latter, and yet free from the defects of both. As applied to a room, it consists of a semicircular apparatus, fitted into an ordinary fireplace, the centre projecting a little beyond the lintel of the mantelpiece. This form, besides giving the fireplace a handsome appearance, serves, as will be seen, a twofold purpose. The grate containing the fire is brought a little further forward than usual. There are two sliding doors opening from the centre to right and left, for modifying, as required, the consumption of fuel and the warmth of the apartment. These doors, when open, expose a large amount of radiating surface, taking in a range of 150° of the room. The products of combustion are purified before escaping to the air, and very little smoke is emitted. On either side of the cast-iron chimney at the back of the fireplace are two caliducts for moderately warming fresh air before entering the room. And here the semicircular form of the apparatus is made to perform its other function. By means of a perforated opening above the fire, and separated from it, the warm fresh air is discharged radially into the room. The tendency of the heated air to ascend is taken advantage of in producing a complete circulation and continual changes, for, as the air passes over and across the room, it becomes gradually cooled, and descends by the

D

walls and windows furthest from the heat. By a simple contrivance, the air which has been vitiated by respiration is returned to the fire, and freed from its impurities. By another arrangement in the system, the vitiated air is conveyed through the ceiling to the fire in a room above. This continual influx of fresh air, and the abstraction, as described, of the air after having been used, besides effectually ventilating and heating the room, serves yet another purpose. The draughts caused by the ordinary fireplace are completely neutralised by Mr Morrison's apparatus. The weight of the new grate is very little more than that of one of the ordinary kind, and it is quite as easily fitted to its place. Mr Morrison intends his system of warming and ventilating to be applied to hospitals, churches, public buildings, and ships, as well as to dwelling-houses; but, especially in its application to rooms, his endeavour has been to realise in this country the mild and salubrious conditions of a warmer climate. The system is applicable to all the apartments of private dwellings. Messrs Kemp & Co., philosophical-instrument makers, Edinburgh, have arranged with the patentee to add a supplementary establishment for constructing and introducing the apparatus in all its applications.

Though cast iron may be readily formed into articles of complex shape, its brittleness sets a limit to its use ; and in the construction of the working parts of machinery, or articles in which great strength and lightness have to be combined, malleable iron must be used. But from the difficulty of working malleable iron, the cost of articles made of it is much greater. The knowledge how to treat the metal, so that, while it might be cast into any shape, it should retain all the qualities of malleable iron, was until recently a desideratum. The possibility of so treating both iron and steel has, we believe, been successfully proved, but as yet the process is kept secret. The conversion of pig-iron into malleable by the "puddling" process was commenced in Scotland about forty years ago, when a number of workmen from England and Wales were brought into Lanarkshire for the purpose of instructing the Scotch ironworkers. The first attempts, however, to establish this branch of trade were not successful, and it was not until 1836 that it was fairly started. There are now nearly four hundred puddling furnaces and fifty rolling-mills in operation, which in 1867 produced 143,800 tons of malleable iron, valued at L.1,006,600. The principal firm in the trade is the Glasgow Iron Company, which has extensive premises at St Rollox, Motherwell, and Wishaw.

The works for the conversion of pig-iron into malleable are nearly all constructed on the same plan. The mill consists of a vast roof supported on iron pillars, so that the sides are open. The puddling furnaces are built at intervals along one or two sides of the mill; and the floor, which is paved with iron plates, is crowded with machinery, a powerful steam-engine occupying the centre.

The work of the puddlers is probably the severest kind of labour voluntarily undertaken by men. The puddling-furnace is a compact structure of firebrick cased in iron. It consists of three parts—the fireplace, the hearth, and the flue. The fireplace is on the left-hand side, and is separated from the hearth, which occupies the central place, by a low wall or ridge. To the right of the hearth is the flue, the entrance to which slopes downward from the hearth, so that, when a fire is lighted in the fireplace, the flame is drawn close over the hearth in its passage to the flue. Each furnace requires two men to work it. One of these is the puddler, who has all the responsibility, and the other his assistant, who performs the portions of the work in which only slight skill is required. The quantity of pig-iron operated upon at a time is about four hundredweight, and is called a charge. One charge is got out of the furnace every two hours, and the work goes on night and day, from one week's end to the other, Sunday excepted—the men taking the night and day shifts by turns. After a charge is withdrawn, the furnace undergoes some slight preparation before another is put in. A coating of "bull dog"—a material prepared from the slag of the furnace—is laid on the hearth, to fortify it against the intense heat. The pig-iron, which has previously been broken into pieces of convenient size, is then thrown in, and the doors of the furnace are closed and sealed up with cinders. Intense heat is then generated; and so fiercely does the fire burn that the flame issues from the top of the chimney, which is upwards of forty feet high. In about a quarter of an hour after the furnace has been sealed, the iron shows signs of melting, and an aperture in the hearth door, about six inches square, is opened. The puddler, whose eyes seem to be proof against a light as dazzling as the sun at noon, looks in at the opening, and determines whether it is time to disturb the iron. So soon as he sees the finer angles of the iron begin to melt, he thrusts in a stout rod of malleable iron, and moves the lumps of metal about, so that the entire mass may be equally heated. If this were not done, the parts which melted first would be burned up and lost, and the quality of what remained deteriorated. The puddler's assistant takes a turn at this part of the work; and during its progress the heat is occasionally moderated by means of the "damper," or by dashing small quantities of water upon the iron

At frequent intervals the puddling bar is withdrawn, and cooled by being dipped into water. The iron dissolves gradually on the hearth, and after a time begins to heave and bubble, innumerable jets of flame bursting forth all over its surface. The desired chemical change is now going on. The hot air from the furnace sweeps over the iron and carries off a great part of the carbon, sulphur, phosphorus, and silicon contained in the pig-iron. Care must be taken to prevent the metal from becoming too fluid; and as soon as it attains a pasty consistency, the heat is moderated. Meantime the puddler uses his rod vigorously; and as the metal begins to "dry," the labour of moving it about is increased. The metal at length seems to curdle and become granular. As it then ceases to give off carbonic oxide, the heat of the furnace is again raised, and the particles of metal begin to adhere together. From this point the chief puddler undertakes and completes the operation. As the metal agglutinates, it becomes very difficult to move. The puddler has to exert himself to the utmost; and he dare not relax his efforts for a single minute, else all the previous labour would be worse than lost. Though the perspiration trickles from his face and arms, and oozes through his scanty clothing, he must toil on. His eye is never removed from watching the contents of the furnace; and the expression of anxiety on his face indicates that the operation has reached a critical point. When the metal has attained a certain degree of consistency, the puddler divides it into five or six heaps. He then works each heap into a "ball" or "bloom" The door of the hearth is opened, and one after the other the balls are drawn out with a large pair of tongs, and dragged over the floor to the "shingling" hammer. As the balls are drawn from the furnace they have a spongy appearance, and slag and other impurities trickle from them. The various operations described occupy, on the average, about two hours, and the quantity of unrefined pig-iron required to make a ton of puddled iron may be stated at from twenty-two to twenty-three cwt. An invention for facilitating and making more perfect the work of the puddlers, has recently been adopted at some of the malleable ironworks. It operates by injecting a current of air at high pressure into the furnace. This is done by making the puddling bar hollow, and affixing to the outer end of it an india-rubber tube, communicating with a powerful air-pump. The patentee is Mr Richardson of Glasgow; and the advantages gained by the contrivance are, that a charge of the furnace can be puddled in fifteen minutes less than the time required by the usual process, and that the iron produced is purer and tougher.

It is the puddler's duty to convey the "balls" from the furnace, and to place them one by one on the anvil of the "shingling" hammer. Before the invention of the steam-hammer, a somewhat clumsy contrivance was used for squeezing the slag out of the puddled iron, and beating it into shape. Now the steam-hammer is everywhere employed for those purposes. When the puddler lays a "ball" on the anvil, he waits to see the result of the first blow, and from it he is enabled to judge of the quality of his work. The "shingler" then steps forward, and takes charge of the "ball." His feet and legs are encased in iron armour, his body is covered by a stout leather apron, and he wears a mask of the same material. One stroke of the hammer makes apparent the use for this warlike attire, for it sends out in every direction jets of liquid fire, which patter against the legs of the workman, and would inflict fearful injuries were they to come in contact with the skin. The manipulation of the ball under the hammer is severe work, and requires great expertness. The "shingler" uses a pair of tongs about four feet in length, and with these seizes the ball and turns it on the anvil every time the hammer ascends. He so manages that the iron assumes the shape of a brick; and the operation occupies only two or three minutes. The "shingler" passes the metal, yet at a white heat, to the "rollers," who pass it through a series of grooves, in a pair of solid iron cylinders. By this means it is drawn into bars of the required size.

The iron produced by the above process is called "puddled-bar," and has to go through another operation before it is suited for even the commoner purposes of the blacksmith. In order to produce what is known in the trade as "common iron," the puddled bars are cut up into short lengths, and a number of these are laid in a heap of sufficient size to make a bar of any stated dimensions. They are then placed in a "reheating furnace," and exposed to a free circulation of heat. In about half an hour the iron becomes heated to what is known as the welding-point, and is then removed and rolled as before. When the rolling is completed, the bars are taken away by boys, and cut to the desired length by means of a circular-saw, which passes through the metal with astonishing rapidity and with a hideous noise. The bars are then straightened on an iron plate, stamped with the maker's name, and allowed to cool. From the moment the iron is taken out of the reheating furnace till the bars are ready for the market, the utmost expedition is required on the part of the workmen; and their operations, especially when witnessed at night, form one of the most interesting sights connected with the manufacture of iron. When a finer quality of iron is required, another welding and rolling are given to it. These repeated heatings, however, entail a considerable loss of material—equal, we believe, to eight or ten per cent. for each heat. In making the best quality of malleable iron, it is usual to refine the pig-iron before putting it into the puddling-furnace. The refining is done in a furnace specially constructed for the purpose; and the process consists in fusing the iron with coke, and thus ridding it of a large proportion of its impurities.

The quantity of malleable iron used in making machinery, building ships, and for other purposes, is immense. From year to year the workers in that material have been called upon to produce heavier pieces of work than formerly; and it is gratifying to find Scotch firms occupying the foremost place among the makers of gigantic smithwork. The heaviest forgings required for the largest war and mercantile vessels afloat have been made at Glasgow. When the Great Eastern was building, it was feared that no firm would be found willing to undertake the forging of her shafts; but the Lance-field Forge Company, of Glasgow, accepted the task, and executed it in a most satisfactory style. The shafts of many large war-ships, mail-steamers, and other vessels, have been made at the splendidly appointed works of this company. The shafts for the steamers of the Cunard, Peninsular and Oriental, and Royal Mail Companies, as well as for the Achilles, Black Prince, Monarch, and other ships of the British navy, and also for the war-ships built by Messrs Napier & Son for foreign governments, were made at Parkhead Forge, Glasgow, by Messrs Rigby & Beardmore. The heaviest piece of work produced at this forge was the crank-shaft of the Monarch—an immense war-ship, recently built at Chatham. When the shaft left the hammer, it weighed thirty-two tons, and when finished it measured 23½ inches in diameter. It was feared that the passage of such a heavy and compact mass over the bridges on the public road would not be safe, and some difficulty was experienced in arranging for its conveyance to Chatham. The North British Railway Company undertook to carry it, and the journey to the railway station and thence was safely accomplished, a special train being run for the purpose.

The Parkhead Forge is an extensive establishment, giving employment to seven hundred men and boys; but in consequence of the heavy nature of the work, the proportion of boys to men is smaller than in other branches of iron manufacture. The buildings cover several acres of ground, and are constructed in a most substantial style. On approaching the entrance to the forge, the visitor is startled by the vibration of the ground under his feet, caused by the incessant blows of the steam-hammers; and a peep inside reveals a scene of extraordinary activity. The rolling-mill is three hundred feet in length and one hundred and fifty in breadth. At one end of the mill are ranged twenty-two puddling furnaces, and half a dozen reheating furnaces. The rolling and other machines are driven by a pair of horizontal engines of 300 horse power. The fly-wheel of the engines is eighteen tons in weight, and makes one hundred revolutions a minute. The steam is supplied by fourteen vertical boilers heated from the puddling furnaces. The iron is first rolled into bars, then cut up, reheated, and either rolled into ship and boiler plates, or into pieces suitable for the forge. A number of years ago the firm devoted attention to making armour plates. Their specimens stood the test of competition with those of English makers most creditably; and but for the want of convenience for carrying the plates—the nearest railway being a mile distant—Messrs Rigby & Beardmore would have obtained a fair share of patronage from our own and other governments. The machines are capable of producing plates eight inches thick, and some of the plates made of that thickness have weighed twelve tons each.

The forge or smithy is nearly as large as the rolling-mill, and its fittings are on a most gigantic scale. There are two steam-cranes, capable of lifting fifty tons each; four, forty tons each; and four, twelve tons each; and these are so arranged that a shaft or other piece of work may be passed from one to the other all over the shop. There are fifteen steam-hammers, varying in weight from seven tons to two. Finished shafts—that is, finished so far as the hammering is concerned—lie about in all directions, and so delicately have these been operated upon by the hammers that turning would seem to be almost superfluous. Yet they are destined, before leaving the place, to be fitted into a lathe and turned with the greatest exactness. In the heating furnaces, and under the hammers, forgings may be seen in various stages of progress, and a glance at these makes plain the whole process of forging. In making a crank-shaft, for instance, a piece of iron eight or ten feet long, and of suitable diameter, is used as a "haft" or handle. At one extremity it is fitted with cross-bars or levers, by which it may be turned on its axis; and the other end is shaped conveniently for having smaller pieces of iron welded to it. The welding end is placed in a furnace, and in about an hour and a-half is raised to a welding heat. The crane by which the iron is moved about is fitted with a chain collar or sling, in the loop of which the iron rests. The collar works in a pulley attached to the chain of the crane, and moves easily, so that the shaft may be readily turned on the anvil. When the proper degree of heat is attained, the stopping of the furnace is removed, the steam-crane put in motion, and the gigantic bolt is swung on to the anvil of the steam-hammer. Several large slabs of iron, similarly heated in another furnace, are then brought out and laid on the "face" of the "haft." At a signal from the head forgeman, the hammer drops upon the glowing mass, and sparks of dazzling brilliancy fly off in all directions. Again and again the hammer descends, the iron meantime being carefully moved about, so that the whole may be worked into a homogeneous mass. Gradually the iron assumes a dull colour, but not before the desired end is obtained. It is then swung back to the furnace, comes forth glowing, has another addition made to its bulk, and so on. The most difficult part of the work is the formation of the crank piece, which is forged solid, and forms a huge square projection on one side of the shaft. When the shaft has acquired the proper dimensions, it is allowed to cool, and the haft piece is cut off to be used again. As the shafts are turned down until a good surface is obtained, an extra inch or so is allowed in the forging. All shafts are made in lengths of about twenty feet, each length having flanged ends, so that they may be firmly united.

For dressing and finishing such huge pieces of iron special and costly appliances are necessary. These are in the machine shop, an apartment one hundred and fifty feet in length, and fifty feet in breadth, both sides of which are lined with turning-lathes, slotting and boring machines, &c., of extraordinary size. One of the turning-lathes is said to be the largest in the world; and some idea of its dimensions and form may be obtained from the fact, that the crank shaft of the Monarch, though weighing thirty-two tons, was turned in it without taxing its capabilities to the utmost. Iron shavings fully one inch broad, and one-eighth of an inch thick, are turned off with apparently as little effort as if the material had been wood, instead of iron. One of the boring machines is sufficiently powerful to drill a hole ten inches in diameter through a solid block of iron, and the largest slotting machine can send off chips a pound or two in weight. When the work leaves this department, it is generally ready for being fitted into its place.

Messrs Rigby & Beardmore pay nearly L.40,000 a-year in wages; and in all departments of the establishment, fifteen thousand tons of iron and sixty thousand tons of coal are used annually.

The "forehands" employed in the operations described above earn much higher wages than any other class engaged in the manufacture of iron; but no one who knows the nature of their work

will say that they are overpaid. When trade is moderately brisk,
a puddler working full time makes from 8s. 6d. to 9s. per " shift "
of twelve hours ; but out of that sum he has to pay his assistant or
" chap " 3s. or 3s. 6d., so that his weekly earnings, supposing him
to work five shifts, are from 25s. to 28s. It follows that when the
puddlers are slack, the men employed in the other branches of iron-
making are equally so. When working full time, shinglers make
about L.4 a-week, and chief rollers £5. All are paid according to
piecework—so much a-ton. The "forehand " shinglers, rollers, and
heaters engage their own assistants, and pay them out of their joint
earnings ; but men are so eager to learn the work, in consequence of
the high rate of remuneration obtained ultimately, that they are
found ready to undertake the subordinate position at low wages, so
that a liberal share falls to the " forehands." The work, as may be
gleaned from the foregoing, is of the most arduous kind, and the
best constitutions cannot stand it long. One effect of the severe
heat and exertion is the creation of a craving for stimulants, such as
beer, which at once cool and support the workmen; and to a certain
extent no man would grudge them these ; but unfortunately, the
craving does not always cease with the work, and the consequence
is that a considerable proportion of them may be set down as being
of irregular habits. When their day's work or night's work is done,
they are too much exhausted to devote attention to anything of the
nature of mental culture, so that they are not so well-informed nor
intelligent as the average of workmen engaged in other occupations.
The forehand forgemen are paid at the rate of from 10s. to 15s.
a-day. They require to exercise great care and skill in the manipulation of heavy forgings. A flaw in the forging of a crank might be
attended by the most disastrous results to life and property. In
order to prevent such a thing, every blow of the hammer has to be
carefully directed, and its effect closely watched. A good forgeman
must know something about the chemistry of iron, and also be well
up in figures. In making machinery, the greatest exactness has to
be observed in the dimensions of the respective parts, and a crank
made at Glasgow must be so nicely finished, that when it is taken to
Chatham or elsewhere it will fit exactly into the bearings prepared
for it.

SHIPBUILDING.

THE shipbuilding trade of Scotland figures largely in the industrial
returns of the country, the value of the vessels of all kinds built
during recent years giving an average of close upon L.3,000,000 per
annum.   Little is known of the early history of the trade, though
it is beyond doubt that vessels were built at both Leith and Aberdeen some time during the fifteenth century.   In the year 1475,
three ships were fitted out at Aberdeen for the service of the king,
the cost being defrayed by the inhabitants of the town.   In the
same year, another ship was furnished with guns, ammunition, &c.,
by the loyal Aberdonians ; and the vessel was manned by twenty-four
young men belonging to the town.   The cost of this ship did not
exceed the equivalent of L.176 in our money.   It is on record that
in 1511 there was built in the vicinity of Leith "ane varie mon-
strous great schip called the Michael."   Lindsay of Pittscottie gives
the following account of this vessel and her armament :—" The
Scottish King bigged a great ship called the ' Great Michael,' which
was the greatest ship, and of most strength, that ever sailed in
England or France ; for this ship was of so great stature, and took
so much timber, that, except Fackland, she wasted all the woods in
Fife ; bye all timber that was gotten out of Norway ; she was so
strong, and of so great length and breadth (all wrights of Scotland,
yea, and many other strangers, were at her device, by the king's com-
mandment, who wrought very busily in her ; but it was a year and a

day ere she was complete)—to wit, she was twelve score feet of length, and thirty-six foot within the sides; she was ten feet thick in the wall, outted jests of oak in her wall, and boards on every side so stark and so thick, that no cannon could go through her. This great ship cumbered Scotland to get her to the sea. From that time she was afloat, and her masts and sails complete, she was counted to the king to be L.30,000 of expenses, with tows and anchors effeiring thereto—by her artillery, which was very great and costly to the king. She bare many cannons, six on every side, with three great bassils, two behind in her deck, and one before, with three hundred shot of small artillery—that is to say, mijand, and battered falcon, and quarter falcon, slings, pestilent serpents, and double dogs, with hagtor and culvering, cors bows and hand bows; she had three hundred mariners to sail her; she had six score gunners to use her artillery; and had one thousand men of warre, bye her captains, skippers, and quartermasters. When this ship passed the sea and was lying in the road of Leith, the king gart shoot a cannon at her, to essay her if she was wight; but I heard say it deared her nocht, and did little skaith."

In Rapin's History of England it is stated that, in 1512, James IV. of Scotland equipped a fleet in which was the largest ship that had up till that time been seen on the sea; and, though it is not so stated, the probability is that this ship was the one built at Leith. In the same year, King Henry VIII. had constructed the largest ship ever known in England. The latter vessel was named the Regent, and her burthen is stated to have been one thousand tons. It is asserted by some writers that the Regent was a copy in every respect of the big ship built by the Scotch King. Those two vessels were much larger than any others then in existence; and England possessed only three ships, in addition to the Regent, which were over four hundred tons. The vessels which traded along the coasts or to the Continent were of small size, and the greatest geographical discoveries were made by men who ventured forth in ships of from forty to sixty tons. The vessels belonging to the port of Leith in 1692 were twenty-nine in number, with an aggregate burthen of one thousand seven hundred and two tons, or an average of fifty-nine tons; and it is probable that all or most of those ships were built at the port.

The shipping trade of the Clyde has sprung from a very small beginning. In a report on the revenue of the Excise and Customs of Scotland, written in 1651, it is said of Glasgow that, "with the exception of coliginers, all the inhabitants are traders—some to Ireland, with small smiddy coals in open boats from four to ten tons, from whence they bring hoops, rungs, barrel-staves, meal, oats, and butter; some to France with plaiding, coals, and herrings, from which the return is salt, pepper, raisins, and prunes; some to Norway for timber." A few traders had by that time ventured as far as Barbadoes, but had not met with success. The mercantile genius of the Glasgow people was stated to be strong; but, in consequence of the shallowness of the Clyde, no vessel of any size could approach within fourteen miles of the city. Their cargoes had therefore to be transhipped into boats and cobles, and thus carried up to town. In the year mentioned, the merchants of Glasgow owned twelve ships, the aggregate burthen of which was nine hundred and fifty-seven tons.

When the Darien Scheme was set afloat, the traders of Glasgow, who had already experienced the advantages of commerce, went heartily into the speculation, many of them embarking their all in the venture. The last expedition in connection with the scheme sailed from Rothesay in September 1669. It consisted of four frigates, conveying twelve hundred emigrants. The unhappy fate of this great national enterprise is well known. The commerce of Glasgow received a shock from which it took many years to recover, and nearly fifty years elapsed before the merchants of the city came to possess any shipping of their own. The Union, from which they hoped no good, opened up new fields of enterprise for them; and as they had no vessels to take advantage of these, they had to charter some belonging to the northern ports of England. After a time, however, they found themselves in a position to have ships built for themselves. The first of these, measuring only sixty tons, was constructed at Greenock, and in 1718 made her first trip to Virginia and Maryland, with which States a trade in tobacco leaf had sprung up. From that time the commerce of Glasgow has enjoyed almost uninterrupted success. In the beginning of the present century, a trade was opened up with the East Indies and other distant countries; and now the commercial connections of the city extend to every land under the sun. The spirit of enterprise which has brought about such happy results in this respect, was early and most successfully applied to home manufactures; and few, if any cities of the world have made such rapid and substantial progress in commerce and industry as Glasgow. No branch of trade which could be made remunerative has been neglected, and not a few branches have been developed into specialties which confer a world-wide reputation on the city and its now noble river.

The steam and sailing vessels built on the Clyde are unsurpassed in strength, speed, or beauty. The impetus given to this trade, by the revival of commerce which followed on the Union, was chiefly confined to the lower reaches of the river, where sufficient water could be obtained to float the vessels. As the deepening of the river proceeded, the shipbuilding trade crept up nearer to Glasgow, and the size of the vessels was increased. While the operations for deepening the Clyde were going on, James Watt was perfecting his improvement on the steam-engine ; and a number of persons in Scotland and America were engaged in solving the problem of propelling vessels by steam. The inventors were successful ; and, though the honour is one that has been much disputed, the Clyde is now generally admitted to have been the cradle of steam navigation. The history of the invention is full of interest ; and as Scotland played a chief part in it, we shall briefly state how it was brought about, drawing mainly from a history of steam navigation compiled by Mr John Timbs.

In the Commissioners of Patents' Museum, at South Kensington, is "the Parent Engine of Steam Navigation," the history of which is briefly as follows :—For some years prior to 1787, Mr Patrick Miller, of Dalswinton, Dumfriesshire, had experimented with double and triple vessels propelled by paddle-wheels, worked by manual labour. In some experiments made in 1786 and 1787, he was assisted by Mr James Taylor, the tutor to his sons ; and at the suggestion of Taylor, it was determined to substitute steam-power for manual labour. For that purpose, early in 1788, Taylor introduced Symington, the eminent engineer, who had, the year before, patented his "new invented steam-engine on principles entirely new;" and Symington applied an engine, constructed according to his invention, to one of Mr Miller's vessels—which is the engine now at South Kensington. In October 1788, the engine, mounted on a frame, was placed upon the deck of a double pleasure-boat, twenty-five feet long and seven feet broad, and connected with two paddle-wheels, one forward and the other abaft the engine, in the space between the hulls of the double boat. This engine propelled the vessel along Dalswinton lake at the speed of five miles an hour. The engine is of the class known in the early history of steam machinery as the "atmospheric engine," in which the piston is raised by the action of steam, and, on a vacuum being produced beneath by the condensation of the steam, is forced down again by the pressure of the atmosphere. Numerous projects had been proposed, and several attempts had been made to propel vessels by steam-power, commencing with an experiment said to have been made in the year 1543; but the whole of the projects and experiments, previous to the application of Symington's engine, proved valueless for any practical use.

The result of the experiments with this engine, and with a larger one subsequently made on the same plan for Mr Miller, demonstrated to Symington that a more simple arrangement of the parts forming a steam-engine was required before steam-power could be practically applied to navigation.

In 1801 Symington was employed by Lord Dundas to construct a steamboat; and having, by his former failures, learned what was required, he availed himself of the great improvements made in the steam-engine by Watt and others, and constructed an improved engine, in combination with a boat and paddle-wheel, on the plan which is now generally adopted. This boat, called the Charlotte Dundas, was the first practical steamboat; and for the novel combination of all the parts Symington obtained letters-patent on the 14th October 1801. The vessel had an engine in which the steam acted on both sides of the piston, and was then discharged from the cylinder into a separate condenser; the rectilinear motion of the piston was converted into rotary motion by a connecting rod and crank; and the crank was united to the axis of Miller's improved paddle-wheel. Thus had Symington the undoubted merit of having combined together for the first time those improvements on which is founded the present system of steam navigation. The speed, when running alone and not towing other boats, was six miles an hour. "The use of this vessel," says Dr Macquorn Rankine, "was abandoned, not from any fault in her construction or working, but because the directors of the Forth and Clyde Canal feared that she would damage its banks. Yet the man in all Britain who possessed, at that time, the greatest practical experience of the working of canals—the Duke of Bridgewater—was not deterred by any such apprehension from ordering, in 1802, eight similar vessels from Symington, to be used on his canal. The death of the Duke of Bridgewater, early in the following year, prevented the execution of that order." It gives an amusing and suggestive insight into the popular view of these surprising changes in the methods of propelling vessels, to remark that a poetical saddler in Kirkintilloch thus described his thoughts when he saw the Charlotte Dundas pass along the canal with two vessels in tow :—

> " When first I saw her in a tether
>   Draw twa sloops after ane anither,
>   Regardless o' the win' an' weather,
>   Athwart her bearin' :
>   I thought frae hell she had come hither
>   A privateerin'."

The widow of Mr Taylor received, in recognition of his efforts to introduce steam navigation, a pension from Government of L.50 per annum; and in 1837 each of his four daughters received a gift of L.50 through Lord Melbourne. About the year 1825, Symington memorialised the Lords of the Treasury, when L.100 was awarded to him from His Majesty's privy purse; and a year or two afterwards a further sum of L.50. The poor inventor hoped that the allowance would be repeated annually, but his hopes were defeated. He received a small sum from the London steam-boat proprietors, and kind relatives contributed to his support in the decline of life. Such was the reward received by the inventor of "the first practical steamboat" in the great country of the steam-engine.

Many attempts have been made, and much misrepresentation used, to obtain for Fulton, the American engineer, the credit of first using steam locomotion on the water. He certainly did not fail to profit by the labours of others. Although Fulton possessed much inventive genius, and had been engaged with Chancellor Livingston, who was at the time Minister for the United States in Paris, in the construction of vessels to be propelled by steam, still he never accomplished anything until after he had seen the vessels of Symington.

Among the persons who had been acquainted with the experiments of Mr Miller and his associates on the Forth, was Mr Henry Bell, of Glasgow, who had been the medium of communication between Fulton and the Scotch coadjutors, and had sent to Fulton drawings of Mr Miller's boat and engines. Some time after, Fulton wrote to Bell to say that he had constructed a boat from the drawings; and this prompted Bell to turn his attention to the introduction of steam navigation in his own country. He accordingly set to work, but had to make several models. At length he put one into the hands of Messrs John Wood & Co., of Port Glasgow, who, from it, built for him a vessel of 40 feet keel and 10 feet 6 inches beam. This vessel he fitted with an engine and paddles, and christened the Comet, from the circumstance of a brilliant comet appearing towards the end of the year 1811, in which she was launched. Bell was enabled to turn his boat to profitable account; for, being a builder, he had erected a bath-house and hotel at Helensburgh, a watering-place on the northern bank of the Clyde, and he employed the Comet to convey passengers across the river, and thus derived a double advantage from it. The Comet began to run in January 1812. She was moved at first by single paddles, and attained a speed of five miles an hour; but Bell substituted wheels, with four paddles of the malt-shovel form. The engine, which was of four-horse power, was made by Messrs Anderson, Campbell, & Co.; and Mr

David Napier, then a workman, was employed in making the boiler.
The Comet was lost in one of the dangerous channels in the West
Highlands. Her engine, after lying in the sea for a number of years,
was recovered, and at the meeting of the British Association in Glasgow in 1840 it was exhibited as a curiosity. Soon after the success
of the Comet had been proved, Mr Hutchison, of Glasgow, had a
steamer built by Mr Thomson, an engineer who had been engaged in
some of Bell's experiments. The vessel was larger than the Comet,
being 58 feet long, 12 feet beam, and 5 feet deep ; engines, ten-horse
power. She was named the Elizabeth, and plied between Greenock
and Glasgow.

In 1813 a Mr Dawson, an Irishman, and Mr Lawrence, of Bristol,
attempted to run a steamer on the Thames, but succumbed to the
opposition of the Thames watermen. The boat was sent soon after
to ply between Seville and San Lucar, in Spain. Another vessel,
the Margery, of about seventy tons, which was built on the Clyde,
was taken south, along the east coast of Scotland. When she
reached the Thames the English fleet was at anchor, and she passed
close by. " The extraordinary apparition," we are told, " excited a
great commotion among officers and men ; none of them had ever
seen a steamer before, and by some of them she was taken for a fire-
ship." She made her first trip from London to Gravesend on the
23d January 1815, and continued to run between the two places
during the following summer, but was frequently laid up for repairs.
The Margery continued for several years to ply as a pioneer steamer
on the Thames. She was followed by another steamer, about seventy-
five tons burthen, with engines of sixteen-horse power, and wheels
nine feet in diameter. This vessel was also built on the Clyde. When
launched she was called the Glasgow, but that name was afterwards
altered to the Thames.

In 1818, so much had the principle of steam navigation spread,
that besides the vessels on the Thames, there were two on the Trent,
four on the Humber, two on the Tyne, one on the Orwell (Harwich),
eighteen on the Clyde, two on the Tay, two at Dundee, six on the
Forth, two at Cork, two on the Mersey, three on the Yare, one on
the Avon, one on the Severn, and two running between Dublin and
Holyhead. There were other steamers in active employment in
Russia, France, Spain, and the Netherlands, and a large number on
the rivers of the United States. Up till this period, although there
had been isolated voyages by sea from one station to another,
no regular passages had been made. The delay which was often
experienced by the sailing packets in traversing the stormy channel

between Holyhead and Dublin suggested the adoption of steam to avoid this loss of time. The first steam-vessel that was employed on the open sea was the Rob Roy, about ninety tons burthen, and thirty horse power, the property of Mr David Napier, of Glasgow. The Rob Roy was appointed to run between Glasgow and Belfast, a passage which she performed during the stormy months of winter, although previously steamers had been out only during the summer season ; and after running for two years there, she was transferred to the Dover and Calais passage as a Government packet. In the following year Mr Napier employed Messrs Wood to build a vessel named the Talbot, of one hundred and eighty tons burthen, with two engines of thirty horse power each. The Talbot was soon after followed by the Ivanhoe, and these were the finest and most complete vessels of the time. They were placed on the Holyhead station, to run between that port and Dublin, and assist the sailing packets which carried the mails ; but such was their speed and regularity that they soon superseded the packets.

The use of iron in shipbuilding was commenced in Scotland in 1818, when the passenger-boat Vulcan was built for the Forth and Clyde Canal Company by Mr Robert Wilson. Two small boats had been built of iron in England before that time, but, with those exceptions, we believe the Vulcan was the first iron vessel constructed. The builder of the Vulcan had great difficulties to contend with, and, in an account of the building of the vessel which he wrote to a friend, he said,—" There was no angle iron in those days, nor any machinery, except an old-fashioned piercing-machine, a cast-iron grooved block to form the ribs, a smith's fire, and one foot knee'd at a heat was considered good work." The vessel was designed by the late Sir John Robison, of Edinburgh, and was so substantially constructed that, we believe, she is still in existence, and doing duty. From time to time, within the past dozen years even, inventors have come forward and patented what they fancied were improvements in the construction of iron ships ; but when the way to prosperity seemed clear before them, an examination of the old Vulcan has shown that they had been forestalled, and consequently the patents became null. Two patents relating to the keels of iron vessels were cancelled when the keel of the canal-boat was examined. The Vulcan was nearly becoming remarkable for another reason than her being the first vessel built of iron. Mr Robert Wilson, Bridgewater Foundry, Patricroft, in the year 1827, when residing in his native place, Dunbar, exhibited to Mr Thomas Wilson, the builder of the Vulcan, a working model of a vessel propelled by a screw at the

E

stern, and also by side paddle-wheels, the whole propelled by clock-work, and so arranged that the side-wheels and stern-propeller could be worked alternately. This model was afterwards exhibited to the Governor and Council of the Forth and Clyde Canal Company, by Mr Thomas Wilson, he having obtained it from the inventor on the understanding that he would advise them to alter the Vulcan passage-boat into a steam-vessel, to be propelled by a screw at the stern. The governor and some of the directors were favourable to the scheme, but the proposal was strenuously resisted by others, who were of opinion that machinery could not be made to supersede horse-power for drawing vessels on the canal. The Cyclops, another canal boat, constructed of iron by Mr Wilson, was converted into a steamer, propelled by a paddle-wheel placed in the stern. The Cyclops steamed between four and five miles an hour, and plied between Port Dundas and Alloa.

Before the invention of steam navigation the shipbuilding trade of the Clyde had attained considerable importance ; and the builders did not neglect the opportunity afforded them by the application of steam to propelling vessels for greatly increasing their trade and effecting improvements in their models. There was a pause, however, while the steam-vessels first constructed were on probation; and up till the year 1830, not more than five thousand tons of steam shipping had been built on the Clyde. From that time the trade rapidly increased. Several companies were formed for running steam-vessels between Glasgow and Liverpool, Dundee and London, and on other routes ; and a spirit of competition arose, which resulted in a great improvement in the form, and increase in the number, size, and power of steam-ships. The directors of the East India Company were induced, by the achievements of the vessels built on the Clyde for other companies, to have two large steamers constructed by Mr Robert Napier of Glasgow. These gave so much satisfaction that, in 1839, Mr Napier received another order, this time for a vessel with engines of four hundred and twenty horse power. Though the practicability of constructing vessels of iron had previously been successfully tested, it was not until 1838 that it was used in the hull of any vessel of large size. In that year Messrs Tod & Macgregor built two iron steamers—the Royal Sovereign and the Royal George—for the trade between Glasgow and Liverpool. It was predicted by many eminent seafaring men that these vessels would prove failures ; but the predictions were not realised, as the steamers were found to possess all the good qualities of wooden ships, besides advantages peculiar to themselves.

The first steam-ship that crossed the Atlantic was the Savannah, which, in 1819, made the passage from New York to Liverpool in twenty-six days; but not until 1838 did a British steamer attempt the voyage. It was not believed that steam-ships could be profitably employed on the route between Britain and New York, in consequence of nearly all their carrying power being, according to the notions that then prevailed, required for the coal which would be consumed on the voyage. In a lecture delivered at Liverpool in December 1835, Dr Lardner is reported to have said that the project of making the voyage direct from Liverpool to New York was perfectly chimerical, and that they might as well talk of making a voyage from New York or Liverpool to the moon. When the project of establishing steam communication between the Old World and the New was started, Valentia was chosen as the most convenient port of departure and arrival. A company was formed, and an Act of Parliament obtained, in 1825, for conveying passengers between Valentia and America by way of Newfoundland; but, after procuring the necessary powers, the company made the perplexing discovery that, though it was easy to get passengers from America to Valentia, it was a different thing to get them to and from London, Edinburgh, Glasgow, and Liverpool, so the scheme was shelved. In 1835 it was started afresh, backed by a railway from London to Liverpool, the post-office packets to Dublin, and the Valentia Railway. The idea of making the voyage by way of Newfoundland, was, however, retained. Nothing came of the scheme after all, as, while it was being agitated, one steamer from London and another from Bristol demonstrated the practicability of making the voyage to New York direct.

A great advance in steam navigation was made about the year 1840, by the formation of what is now known as the Cunard Company, which was originated by Mr (now Sir) Samuel Cunard, who consulted with Mr Robert Napier, and, along with him, formed this now celebrated company, the larger number of whose original members were eminent and wealthy citizens of Glasgow. To Mr Napier was committed the contract for the hulls and engines of the first four vessels. The hulls were constructed for him, and under his supervision, by Mr John Wood, Mr Charles Wood, Messrs Steele, and Mr Robert Duncan respectively, all being fitted with engines of four hundred horse power made by Mr Napier.

An immense amount of capital has been embarked in the steam-shipping trade during the past twenty years. Lines of steamers run to and from all the principal ports of the world, and the most formidable competitors of the railways in this country are the coasting steamers. As the size of the vessels has been increased, a considerable improvement has been made in the matter of speed ; and in the case of those destined for carrying passengers, the fittings have come to be of the most superb and luxurious description. The Cunard, Inman, and Oriental vessels are floating palaces. Most of them, and many others of the same class, have been built on the Clyde.

During the seven years from 1846 till 1852 inclusive, the number of steam-vessels built on the Clyde was—14 with wood hulls, 233 with iron hulls—total, 247 ; of which 141 were paddles, and 106 screws. The tonnage of the wooden steamers amounted to 18,331, and of the iron to 129,273; the horse power of the engines in the wooden hulls was 6739, and in the iron hulls 31,593. The total value of the vessels and their engines was L.4,331,362, which gives an average of L.618,766 a-year for this branch of trade alone. In 1861, 81 steamers were built, the aggregate tonnage of which was 60,185 ; and the horse power of the engines 12,493; value of hull and fittings, L.1,252,300; value of engines, L.456,800. The number of vessels of all sorts built in the following year was 122, with a tonnage of 69,969. The following are the figures for the six succeeding years :—1863—170 ships, 124,000 tons; 1864—220 ships, 184,000 tons ; 1865—267 ships, 153,300 tons ; 1866—247 ships, 129,989 tons; 1867—241 ships, 114,598 tons; 1868—227 ships, 174,978 tons. At the close of 1868 there were orders in hand for 123 ships of 129,400 tons.

In 1864 the Clyde shipbuilders did a lucrative business in constructing swift and handy steamers for running the blockade of the ports of the Southern States of America. A serious interruption to the trade was caused in 1866 by a lock-out of the workmen, consequent on a partial strike, made to enforce what the employers considered an unreasonable demand on the part of the men.

As exhibiting the extent of the undertakings of individual firms, the number and tonnage of vessels which a few of them had orders for at the beginning of 1868 are given :—Messrs Napier and Sons, four iron-clad ships of war—10,636 tons, and 2140 horse power ; Messrs Caird & Co., six steamers, each of 3000 tons and 600 horse power—18,000 tons, 3600 horse power ; Messrs Denny Brothers, two composite gunboats, one twin-screw armour-clad war-ship, and five screw-steamers—10,520 tons, and 1970 horse power; Messrs Randolph, Elder, & Co., eight screw-steamers, two iron ships, and two lighters—10,510 tons, and 1310 horse power; Messrs J. and G. Thomson, three screw-steamers, one paddle-steamer, one gun-boat for the British Government, and one iron ship—7700 tons, and 1020 horse power.

The Clyde owes much of its prosperity to the invention of the dredging-machine, and the shipbuilders on its banks do a considerable trade in building dredgers, and other vessels and appliances for improving navigation. In 1866 Messrs Randolph, Elder, & Co. built two floating graving-docks of iron, one for the French Government, to be stationed at Sargon in Cochin-China, and the other for a company at Callao. Each dock was three hundred feet long, seventy-six feet broad, and forty feet deep in the walls. The quantity of mallable iron used in each was three thousand tons, and the cost L.100,000. This kind of dock is a most ingenious and useful contrivance. It is not always possible or convenient to construct a permanent dock, and the floating-dock, while answering all the purposes of a stone structure, may be moved about from one place to another as required. The floating-dock has hollow and water-tight sides of sufficient size to give a floating power equal to its own weight and that of the vessel placed in it. When a vessel is to be put into the dock, a series of valves in the sides of the latter are opened, and the water is allowed to rush in until the dock is sunk sufficiently low to admit the vessel. When the vessel has been floated in, the powerful steam-pumps with which the dock is fitted are set in motion. The buoyant powers of the structure are thus gradually restored, and in three hours the vessel is left dry.

A novel application of steam power to shipping has recently been made by one of the Clyde builders. Mr Robert Duncan, shipbuilder, Port-Glasgow, has obtained a patent for supplying auxiliary steam power to sailing vessels for carrying them through those latitudes where calms prevail, on voyages to the East Indies, China, and Australia. The auxiliary consists of a steam-launch, which is carried on board the sailing vessel so long as the winds are favourable ; but on entering a calm region steam is got up in the launch, and she is sent to do duty as a tug, not a-head of the vessel as is usually the case, but alongside. The first vessel furnished with an aid of this kind was the ship Niagara, of Port-Glasgow, which sailed for Melbourne in September 1867. The Niagara is one of the largest vessels registered in the Clyde, and had on board a heavy cargo of machinery. A letter from the captain thus records the first trial of the auxiliary steamer:—" I arrived at Melbourne on the 20th December, after a passage of ninety-six days, everything in good condition, and all in good health. I have had but one trial of the steamer, and cannot say it was under the most satisfactory circumstances, yet in some respects it was. I have proved that it is quite practicable to tow alongside with her, even in a pretty heavy sea. I had not on the passage what might be called a calm, but a good deal of light weather.   Our nearest approach to a calm was on the 9th October (lat. 9° N. 27° W.)   There were heavy rollers coming in from the southward, making us tumble about a good deal.   A ship that had been in company with us for several days had, in the light airs, stolen ahead of us nearly hull down.   I had many a look at her, and then at the boat, but feared to put her out, the sea being so heavy. At last, out she went, with six sailors in the stern for ballast, and the carpenter as engineer, and 'let her rip.'   She pulled like a fiend; and was, from the tightening of the forward and aft check and tow lines when caught by the sea, twisted in every direction.   Yet she took us on end.   At noon, when we set her going, the ship was barely making half-a-knot; and till dark, when we took her in, we were steaming fully two knots an-hour.   We passed our friend ahead, and left him at sundown about three miles astern."

No name is more widely known in connection with the shipbuilding trade of the Clyde than that of the firm of Messrs Robert Napier & Sons.   Steam navigation owes much to the founder of this firm, and some of his relatives.   They took an early interest in the subject, and bestowed no small amount of fostering care on it.   It was in connection with the making and improving of marine engines that the firm first acquired a reputation; and without losing any of their pre-eminence in that department, they soon came to be equally famous as builders of steam-vessels.   About thirty years ago they had received Government patronage, and since that time they have executed numerous commissions for the naval authorities, not only of Britain, but of France, Russia, Turkey, Denmark, and Holland.

In 1834 the Admiralty undertook experiments to ascertain the resistance of iron plates to shot.   The results led to the conclusion that iron was not suited for the construction of war-ships.   The question was revived from time to time, however, and during the Crimean war was brought into prominent notice by the French Emperor having some iron-plated floating batteries constructed.   Our Government followed the Imperial example, and in April 1855 the iron-clad floating-battery Thunder was launched.   The vessel had a wood hull covered with iron plates.   Next year, three batteries, composed entirely of iron, were launched; but peace having been concluded before the vessels were ready, they were not called into use.   Two of them, indeed, were never completed for sea; and as some experiments, made soon after they were built, showed that such vessels could never be of effective service, they were condemned. One of the batteries, the Erebus, was built by Messrs Napier &

Sons, and the engines of that vessel, and also of the Terror, were made by them. This was the first experience of the firm in the construction of ironclads. In 1859 they undertook the building of the Black Prince—one of the largest and finest vessels in our iron navy. In September 1862 they launched the Hector, a powerful iron-plated ship on the ram principle; and in the following year they completed the Rolf Krake turret-ship for the Danish Government. The latter became famous by her achievements during the Dano-German war, and was the first turret-ship ever engaged in actual warfare. Three magnificent frigates of 4221 tons each, for the Turkish Government, were their next commissions in this line; and in the spring of 1868—to which period what follows with reference to Messrs Napier and their operations refers—they had on hand no fewer than four ironclad war-ships. Some of the largest and finest mercantile vessels afloat have been built by Messrs Napier. In their building-yard, engine-work, and foundry, they employ from three thousand to four thousand men and boys; and the various departments contain perhaps the finest collection of machine tools in existence.

The building-yard is at Govan. One-half of the workmen are employed there, and the remainder at the engine-work of the firm at Lancefield, and their foundry in Washington Street, Glasgow. The workshops and offices at Govan cover a large extent of ground. In order to become acquainted with the various operations that go on in the place, the visitor must first enter the "drawing loft." The draughtsmen delineate on paper the shape and dimensions of every rib and plate required in the construction of a vessel, and work out the internal arrangements much in the same way that an architect lays off the rooms in a house. The working drawings are passed to the moulders, whose "loft" is so large that the full-sized outlines of a vessel of 5000 tons may be drawn on the floor, which is merely a gigantic blackboard. There the drawings are enlarged to the full dimensions in chalk, so that the form of each frame or rib, and the dimensions and curve of each plate, may be ascertained exactly. Adjoining the moulding loft is a workshop, the floor of which is paved with blocks of cast iron, pierced at regular intervals with holes about an inch in diameter. On this floor full-sized outlines of the frames of the vessel are drawn in chalk. With the exception of a few amidships, no two ribs are exactly alike, so that much care and no small amount of skill are required on the part of the workmen in this department. The ribs are formed of angle-iron—that is, iron having a section like the letter L. After the floor has been prepared by placing a series of pegs in the holes bordering on the chalk line, the iron bar intended for the rib is taken out of the furnace in which it has been heating, and, by pressure against the pegs and some hammering, is brought to the required shape with great facility. After the frames and plates have been shaped, they are taken to the punching or drilling machines, by which the holes for the rivets are made. The keel, which is composed of a strong beam of malleable iron, having in the meantime been laid down, the ribs are fastened to it, and covered over with the plates. The latter are secured to each other and to the frame-work of the ship by rivets inserted and clenched at a white heat.

In describing how an iron merchant-ship is built, little would remain to be added to what has been already written; but the work which Messrs Napier have on hand at present is of a somewhat different kind, and deserves more than a passing notice. They are building two large armour-plated ships of war for the British and two for the Dutch navy. One of the Dutch vessels is a ram and the other a monitor, but both have a turret amidships. The ram is a fine-looking vessel of 1473 tons, builders' measurement. She measures 205 feet in length, forty in breadth, and twenty-four in depth. Her sides, from the deck to a depth of three feet below the water-line, are protected by a belt of iron, six inches thick amidships, and tapering to four and a-half inches at the stem, and three inches at the stern. The plates are backed by ten inches of teak, within which is an iron lining half an inch thick. The turret is twenty-two feet in diameter, and rises four feet above the deck. It is composed of eight-inch plates, backed by twelve inches of teak. Below the level of the deck the turret consists of an iron framework supported on wheels, which run upon a circular rail laid eight feet below the deck. In order to protect the framework and mechanism, a wall of eight-inch plates is erected round the circular "well" in which the turret stands. The stem of the vessel is of immense strength, and projects into a knuckle in the central part. It consists of one huge piece of malleable iron, weighing about six tons. The monitor is a vessel of 1613 tons, though she is eighteen feet shorter than the ram, and not more than half as deep. She is, however, four feet broader, and the lines throughout are fuller. A belt of armour plates, five and a-half inches thick, and five feet deep, passes round her. The turret is of the same dimensions and strength as that of the ram, but is somewhat differently constructed below the deck. Both vessels are fitted with twin-screw propellers.

The rams being built for the British navy are much larger than the Dutch vessels. They are named respectively the Invincible and the Audacious, and will be exactly alike in every respect, their dimensions being—length, 280 feet; breadth, fifty-four feet; and depth, twenty-four feet. Their tonnage will be 3775, and the horse power of their engines 800. They will be propelled by twin screws, and will in addition be fully rigged. The form of the vessels and mode of construction is the latest adopted by Government. One of the vessels is sufficiently far advanced to enable one to form some idea of their great strength and buoyancy. The construction is what is known as cellular—that is, the vessel has an outside and an inside "skin,' the space between which is divided by ribs and longitudinal webs into a series of water-tight cells, from three to four feet square. The advantage of this is, that, though the vessel should sustain injury to her external skin, she shall yet remain as buoyant as ever, and can be made to leak only when both the outside and inside plates are pierced. There is no keel, according to the common understanding of the term; but running from stem to stern, along the central line of the vessel, is a vertical plate of steel, fully three feet in depth, and half an inch thick. This plate rests on, and is securely rivetted to, the outside plating; and branching off to the right and left from it are the ribs of the vessel. These are composed of iron plates of nearly the same depth as the central web, and are about four feet apart. The ribs are intersected by longitudinal plates, and the structure is further strengthened by two webs of steel similar to the central one, being carried along each side of the ship. The external plating of the bottom is three-quarters of an inch thick. At a point about three feet below the water line of the vessel, the cellular structure terminates, and the armour-plating commences. The armour is eight inches thick amidships, tapering to six inches forward and aft, and is backed by 10 inches of teak. On the centre of the deck a square battery, protected by armour-plates, will be constructed. In each vessel about 890 tons of armour-plates and about 2400 tons of iron of other kinds will be used.

The operations which require to be performed in the construction of a vessel are varied, and each requires special skill on the part of the workmen. A dozen distinct branches of trade, at least, are represented in the building of a ship, not counting the men engaged in the production of the raw material; and labourers and boys are employed to do parts of the work which do not require any special training. The following are the average rates of wages in Messrs Napiers' yard, according to a return recently made to the Board of Trade—Carpenters, 26s. 6d. per week of fifty-eight hours; painters,

26s. 6d.; joiners, 26s.; smiths, 26s.; fitters, 25s.; engineers, 24s.; sawyers, 24s.; riveters, 19s.; caulkers, 19s.; strikers, 16s.; chippers, 14s. 6d.; drillers, 13s. 6d.; lads and boys, 7s. to 14s.; labourers, 14s.

The number of persons employed in the shipbuilding trade on the Clyde is estimated at from eighteen thousand to twenty thousand.

Both at Leith and at Newhaven an extensive shipbuilding trade was carried on at various periods, but for a long time past no ships have been built at the latter place, and at Leith the trade has declined considerably in recent years. James IV. established a royal dockyard in Newhaven, where was also a manufactory for cables and ropes in 1506. The war-ship Great Michael, of which mention has been already made, was built at Newhaven. In 1544 the shipping of Leith was represented to be prosperous. After the Union, a line of battle ship, the Fury, was built at the old Sandport, the site on which the Custom House now stands. Within the recollection of many persons, shipbuilding was one of the most important branches of industry carried on at Leith. In 1840 two steamers, larger than any then afloat, were contracted for, and successively launched by Messrs Menzies. About the same time other ships of the largest size were built at Leith, which led many to suppose that the port would keep the lead in shipbuilding. Contrary to expectation, the business has, as already stated, gradually declined, the well-known character of the Aberdeen "clippers," and the celebrity of Clyde-built ships having diverted it partly to the north, but more particularly to the west. The change threw many of the Leith carpenters out of work, and compelled them to seek employment elsewhere. A few years ago the well-known firm of Messrs S. & H. Morton, makers of the patent slips, of more than European reputation, began to build iron ships; but after completing a few steamers, a sailing ship, and one or two dredgers, the trade came to a temporary stand. It is right, however, to say that the business of shipbuilding has not been abandoned by the firm, but is conducted by them in conjunction with the other departments of their trade. Though the building of new ships has not been carried on to any extent recently, a considerable number of ship-carpenters are employed in the port repairing vessels, some afloat, and others in dry docks. Artisans of this class are more in demand during the winter and spring, in reclassing and otherwise improving ships laid up during the winter months, and in overhauling vessels arriving from long voyages. There are six graving docks, all of which are generally occupied. The last constructed (the Prince of Wales Graving Dock) is capable of receiving the largest ship in the merchant service, except the Great Eastern.

A number of vessels have been built at Granton, and since the construction of the patent slip there in 1852, a considerable trade in repairing ships of all kinds, but chiefly steamers of a large size, has been carried on.

Aberdeen is one of the oldest shipping-places in the kingdom. The fitting out of war-ships by the loyal inhabitants of the city in the fifteenth century has already been mentioned, and much might be added about the early maritime relations of the port. Before the invention of steam navigation, the coasting trade was carried on by smacks, small vessels which made tedious voyages, and met with frequent mishaps. A voyage in one of those vessels from Leith to Wick was regarded, especially in the winter time, as being a much more perilous undertaking than a voyage to America would be at the present time. It was no uncommon thing for the vessels to knock about the coast for weeks without making fifty miles of progress. When a gale came on, they ran into the nearest port, and did not venture out again until the storm had passed, and the direction of the wind favoured them. The building and repair of the smacks formed a considerable item in the shipbuilding trade of the country up till thirty or forty years ago, and Aberdeen had not only a fair share of business in that line, but acquired a celebrity for producing fast-sailing and strong vessels. At the beginning of the century there were as many building yards at Aberdeen as there are at present, and the Halls and the Duthies were laying the foundation for that reputation in the trade which their descendants have for some time enjoyed. The early builders had many difficulties to contend with, the chief of which was the want of convenient ground to build their vessels upon. There were no building-yards nor slips, and the work was carried on upon the beach. The extension of commerce, and the consequent increased demand for ships, induced the builders at Aberdeen, as elsewhere, to construct vessels of a larger size than formerly. The year 1816 was a memorable one in the trade, as in it was launched the Castle Forbes, the first vessel built expressly for the Indian trade, and the largest that had been built at the port up till that time. The Castle Forbes was a local wonder on account of her size, though she measured only 439 tons. About twenty ships, having a gross tonnage of 2770, were launched in 1817, and next year twenty-two, measuring 3300 tons, were built.

Since the time referred to the trade has been much extended, chiefly by the enterprise of the late Mr Alexander Hall, who introduced the "clipper" mould of vessels. Until about thirty years ago ships were built according to a conventional model, which would appear to have been held sacred against attempts at improvement. Bluff bows, a full stern, heavy sides, and massive rigging, were the characteristics of this ideal of the shipbuilders. With the increase of commerce, however, swift-sailing vessels came to be demanded, and the old notions gave way to the requirements of the times. It did not need a profound knowledge of natural philosophy to discover that the speed of a vessel might be increased by making her bows more acute; but though the fact could not fail to be known, it was acted upon only to a limited extent. Mr Hall, who began shipbuilding about seventy-five years ago, was a most energetic man, and came to have an extensive business in the construction of vessels for the Indian and other branches of foreign trade. He paid great attention to the forms of his vessels, and having come to appreciate the value of the sharp-bowed or "clipper" model, he, in the year 1839, built the Scottish Maid, a vessel of 142 tons, and in her demonstrated the advantages of sharp lines. The vessel attracted much attention, and soon afterwards the Aberdeen shipbuilders became famous for their "clippers." The shipping firms engaged in the Australian emigration trade got a considerable number of vessels built at that port. Mr Hall was succeeded in business by his sons. Perhaps the best known vessel that they have built was the Schomberg, completed in 1854, for Messrs James Baines & Co., of Liverpool. A description of this vessel, illustrated by diagrams, appears under the article "Shipbuilding," in the "Encyclopædia Britannica." Constructed specially for the Australian passenger trade, the Schomberg was built and fitted up with the best materials, and when she was ready for sea was one of the finest as well as largest vessels afloat. Her length was 262 feet; breadth, 45 feet; depth, 30 feet; and tonnage, 2600. The frames of the vessel were of British oak, and the planking consisted of four layers of Scotch larch, each $2\frac{1}{2}$ inches thick. The first two layers were fixed in a diagonal position, passing down one side of the vessel and up the other, beneath the inside keel. The third layer was put on in a perpendicular position, and also passed under the vessel; and over this the outer layer was fixed horizontally. By arranging the planking in the way described, great strength was obtained. The vessel, which cost L.42,000, sailed for Australia in 1854, and was unfortunately wrecked on Cape Otway, on the 184th day after leaving Liverpool. One or two vessels of nearly similar size and build have since been turned out, but these began their career and have been pursuing it without attracting special attention. The general substitution of iron for wood as a material for building ships of a large size has had its effect upon the trade of Aberdeen, and now the achievements of the "clippers" built there have ceased to be spoken of in terms of wonder. The "clipper" form is almost universal, and has reached its highest development in the China traders built on the Clyde, the homeward voyages of which are among the most interesting nautical events of the year. The Aberdeen builders have not, however, fallen off in prosperity, as they have complied with the requirements of the day, and taken to building ships of iron, and of a combination of wood and iron. An extensive trade is also done in repairing wooden vessels, for which there are special facilities at all the yards. The three principal shipbuilding firms—Messrs Alexander Hall & Sons, Messrs John Duthie & Sons, and Messrs Walter Hood & Co.—have their premises at Footdee, and one or two small yards are on the Inches. The tonnage of the vessels built in 1863 was 1230 ; in 1865, 9701 ; in 1866, 9224 ; and in 1867, 12,112. Upwards of 1000 persons are employed in the trade. Carpenters receive 22s. a-week, working fifty-seven hours. Apprentices are taken at the age of fourteen years, and their wages average 7s. a-week over the term of five years which they require to serve. Most of the journeymen mechanics are married, and occupy houses in the neighbourhood of the yards.

Shipbuilding has long formed an important branch of industry at Dundee, and even at the beginning of the present century the number of vessels built there for the coasting and over-sea trades was large. The size of the vessels went on increasing until about 1856, when wooden shipbuilding may be said to have reached the height of its prosperity. In that year the Eastern Monarch was built by Messrs Alexander Stephen & Son. The vessel measured 1848 tons, was classed fourteen years A1 in Lloyd's Register, and was one of the largest, if not the largest, of her class then afloat. At that time there were six firms engaged in building timber vessels, whereas there are now only two which are so exclusively. A considerable number of vessels are also built on the Tay, at Perth, Newburgh, and Tayport. Iron shipbuilding was introduced at Dundee in 1838, when Messrs James Carmichael & Co. built an iron paddle-steamer named the Caledonia, intended for the river traffic between Dundee and Perth. A small iron schooner was also built by this firm. These vessels attracted great attention, there being few iron vessels then afloat. No other iron ships were built till 1840, when several iron paddle-steamers were constructed by Mr P. Borrie. Between 1842 and 1854, none but wooden vessels were built in Dundee; but in the interval the iron shipbuilding trade had been taken up at other ports, so that although Dundee was early in the field, that advantage was lost. In the latter year the firm of Messrs Gourlay Brothers & Co. began to build iron vessels, and the trade has since steadily increased. The largest iron ship built at Dundee is named the Dundee, and measures 1295 tons register. She was built by the Messrs Gourlay, and is owned by Messrs Gibson Brothers & Co., of Dundee.

Steam shipbuilding was begun at Dundee in 1823, when a paddle-vessel named the Hero was built for the traffic between Dundee and Perth; and in 1834 this branch of trade was energetically taken up by Mr Peter Borrie in conjunction with Mr Thomas Adamson. The steamers were built of wood, but in 1840, as has been mentioned, Mr Borrie began to construct iron steamers. The first screw-steamer built was launched from the yard of Mr John Brown in 1851. Since Messrs Gourlay began to build iron ships, they have turned out a large number of steamers, many of which have been of considerable value, fitted for carrying mails and passengers. Until the year 1865, all the vessels built in Dundee had been constructed either of wood or iron, but in that year Messrs Stephen & Son began to build ships with a combination of these materials. Such vessels are known as composite, the frames, keelson, stringers, tie plates, and beams being of iron, and the planking, keel, stern, and stem posts of wood. In 1868 there were five shipbuilding firms in Dundee—viz., Messrs Stephen & Son, Messrs Brown & Simpson, Dundee Shipbuilding Company, Tay Shipping Company, and Messrs Gourlay Brothers & Co. The total tons of all sorts of vessels launched and on hand from 1861 till June 1867, was 41,564 tons, representing a value of L.627,000, or L.104,500 annually.

During the past four or five years a thriving shipbuilding business has been carried on at Abden, near Kinghorn, by Mr John Key, who has turned out a number of fine steamers of large size. One of these, recently launched, is a vessel of 2000 tons, and 400 horse power, for the Peninsular and Oriental Company. Her dimensions are :— Length, 280 feet; breadth, thirty-six feet; depth, twenty-eight feet. She is fitted up in a splendid style for 105 first-class passengers, and fifty second-class. Before commencing shipbuilding, Mr Key had an extensive engineering business at Kirkcaldy. His marine engines are well known for many good qualities, and for them he has obtained orders from both the British and the French Governments. Mr Key employs about 200 men in his foundry, and 350 in his shipbuilding establishment. The River Tay, the first iron steam vessel built specially for the whaling trade, was constructed by Mr Key in 1868. The vessel is the property of Messrs Gilroy Brothers & Co., Dundee, and her dimensions are :—Length, 145 feet; breadth, thirty feet; and depth, eighteen feet six inches; she is barque-rigged, and measures 608 tons; she is divided into six water-tight compartments, and her hull is constructed to withstand the pressure of ice. There is a considerable shipbuilding business at Inverkeithing; and at Banff, Inverness, Perth, Peterhead, and elsewhere, numbers of small wooden vessels, intended chiefly for the coasting-trade, are built. Exclusive of those built on the Clyde, and at Aberdeen and Dundee, the number of vessels built in Scotland in 1867 was 79 of wood and 4 of iron, the aggregate tonnage of which was 13,237.

The following figures will illustrate the growth of Scotch shipping: —In 1656, the number of vessels belonging to Scotch ports was 137, measuring in the aggregate 5736 tons, which gives an average of about 42 tons. In 1760, there were 999 vessels, of 53,913 tons in the aggregate, or an average of 54 tons. An immense increase took place in the forty years following; and in 1800 there were 2415 ships, measuring 171,728 tons, averaging a fraction over 71 tons, and employing 14,820 seamen. The number and size of the vessels went on increasing, and on 31st December 1840, there were 3479 ships of all kinds, the aggregate tonnage of which was 429,204, the average being over 123 tons, and the number of seamen 28,428. Ten years later the numbers were—Sailing ships, 3432; aggregate tonnage, 491,395; average, 143 tons; steamships, 169; aggregate tonnage, 30,827; average, a fraction over 182 tons. During the next decade, a great change took place in the size of the ships, consequent on the extension of foreign trade and the improvement of harbours and docks. The total number of sailing ships in 1860 was 3172, being 260 fewer than in 1850; but the tonnage showed an increase of 60,817, so that while the average tonnage in 1850 was 143, in 1860 it was nearly 175. The number of steam-vessels had increased in 1860 to 314, with an aggregate tonnage of 71,579, the average being 228. On 31st December 1867, the number of sailing vessels of and under 50 tons registered in Scotland was 1007, with an aggregate tonnage of 30,604; above 50 tons, 1935 vessels, with a tonnage of 637,824. The number of steam-vessels of and under 50 tons was 138, with a tonnage of 3740; above 50 tons, 363; tonnage, 148,489.

RAILWAYS.

RAILWAYS, strictly speaking, do not form a branch of industry; and
yet no record of the industrial progress of the country would be complete without some reference to them.   They are most important aids
to the convenience and enjoyments of civilised life, and exercise a
fostering influence on all arts.   No better proof of their importance
could be given than is afforded by their rapid extension, and the
amount of capital invested in them.   It is only forty-three years
since the sanction of Parliament was given to the construction of the
first public railway worked by locomotives in Britain—the Liverpool
and Manchester line ; and at the present time there are upwards of
14,000 miles of railway in the country, the annual receipts of
which amount to nearly L.40,000,000.   Few towns of any note
are beyond convenient distance of the iron road.   It has been carried
across plains, through valleys, beneath hills, and over rivers, no
natural difficulties being allowed to stand in the way of its extension
to the centres of population and trade.   Night and day, trains rush
to and fro incessantly—this laden with passengers, that with cattle,
and the other with goods or minerals.   The scream of the engine-
whistle and the rattle of the wheels on the rails, are sounds familiar
to most ears ; and the safety and precision with which the traffic is
conducted, are known by experience to nine-tenths of the population.
The number of persons who travel by rail in the United Kingdom,
is close upon three hundred millions annually.

Though the term "railway" is now employed almost exclusively to designate the whole system and appliances of a firm or company which conveys passengers and goods by steam-power over a road laid with rails, there were railways long before the locomotive was invented. In the early years of the seventeenth century, wooden rails were laid down on the roads leading from some of the coal pits at Newcastle to the quays, and for more than a hundred years no attempt was made to improve upon these, except to the extent of fixing thin plates of iron on the upper side of the wood wheel-track. The first rails made wholly of iron were cast at Colebrook Dale Iron-works in 1767. These were found to possess such a decided advantage over the wooden rails, that they came into general use at the collieries, and ingenious men set about improving their shape and extending their use. The idea of laying rails along the public high-ways had not yet dawned on the mind of any one, though in several quarters wheel-tracks formed of stone were in existence, and had been known in Italy for centuries. In the year 1808 Parliamentary powers were obtained for the construction of the first public railway in Scotland. This was a tramroad, nine and a-half miles in length, extending from Kilmarnock to Troon. The rails, of which there were two lines, were of cast iron, fixed in stone blocks. The railway cost L.50,000, and was opened for traffic in 1812, the carriages being drawn by horses. A few years afterwards an attempt was made to use a locomotive on the line, but without success. The Troon Railway was constructed at the expense of the Duke of Portland, for the improvement of his Ayrshire estates. Having been adapted to locomotive traffic, it is now leased and worked by the Glasgow and South-Western Company, and, in proportion to its mileage, is the most remunerative line in Scotland. The Carron Iron Company early established a railway in connection with their works, and thereby reduced their carrying expenses from L.1200 to L.300 a-month. Rails were also laid down at the principal collieries in Mid-Lothian, Fife, Lanark, and Ayr, a number of years before locomotives were introduced. It was proposed to form a railway of the same kind from Glasgow to Berwick in 1810, and the ground was surveyed by Telford, who estimated the cost at L.2926 per mile; but the work was never commenced.

The formation of a railway from Edinburgh to Dalkeith was begun in 1826, and the line opened for traffic in 1831, the late Mr James Jardine of Edinburgh being the engineer. The railway is still in existence, but has undergone a great change. It was originally

F

constructed for the purpose of conveying coal, manure, and other heavy material, and with that view branches were sent off to the principal coal-fields of Mid-Lothian, and also to Leith and Mussel-burgh; but passenger traffic soon became the chief source of profit. The railway was formed of cast-iron plates of what is known as the fish-bellied pattern, and up till 1845, when it was purchased by the North British Railway Company, was worked by horses. The length of the line and branches was fourteen miles; and so numer-ous were the curves, that eleven miles had to be travelled in order to get to Dalkeith from Edinburgh. Towards the close of its "horsey" days, when railways worked by locomotives became common, this railway, with its lumbering carriages, slow-paced steeds, and noisy officials, was laughed at as an old-fashioned thing; but many per-sons have pleasant recollections of holiday trips made over the line. Then, as now, people took advantage of the Fast Days to spend a few hours outside the city, and it was no uncommon thing for the Dal-keith Railway to bear away four or five thousand pleasure-seekers on such occasions. The Musselburgh Races were also a fruitful source of revenue to the line. The passenger carriages were a sort of hybrid between the old-fashioned stage-coach and the modern omnibus, and in summer the outside seats were the most popular.

Mr Robert Chambers, in one of his essays, gives a sketch of this line, under the name of the "Innocent Railway," in which he says :—

"On arriving at the St Leonard's depôt—about the spot where Scott locates the Deans family—you are at once ushered into a great wooden carriage, where already perhaps two or three young families, under the care of their respective mammas, have taken up their quarters. But probably you prefer an outside seat—for there are outsides on the Innocent Railway—and so you get mounted up in front beside the driver, or else upon a similar seat behind. Your com-panion is perhaps a farm-servant, or a sailor, or one of those numer-ous indescribable blue-and-drab men who live about Dalkeith, and have a great deal to say about markets. An open carriage, full of fishwomen from Fisherrow, is placed judiciously in the rear; and there they sit, smoking their pipes or counting their money in their tenfold laps—the labours of the day all past—nothing now to be done but to cruise home by the Innocent Railway, 'in maiden meditation fancy free.' Singly, and in groups, come up the passen-gers, country-people most of them, with a great tendency to cotton umbrellas and bundles, but also a sprinkling of more lady and gentleman-like personages. There being only one set of carriages with one set of charges, the conductor makes an eye selection of passengers for a certain set of seats, and contrives to gratify most without offending any. The carriage begins to move. But even after a movement has commenced, you can hardly be said to have taken leave of the station. There is always a woman with some children seen running after the carriages, flagrant and sudorific, in a needless fright at the idea of being left behind, and who has to be taken in, juveniles and all, during the pause which is made before descending the tunnel. This reminds me, by-the-by, to say that nobody is ever too late for the Innocent Railway. One day we had started from Fisherrow up the inclined plane, when a washerwoman, with a huge bundle of clothes upon her back, was seen making after us along the line, occasionally waving a hand, in the hope of its prevailing upon the conductor to stop. We thought the poor woman had no sort of chance of making out her passage ; but, wonderful to say, she overtook us, burden and all, at a place where a short pause is made a mile and a half forward. The Innocent Railway has a great consideration for such of the dilatory as heroically persevere. The first pause, while the rope is fixing for the tunnel descent, suffices to take in the perspiring female and family. There is a second stoppage—quite leisurely—at the bottom, to detach the rope, and yoke the horses to their respective carriages. Off they then go, trotting at a brisk pace past Duddingston Loch ; but we have not advanced above a quarter of a mile, when a lady with a parasol and ten bandboxes is seen waiting for us at a cross-road ; and there is, of course, a pause to get her taken in. This accomplished, on we go again ; but lo, ere we have gone another mile, we have to stop at another cross-road to let off a farmer. Once more in motion, we advance rather briskly—that is, at the rate of about eight miles an hour—in order to make up for lost time ; but this has not lasted half a mile, when we meet the carriages proceeding to town, and have to stop, in order that the drivers may pass some message in the one or the other direction. Such are the incidents which mark a passage by the Innocent Railway. . . . A few more minutes bring us into the station at Fisherrow. The passengers land in a place like a farm-yard, where ducks and hens, and a lounging dog, and a cottager's children, are quietly going about their usual avocations, as if undreaming that they are within fifty miles of such a thing as machinery. And so ends the journey of exactly four miles and three-quarters by the Innocent Railway. On consulting your watch, you find it has required exactly forty minutes. And now, my co-mates, I would ask if a railway of this simple and primitive character be not something infinitely better than your whisking locomotive lines, where you never have leisure to look a moment about you? There cannot, in my opinion, be a shadow of a comparison between the two. By the Innocent Railway you never feel in the least jeopardy; your journey is one of incident and adventure; you can examine the crops as you go along; you have time to hear the news from your companions; and the by-play of the officials is a source of never-failing amusement. In the very contemplation of the innocence of the railway you find your heart rejoiced. Only think of a railway having a board at all the stations forbidding the drivers to stop by the way to feed their horses, under a penalty of half-a-crown—the 'way' being altogether only a few miles! Just conceive a railway where the carriages have barefooted boys to come off and run on in advance to change the switches! Or imagine any other railway on earth where such a circumstance as the following could take place. During the pause of a Mussel-burgh *up*-train at the bottom of the tunnel, a quiet-looking man, seated on the back of the carriage, said to a friend whom he recog-nised on the front of the next behind, 'Is the charge for this rail-way raised lately?' 'No.' 'Why, I have paid sixpence.' 'You should only have paid fourpence.' The inquiring party asked for an explanation of the driver, who came up at the moment. An answer was given in a voice that made the quiet man shrink up into half the space—'Didna I tell you at Fisherrow that I couldna gie ye change till we got up to the toon-n-n!'"

Railways, as they now exist, are the result of many years of experiment and much anxiety and cost; and there has been perhaps more controversy in connection with the claims of railway and locomotive inventors than in the case of any other mechanical con-trivance. The rail was invented more than a century before the steam-carriage, yet, singularly enough, the contrivers of the first loco-motive did not think of using it on a railway. James Watt has recorded that, in 1759, his friend Dr Robison, who was then a student at Glasgow College, suggested that the steam-engine might be employed in moving wheeled carriages on the highways. Watt does not seem to have acted on the hint until the year 1784, when he took out a patent for an adaptation of the steam-engine to the propulsion of land carriages. He apparently had not much hope that anything could be achieved by such a contrivance, for he stated that "a carriage for two persons might be moved with a cylinder of seven inches in diameter when the piston had a stroke of one foot, and made sixty strokes in a minute." So little did he regard his invention, and so averse was he to the use of high-pressure steam, that he never built a steam-carriage; but his friend and assistant, Mr William Murdoch, constructed, in 1784, a working model of a locomotive which, though only fifteen inches in length, attained a speed of six or eight miles an hour. This was the first locomotive in Britain, and it is preserved in the Patent Museum. In 1802 Messrs Trevithick & Vivian, of Camborne, in Cornwall, patented a steam carriage for common roads, and two years afterwards they constructed a locomotive for the Merthyr Tydvil Railway. This was the first steam-engine applied to locomotive purposes in Britain, and the leading features of it were essentially the same as those of the railway engines of the present time. For twenty years after, however, little progress was made in working out or extending the use of steam-engines on the railways. A number of machines had been devised, but one after the other they were set aside. In 1814 George Stephenson made a locomotive for the Killingworth Colliery Railway. It could draw thirty tons at the rate of four miles an hour, and was regarded as a great step in advance. On the Stockton and Darlington Railway, opened in 1825, of which Stephenson was engineer, an engine of the same kind was used. This engine may be seen at Darlington Station, where it has been set upon a pedestal. The number of cranks and rods about the machine give it a complicated appearance, and it looks odd in contrast with the engines that have superseded it. When the Manchester and Liverpool Railway was being constructed in the years 1826–29, so little was known either as to the capabilities of railways, or the most advantageous mode of working them, that the directors and engineers had some difficulty in deciding whether the line should be worked by fixed engines or by locomotives. It was ultimately decided to use locomotives, and the directors offered a premium of L.500 for the best locomotive that could be produced in accordance with the following conditions :— That the chimney should emit no smoke, that the engine should be on springs, that it should not weigh more than six tons, or four tons and a half if it had only four wheels, that it should be able to draw three times its own weight, and not cost more than L.500. Four engines were entered to compete for the prize, and the trial of these, on the 15th September 1830, was one of the most interesting incidents in the history of railways. George Stephenson's "Rocket" won the day. It drew three times its own weight, or twelve tons fifteen cwt., at an average speed of fourteen miles an hour, and attained a maximum velocity of twenty-nine miles an hour.

Before the experiments on the Liverpool and Manchester Railway, few engineers would admit the possibility of a locomotive engine attaining a speed of over ten miles an hour. The theory of friction and velocity as bearing on the matter was little known, and nobody seemed to think its study to be essential in developing the new mode of locomotion. Practical men engaged in forming the earlier railways and constructing the locomotives had very indistinct ideas as to the ultimate result of their work. They neglected first principles, and consequently wrought at a great disadvantage. Though George Stephenson is said to have predicted that there was no limit to the speed of a locomotive engine, there is no proof that he was acquainted with the principles referred to ; rather it is probable that he spoke in that spirit of enthusiasm natural to inventors.

Towards the close of last century certain experiments were made to discover the laws of friction and velocity, and the result of these, though bearing directly on the working of railways, was entirely overlooked by the promulgators of steam locomotion, and the writers on the subjects of roads and railways. In December 1824, when engineers and mechanicians were uniting their efforts to the production of railways and locomotives, with which a distance of ten miles an hour might be accomplished, the late Mr Charles Maclaren wrote a series of papers in the "Scotsman," in which he drew attention to the experiments referred to, and demonstrated in the clearest manner that the friction of a sliding or rolling body is the same at all velocities, and that a speed of twenty miles or more might be realised on railways. Mr Maclaren's essays attracted much attention in the scientific world, and threw a new light on the labours of the railway engineers. The papers were reprinted in various forms, and obtained a wide circulation in Britain and America, and also on the Continent—having been translated into French and German. The editor of the "Mechanic's Magazine," in commenting on the competition of locomotives on the Liverpool and Manchester Railway in 1829, prefaces an extract from one of the papers by the remark that—"The 'Scotsman' had the honour, four years ago, of first bringing forcibly under public notice the advantages derivable from locomotive carriages on railways." In 1851 the "Economist," in referring back to Mr Maclaren's papers, said of them that "they prepared the way for the success of railway projectors."

Not only did Mr Maclaren anticipate the achievements of railways in the matter of speed, but he foreshadowed their general utility and the effect they would produce on society. Mr Maclaren thus urged the importance of a high rate of speed on railways :—
"In speaking of twenty miles an hour, it is not meant that this

velocity will be found practicable at first, or even that it should be attempted. No complex invention can be perfect at the moment of its birth ; and our object at the present time should be to make the best use of our present means. Every man who knows anything of the history of the arts will readily believe that railroads and loco-motive engines have yet to receive many improvements. His vision must be narrow who considers the results of the first rude trials as fixing the boundaries of the new power thus put into our hands, and he must see far indeed who can define its ultimate limits. Mechanical skill has accomplished a hundred things at the present day which the practical men of the preceding generation would have derided as chimeras. In proportion as the mechanism of the rail-road and the engine is perfected, the engineer will feel his way towards a more rapid rate of movement ; for it is probable that all the advantages of the locomotive engine will be found to depend on the practicability of employing a high velocity."

Mr Maclaren had also formed a pretty sound idea of what was necessary in the passenger carriages of railways, and it is a matter of regret that some of his suggestions were not acted upon. The passenger carriages on the American railways are, in several important respects, exactly what is here proposed :—" In the construction of the steam-coach, the object should be to unite the highest practical velocity with as many comforts and accommodations as possible. With this view, perhaps, a form analogous to that of the steam-boat and track-boat would be the best. It might, for instance, consist of a gallery seven feet high, eight wide, and 100 feet in length, formed into ten separate chambers ten feet long each, connected with each other by joints working horizontally, to allow the train to bend where the road turned. A narrow covered footway, suspended on the outside over the wheels on one side, would serve as a common means of communication for the whole. On the other side might be outside seats, to be used in fine weather. The top, surrounded by a rail, might also be a sitting-place or promenade, like the deck of a track-boat. Two of the ten rooms might be set apart for cook-ing, stores, and various accommodations ; the other eight would lodge 100 passengers, whose weight, with that of their luggage, might be twelve tons. The coach itself might be twelve tons more ; and that of the locomotive machine eight tons, added to these, would make the whole thirty-two tons. Each of the short galleries would rest upon four wheels, and the whole would form one continuous vehicle."

He thus estimated the advantages to nations and to individuals, which would result from the extension and perfection of railways :—

" When the steam-coach is brought fully into use, practice will teach us many things respecting it, of which theory leaves us ignorant. With the facilities for rapid motion which it will afford, however, there is nothing very extravagant in expecting to see the present extreme rate of travelling [ten miles per hour] doubled. We shall then be carried at the rate of 400 miles a-day, with all the ease we now enjoy in a steam-boat, but without the annoyance of sea-sickness, or the danger of being burned or drowned. It is impossible to anticipate the effects of such an extraordinary facility of communication, when generally introduced. From Calais to Petersburg, or Constantinople for instance, would be but a journey of five days; and the tour of Europe might be accomplished in a shorter time than our grandfathers took to travel to London and home again. The Americans, with their characteristic ardour for improvement, are now collecting information about railways and locomotive machines in England; and to them these inventions will prove of inestimable value. It is pleasing indeed to think, that at the moment when the gigantic Republics of the new world are starting into existence, the inventive genius of man is creating new moral and mechanical powers to cement and bind their vast and distant members together, and to give the human race the benefits of a more extended and perfect civilisation. Nor ought we to over-look the additional security which an opulent and highly improved country will in future derive from the facility of its internal means of communication. Were a foreign enemy, for instance, to invade England, 500 steam-waggons could convey 50,000 armed men in one day to the point assailed; and within one week it would be easy, by the same means, to collect half-a-million at one spot, all quite fresh and fit for action. We cannot scan the future march of improvement; and it would be rash to say that even a higher velocity than twenty miles an hour may not be found appli-cable. Tiberius travelled 200 miles in two days, and this was reckoned an extraordinary effort. But in our times a shopkeeper or mechanic travels twice as fast as the Roman Emperor, and twenty years hence he may probably travel with a speed that would leave the fleetest courser behind. Such a new power of locomotion cannot be introduced without working a vast change in the state of society. With so great a facility and celerity of communication, the provincial towns of an empire would become so many suburbs of the metropolis —or rather the effect would be similar to that of collecting the whole inhabitants into one city. Commodities, inventions, discoveries, opinions, would circulate with a rapidity hitherto unknown, and, above all, the intercourse of man with man, nation with nation, and province with province, would be prodigiously increased."

Though Mr Maclaren's papers made considerable stir in the scientific world, they were evidently regarded with a jealous eye by men directly connected with railways, and it was only after his predictions had been realised that the correctness of his reasoning was admitted ; but then, also, men came forward and disputed the honour with him, though they had not the slightest proof to show that they were entitled to it. When Mr Maclaren was ridiculed for his views, those men were silent ; but when he came to be praised, they claimed the praise, and affected to feel honoured by the ridicule. Mr Nicholas Wood, of Killingworth, published a book on railways in the year after Mr Maclaren's essays appeared, and though he was strongly in favour of locomotives, he said, with evident allusion to Mr Maclaren :—" It is far from my wish to promulgate to the world that the ridiculous expectations, or rather professions, of the enthusiastic speculatist will be realised, and that we shall see engines travelling at the rate of twelve, sixteen, eighteen, twenty miles an hour. Nothing could do more harm towards their general adoption and improvement than the promulgation of such nonsense ! " It would be superfluous to point out how the "ridiculous expectations of the enthusiastic speculatist" have been more than realised.

While one set of inventors devoted their attention to perfecting a locomotive engine for railways, another set were busy devising steam-coaches to run on common roads. The latter met with many discouraging failures, and though a number of ingenious and costly machines were got to work tolerably under favourable circumstances, there were obstacles to their introduction to general use which could not be overcome. Among the Scotch mechanicians who devoted attention to the subject were Messrs T. Burstall and John Hill of Leith, who produced a road steamer, in the construction of which many improvements on previous inventions were introduced. About the year 1829 Mr Goldsworthy Gurney brought out a steam-carriage which met with considerable favour, though it presented few features not embraced in other machines of the kind. In 1831, one of these carriages ran on the road between Gloucester and Cheltenham. Obstructions were thrown in the way of introducing such machines, however, and Mr Gurney petitioned Parliament, and a committee of inquiry was appointed. The report of the committee was favourable to Mr Gurney ; and the success of his carriage led, in 1834, to the formation in Scotland of a company to run steam-carriages on the common roads. The road between Glasgow and

Paisley was the first chosen for the operations of the company; but the steam-coaches had not been long in operation when the boiler of one of them exploded, killed a number of persons, and put an end for the time to all attempts at steam locomotion on the turnpike roads.

At present the only steam-engines on the roads are traction engines for drawing heavy loads. Some of these are by Scotch inventors, and the most perfect, perhaps, that has yet been produced is a "road-steamer" invented by Mr R. W. Thomson, C.E., Edinburgh. Mr Thomson found that by fitting the wheels of a traction engine with stout tires of vulcanised indiarubber, a good gripping power could be obtained without the use of spikes or other appliances usually attached to traction engines. The "road-steamer" has a boiler of peculiar construction and extraordinary power, also invented by Mr Thomson, and altogether the machine is the most compact, powerful, and easily worked that has yet been put forward. Contrary to what might be expected, the tires withstand contact with the roughest road and sharpest stones without injury, while on soft ground or sand the weight is so distributed by the indiarubber that the wheels sink only an inch or two. The first engine of this kind made has been sent to Java; but before being shipped it was severely tested. A waggon laden with ten tons of flour was drawn with the greatest ease along the steepest streets in Leith, and the gentlemen invited to witness the experiment expressed the highest approval of Mr Thomson's invention, and congratulated him on having solved what had long been a most difficult problem with mechanicians.

The first promoters of railways do not seem to have reckoned much on the carrying of passengers as a source of revenue, for we find that of the fifty-three railways in Britain, for the construction of which Parliamentary powers had been obtained prior to 1830, only fifteen undertook the conveyance of passengers. Nearly all were connected with mines or quarries, and were worked either by horses or by fixed engines. Ten of the fifty-three railways sanctioned were Scotch, but of these two were abandoned. The aggregate length of the others was ninety-seven and a-half miles, and the total of the original capital L.469,705. During the ten years from 1830 to 1840, eighty-one railway bills were passed, of which twelve related to Scotland; but one of the proposed lines was abandoned. The length of those proceeded with was 191¼ miles, and the capital L.3,122,133. The most important were the Dundee and Arbroath; Arbroath and Forfar; Glasgow, Paisley, and Greenock; Glasgow, Paisley, Kilmarnock, and Ayr; and Edinburgh and Glasgow. The next decade was marked by a great extension of railways, and the union of numerous minor lines into systems. Enterprising promoters of new railways filled the public mind with golden dreams, and money was readily poured out on the most unpromising schemes. A railway speculation fever spread over the country, which reached its height in 1845 and 1846. In the former year, 225 railway bills were laid before Parliament; but of these only 120 were passed. Among the latter were fifteen relating to Scotland, thirteen being for new railways, and two for alterations on existing lines. The total length of the railways sanctioned was 436½ miles, and the original capital L.6,424,000, with borrowing powers to the extent of L.2,140,331. Seven of the railways were over twenty miles in length—namely, Aberdeen, Caledonian, Dundee and Perth, Edinburgh and Hawick, Edinburgh and Northern, Scottish Central, and Scottish Midland. In 1846 between 500 and 600 railway bills were brought before Parliament, and of these 272 were passed, many of them in a reduced or modified shape. Of the bills passed, fifty-eight related to Scotland. Only a few were for the construction of new independent lines, the remainder being for extensions, alterations, amalgamations, and deviations. The formation of 400 additional miles of railway in Scotland was, however, sanctioned, the original capital of which was L.11,362,980, with borrowing powers to the extent of L.3,736,464.

Popular excitement was most intense on financial matters during the years referred to. Many fortunes were made, and many wrecked, amid the vicissitudes of the railway mania: and the domestic history of the period is marked by numerous incidents of a painful nature. The spirit of the time was admirably satirised by the late Professor Aytoun, in his famous sketch of "How we got up the Glenmutchkin Railway, and how we got out of it." As affording an indication of the extent to which the speculative mania was carried, it may be mentioned, that in the end of 1845, proposals for no fewer than 620 lines of railway were before the public, the capital required for the construction of which was L.563,203,000. In addition to these lines, there were 643 schemes afloat, the prospectuses of which had not been registered. Of the capital subscribed for the railway bills presented to Parliament in 1846, L.121,255,374 was subscribed in sums of L.2000 and upwards. Tempted by plausible prospectuses, and glamoured by the fair speech of designing "promoters," thousands of persons embarked their all in the purchase of shares in railways which never did, and never could get beyond paper; and the natural result followed—that the simple trusting ones were in many cases reduced from affluence to poverty. The railway mania was the golden age of swindlers; but it cast a shadow on many homes, and cut short the career to prosperity and distinction of many a father and son. Individuals suffered, but the country has profited immensely by the energetic manner in which railways were undertaken and completed by the enthusiasts of twenty years ago, though probably not to the extent it would have done had the vast sums of money invested been judiciously expended. In Scotland, as elsewhere, a check was given to the extension of railways by the disastrous results of over-speculation; but as soon as a degree of confidence in railway investments was restored in the public mind, the work was resumed, and continued to make steady progress until a year or two ago, when the finances of several of the great companies got into confusion, and certain awkward revelations were made which tended to shake public confidence once more. Several extensions which had been sanctioned by Parliament were suspended after operations for their execution had been begun; and, perhaps, at no time during the past twenty years has there been less work on hand in the way of railway making than at present.

At the date of writing, the latest official returns relating to railways refer to the state of matters as existing on 31st December 1866. At that date, there were in Scotland forty-eight railways, the aggregate length of which was 2244 miles. The authorised capital was L.50,104,794 by shares, and L.17,024,623 by loans—total, L.67,129,417. The amount paid up on shares and on debenture loans outstanding at the date of the return, was L.53,078,798. Of the forty-eight railways, all, except three, are either leased or worked by one or other of the following companies:—Caledonian, Glasgow and South-Western, Great North of Scotland, Highland, North British, of each of which a brief account in alphabetical order is subjoined.

The Caledonian railway was projected about the year 1840; but the bill for its formation was severely contested during several sessions, and did not receive the royal assent until 31st July 1845. The original line was 137¼ miles in length, and comprised a great fork from Edinburgh to Carnwath, a great fork from the north side of Glasgow to Carnwath, a branch from the Glasgow fork at Motherwell to the south side of Glasgow, with a subordinate branch to Hamilton, a branch from the same fork in the vicinity of Gartsherrie to the Scottish Central Railway near Castlecary, and a main trunk extending from Carnwath to Carlisle. The act of incorporation authorised the company to raise L.2,100,000 in shares of L.50 each, and to borrow a sum of L.700,000. The estimated cost of the railway was L.2,100,000. The Scottish Central, Scottish Midland, Scottish North-Eastern, and several other railways, have been amalgamated with the Caledonian. The company further hold in lease the Alyth and the Arbroath and Forfar railways; while the Bushby, Crieff and Methven Junction, Greenock and Wemyss Bay, Montrose and Bervie, and Portpatrick railways, are worked by them. The total length is 673 miles. The authorised capital of the conjoint railways at 31st December 1866 was L.17,429,181 by shares, and L.5,826,357 by loans—total, L.23,255,538. The amount paid up on shares, and on debenture loans outstanding at 31st January 1867, was L.20,315,652. In 1866, the receipts from all sources of traffic amounted to L.1,784,717, of which sum L.638,376 was derived from passengers, and L.1,146,341 from goods and live stock. The number of passengers, not including 7724 holders of season and periodical tickets, was 9,127,203, carried in 113,512 trains, which travelled in the aggregate 2,699,330 miles. 900,000 head of live stock, 5,691,129 tons of minerals, and 1,830,759 tons of general merchandise, were carried in 136,841 trains, which travelled 3,976,179 miles. The traffic was carried on by means of 479 locomotives, 1068 passenger carriages and luggage-vans, and 13,505 goods and other waggons.

In 1850 a number of lines in the south-west of Scotland were amalgamated under the title of the Glasgow and South-Western Railway. The main line extends from Glasgow by way of Paisley, Kilmarnock, and Dumfries, to a junction with the Caledonian Railway near Gretna. There are, besides, a number of branches. The total length is 254 miles, and the authorised capital L.8,015,100, of which L.6,234,600 may be raised by shares, and L.1,780,500 by loans. At 31st January 1867, L.6,287,311 had been paid up on shares and on debenture loans. The receipts from all sources of traffic in 1866 were L.570,805, of which sum L.189,040 was from passengers, and L.381,765 from goods and live stock. The number of passengers, exclusive of 780 season-ticket holders, was 2,862,928, carried in 40,283 trains, which travelled in the aggregate 1,099,237 miles. 2,755,305 tons of minerals, and 426,131 tons of general merchandise, were carried in 75,395 trains, which traversed 1,855,085 miles. The traffic was carried on by 152 locomotives, 401 passenger carriages and vans, and 56,691 goods and other waggons.

As originally authorised by Parliament in 1846, the Great North of Scotland Railway was to embrace a line from Aberdeen to Inverness, with branches to Banff, Portsoy, and Burghead, the total length being 138¼ miles.   It was to have formed one undertaking, with a line from Aberdeen into Forfarshire, which had been sanctioned in the preceding year.   From various causes, however, the scheme was not carried out in its integrity—indeed, only a small portion of this line was constructed under the original proprietary ; but lines which were formed as separate undertakings in the district have been amalgamated with it, and the Great North of Scotland is now a much more extensive concern than its original promoters contemplated.   The more important railways that have been amalgamated with the Great North are the Banffshire, Strathspey, Formartin and Buchan, and Deeside.   The total length is 289 miles, of which 284 miles have only a single line of rails.   The authorised capital of the conjoint railways is L.3,080,393 by shares, and L.1,003,019 by loans—total, L.4,083,412.   At 31st January 1867, there has been paid up on shares and on debenture loans outstanding L.3,638,778. The receipts from all sources of traffic in 1866 were L.172,339 ; of which sum L.87,342 was from passengers, and L.84,997 from goods and live stock.   The number of passengers, exclusive of 4536 season-ticket holders, was 1,736,246, carried by 31,247 trains, which travelled in the aggregate 624,124 miles.   207,893 tons of minerals, and 313,345 tons of general merchandise were carried in 10,382 trains, which traversed 261,643 miles.   The rolling-stock consisted of 54 locomotives, 200 passenger carriages and vans, and 1453 goods and other waggons.

The Highland Railway comprises several undertakings, which by gradual amalgamation became in 1865 one system under the present title.   The first portion of this important system was a line from Inverness to Nairn, which was opened in November 1855. That was followed by the Inverness and Aberdeen Junction, which extended from Nairn to Keith—the northern terminus of the Great North of Scotland Railway—and was opened throughout in August 1858.   The Inverness and Nairn was amalgamated with this line in 1861.   In the following year, the Inverness and Ross-shire Railway was opened from Inverness to Dingwall, and in 1863 from Dingwall to Invergordon.   The Ross-shire line was amalgamated with the Inverness and Aberdeen Junction in 1862, and the next year an extension from Invergordon to Bonar Bridge was commenced.   This, with a branch to Burghead, which was opened at the end of 1862, completed the system of the Inverness and Aberdeen Junction Company.   In 1863 the direct Inverness and Perth Junction Railway was opened.   It consisted of a line from Forres to Dunkeld, where it joined the Perth and Dunkeld Railway. The latter was amalgamated with the Inverness and Perth line in the same year. A branch to Aberfeldy was made in 1864, which completed the line of the Inverness and Perth Company. The Inverness and Aberdeen Junction Company worked the Inverness and Perth line, and by the Amalgamation Act of 1865 these two undertakings became the Highland Railway. The total length of the system is 246 miles, 239 of which are single. The Findhorn Railway, a short line of $3\frac{1}{4}$ miles, is worked by the Highland Company. The authorised capital of the conjoined railways at 31st December 1866 was L.2,338,000 by shares, and L.703,880 by loans—total, L.3,041,880. The amount paid up on shares and debenture loans at that date was L.2,285,012. The receipts from all sources of traffic in 1866 were L.190,193, of which sum L.108,219 was from passengers, and L.81,974 from goods and live-stock. The number of passengers, exclusive of 923 season-ticket holders, was 946,461, who were conveyed in 15,059 trains, which travelled in the aggregate 522,592 miles, 102,496 tons of minerals, and 146,131 tons of general merchandise were carried in 3875 trains, which travelled in the aggregate 364,599 miles. The rolling stock consisted of 55 locomotives, 176 passenger carriages, 1169 waggons, &c.

The North British Railway is the longest in Scotland—measuring over all 735 miles. It extends from Perth and Dundee on the north, to Carlisle, Silloth, and Newcastle on the south, and passes across the country from Helensburgh to Berwick, sending out numerous branches and loops in its course. The railway originally consisted of a line from Edinburgh to Berwick, measuring fifty-eight miles, with a branch to Haddington four miles in length. The Edinburgh and Dalkeith Railway was purchased by the Company in 1845, adapted to locomotive traffic, and connected with the main line. A company had been formed, and powers obtained for the construction of a railway from Edinburgh to Hawick; and in 1845 the powers of that company were purchased by the North British, who next year got a bill passed, to enable them to send out branches from their main line to Tranent, Cockenzie, North Berwick, and Dunse; and from their Hawick line branches to Selkirk, Kelso, and Jedburgh. The main line was opened on the 18th June 1846. Further powers were obtained in the following year, and by that time the company had either constructed or held authority for a total length of 163 miles of railway. Branches to Musselburgh and Peebles were the next works undertaken. The latter of these was opened in June 1855. Since that time numerous additions have been made by new works and amalgamations, and at present the company hold in lease the Carlisle and Silloth Bay, Edinburgh and Bathgate, Peebles and Port-Carlisle Railways, while they work the Berwickshire Devon Valley, Glasgow and Milngavie Junction, Leslie, and St Andrews Railways. The authorised capital of the entire system at 31st December 1866 was L.16,687,620 by shares, and L.6,266,467 by loans—total, L.22,954,087. At 31st January 1867, there had been paid up on shares and on debenture loans L.19,178,407. The receipts from all sources in 1866 were L.1,374,702, of which sum L.561,185 was from passengers, and L.813,517 from goods and live stock. The number of passengers, exclusive of 6401 season-ticket holders, was 8,196,291, carried in 158,117 trains, which travelled in the aggregate 2,577,614 miles. 4,118,943 tons of minerals, and 1,539,506 tons of general merchandise were carried in 181,839 trains, which traversed 3,571,335 miles. The rolling stock was—Locomotives, 367; passenger-carriages and vans, 1261; waggons, 16,277; other vehicles, 159.

The only railways not belonging to or worked by the five companies above mentioned are the Forth and Clyde Junction (thirty miles); the Leven and East of Fife (nineteen miles); and the Drumpeller Railway, which belongs to the Forth and Clyde Navigation Company (two miles). The capital of the Forth and Clyde Junction is L.192,000 by shares, and L.64,000 by loan—total, L.256,000. The total paid up on shares and debentures at 31st December 1866, was L.250,051. The receipts from all sources were L.17,168, of which sum L.5381 was from passengers, and L.11,787 from goods and live stock. The number of passengers, exclusive of 72 season-ticket holders, was 92,243, carried in 1387 trains, which travelled in the aggregate 41,612 miles. 60,000 tons of minerals, and 32,241 tons of general merchandise, were carried in 1092 trains, which traversed 32,752 miles. The rolling stock consisted of 4 locomotives, 14 passenger carriages and vans, and 289 waggons. The capital of the Leven and East of Fife Railway is L.130,000 by shares, and L.43,300 by loans—total, L.170,300. The paid up shares and debentures on loans were L.136,170. The receipts from all sources were L.15,030, of which sum L.6592 was from passengers, and L.8438 from goods and live stock. The number of passengers, exclusive of 131 holders of season-tickets, was 121,027, carried in 2584 trains, which travelled in the aggregate 54,507 miles. The number of goods trains is not stated, but 19,687 tons of minerals and 48,429 tons of general merchandise, were carried. The rolling stock consisted of 3 locomotives, 7 passenger carriages, and 168 waggons and vans. The Drumpeller Railway, which carries no passengers, conveyed 239,867 tons of coal, the revenue from which was L.2177.

It will be seen from the above figures that the number of passengers who travelled on the railways of Scotland in 1866 was 23,082,369, exclusive of 20,567 season-ticket holders. The other traffic comprised 345,430 cattle, 1,788,321 sheep, and 82,230 pigs; 13,195,851 tons of coal and other minerals; 4,336,512 tons of general merchandise. 771,613 trains of all kinds were run, and the aggregate distance traversed was 17,680,579 miles. The receipts from passenger traffic amounted to L.1,596,135, and from goods and live stock to L.2,530,996—total, L.4,127,131. Under the head of working expenditure the following facts are stated:—The maintenance of way and works of the Scotch railways in 1866 cost L.387,425; locomotive power, L.587,195; repairs and renewals of carriages and waggons, L.142,280; traffic charges (coaches and merchandise), L.519,053; rates and taxes, L.71,872; Government duty, L.33,911; compensation for personal injury, &c., L.16,989; compensation for damages and loss of goods, L.19,829; legal and Parliamentary expenses, L.34,038; miscellaneous expenses, L.200,494—making a total working expenditure of L.2,013,087, representing an increase of L.234,754 as compared with the preceding year. The proportion per cent. of expenditure to total receipts was 49. There is some difficulty in ascertaining the number of persons engaged about railways, but a careful calculation leads to the conclusion that not fewer than 30,000 persons are so employed in Scotland.

Of the works in progress, the Callander and Oban Railway is the most extensive. Powers were obtained in 1865 for the construction of this line, which is to extend from Callander to the town of Oban, a distance of 70¾ miles. The authorised capital is L.600,000 in shares, and L.200,000 on loan. An arrangement was made with the Scottish Central Company (now merged in the Caledonian) that they should subscribe L.200,000 to the undertaking, and also work the line. In 1867 the Caledonian Company were empowered to construct new lines, seven and a-half miles in length, in substitution of a portion of a line from Dundee to Forfar, for which an Act was obtained by the Scottish North-Eastern in 1864. The Caledonian Company have also, with a view of rendering their system more efficient, constructed a number of short branches in Lanarkshire and Mid-Lothian. Other works were contemplated by this company, but in consequence of financial complications and for other reasons, powers were obtained in 1867 for the abandonment of certain branches, extension of time for the construction of other authorised works, power to raise additional money, and alteration of application of monies and terms of issue of certain unissued share and loan capital. In 1868 the North British Company also obtained powers to relinquish certain works. The Glasgow and South-Western Company have, during the past year or two, been forming a number of new branches in Ayrshire.

The Sutherland Railway, extending from the Bonar Bridge station of the Highland Railway to Brora, a distance of $32\frac{1}{4}$ miles, was completed in 1868, and opened as far as Golspie (27 miles). The capital of this line is L.180,000 in shares, and L.60,000 on loan. The traffic is being worked by the Highland Company, who subscribe L.15,000 to the undertaking. A few years ago, a number of capitalists in Caithness thought that the iron road might be advantageously extended to John o'Groat's, and a spirited movement was made to get up a railway leading from the Sutherland line at Brora to Wick, and thence to Thurso. It was decided to promulgate the Wick and Thurso section first, and an Act was passed in 1866 authorising the construction of that section, $21\frac{3}{4}$ miles in length, the capital to consist of L.130,000 in L.10 shares, and L.43,000 on loan. The Highland Company undertook the working of the line, and subscribed L.10,000. The scheme, however, has not received sufficient support to warrant the commencement of operations for the formation of the line.

A necessary part of the organisation of a railway of any extent is an engineering and carriage building establishment, at which the rolling stock may be made or repaired. It is also necessary, on an extensive railway, to have at convenient stations workshops at which slight repairs may be executed on locomotives and carriages. Thus, the North British Company have a great central establishment at Cowlairs, and workshops at St Margaret's, Coatbridge, Stirling, Haymarket, Burntisland, Hawick, Berwick, and Carlisle. The chief establishment of the Caledonian is at St Rollox, and the minor workshops at the Southside Station, Glasgow, at Greenock, Motherwell, Gartsherrie, Carstairs, Stirling, Perth, Edinburgh, and Carlisle. The chief workshops of the Glasgow and South-Western Company are at Kilmarnock; those of the Highland Company at Inverness; and those of the Great North of Scotland at Aberdeen. The North British workshops at Cowlairs are the most extensive and completely appointed of the kind in Scotland, and a brief description of them will give an idea of the kind of work carried on, and also of the vast expense of maintaining in a state of efficiency the rolling stock of a railway.

Before the Edinburgh and Glasgow Railway Company was amalgamated with the North British, the chief workshops of the latter were at St Margaret's, in the neighbourhood of Edinburgh ; but the Edinburgh and Glasgow Company's workshops at Cowlairs occupied a more convenient situation ; and they have been extended and adapted to answer the purposes of a central establishment for the entire North British system, and all the heavier repairs are done there. A process of centralisation has been going on, and a reduction has been made in the number of persons employed in the district workshops, at which, except in one or two cases, only what are designated "running repairs" are now made. The Cowlairs works cover several acres of ground, and are arranged in two departments— one for making and repairing locomotives and tenders, and the other for building and repairing carriages and waggons. In the former about 800 men and boys are employed, and in the latter between 400 and 500. About twenty per cent. of the locomotives owned by any railway company are usually undergoing repairs at a time, so that there are rarely fewer than from forty to fifty engines in the "hospital" at Cowlairs. Some engines are brought in which require only a few trifling repairs, and are turned out cured in a day or two ; others give way in a vital part, and it requires weeks of work to put them right ; a third class are the aged and debilitated, which can only be set going again by being fitted with new boilers, cylinders, &c. In addition to the invalids, one or two new locomotives are generally on hand. Only the best material will stand in a railway engine, and that material can be properly operated upon only by efficient workmen. A first-class locomotive may indeed be said to be the greatest achievement of the mechanical engineer.

Like other adjuncts of railways, the locomotive has had a gradual growth to perfection. George Stephenson's "Rocket" was a clumsy toy compared with an express engine of our day. The "Rocket," with its tender, weighed seven tons nine cwts., and was capable of drawing two waggons, weighing nine and a-half tons, at an average speed of fourteen miles an hour. One of the North British express locomotives weighs, with its loaded tender, about forty tons, and exerting a force equal to 750 horse power, can convey a load of upwards of 200 tons at the rate of forty miles an hour. One of these engines is calculated to run about 100,000 miles without requiring repairs, except of a trivial kind; and the average distance run by each engine at work is 130 miles a-day. A locomotive does not break down at once in all its parts; some portions require frequent renewal, while others continue good for a number of years. Generally, the first important part of the engine that gives way is the crank-axle, which, though of great strength, is pretty sure to break after accomplishing a certain amount of work. The average distance run by an engine on the North British system before the crank-axle gives in is from 80,000 to 100,000 miles; but before that mileage is accomplished, certain working parts of the engine and their bearings have to be renewed, and once in twelve months or so the tires of the wheels have to be "turned up" on the bearing surface. The deterioration in the value of a locomotive is estimated to be three-halfpence for each mile run. A locomotive of the best kind costs upwards of L.2000, and at the end of three or four years' service requires renovations to the extent of L.200 or L.300. It will thus be seen that the maintenance of the locomotive power of a railway amounts to a large proportion of the working expenses. No less a sum than L.100,000 is expended annually on the repair and renewal of engines at the North British Company's workshops. The cost of new stock, and the running expenses, amount to an additional annual sum of L.129,000.

Excepting the cylinders, axle-boxes, and fire-bars, there is little or no cast iron used in the construction of a locomotive, while in the passenger carriages and goods waggons only the axle-boxes and buffer cylinders are made of that material. The foundry department at Cowlairs is on a considerable scale, and in it is made all the cast-iron work required for the company's locomotives, steamboats, carriages, and waggons, as well as a considerable quantity of castings for the permanent way. About fifty men are employed in the iron and brass foundries; but their operations do not call for special notice, all the work being of a simple kind. A large quantity of malleable iron is used, and, with the exception of the crank-axles, all the forgings are made at the works. The smithy is an immense place, containing upwards of sixty fires, and having among its fittings four steam hammers, which are kept going constantly. A great number of bolts and rivets are required, and these are turned out by the workmen at a rapid rate. The bolts are screwed at machines attended by boys, who are paid by piece-work, and make excellent wages. One little fellow about twelve years of age is so expert that he makes ten or eleven shillings a-week. In an adjoining place springs are made. The turning and fitting shops are abundantly supplied with all the appliances of a first-class engineering establishment, and there appears to be no end to the variety of operations that are carried on in them. Upwards of 5000 separate pieces of metal are used in the construction of a locomotive ; and the making, adjusting, and uniting of these entails, as may be supposed, a vast expenditure of pains-taking labour. The tires of the engine wheels are now for the most part made of steel by an ingenious process which dispenses with welding, and so lessens risk of breaking. None of the engine tires are made by the company, it being found most profitable and con-venient to obtain these, as well as the crank-axles, from firms who devote special attention to their production.

In the boiler-shop about 120 men and boys are employed. The boilers, with the exception of the inner shell of the fire-box, are made of the best iron, in plates half an inch thick. In consequence of the intense heat of the furnace, the fire-box is made of copper of the same thickness as the iron. The boiler plates, after being shaped, punched, and bent, are riveted together by a machine which is capable of doing as much work in two hours as half a dozen men could accomplish in a day, and in a much superior style. Two men and a boy are required to work the machine.

When the parts of an engine are ready to be put together, they are taken to the erecting-shop, in which a special class of workmen called "erecters" are employed. There the engine is completed, and steam got up, and thence, radiant in paint and polished brass, it goes forth a thing of beauty and of strength, ready to do good service alike to prince and to peasant.

The carriage building department comes next under notice. There huge logs of timber are converted into carriages, waggons, and vans, by the hands of upwards of 400 workmen, aided by a large assortment of beautiful machines. The logs are conveyed by rail to the saw-mill, where they are cut by vertical and circular steam-saws into planks of the required dimensions. The planks are piled in a drying-shed, and, after remaining there a certain time, are taken to the cutting-out shop, where they are planed, moulded, morticed, tenoned, and bored by machines. Every piece is fashioned accord-ing to a standard pattern, and little skill is required on the part of the workmen. They have to make scarcely a single measurement or calculation, but simply to mark the wood according to the patterns and place it on the machines.

When the wood leaves the cutting-out shop it is returned to the drying-shed, where it remains until required by the carriage builders. The latter occupy a vast range of workshops, in which carriages in all stages of completion may be seen.   The frames of the carriages are of oak, and the planking of fir ; but in the first-class carriages a good deal of teak is used.   There is in all classes of carriages a considerable quantity of iron work, which is brought from the smithy in a finished state.   The carriage and waggon builders have everything prepared to hand, and they have simply to put the materials together.   They are paid according to piece-work, and generally two or four work together and contract to build a carriage or waggon for a certain sum.   The building of goods and cattle waggons is a coarser kind of work ; but for these the wood is prepared in the same way as for passenger carriages.   The working power is equal to producing fifty waggons and six passenger carriages a-month.   In the finishing department women are employed in making the trimmings of first-class carriages.   The painting-shop is on a scale of vastness commensurate with the other parts of the establishment.   In it the carriages are painted and varnished, and when they leave it are ready for use on the line.

So far as practicable, piece-work is the rule at Cowlairs, and is attended with the most satisfactory results to employers and employed.   When piece-work was first proposed, some of the men demurred, until they discovered that they could thereby increase their wages by a few shillings a-week ; and, in certain cases, men are making thirty per cent. more money than they received for the same number of hours when paid according to time.   Fifty-eight hours a-week is the working time throughout the establishment, and the average rate of wages is as follows :—Locomotive department—moulders, turners, and boiler-makers, 27s. a-week ; smiths, fitters, and erecters, 26s. ; machine attendants, 20s. ; boiler-makers' assistants, 18s. ; boys, from 4s. to 10s.   Carriage building department—carriage builders and joiners, 24s. to 26s. a-week ; painters, 24s. ; machine attendants, 15s.

After a railway is made and opened for traffic, it becomes of the utmost importance to pay close attention to the permanent way ; for, notwithstanding the perfection which has been attained in the making and laying of rails, fractures and displacements are not impossible, and are terribly dangerous things.   There is a staff of officials whose sole business is to look after the line.   These immediately repair any defect, or give warning in cases of danger.   The inspector of the permanent way has got under him a certain number of sub-inspectors, to each of whom a section of the railway, generally from twenty to thirty miles long, is intrusted, and who is responsible for maintaining it in a safe working condition. Each section is divided into portions from two to three miles in length, and to each of these four or five platelayers are allotted. One of the platelayers acts as foreman, and is responsible for his portion of the line, every yard of which he has to examine carefully twice a-day. His men traverse the line after him, replace fractured chairs, turn or remove bad rails, trim the ballast, and repair the fences. When a rail has to be turned or replaced one of the men is sent along the line with a signal to stop any train that may approach before the operation is completed. The platelayers are chiefly drawn from the agricultural or labouring class. The foremen receive 18s. a-week, and the assistants 15s.

In conducting the traffic two sets of officials are engaged—one having charge of, and accompanying the trains, and the other attending at the offices or stations. Of the travelling officials the engine-driver is the most important, for on him the safety and punctuality of the train chiefly depends. A keen eye, a steady hand, and a clear judgment are essential qualities in a driver, and to these must be added the power of close application to duty. The guard having got the passengers and luggage on board the train, and proclaimed "all right," the driver releases the breaks of his engine and turns on the steam. Immediately the driving wheels revolve, and the train moves off. The controlling appliances are concentrated at the left-hand side of the engine, and there the driver stands with his hands on the levers which regulate the steam and the draught of the fire. His eyes are always forward, except when they glance occasionally at the steam and water gauges. He must keep a sharp look-out for signals, and at the same time work the engine so as to maintain a steady pace. In descending an incline the steam is shut off, the momentum of the train being sufficient to carry it along without assistance from the engine. A careless or incompetent driver generally makes the train travel by fits and starts. He expends his steam injudiciously on the level parts of the line, and makes no provision for the extra effort required from the engine in ascending a slope. Time is thus lost; and to make amends the fire is urged, and the passengers are by-and-by startled by a series of spurts which are not only dangerous, but most destructive to the engine and rails. Each driver has an assistant or stoker, whose duty is to prepare the engine for work, keep the fire up, work the breaks, and make himself generally useful about the engine. The stoker's chief care is to keep up a good supply of steam, and prevent the water from falling too low in the boiler. In approaching a station, the driver shuts off the steam and applies the breaks gradually, so as to stop the train at the proper point, and in doing that he has to take into account the state of the rails ; if they be wet, he must apply the breaks sooner than if they were dry, for when the rails are wet a train will run a considerable distance on a level even though the breaks should be full on.   There are many other little niceties in driving an engine ; and, though railway travellers generally may not be aware of the fact, the comfort of a journey depends a good deal on the competency of the driver.   A number of the drivers have served an apprenticeship at the engineering trade, but the greater proportion of them have merely had a course of training as cleaners and stokers.   Men entering the service as cleaners must be able to read and write.   Their duties are to clean the engines after they come in from their spell of duty, and to make themselves useful in other ways.   According to capability, the cleaners are promoted to be stokers, and have their wages increased to from 16s. to 20s. a-week; but no cleaner is so promoted until he is nineteen years of age.   In course of time the stoker becomes a driver if he shows sufficient ability for that responsible post.   Some men get through the preliminary grades in four or five years, while others, if they be promoted at all, are so only after seven or eight years' probation.   Drivers receive from 4s. 6d. to 7s. a-day, according to the nature of the work in which they are engaged.   On all trains there is a guard whose duty is to look after the freight —in the case of goods trains, to see that everything is properly stowed, and to pick up and let off waggons from or to certain places on the route; and in the case of passenger trains, to see that the passengers are safely on board before starting the train, take charge of the luggage, see that sufficient time is allowed for passengers to alight at intermediate stations, and pay polite attention to any one asking information respecting the train.   The guards are selected from the station porters, and begin by taking charge of goods trains.   Their wages range from 18s. to 30s. a-week.

At the more important stations are concentrated large numbers of officials—such as managers of departments, clerks, and accountants. The most numerous class are the porters, who look after the loading of trains, passengers' luggage, and so forth.   Their wages range from 15s. to 18s. a-week.   Each station is in charge of a stationmaster, who has to see to the proper working of the traffic in his district, keep a set of books relating thereto, and superintend and pay all the subordinates on the portion of the line over which his authority extends.   The office of stationmaster at the centres of traffic is an important and responsible one, and those who hold it must have considerable powers of administration. In addition to the officials enumerated, there are signalmen, pointsmen, telegraphists, greasers, lamp-trimmers, and others. The signalmen and pointsmen require to exercise great watchfulness and care, as the slightest blunder on their part might be attended with serious results. Their wage is 18s. a-week, but their work is light. Neglect of duty on the part of any of the officials is punished by the infliction of a fine, or by immediate dismissal. As a rule, they are well treated, and have every inducement to attend to their work. It has come out in connection with accidents on English railways, that some of the subordinates have been kept on constant duty for an excessively long time, amounting in some cases to sixteen and even eighteen hours in a day; but the *employés* on Scotch railways are not worse off with regard to the time they have to work than men engaged in other departments of labour. Several benefit societies exist among them. Most of the drivers and stokers are members of the Locomotive Engine-Drivers' and Firemen's Amalgamated Benefit Society. Members pay sixpence a-week, and when sick or injured receive 10s. a-week. An additional payment of 2s. 6d. a-quarter entitles a member to a sum of L.70 in the event of his being injured to an extent which prevents him from again following the occupation of a driver or fireman.

COACHMAKING.

COACHES were introduced into Britain in the sixteenth century, and
the event is thus recorded by Taylor, the "water poet," who wrote
in 1623:—"In the yeare 1564, one William Boomen, a Dutchman,
brought first the use of coaches hither, and the said Boomen was
Queen Elizabeth's coachman. A coach was a strange monster in those
days, and the sight of them put both horse and man to amazement;
some said it was a great crab-shell brought out of China, and some
imagined it to be one of the Pagan temples in which the cannibals
adore the devil. The mischiefs that hath been done by them are
not to be numbered, as breaking of legges and armes, overthrown
downe hills, over bridges, running over children, lame and old
people." A great obstacle to the use of coaches was the want of
suitable roads; but we find that, so early as 1605, covered waggons
were employed for the conveyance of passengers and goods between
London, Canterbury, and other large towns.

The first public coaches in Scotland were placed on the road
between Edinburgh and Leith in 1610, by Mr Henry Anderson, a
native of Stralsund, in Pomerania—who, on condition of obtaining
a royal patent conferring on him the exclusive privilege of running
coaches between the two places for a period of fifteen years, brought
from his native country coaches and waggons, with horses to draw,
and servants to attend them. The fare was fixed at twopence for
each passenger.

At the close of the seventeenth century, coaches and chariots had
become fashionable with the Scotch nobility, but were chiefly used

in town. When the Duke of Queensberry came to Edinburgh as King's Commissioner in 1700, he was met by a train of forty coaches, most of which were drawn by six horses. In 1673 there were twenty hackney coaches in Edinburgh; but these were not managed in a way to make them popular, and the number gradually decreased, until in 1778 there were only nine registered hackney coaches in the city. The sedan chairs were formidable opponents of the coaches; and at the last-mentioned date there were 188 chairs for hire in Edinburgh, besides fifty private ones. In the course of time, however, hackney coaches became popular, and the number of chairs gradually decreased, until, about twenty years ago, they went entirely out of use. A fine sedan chair is preserved in the Edinburgh Antiquarian Museum. It is simply a box about two and a-half feet square and five feet high, fitted with a seat and a glass front. Two poles, attached to the sides, formed handles by which the chair could be carried by one man walking before and another behind. The chairs were usually borne by stout Highlanders. The use of a chair could be obtained for a day, from ten in the morning till twelve at night, for 7s. 6d.; but the fares for short journeys were higher than the cab fares of the present day.

Stage-coaches were introduced into England in 1658—at least the earliest public notification of that mode of travelling was made in that year. Twenty years later, the Provost and Magistrates of Glasgow entered into an arrangement with Mr William Hume of Edinburgh, that he should run a coach once a-week between the two cities. This was the first stage-coach in Scotland, for the Edinburgh and Leith coaches scarcely came under that designation. It is not stated how long the coach between Edinburgh and Glasgow continued; but prior to 1763 it had ceased to run, and in that year a heavy coach, drawn by four horses in good weather, and six in bad, ran three times a-week between the two places. Subsequently the coach was run daily, and took from eleven to twelve hours on the road. Lighter vehicles were afterwards introduced, and the journey came to be accomplished in six hours.

Up till the middle of last century there was no stage-coach on the route from Edinburgh to London. When a traveller wished to make the journey, it was no uncommon thing for him to advertise for a companion to share a post-chaise. In 1753 a stage-coach was running between the two capitals, and next year the following advertisement regarding it appeared in one of the Edinburgh newspapers :—

"The Edinburgh stage-coach, for the better accommodation of

passengers, will be altered to a new genteel two-end glass coach machine, hung on steel springs, exceeding light and easy, to go in ten days in summer and twelve in winter, to set out the first Tuesday in March, and continue it from Hosea Eastgate's, the Coach and Horses, in Dean Street, Soho, London, and from John Somerville's, in the Canongate, Edinburgh, every other Tuesday, and meet at Burrowbridge on Saturday night, and set out from thence on Monday morning, and get to London and Edinburgh on Friday. In winter to set out from London and Edinburgh every other Monday, and to go to Burrowbridge on Saturday night; and to get out from thence on Monday morning, and get to London and Edinburgh on Saturday night. Performed, if God permits, by your dutiful servant, HOSEA EASTGATE."

Glasgow did not possess means of direct communication with London until 1788, when a coach was started to carry the mails and passengers. The arrival of the first coach from London was an event of much interest in Glasgow, and a large number of the citizens turned out on horseback to welcome it. So little were people disposed to travel in those days, that for many years there was not a sufficient number of passengers to make the coach remunerative to the contractors. The coach accomplished the journey in sixty-three hours. After Glasgow and Edinburgh had been placed in communication by coaches, vehicles were run to places in the neighbourhood of both cities; and the journeys of these were gradually lengthened, and the number of coaches increased, until in the beginning of the present century regular communication was maintained between all parts of the country with a frequency proportioned to the importance of the respective towns. Macadam and Telford came opportunely on the scene, and, by improving and extending the roads, gave a great impetus to traffic. In 1825 eight royal mail-coaches, and upwards of fifty stage-coaches, started from Edinburgh every day. Of the stage-coaches, ten ran to Glasgow and six to London. There were in addition local coaches to such places as Newhaven, Leith, and Portobello, and carriers to every town and village of any consequence between Wigtown and Nairn.

Now that the surface of Scotland is traversed by a network of roads of the best kind, the difficulties of travelling in the early days of coaches are apt to be under-estimated. As a specimen of the troubles arising from bad roads, may be mentioned the case of the Marquis of Downshire, who, in travelling through Galloway in the middle of last century, took with him what was then considered to be a necessary part of his retinue—namely, a staff of labourers with their tools to smooth the way and get the coach out of ruts. Yet such was the nature of the road, that when the coach got to the Carse of Slakes, a hill three miles from the village of Creetown, it came to a dead halt; his lordship had to send his servants away, and he and his family passed the night in their coach on the hillside. So late as 1780, it was necessary, in some quarters, to have the carriage attendants provided with axes with which to clear a way through the woods. For other reasons than the absence or badness of roads, travelling in the early days of coaching was far from being a pleasant thing. The vehicles were clumsy, badly constructed, and without springs. Accidents were of frequent occurrence; and the number of persons killed or injured was much greater, in proportion to the number of travellers, than is the case in the present day, notwithstanding the popular notion as to the dangers of railways. Upwards of twenty-three millions of passengers travelled on the Scotch railways alone in the year 1866; and of these only five were killed—two of them while incautiously crossing the rails in front of advancing trains, and two while getting out of trains before they had been brought to a stand. In 1806, a parliamentary committee was appointed to consider, among other things, the act limiting the number of passengers to be carried by stage-coaches. It was stated in evidence before the committee, that "accidents are continually happening in one part of the kingdom or another—indeed, scarce a week passes without some of the coaches breaking down, and often killing the unfortunate passengers."

The first coach-making establishment in Scotland was set up in Edinburgh about the year 1696; but for a considerable time the only work done, beyond repairing the coaches brought from London, was the making of a few clumsy carriages. In 1738 Mr John Home, who had carried on the business of coachmaker for several years previously, went to London, and received instruction in the trade. On returning to Edinburgh he brought with him a supply of tools, and set about conducting his business in a new style. There had hitherto been no division of labour in making a coach; but Mr Home allotted to different workmen the fashioning of the various parts of a carriage. Thus the men became expert at their parts, and the result was a great improvement in their productions, while the Scotch nobility and gentry, with whom chaises had become fashionable, instead of bringing their vehicles from London or Paris, as formerly, had them made in Edinburgh.

A letter on the progress of Edinburgh, published in 1783, says:—
" Coaches and chaises are constructed as elegantly in Edinburgh as

anywhere in Europe. Many are yearly exported to St Petersburg and the cities in the Baltic; and there was lately an order from Paris to one coachmaker in Edinburgh for 1000 crane-necked carriages, to be executed in three years." A number of carriages were exported from Leith to the West Indies in 1766, and in subsequent years there was also a large exportation to Holland, Russia, France, and Poland. The annual value of the carriages exported from Leith was stated in 1778 to be L.2200. As the manufactures and commerce of the country increased, and wealthy people became more numerous, the use of carriages, of course, became more common, and stage-coaches also increased in number. Though railways have superseded stage travelling except in a few remote districts, they have not acted unfavourably on the coachmaking trade; rather the contrary, for, with other causes adding to the prosperity of the community, they have helped to multiply the persons who can afford to keep carriages, while for their own service a large number of vehicles are required.

The number of carriages assessed under schedule D in Scotland is over 25,000, and the gross amount of duty charged is L.33,000. In addition to these, there are about 2500 licensed hackney coaches, 400 stage-coaches, and 1800 other vehicles, which are exempted from taxation. Drawn up in a continuous line, with eight yards allowed to each, the whole of the carriages, with the horses employed to draw them, would form a procession about 136 miles in length. Time has come to be so valuable with people in business, that few journeys of more than a mile or two are made on foot. The main thoroughfares of Edinburgh and Glasgow are traversed at frequent intervals by splendid omnibuses, which for an almost nominal sum convey passengers from or to any part of a route extending to two or three miles. Then there are hundreds of hackney coaches, or "cabs," stationed in convenient localities. In making and maintaining these vehicles many men are employed. There are fourteen coachmaking establishments in Edinburgh, several of them of considerable extent, and all turning out work of the best description. Indeed, no coachmakers in the world produce carriages which for comfort, strength, or elegance, surpass those made in that city. The "cabs" of Edinburgh are superior to any to be found elsewhere, while the city omnibuses are extremely comfortable in all their appointments.

The coachmakers of Edinburgh are chiefly engaged in constructing private carriages; but at the same time they turn out a large number of vehicles which do not fall under that designation. They have customers in all quarters of the world, and their handiwork is admired wherever it is seen. The coachmaking trade of Scotland employs upwards of 2000 persons. The largest establishment in Edinburgh is that of Messrs J. & W. Croall, York Lane, in which about 100 workmen are employed ; but Messrs James Macnee & Co.'s works at Fountainbridge are also of considerable extent. These firms have always splendid carriages on exhibition in their show-saloons. Conspicuous by their size and richness of style are the four-in-hand "drags," much in fashion among the members of the upper ten thousand who attend race meetings. Gaudily painted and expensively equipped carriages, such as young noblemen delight to possess, next arrest the eye ; and in the glitter of these, the quiet but genteel " brougham " of the professional man looks excessively grave. Then there are the " landau sociable," which has to some extent supplanted the " phaeton ; " the clarence, with its glass front and sides, which afford shelter while they do not interrupt the view; elegant pony carriages for fair charioteers ; and a host of other vehicles adapted to all requirements, and suited to all ranks. The prices usually range from L.60 to L.300 ; but there is practically no limit to the amount that may be expended on the fittings and decorations of a carriage. Probably the most costly ever constructed is Her Majesty's state-coach, which was made for George III. in 1762. The cost of the coach was L.7562, of which sum the coachmaker received L.1673 ; the carver, L.2500 ; gilder, L.933 ; painter, L.315 ; laceman, L.737 ; chaser, L.665 ; and harnessmaker, L.385. Some alterations have been made on the coach during Her Majesty's reign, but, in the main, it retains its original character.

Strength, lightness, and elegance, combined with suitable accommodation and easy springs, are the objects to which the coachmakers have to pay chief attention, so that the material used must be carefully selected and judiciously combined. In the construction of a carriage six distinct trades are directly concerned, and contributions from as many more are required. Take a "brougham," for instance, and trace it through the various stages of construction. As in building a house or a ship, the first thing to be done is to prepare a design. That is usually done by the foreman of the establishment, who makes a full-sized chalk drawing of the proposed vehicle on a black board. The different kinds of carriages derive their names from some peculiar arrangement of the more important parts, but carriages of the same designation may differ widely in their details. Persons ordering carriages are allowed an opportunity of inspecting the design and suggesting alterations thereon, and the result is that it is rare to find two carriages exactly alike. After the chalk drawing has been approved of, operations are commenced. The body-makers take measurements of the upper or principal parts of the drawing, and forthwith begin to make that part, to which their attention is exclusively confined. Equally distinct are the occupations of the carriage-makers, the wheelwrights, and the smiths. The carriage-makers construct the framework on which the body of the carriage rests, and the pole or shafts. The wheelwrights are solely occupied in making the wheels. The amount of smith work required for a carriage is considerable, and some of the pieces are exceedingly complicated in shape. The woods chiefly employed in coachmaking are ash, mahogany, and oak, and these must be thoroughly "seasoned." Ash strengthened with iron is used in the framework of the "body." Straight lines are avoided as much as possible in the construction of carriages, and the consequence is that the body-maker has to bend and shape his wood to a great variety of curves. After the framework is completed, the sides and ends, except the spaces for the doors and glass front, are boarded in by panels of mahogany, which are brought to the required curve by being damped on one side and exposed to heat on the other. As the strength of the carriage does not depend on the panels, and as lightness is a great point in carriage-building, they are made quite thin. The roof is composed of the same material, and the floor is planked with fir. The "carriage," or that part of the vehicle which unites the fore and hind wheels, and on which the "body" is supported, is made of a combination of ash, elm, and iron. The carriage-maker shapes and fits together all the pieces of wood for the "carriage," and hands them over to the smith, who makes and fixes on the iron parts. Meantime the wheelwright has been busy with the wheels, in which three kinds of wood are used. The nave, or centre, is made of elm, the spokes of oak, and the felloes or rim of ash ; and all are firmly bound together by a stout hoop of iron. Beneath the carriage and the body the springs are introduced. These are delicately fashioned in fine steel, and are used in a variety of forms.

When the operations above described are completed, the painters and trimmers execute their part of the work. In the best class of broughams, however, a piece of currying work has to be done before painting can be proceeded with. The roof and upper part of the back and sides are covered with a hide of leather, which is so manipulated, that without a seam it covers the parts mentioned, imparting strength and rendering the carriage waterproof. A less expensive and more common mode of effecting this object is to use fustian, or "moleskin," instead of leather. The fustian makes a good ground for painting upon, and is very durable—while it does not, as some hides have been found to do, exude oil, which, finding its way through the paint and varnish, spoils the appearance of the carriage. After several coats of "priming," a number of coats of "filling" are laid on. Each coat is allowed to dry thoroughly before another is added. Five or six coats of paint of the colour which the coach is ultimately to have are next applied, the entire surface being smoothed and polished from time to time, until a beautiful finish is obtained. Half a dozen coats of copal varnish are then laid over all. The varnishing has to be done in an apartment from which dust and flies are carefully excluded. In all, twenty-five coats of paint and varnish are required. Most carriages are decorated on the wheels, shafts, and other parts, by fine lines of a light colour. These are executed before the varnishing is done, and so are the armorial bearings or monograms, which few carriages are without now-a-days. The heraldic painting is done by a superior tradesman, and some specimens of this kind of work are remarkable for clearness of outline and vividness of colour. When the painting is completed, the carriage is put together and passed to the trimmers, of whom there are two classes—one doing the upholstery work for the interior, and the other the "blackwork" or leather fittings. In connection with a "brougham" little service is required from the black-trimmers; but in the case of a "landau sociable," or other kind of hooded carriage, they have a good deal to do. The trimmings of the inside are composed of various materials, according to the price to be paid for the carriage—Spanish cloths, plain and embossed silks, embossed leather, lace of various kinds, &c. The metallic beading, door-handles, and other decorations of the kind, are obtained from manufacturers who devote special attention to their production.

Omnibuses and stage-coaches are fashioned much in the same manner as carriages, only they are made heavier and stronger. The principal builders of these in Edinburgh are Messrs J. Croall & Sons and Messrs Carse & Co. Carts, waggons, and the like, are made by a distinct class of workmen. Cartwrights are to be found in nearly all the towns and villages in the country, but a considerable proportion of the county towns even are without a coachmaker. Edinburgh, Glasgow, Aberdeen, Perth, and Stirling, are the chief seats of the coachmaking trade in Scotland, and from these towns a large number of carriages are annually exported.

H

Trade-unions have not interposed to produce disagreements between employers and employed in the coachmaking trade, and probably in no other branch of industry are the merits and remuneration of the men so nicely adjusted. In all departments the wages, on the average, are equal to the highest paid to workers in wood and iron; but then some men, by exercising greater skill and expertness, are able to make nearly twice as much money as others. There is, consequently, a considerable difference between the highest and the lowest wages earned in each department. So much of the work is done by "piece" as can be satisfactorily reckoned, and the remainder according to a time-scale. The following are the current rates of wages, earned by piece-work and otherwise, in the shops of the leading firms in Edinburgh :—Bodymakers, 20s. to 40s. a-week ; carriage makers, 20s. to 27s. ; smiths, 17s. to 40s. ; wheelwrights, 24s. to 28s. ; painters, 19s. to 27s. ; trimmers, 20s. to 27s.

MANUFACTURE OF PLATE AND JEWELLERY.

MANY centuries before coal or iron or any other mineral was known to them, the inhabitants of Scotland were acquainted with gold. Found in the river-beds of their own rugged country, the precious metal was to them an object of delight; and with the aid of stone hammers they formed it into rude ornaments for the decoration of their persons. Antiquarian research has brought to light many curious and interesting facts relating to the use of both gold and silver in this country in prehistoric times, and numerous articles of ornament fashioned in these metals are preserved in the Museum of the Society of Antiquaries in Edinburgh. Coming down to historic times, we find that the Scotch had such a love for trinkets, that the Norsemen in their sagas reproachfully characterised them as "forlorn wearers of rings." In the records of the twelfth century, there is distinct evidence that gold was found in Scotland; for David I. conveyed by charter to the abbey of Dunfermline one-tenth part of all the precious metal found in Fife and Kinross. If one-tenth of the gold found in these counties was considered a fit gift from a king to a favourite ecclesiastical institution, we may conclude that the entire quantity obtained was considerable. Sir David Lindsay, the tutor of James V., in recounting the advantages of Scotland, says:—" Of everilk mettell we have the riche mynis, baith gold, silver, and stanes precious." James IV. had opened gold mines at

Leadhills, and the search for the precious metal was continued by James V., who obtained the service of foreign miners, and conducted the operations in a more systematic manner than formerly. It is said that his enterprise was rewarded by obtaining gold to the value of L.300,000. At a later period gold was got at Wanlockhead by a Dutchman named Greig. The gold found by him was made into a basin capable of holding an English gallon of liquid. The basin was filled with coins also made of Scotch gold, and presented by the Regent Morton to the King of France. One of the Earls of Hopetoun caused a search for gold to be made in the same locality at a later period; but the expenses were greater than the value of the metal found, and the adventure was abandoned, though not before sufficient gold had been obtained to form a small piece of plate. It is recorded by Boethius, Buchanan, and others, that gold was at one time found in remunerative quantities in Glengabber, a tributary of Megget Water in Peeblesshire. Mention is made of a nugget weighing thirty ounces having been found by some of the early miners. The largest piece ever got at Wanlockhead weighs four or five ounces, and is preserved in the British Museum. A grand scheme for the formation of a mining company to search for gold in Scotland was submitted to the Council of Queen Elizabeth. The company was to consist of twenty-four landed gentlemen, each of whom was to pay L.300 in support of the company. The prospect of success not being considered strong enough to induce gentlemen to embark in the scheme, it was suggested that each shareholder should be knighted, and called the " Knight of the Golden Mine," or the " Golden Knight," but the company never was formed.

Geologists have at various times expressed a belief that gold exists in Scotland in considerable quantities. At a conversazione of the Royal Society of Edinburgh, which was held in February 1863, Dr Lauder Lindsay read a paper on the gold fields of Scotland. He pointed out the general resemblance of the auriferous slates of Otago, in New Zealand, to the metamorphic slates of the Grampians, and discussed the question of the probable diffusion of gold in the Silurian slates and their derived " drifts," or alluvium in Scotland. Nuggets, he showed, had already been found in Leadhills, Tweeddale, Breadalbane, and in Sutherland. In two other quarters Dr Lindsay tried to excite interest in this important social question, namely, at the British Association for the Advancement of Science in 1867, and at the Geological Society of Edinburgh in 1867 and 1868. In the "Transactions" of the latter for 1868, will be found his paper on "The Gold and Gold Fields of Scotland."

Dr Lindsay, it appears, made a personal survey of the gold fields of New Zealand, and it was when visiting these that he was struck with the similarity, as respects physical geography and geology, between the two countries; and on his return to Scotland surveyed it afresh, to see how far such a parallelism really existed. He found the rocks of a great part of Scotland similar to those of most auriferous countries, and came to the conclusion that gold is much more extensively and generally diffused over Scotland than had been hitherto supposed. He points particularly to the greater part of the counties of Sutherland and Ross, of Inverness and Argyll, north of the Caledonian Canal; and to parts of Aberdeen, Banff, Kincardine, Perth, Forfar, Stirling, and Dumbarton, north of the Tay; also to various counties south of the Forth, where gold has been found or may be looked for. Geologically, the area of the Scottish gold fields corresponds to that occupied by the Lower Silurian strata and their drifts; in the south, represented by the greywackes and graptolitic slates of the Lowthers; in the north, by the micaceous schists of the Grampians. But it is not necessarily confined to the Silurian area, and has been found in other countries in rocks of many different characters and ages. It is obvious, however, that the extent to which it occurs in Scotland can only be determined by systematic investigation, equivalent at least to the " prospecting" of gold-diggers, and Dr Lindsay suggests that the investigation should form part of the duties of the staff of the National Geological Survey.

Actual discoveries of gold have been made at different times in the following localities:—

Sutherlandshire—Kildonan on Helmsdale Water.

Perthshire—1. Breadalbane—area of Loch Tay and head waters of the Tay (Tyndrum and Taymouth). 2. Upper Strathearn—area of Loch Earn and head waters of the Earn (Glen Lednock—streams falling from the north into Loch Earn; Ardvoirlich, south side of Loch Earn; Glenturrit). 3. Glenalmond—Glenquoich and other valleys of the Grampians.

Forfarshire—Clova district—" Braes of Angus," Edzell and Glenesk.

Aberdeenshire—area of the Dee (Braemar, Invercauld, coast about Aberdeen).

Argyllshire—Dunoon.

Lanarkshire—Head waters of the Clyde, including the rich Crawford Moor or Leadhills district (Elvan water, Glengonner, Glencaple, Mennlock and Wenlock, Short Cleuch, Lamington Burn).

Peeblesshire—Head waters of the Tweed (Manor water, which

flows north to the Tweed; Megget water, which flows south to St Mary's Loch; various feeders of the Yarrow; Glengabber).

Dumfriesshire—Head waters of the Annan (Moffatdale—streams falling into Moffat water; Hartfell range above Dobbs Linn).

A gentlemen belonging to Sutherlandshire, while engaged digging gold in Australia or New Zealand, was struck with the likeness which the geological formation bore to that of his native county. Having returned home, he set out in the end of last year to visit the quartz reefs of Kildonan, and was not long in satisfying himself that the rocks were not only the same in appearance, but that they also contained gold. The report that gold was to be found in their neighbourhood created a deal of excitement among the natives of the district, who soon flocked to the strath of Kildonan, and began gold-washing. Their labours so far have been confined to the bed of the stream; and the quantity of gold obtained has not been very encouraging; but experienced men are confident, that were permission obtained to conduct mining operations in a systematic style, the results would be highly remunerative. The superior of the soil is having the ground tested, and should he be satisfied that it would be advantageous to both the diggers and himself, the ground will be let off into "claims," so that there is a probability of extraordinary scenes being witnessed in the lonely glens of Sutherlandshire, if not in other quarters.

There are no silver mines in Scotland; but the lead obtained from the mines at Wanlockhead contains a proportion of silver, which is extracted. The quantity of silver thus obtained is from 6000 to 8000 ounces a-year, worth from L.1500 to L.2000. The silver produced from lead ore in the United Kingdom amounts to about 700,000 ounces a-year, and a very small quantity is produced from native silver ore. From official returns, it appears that from 500 to 5000 ounces of gold per annum are produced by mines in the United Kingdom.

A love for finery seems to have been a conspicuous trait in the character of several of the early rulers of Scotland; and when trade was opened with some of the Continental countries in the twelfth century, among the first things imported were vessels of gold and silver, armour, &c.; and a great show of these articles was made at the court, and among the nobility. In those days the churchmen were the great masters of the necessary and ornamental arts, and were so jealous of their skill, that they could not afford to allow foreigners to have the sole privilege of supplying such articles as plate and jewellery; and accordingly they turned their attention to working in the precious metals. They became goldsmiths, jewellers, and lapidaries, and soon succeeded in making articles which competed with some measure of success against foreign productions. This was the beginning of the art of working in gold and silver in Scotland. In the course of the thirteenth and fourteenth centuries the trade assumed considerable importance. Ladies of the court and nobility wore tiaras, girdles, brooches, and earrings of gold and silver, set with native pearls and precious stones; while the armour and horse-trappings of the gentlemen were, according to some accounts, most gorgeously decorated with the same materials. The plate used in the churches was of the most superb description. Though the native goldsmiths and jewellers had attained great expertness in their art, their handiwork was not sufficiently grand for the taste of some of the nobles, who obtained splendid specimens of armour, jewellery, and gold and silver work, from Italy and Flanders.

Some idea of the kind and quantity of plate and jewellery in the possession of old Scotch families may be obtained from an inventory of the jewels and valuables of the Campbells of Glenurchy, drawn up in 1640. Among the articles were a target of enamelled gold, set with diamonds, topazes, rubies, and sapphires—a gift from King James V.; a round jewel of gold, set with precious stones, among which were twenty-nine diamonds, and four great rubies—the gift of Queen Anne; a gold ring, set with a diamond shaped like a heart, and other diamonds; a silver brooch, set with precious stones; sixty-six gold buttons; twelve silver plates; four great silver chargers; two silver basins and jugs, partly gilt; one dozen silver trenchers; one dozen silver saucers; a great silver cup, partly gilt, and bearing the arms and names of the Laird of Duntrons; seven other silver goblets and cups, partly gilt; a great silver cup, with lid partly gilt, and ornamented with raised work; an engraved silver cup; three silver jugs for vinegar; three silver salt cellars; two bowls, with silver lips and feet; eleven plain silver spoons, with the laird's name on them; six silver spoons, with "round knapit endis overgilt;" and thirty-eight other silver spoons. It is evident that several of the articles mentioned were of foreign manufacture; but there is ground for supposing that a considerable number were home-made, and, moreover, that similar collections of plate were in the possession of other Scotch families at the date of the above inventory.

About a century before that time one of the most important

trades in Edinburgh was that of the goldsmith ; and the city possesses in George Heriot's Hospital a substantial token of the prosperity which rewarded some of the workers in the precious metals.   George Heriot succeeded his father in the business of goldsmith and jeweller, and in 1597 was appointed goldsmith to Anne of Denmark, the queen of James VI.   Subsequently he received the appointment of goldsmith and jeweller to the King.   Upon His Majesty's accession to the throne of England, Heriot, whose skill in his trade would appear to have been remarkable for those days, accompanied the king.   He died in London in 1624, leaving a fortune of L.50,000, which he had accumulated in thirty-eight years ; and nearly half of that amount was bequeathed in trust to the Town Council and ministers of Edinburgh, for the purpose of building an Hospital in Edinburgh for the maintenance and education of indigent children, the sons of burgesses of the city.

The trade of the goldsmith was made the subject of legislative enactments in very early times.   Gold and silver in a state of purity would be too soft and ductile to be used in the manufacture of plate or coins, and, accordingly, it is necessary to add a certain proportion of the baser metals.   The minimum quantity of alloy required is well known ; but as great facilities existed for the workers in the metals unduly increasing that quantity, and as they had availed themselves of those facilities to an extent warranting the interference of the State, a law was passed in 1238 which prohibited the use of gold of less than a certain standard of fineness, or of silver of a lower standard than the coin of the realm.   The mode of testing the quality of gold and silver was by means of the " touchstone," a black stone of close fine grain, on which the article to be tested was rubbed, the quality of metal being determined by the shades of colour presented by the metal which adhered to the stone.   The assaying or testing of the precious metals was a privilege conferred on the Goldsmiths' Company of England in the year 1300 by an Act of Edward I. The wardens of the craft were empowered to go from shop to shop to see that no inferior gold was used in making plate or jewellery.   All that came up to the standard of purity was then stamped with a leopard's head, while the inferior metal was forfeited to the king. Honesty would appear to have been at a discount in those days ; for frequent reference is made to deceptions—such as that practised by the cutlers, " who covered tin with silver so subtilely and with such sleight that the same could not be discerned and severed from the tin, and by that means they sold the tin so covered for fine silver." The preamble of an Act passed in 1379 sets forth that the gold and silver worked by English goldsmiths was oftentimes "less fine than it ought to be." An Act of Henry IV., dated 1403, recites "that many fraudulent artificers do daily make locks, &c., of copper and atten, and the same do over gild and silver like to gold and silver, to the great deceit, loss, and hindrance of the common people, and the wasting of gold and silver." Persons continuing such practices were made liable to a heavy penalty. It appears, however, that ornaments for the church might be made of gilded or silvered copper, provided some part of the copper was left exposed, to show that the article was not solid. A number of Acts were subsequently passed to regulate the trade in England.

The workers in the precious metals in Scotland seem to have been afflicted with weaknesses similar to those which caused their English brethren to be the subjects of so much legislation. In 1457 a statute was enacted for "the reformation of gold and silver wrought by goldsmiths in Scotland ; and, to eschew the deceiving done to the king's lieges, there shall be ordained in each burgh where goldsmiths work one understanding and cunning man of good conscience, who shall be deacon of the craft ; and when work is brought to the goldsmith, and it be gold, he shall give it forth again in work no worse than twenty grains, and silver eleven grains fine, and he shall take his work to the deacon of the craft, that he may examine that it be fine as above written, and the said deacon shall set his mark and token thereto, together with the said goldsmith's ; and when there is no goldsmith but one in the town, he shall show that work, tokened with his own mark, to the head officers of the town, who shall have a mark in like manner ordained therefor, and shall be set to the same work." It had evidently been found difficult to resist the temptation to deceive, for in 1555, "forasmuch as there was great fraud," it was enacted that no goldsmith should "make in work nor set forth either his own or other men's silver under the just fineness of elevenpenny fine, under the pain of death and confiscation of all their goods and movables ; also, that no goldsmith set forth either his own, or other men's gold under the just fineness of twenty-two carats fine, under the pain aforesaid."

About the earliest incorporated trade in Edinburgh was that of the hammermen, under which term were included goldsmiths, blacksmiths, saddlers, cutlers, and armourers. Other branches were subsequently added, but in 1586 the goldsmiths were formed into a separate company. By the articles of the company apprentices were ordained to serve for a term of seven years, and masters were obliged to serve a regular apprenticeship and three years over and above, to make them more perfect in their trade. They were, moreover, bound to give to the deacon of the craft proof of their skill in working and knowledge of the fineness of metals, &c. Only those admitted to the company by the deacon and master were to work, melt, or break down, or sell any gold or silver work, under penalty of twenty pounds or imprisonment. In 1687 the company was incorporated by a charter granted by James VII., and obtained additional powers for regulating the trade. According to the terms of the charter, those powers were granted " because the art and science of goldsmiths, for the most part, is exercised in the city of Edinburgh, to which our subjects frequently resort, because it is the seat of our supreme Parliament, and of the other Supreme Courts, and there are few goldsmiths in other cities." In virtue of the powers conferred on it, the company, from the date of its formation, tested and stamped all the plate and jewellery made in Scotland. The first stamp used was a castle, consisting of three towers, the central one being higher than the others. In 1681 a letter representing the date was stamped on as well as the castle. The letter **a** indicates that the article bearing it was made in the year between 29th September 1681 and the same day in 1682—the other letters of the alphabet, omitting j and w, representing the succeeding twenty-three years. Each piece bore, in addition to the castle and date-letter, the assay-master's initials, and the maker's initials. Seven alphabets of a different type have been exhausted in recording the dates ; and the letter of the eighth alphabet, for 1869, is an Egyptian capital **M.** In 1759 the standard mark of a thistle was substituted for the assay-master's initials, and is still continued. In 1784 a " duty-mark " was added, the form being the head of the sovereign.

The silver mace of the city of Edinburgh is dated 1617 ; the High Church plate, 1643 ; Newbattle Church plate, 1646. Other towns in Scotland seem to have availed themselves of the early Acts of Parliament, and used their own "town marks." The plate of the parish church of St Andrews bears date 1671, and is marked with a St Andrew cross ; and the plate of the West Church of Perth, dated 1771, bears, in addition to the Edinburgh marks, the town symbol of a spread eagle. Glasgow was not made an assay town until 1819. The marks used on the plate stamped at Glasgow are a lion rampant, the arms of the city, the maker's initials, the date letter, and the sovereign's head.

Few occupations afford such a wide field for the exercise of artistic taste as those of the goldsmith, silversmith, and jeweller. The variety of articles fashioned by them is very extensive, and practically there is no limit to the number of designs that may be employed. Sometimes the task of the goldsmith or silversmith is to produce an article of a purely ornamental character, but more commonly use and ornament are combined. With the jeweller ornament is the chief object, and he is, accordingly, less restricted in working out his designs. Only in a few cases is the vendor of plate and jewellery the actual maker, and this is one of the peculiarities of the trade. There are a number of "small masters" in each department, who occupy shops of their own, and work for the merchants by contract or otherwise, none of them being bound to work exclusively for one merchant.

In Edinburgh the workshops of the small masters are situated in out-of-the-way lanes in the New Town, and, like those in which baser metals are dealt with, are dingy, smoke-begrimed places. A brief description of the making of a teapot will suffice to convey an idea of the silversmith's occupation. There are two ways in which the work may be accomplished—an old and a new. According to the old method, the bowl or body of the teapot is made of one piece of silver hammered up from the flat. The silver is first rolled out into a sheet about the thickness of a shilling. A piece of the required size having been cut, it is hammered on a block of hardwood in the surface of which a smooth saucer-like hollow has been formed. In that way the metal is brought to the shape of a bowl. It is then taken to an anvil, and by skilful hammering the rim of the bowl is gradually contracted until the vessel is almost of a globular shape, with an opening three or four inches in diameter on what is to be its upper end. As the striking tends to harden the silver, the vessel is annealed several times during the hammering process. This is done by bringing it to a red heat and allowing it to cool gradually. When the body of the teapot is brought to nearly the required shape by the first hammering, it is planished or made smooth by a slightly different process. Great care must be taken to hammer the vessel equally, else some parts might become thin or be broken through completely. The lid, spout, and handle are next made, and a hoop of metal is soldered on the lower part to form a base or foot. The spout is stamped out of a sheet of metal by means of dies, and is made in two halves which are soldered together. The handle is made in the same manner. An ornamental form is thus obtained without much labour. It depends on the design whether the spout and handle should be attached or not before the chasing is done on the body of the teapot. All the parts are put together for engraving, as the annealing in soldering destroys the bright cut of the engraving. To the first processes in the mode of making teapots that has just been described, there are various technical objections, the principal being that the blisters to which silver is liable become cracks under the hammer, and must be soldered, and that it is difficult to keep an equal thickness of metal all through the article. The improved way is to have the silver rolled thinner, so that on being annealed all the blisters show themselves. A sound piece being selected, a cylinder is formed, from which the body of the vessel is formed partly be expanding and partly by contracting. The expansion is done with a small bullet hammer on a sand-bag, and the contraction on variously formed beak-irons by hammers. The improved mode of planishing is to have the hammer face padded with a kind of cloth, outside which is a piece of fine steel clock-spring. This finishes the surface with almost a burnished effect. Sugar basins are still made by the process first described.

The chasing of silver is a highly artistic occupation, and on the manner in which it is executed the beauty and value of an article chiefly depend. The chaser begins by drawing the design on the silver with a hard lead pencil. Where parts have to be brought out in high relief—such as figures, festoons of flowers, or bunches of fruit—the metal is struck out from the inside by means of a " snarling-iron." No attempt is made at this stage to produce anything approaching a likeness of the fruit or flowers, the object being merely to raise the metal over a space and to a height sufficient to admit of the forms being produced by manipulation from the outside. The vessel is then filled with pitch, a substance sufficiently consistent to preserve it from losing shape under the punches of the chaser, and yet not too hard to prevent the necessary indentations being made. A vessel chased in high relief looks to the uninitiated as if the work had been done by using dies or punches from the inside, whereas, except in such cases as have been mentioned, the chasing is done entirely from the outside. Having completed the operations described, the chaser rests the vessel on a circular cushion, and begins the punching. He first goes over the outline of a small patch of the design, and then fills in the details. His tools are small steel punches and a hammer ; but such variety is there in the details of patterns, that not fewer than about three hundred punches are required to form a complete set. The groundwork of chasing is usually rough or "matted," and that part is done with punches having chequered faces in all degrees of fineness. The rapidity with which the most elaborate designs are worked out is surprising.

The occupation is light, but can be followed only by persons possessing artistic taste and skill. After the chasing is completed the parts are soldered together, and the article is then ready for "finishing." The process of finishing plate—a department of work by itself—is as follows :—After the article is finally annealed, and allowed to lie a certain time in "pickle" (water with a little sulphuric acid in it), the surface is scoured or rubbed with a lump of pumice stone and water. The vessel is then taken to the "mill" to be brushed with rottenstone and oil, after which it is either flat chased or engraved. Engraved work is finished by being hand polished with rouge. Chased work, both repoussé and flat-chased, is whitened by being smeared over with a mixture of borax and saltpetre, annealed, and laid in a bath of "pickle." The chased parts are brushed at a lathe with a circular brush made of very fine brass wire, charged with sour beer, the smooth parts being burnished or polished with the finger and rouge.

Silver engraving is a branch distinct from chasing. The engraver draws his designs on the silver, and cuts out the lines, thus removing a portion of the metal, whereas the chaser produces his effects without reducing the material. In some of the heavier kinds of plate—such as candelabra, centre-pieces, and tea-urns—the ornaments are formed by casting. Effigies of animals, detached shields, trunks of trees, &c., are formed in that way. The models are made in wax, and from these moulds are taken in sand, much in the same way as for castings in brass or iron; but the sand leaves a rough surface, and the chaser has to give the castings a finishing touch. Articles of electro-plate are usually made of German silver, and treated in the same way as silver through all the preliminary stages; but after the chasing has been done, the vessels are exposed to the action of a battery, and thus coated with pure silver. Sometimes articles made of silver and electro-plate are either wholly or partially plated with gold by the electro process; when partially plated they are said to be " parcel gilt."

The making of spoons and forks was at one time an extensive branch of the silversmith trade in Edinburgh, but now there are only two workshops in which these articles are produced. It appears that the profits on spoons and forks are small, and hence there is no inducement to enter into competition with the manufacturers in London, who have extensive establishments in which machinery is applied to most parts of the work. As made in the old fashioned but thoroughly substantial way, spoons are first forged, then stamped by means of dies, and afterwards filed and polished by hand.

The assay office of the Goldsmiths' Hall, on the South Bridge, Edinburgh, is open on alternate days, when articles of gold or silver that require to be guaranteed by the stamp of genuineness are sent in and assayed. The assay-master scrapes a small quantity of metal off each article, and submits it to a test, in order to ascertain the quality. The duty charged on each ounce of gold-plate is 17s. 6d., and on silver-plate 1s. 6d.

Messrs Mackay, Cunningham, & Co., Her Majesty's goldsmiths for Scotland, have done much to improve and extend the art of working in gold and silver in Scotland. Liberal encouragement has been given to native talent, and the result has gone to prove that the highest class of plate and jewellery can be produced in Edinburgh as readily as in London or elsewhere, and that, if there be any difference in excellence of workmanship, it is in favour of Edinburgh. Designers and modellers of great ability are now regularly employed in the trade; and at least one hand, whose fame in another department of art will live through many generations, has lent assistance to establish a genuine reputation for the productions of the Scottish goldsmiths. One or two of the more important works recently executed by Messrs Mackay, Cunningham, & Co. are worthy of mention. Remarkable for its size and novelty of style is a trophy for the officers' mess of the 92d Regiment. It is in the form of a triangular obelisk of granite, rising from a silver pedestal, which rests upon a broad base of granite, the total height being about three feet. Standing on the angles of the base are three figures, representing an officer, a sergeant, and a piper, all in the full uniform of the regiment. The figures, which are eight inches high, are in frosted silver, and have been beautifully modelled. They were the last work executed by the late Mr William Beattie, whose skill as a modeller is well known in artistic circles. The sides of the pedestal bear in high relief the crest and badge of the regiment —a stag's head and a wreath of ivy. On each of the angles of the pedestal is a sphinx in frosted silver, and the shaft of the obelisk is girt at intervals by bands of silver, on which are emblazoned the names of the more important actions in which the regiment has participated. Both in conception and execution the trophy is a noble piece of plate, and will no doubt be cherished as such by the owners. A beautiful epergne or centre ornament of silver next claims attention. It has been made for a famous breeder of Leicester sheep, and consists of an oak tree denuded of its upper part, but retaining a few of its branches, which support a crystal fruit-dish. On a grass plot surrounding the trunk of the oak are a group of Leicester sheep, which have been finely modelled by Mr Gourlay Steell, R.S.A. The base is about three inches in depth, and is surrounded by a series of circular panels or recesses in which are disposed the medals, about twenty in number, won by the owner at various agricultural shows. A communion service for the congregation of the Rev. Dr Candlish shows a great advance on the common style of Presbyterian church plate.

Equally worthy of mention are the efforts made to improve the trade by Messrs William Marshall & Co. of Edinburgh, who have devoted much attention to the production of plate and jewellery of a highly artistic kind. Mr J. D. Marshall has laboured most assiduously for twenty years past in working out designs, which, while peculiarly suitable for production in the precious metals, have had the effect of creating a distinct character and celebrity for Scotch jewellery. He has studied the national antiquities to good purpose, and has borrowed hints from the most unlikely quarters. The enamelled and engraved jewellery of a runic type received its first development from Mr Marshall, and his designs in that class number many hundreds. He has also applied himself most successfully to designs for plate. The saloon of the firm contains a collection of native workmanship which would do credit to any country; and that its merit is recognised beyond the borders, is attested by the honours they have won at the London, Paris, and other Exhibitions. Messrs Marshall & Co. are the most extensive makers of plate and jewellery in Scotland.

No ornamental art enjoys a wider patronage than that of the jeweller. His productions grace the brow of royalty, and form an object of pride with the poorest domestic, he is intrusted with the "setting" of gems worth a hundred fortunes, and has to exercise his ingenuity to produce trinkets for the million. In jewellery "a thing of beauty" is not "a joy for ever," nor would the jeweller wish it were so. With changes of fashion in dress come changes of fashion in jewels, and there is thus a constant demand for new designs. To meet that demand, gold, silver, and gems are combined in ever-varying styles. The manufacture of jewellery, as already stated, was early practised in Scotland, and for many years past the "pebble jewellery" made in this country has been much in demand at home and abroad. The style has been copied by the English manufacturers, who, by using an inferior quality of materials, have prevented the Scotch makers from reaping the full benefit of this branch of their trade, for the exercise of which the abundant supply of fine pebbles to be obtained in Scotland gives them peculiar facilities.   In Edinburgh great attention has been paid to the manufacture of pebble jewellery, and a degree of excellence has been attained which it would be almost impossible to surpass.   Some of the early work in pebbles was very coarse and inartistic; the stones were roughly cut, and arranged without regard to shades of colour; but now the utmost care is taken in the cutting and arranging of the pebbles, and beautiful effects are thereby produced.

The gold used by jewellers is always alloyed with certain proportions of pure silver and the finest copper, according to the quality desired.   The legal standard of silver is 11 oz. 2 dwt. of fine silver in the pound troy, the balance being copper.   The jeweller melts his metals in a crucible, and casts them into ingots about two inches broad, three inches long, and one-eighth of an inch thick. The ingots are reduced to any degree of thinness by being passed between steel rollers.   The sheets or plates of metal thus produced are intrusted to a workman, who, guided by drawings or models, clips out the pieces required for the various articles to be made.   The pieces are given, along with the designs, to other workmen, who put them together.   These men are seated at large tables, round the sides of which are a series of semicircular recesses, each recess being occupied by a workman.   After the pieces are brought to the exact size required, they are soldered together by means of a blow-pipe.   Articles of an ornate character—such as brooches and bracelets—covered with designs in filigree work, or inlaid with pebbles, require great nicety of manipulation, and the number of parts which go to compose some of them is immense.   Pebble bracelets of a finely worked geometrical pattern are made in which there are no fewer than 160 pieces of stone.   In making an article which is to be inlaid with pebbles, such as a brooch, the jeweller forms a back or foundation, to which a plate, pierced with apertures for the pebbles, is fixed, a convenient space being left between the two plates.   At this stage the work is passed to the lapidary, who cuts and fixes the pebbles. The stones are first cut with a revolving disc of iron, charged with diamond dust and oil, and roughly shaped with a pair of pincers. Each piece is then taken in succession, and attached to a "cement-stick"—a small piece of wood with a quantity of strong cement on one end.   Held in that way, the stone is ground to the required shape on a revolving disc of lead, charged with emery and water. When all the pieces are brought to the shape of the apertures designed for them, they are set in with shellac.   The outer surface has up till this time been left rough; but, after the cement has hardened, the lapidary takes the brooch in his hand, and manipulates it on the grinding disc until the stone is reduced to the level of the metal which surrounds it. The surface is next polished on a disc of tin, charged with rotten-stone and water, and the brooch is returned to the jeweller. Usually pebble brooches have in the centre a " cairngorm," or what is commonly supposed to be one. The cairngorms are not "set" until the work on the other parts of the brooch is all but completed. The exposed surface of the metal on the face of the brooch is usually relieved by engraved scroll-work. Enamelled jewellery has recently come into fashion to some extent, and fine specimens have been produced, the runic patterns especially being very pretty.

The lapidaries obtain their pebbles from various quarters of the country. Aberdeenshire furnishes agates, beryls, and the famous Cairngorm crystal ; and in the parish of Leslie, in the same county, is found a beautiful amianthus, which is fashioned into snuff-boxes, &c. Ayrshire furnishes agates and jaspers ; Perthshire, bloodstone and a variety of others; Forfarshire, jaspers; and Mid-Lothian, the Pentland pebble and the Arthur Seat jasper. Amethysts were once abundant in Scotland, but they have now become scarce. At Elie, in Fifeshire, garnets are found. Then there are the Scotch pearls, so much valued for their size and beauty, though inferior in some respects to the Oriental kind, being more opaque. With such a variety of material, the Scotch jewellers have great facilities for producing multitudinous designs, and they seem to be improving their opportunity.

As might be expected, the silversmiths and jewellers are an intelligent class of workmen, and nearly all of them are or have been students in the School of Design. Their occupation being, however, to a great extent simply mechanical, their wages are not higher than those of skilled workmen in other trades which fall under that designation. Lapidaries serve an apprenticeship of six years, and jewellers, silversmiths, and silverchasers of seven years. Silversmiths and jewellers generally receive from 18s. to 32s. a-week, and lapidaries, 24s. ; but in exceptional cases higher rates are earned. About two years ago the men made a successful movement for the reduction of their hours of labour to fifty-seven a-week ; but, without any pressure on their part, a considerable advance has been made on the rate of wages within the last few years.

Besides being used in the manufacture of plate and jewellery, gold and silver are extensively employed in decorating various articles, such as picture-frames and other articles of furniture, books, carved work, &c. For such purposes the metals are hammered out into exceedingly thin plates or leaves. Gold-beating, as the process of making those leaves is called, is an art of great

I

antiquity ; and it would seem that gilded articles were so fashionable at one time in this country, that it became necessary for the State to interfere to prevent the precious metal from being wasted in such a way.　About the year 1619, a statute of James I. enacted that, " the better to prevent the unnecessary and excessive vent of gold and silver foliate (*i.e.*, leaf) within this realm, none such shall from henceforth be wrought or used in any building, ceiling, wainscot, bedstead, chairs, stools, clothes, or any other ornament whatsoever, except it be armour or weapons, or in arms or ensigns of honour at funerals or monuments of the dead." Gold-beating was introduced into Scotland in the year 1805 by the late Mr Wright, of the firm of Ross & Wright, Calton Street, Edinburgh.　Mr Wright was presented with a complete set of working apparatus by the Highland Society for having introduced the art, and he was appointed gold-beater to his Majesty King George IV. The trade has not thriven in Scotland however, there being but two or three gold-beaters' shops north of the Tweed, and in these only a few men are employed.　Gold or silver leaf is made by first rolling out the metal into thin plates, and then hammering them between layers of prepared ox-gut, called "goldbeaters' skin."　The gold leaves are made so thin that it would require 300,000 of them laid one on the top of the other to make the thickness an inch.　The leaves measure 3·3 inches square, and 2000 of them are produced from a piece of gold weighing four pennyweights less than an ounce. Machinery has recently been applied to supersede the arduous manual labour of gold-beating.

Seal-engraving is an art akin to jewel-making, and merits a passing notice.　The practice of using gummed envelopes has, by superseding wax, gone far to extinguish the occupation of the seal-engraver. Not many years ago a massive seal, bearing the crest of the wearer —if he were fortunate enough to have one, or his initials if he could not claim heraldic privileges—was invariably suspended on the watch-guards of gentlemen ; and ladies carried daintily got-up seals, with which they impressed emblems of love on the gaudily coloured and perfumed wax which preserved the contents of their *billets-doux* from the glance of profane eyes.　Wax and seals have had their day ; but signet-rings are still in fashion, and keep the lathes of the engravers from coming to a dead stand.　Engraving on gems is one of the nicest artistic occupations.　It is easy for workers in metals to repair flaws or imperfections, but the seal-engraver has no facilities for doing so.　If he makes a blunder the gem is ruined, and his labour is lost.　He begins operations by fixing the gem on a convenient handle, and then draws the design upon it with a brass needle. The engraving is done by means of fine tools resembling drills, to which a rapid revolving motion is given in a small lathe. The tools are dipped from time to time into a composition of diamond dust and olive oil, and the operator holds the gem in his hand and applies it to the tools. So fine is the work generally that a powerful eyeglass has to be used, and so slow is the process of cutting, that a whole day is required for the engraving of a ribbon and motto.

The number of persons working in gold, silver, and precious stones in Scotland is little short of two thousand, and a large proportion of these are employed in Edinburgh. Plate and jewellery being articles of luxury, the demand for them fluctuates according to the prosperity of the country—a fact clearly brought out by the returns of the quantities of gold and silver used in each year. The high price of gold-plate puts it beyond the attainment of all save a select few in the highest ranks of society ; but for those who have the desire, without the means, to possess real articles, electro-plate forms a passable substitute—though half the charm of the possession is lost in the knowledge that the beauty of the articles is only skin deep, and that the skin is a very thin one. Silver-plate has become common among the middle-class of the population, and articles in electro-plate are in great demand. Some of the latter are beautifully got up, but the ornamentation generally is not so finely executed as in the case of solid silver work. Only a small quantity of plated goods is made in Scotland, and a considerable portion of the other work is done to order.

For a number of years past the silversmith and jeweller trades have been extending in Edinburgh, and there are indications that they will increase still further. There are upwards of thirty master jewellers in the city, who employ from half a dozen to thirty men each. All the work done is of a superior kind, no attempt being made to vie with Birmingham in the production of cheap and showy articles, the beauty of which is as transient as that of a flower. The city is not likely to become a manufacturing centre in the common meaning of the term, nor in some respects would that be desirable. It is, however, well adapted to become a seat of light, artistic occupations, and many such are carried on in it. No city in Britain possesses a better School of Design, and it is gratifying to know that it is largely taken advantage of. Workmen trained in Edinburgh are highly valued by the London manufacturers of plate and jewellery, and some of the best work done in the metropolis is by their hands.

MISCELLANEOUS MANUFACTURES IN METALS.

In making steam-engines, mill-work, and general machinery the engineers of Scotland have acquired considerable repute. When the possibility of propelling vessels by steam was successfully tested on the Clyde, the enterprising mechanicians of the west did not neglect to improve the occasion. As soon as the demand for steam-vessels arose, they were ready to supply the motive power, and in that particular branch of work they have no superiors. The impetus given to the industrial arts by Watt's improvements on the steam-engine was accompanied by a demand for machinery to take the place of tardy hand-labour, and the inventive faculties of ingenious men were stimulated to the production of an endless variety of appliances for economising time and labour. The manufacture of these has engaged a large share of the attention of Scotch machine makers, and their handiwork is to be found in every important seat of industry in the world. Steam-engines of all kinds, locomotives, and printing, spinning, and weaving machinery, may be mentioned as among the more important productions of our engineering establishments.

No statement of the number of persons employed in the trade exists; but there cannot be fewer than 12,000 to 15,000 persons directly engaged in making engines and machines in Scotland. It is impossible to ascertain the value of the engines and machinery made. The annual export of machinery and millwork is valued at about half a million sterling. In 1866 it was L.414,810, the following being the amount from the different ports :—Glasgow, L.245,008 ; Leith, L.133,518 ; Greenock, L.17,529 ; Grangemouth, L.6024 ; Ardrossan and Troon, L.1355 ; Dundee, L.7137 ; other ports, L.4239.   This, of course, does not include the value of engines of steamers built for ports beyond Scotland.   The value of the engines of the steam vessels built on the Clyde is about L.400,000 annually ; and the value of the proportion of the above for ports beyond Scotland generally reaches about L.300,000. An extensive trade is done in the manufacture of engineers' tools ; yet many of the principal tools in the large engineering establishments bear the names of English makers.

Glasgow produces almost every article made of steel or iron, ranging between and including needles and 6000-ton iron-clad ships of war.   The immense engineering establishments which abound in the city present many interesting sights, and illustrate in an extraordinary degree the capabilities of machinery to supplant human labour.   The workmen are little more than passive agents.   They assign work to the machines, and stand by to see that it is properly accomplished.   Turning, boring, planing, and such operations are done by ingeniously-constructed mechanism, which performs the work with unfailing accuracy and astonishing speed.   A bare enumeration of the variety of work carried on in the engineering and machine shops would fill a book, and the information would be of interest to few persons.

Without further allusion to the general trade, mention may be made in passing of one of the most recently established branches of machine-making.   During the past ten years sewing-machines have been introduced into most manufactories and workshops devoted to the production of articles of clothing, and are gradually coming to be regarded as indispensable in the equipment of a house.   The sewing-machine is a decided favourite with the ladies, and mothers who have to bring up large families on limited incomes find in it a "friend indeed."   There are now many varieties of machines, the productions of the different patentees or makers having some peculiarity of form or movement to distinguish them.   In various parts of Britain there are establishments devoted to the manufacture of the machines.   Several of these are in Glasgow, and there is one in Edinburgh.   The most extensive makers of sewing-machines in Scotland are Messrs R. E. Simpson & Co., Maxwell Street, Glasgow, who acquired patent rights for their machines in 1859, and have gone on turning them out in increasing numbers.   The chief characteristics of the "Simpson" machines are the extreme simplicity of the mechanism, and the high speed at which they may be driven without danger of getting out of order. These qualities are most important, for nothing can be more annoying than having to suspend work at short intervals in order to have repairs made. Messrs Simpson & Co. employ about eighty workpeople, who produce from 300 to 400 machines a-month. There are indications that this branch of machine-making will undergo great extension. Already from 300 to 400 "Simpson" machines may be seen at work in individual manufactories of shirts, &c. Messrs Simpson & Co. obtained a medal at the Paris Exhibition of 1867 for the excellency of their sewing-machines. Since then they have established extensive agencies in Paris, Hamburg, Vienna, Brussels, &c.

A large proportion of the men belonging to the engineering trades are members of " The Amalgamated Society of Engineers, Millwrights, Smiths, and Pattern Makers," which is the most powerful trades-union in the country. It was formed in 1851 by, as its name indicates, the amalgamation of separate societies existing throughout the country. The number of original members was 11,829 ; and their collective funds, at the end of the first year, showed a balance in favour of the society of L.21,705. At the close of 1866, the members numbered 33,007, and the balance was L.138,113. The amount contributed to the various funds of the society in the sixteen years was L.535,573. The expenditure in 1866 was L.60,448, 5s. 4d., or L.1, 16s. 7¼d. per member. The cost of management for the year was L.9313, or about 15 per cent. of the entire disbursements. The society had 33 branches in Scotland in 1866, embracing 3105 members, whose portion of the balance amounted to L.13,589. The annual report of the society for 1866 forms a volume of upwards of 400 pages. It affords no clue, however, to the amount expended in promoting objects relating to work and wages, unless that be represented by the first item in the accounts —namely, "Donations, contingent fund, sending members to situations, and beds for non-free members, L.22,782, 8s. 2d."

Though it is no part of the plan of this book to discuss trades-unions, a fact or two bearing on that subject may be here introduced. According to indubitable authority the men engaged in machine-making are, generally, better educated than they were five-and-twenty years ago, yet such is the effect of the "levelling" principle of trades-unions, that fewer men are now to be met with who show a superior knowledge of their business. Indeed, one employer recently stated that out of from 200 to 300 men in his employment he did not know one who would be able to take the foreman's place in the event of its becoming vacant—whereas, before trades-unions became so fashionable with the men, he could, from a smaller number in his workshops, select at least a dozen fit for the superior post.

The term of apprenticeship in the engineering trade is five years, the working hours are fifty-seven a-week, and the wages of journeymen from 20s. to 28s.

Intimately related to the engineers are the ironmoulders; but they have not joined the Amalgamated Society, preferring, it would seem, to hold by their own union. For a considerable time past, the relations between the union moulders and their employers have not been of a harmonious kind. About two years ago the masters found it necessary to associate, so that they might be able to meet the action of their workmen; and in order to protect their own interests they had, in 1868, to close their works against members of the union. That step, of course, embittered the feeling between the employers and employed; but rules so arbitrary as the following, which were applied by the moulders' union to jobbing foundries, could not be submitted to:—" No apprentice above fourteen years of age can be admitted into a foundry, and the apprenticeship must not be for a less period than seven years; no more apprentices can be admitted to a foundry than in the proportion of one to every three journeymen; non-union men cannot be employed; labourers cannot be permitted to do unskilled work in the moulding department; piece-work is prohibited." There are upwards of 3200 men in the union. Before the masters took action to protect themselves, men working on time were paid 28s. 6d. to 30s. a-week of fifty-seven hours. Piece-men were allowed by the rules of their union to earn from 5s. to 7s. a-day, but not more, though an expert hand could nearly double that amount. No man was allowed to work overtime; and though at the hour for stopping work for the day metal should be within ten minutes of being ready to pour into the moulds, the men would walk away, and leave their masters to do with the metal as they liked. This rule against overtime caused serious loss in cases of accidents to machinery in factories, as though an hour's extra work would sometimes be sufficient to make all right, it would not be conceded, and the result was that the machinery and the hundreds of operatives who attended it had to wait the time of the moulders. With men who upheld and submitted to such rules it could not be expected that the public would sympathise; and when the masters took decided measures to free themselves from the operation of the moulders' union, they had the moral support of all reasonable men. In the spring of 1868 the union moulders, after being locked out for more than a month, agreed to accept the terms of the masters, which were as follow:— "That the employer shall be the judge of the kind of hands, and of the number and age of the apprentices he shall introduce into his foundry, and of the kind of work, whether piece-work or otherwise, at which labourers or other hands shall be employed."

Manufactures in copper, brass, &c., come next under notice. Traces of copper ore have been found in most counties in Scotland, and it would appear that from very early times attempts have been made to work the veins; but only in a few cases with success. The census of 1861 showed that there were one copper mine proprietor, and forty-one copper miners in the country; but the quantity of metal obtained at any time has not been great. In 1862 the mines of Scotland yielded 173 tons of ore, from which ten tons of fine copper, valued at L.1060, were extracted. In 1864 the quantity of ore got was only 14 tons, which yielded two tons of fine copper, valued at L.242. Though the supply of native copper is limited, a considerable quantity of the metal is manufactured in Scotland, both for home use and exportation. The value of the quantity sent abroad in 1866 was L.36,385. The largest articles made of copper are vacuum pans for sugar refineries, boilers for dye-works, fire-boxes for locomotives, and steam-pipes for marine engines. The metal is easily worked, and possesses properties which render it of great value in the arts. Upwards of 500 persons are engaged in working copper in Scotland, and these are chiefly employed in Glasgow. The wages range from 23s. to 27s. a-week.

Workers in brass are more numerous than the coppersmiths; but though the alloy is applied to a great variety of purposes both of use and of ornament, the labours of the Scotch artisans are chiefly confined to converting it into portions of machinery, and gas and water fittings. There is a brass-foundry in connection with every engineering establishment of any extent, the quantity of brass-work required for a locomotive or first-class marine engine being considerable. Four extensive brass-foundries in Edinburgh are chiefly engaged in making "plumber-work." There are upwards of 800 brass-founders, fitters, finishers, &c., in Edinburgh; in Glasgow, there are about 1000; and in all Scotland close upon 2500.

The brass-foundry of Messrs Milne and Son, Milton House, Canongate, Edinburgh, which embraces an extensive manufactory of gas metres, is the most extensive establishment of the kind in Scotland. It covers about an acre of ground, and upwards of 350 persons are employed.   The brass-foundry and fitting-shops occupy an extensive range of buildings adjoining Milton House.   In the foundry, taps, valves, joints, couplings, and the other portions of "plumber-work" made of brass are cast.   The moulds are formed in sand, and, as the work is easy, a good deal of it is done by boys, who earn high wages.   On leaving the foundry the castings are taken to the turners, who smooth them down and cut screws on such pieces as require them.   Gas pendants and brackets are made by the fitters, for whom the materials are partly prepared by the founders and turners.   All the parts that may require to be taken separate at any time are screwed together, and the others are soldered.   In some cases the ornaments are cast, but more commonly they are struck by dies out of thin plates of metal.   The finishers take up the work from the fitters, and burnish, lacquer, or bronze it.   The framework of crystal gasaliers is electro-plated with silver and highly polished, so that the metal becomes almost invisible among the drops and prisms, and the lustre looks as if it were composed entirely of crystal. Messrs Milne & Son do a considerable trade in making lanterns and apparatus for lighthouses.   One of their latest works of this kind was an earthquake-resisting lighthouse for Japan.   The frames of lighthouse lanterns are chiefly made of gun-metal, and the domes of copper.   They thus come to be very expensive, but their durability amply compensates for the cost.   Lighthouse fittings and machinery have been brought to great perfection under the auspices of the Northern Lights Commissioners, who showed a magnificent collection of apparatus at the Paris Exhibition.

The gas-meter factory is a large building three storeys in height. On the ground-floor the iron cases of the meters—which are cast elsewhere—are dressed and drilled.   They are then raised to the floor above, where the drums, indices, and other parts are fitted, and the meters made ready for use.   The upper floor is occupied by the tinsmiths, who make the drums and floats.   Block tin only is used in those parts, and the metal is converted into sheets of any required thinness by means of a powerful rolling mill.   From 1000 to 1500 meters are turned out every month.   Most of them are for houses; but a considerable number of large ones for use in factories and public buildings are also made.   Some of the latter are five feet in height, and are capable of supplying 600 burners ; but even these are not the largest manufactured.   Messrs Milne have made a umber of what are called "station-meters," used in gas-works fo r measuring the quantity of gas made.   One recently constructed was a cube of fourteen feet.   Station-meters are usually ornamented in front by pilasters, pediments, and other architectural details, and have in the centre a series of indices and a time-piece. The firm send large numbers of meters of all sizes to Australia, South America, and other foreign parts.

In another department gas-burners are made, chiefly of the "fish-tail" kind, which was invented many years ago by Mr J. B. Neilson, of Glasgow. They are fashioned from rods of fine cast-iron. The rods, which are a foot in length, are first turned smooth, and then cut into lengths. The pieces thus formed are drilled to a certain depth by boring machines which operate on six at a time. So far the work is done by women, who display great expertness in their respective parts. The next operation is to give the burners a final turning by which they are tapered towards the ends, and have a hollow formed in the centre of the top. The lines which indicate the size or number, and at the same time serve to ornament the burners, are also cut at this time. The holes in the top, through which the gas flows, are then drilled; and, after being tested, the burners are ready for the market. Several millions of burners are made annually in this workshop, some of them of a more complex kind than the "fish-tail."

A separate department of the establishment is devoted to the manufacture of complete sets of gas-making apparatus for use in private residences or factories. These are produced of various sizes, capable of supplying from twenty to an indefinite number of lights. It is found that a mansion can be much better lighted by gas made by the apparatus supplied by Messrs Milne than by the usual means of illumination used in country districts; and the advantage of cheapness in favour of gas is generally about twenty-five per cent.

Veins of lead ore exist and have been worked in half the counties of Scotland; but only in a few cases have the returns been sufficient to induce the continuation of mining operations for any length of time. Half a dozen mines are at present in operation, and the total produce ranges from 1200 to 1500 tons of pure lead per annum. The principal mines are at Leadhills in Lanarkshire and Wanlockhead in Dumfriesshire—the former belonging to the Earl of Hopetoun, and the latter to the Duke of Buccleuch. It is thought probable that lead was dug at Leadhills as early as the time of the Roman dominion; at all events, it is known that an important vein was discovered and worked in the year 1517. In the beginning of the present century the lead-miners enjoyed a period of great prosperity, and those at Leadhills alone turned out 1400 tons of metal annually, which at the price then current was worth L.45,000. The works are now carried on by a company, who are bound to pay to the proprietor every sixth bar of lead produced. The mines at Wanlockhead are about a mile and a half south from Leadhills, and are the most productive in the country. The foreign miners who were employed by James V. to seek for gold in the locality discovered the veins of lead ; but the metal was too base to merit royal attention, and the mines were not opened until about the year 1680, when Sir James Stampfield began operations on a small scale. From that time to the present the mines have been worked with little interruption, but under the auspices of several different companies and individual adventurers. The last company was formed in 1755, and obtained leases of the whole mines for successive periods down till 1842. This company carried on the work of mining with great energy, and succeeded in discovering new and rich veins of ore. They also applied steam-power to keep the workings clear of water. In one year the metal raised brought L.47,000. When the last lease of the company expired in 1842, the Duke of Buccleuch took the working of the mines into his own hands, and has introduced improved apparatus for smelting the ore and extracting the silver from it. There are five principal veins of ore in the mines, and these have been worked to a depth of from 70 to 140 fathoms. In 1861 there were 538 lead-miners in Scotland. The population of the villages of Leadhills and Wanlockhead numbers 1600, all of whom are dependent on the lead-mines. Both villages are 1300 feet above the sea, and are about the dreariest inhabited places in the country. The miners and their families are, nevertheless, a cheerful and contented class of people. They are well supplied with churches and schools, and have a library of considerable extent.

Lead is chiefly used in the shape of sheets for covering roofs or of pipes for conveying water and gas. Its conversion into these forms is a special branch of trade, which is carried on in only three or four establishments in Scotland. The largest is that of Messrs T. B. Campbell & Co., Leith, in which several thousand tons of lead are worked up in the course of a year, besides a considerable quantity of tin and zinc. Machinery is extensively applied in the manufactory, and the fifty workmen employed are chiefly engaged in tending machines. The rolling-mill, in which the lead is formed into sheets, is about fifty yards in length ; and the centre of the floor, from end to end, is occupied by a large bench or framework fitted with wooden rollers. In the centre of two divisions of this bench, each forty feet

in length, are erected a pair of massive iron rollers, the distance
between which is nicely regulated by screws and an index. The
lead is prepared for rolling by melting down six tons of bars at a
time, and running the metal off into an iron mould, so as to form it
into slabs seven feet square and about six inches thick. One of
those slabs is laid upon the rolling bench, and passed between the
iron rollers. The rollers are fitted with reversing gear, and every
time the slab passes from one side to the other its thickness is
reduced and its length increased. When the sheet extends to the
length of one division of the bench, or forty feet, it is cut up into
convenient pieces, to be further reduced in thickness by the same
process. Sheet-lead, weighing from 3 ℔ to 8 ℔ per square foot, is
the kind commonly used, but for special purposes heavier sheets are
made. The rollers are driven by an engine of 20-horse power, and
are capable of turning out eight tons of sheet-lead a day.

The making of lead or tin gas-tubing is carried on in another part
of the establishment, and is an interesting process. It is done by
means of hydraulic presses, which force the metal over a die of
peculiar construction. In the centre of the lower part or sole of the
press, which is composed of a huge mass of iron, is a circular chamber,
in the middle of which is fixed a short rod of steel, of similar dimensions to the internal diameter of the tube to be made. The chamber
is filled with molten metal, and a piston which fits exactly into the
chamber, and is attached to the upper part of the press, is forced
down upon the metal. The piston is pierced through perpendicularly
by a hole of the same diameter as the outside of the tube, and the
steel rod fixed in the lower part of the press enters this hole as the
piston descends. It will thus be seen that there is between the
sides of the bore and the rod a space equal to the substance of the
tube, and that, when pressure is applied to the surface of the molten
metal, it can escape only by passing up through that space. That is
exactly what takes place, and the tube comes forth perfectly formed
from the upper end of the piston. The chambers in the presses are
of various sizes, those for lead pipes containing from two hundredweight to four hundredweight of molten metal, and those for composition and tin tubes smaller quantities. Some of the presses are
capable of making lead tubes five inches in diameter. Tin tubes of
more than one and a quarter inch are made by "drawing," on a
drawbench, by means of an endless chain. A cylinder of metal,
eighteen inches in length, of suitable diameter, with a hole through
the centre of it, is taken by the drawer, who inserts a mandrel into
it, and draws it through a series of dies of gradually diminishing

diameter. The " ingot " is thus extended from a length of eighteen inches to nearly as many feet.

Zinc has, by its cheapness and lightness as compared with lead, come extensively into use for covering roofs, and making rain-pipes and ridging ; and Messrs Campbell & Co. have a considerable trade in the manufacture of the metal for those purposes. The zinc is imported in large sheets, which are cut into stripes of the required dimensions, and these are passed through machines which give them any shape that may be desired. Thus a stripe is drawn through a die on a machine, and comes out folded up into tubeform, with its edges bent respectively inward and outward. A fresh die is placed in the machine, and a mandrel in the now half-formed tube, which is sent through again. This time it comes out with the edges of the plate firmly locked together in a water-tight joint, and the whole finished more perfectly than it could be by any workman in as many hours as the minutes which it occupied the machine. Ridging pieces require but one operation, and are produced with great rapidity.

One of the most important purposes to which lead is applïed is the manufacture of printing types, in which way it was used upwards of four centuries ago. The extraordinary demand for books which has sprung up within recent times has given an impetus to the trade of the typefounder as well as to that of the publisher, and he has been encouraged to improve his productions until a degree of excellence has been attained which leaves little to be desired. The beauty of the letters now manufactured far surpasses that of the best made in any previous period in the history of typefounding, while their variety is being daily increased. In 1778 most of the types used in Scotland were made in Glasgow, and there was then only one typefounder in Edinburgh. The trade afterwards came almost entirely to Edinburgh, and a few years ago there were several typefounders in the city. Typefounding is now almost exclusively in the hands of two firms—Messrs Miller & Richard, Edinburgh, Her Majesty's Letter-Founders for Scotland; and Messrs James Marr & Co., also in Edinburgh. The former, which is the most extensive firm, commenced operations more than half-a-century ago, and has always had a reputation for producing elegant and durable types—a reputation acquired by devoting great care to the designing of styles, maintaining a superior class of workmen, and applying machinery wherever it was available. The foundry of the firm, which occupies an extensive range of buildings between Nicolson Street and Potterrow, presents a most interesting sight.

Upwards of 500 men and boys are employed in it, and these are aided by more than a hundred beautiful and ingeniously devised machines, set in motion by two steam-engines of 40 horse-power.

When it is desired to produce a "fount" or set of types of a new style or form, the first operation is to cut a set of punches. These are made of the finest steel, and the cutting is an operation of great nicety. There must be a punch for every letter, figure, point, and reference mark. In the case of type used in newspapers, there are usually five alphabets of each size—namely, ROMAN CAPITALS, SMALL CAPITALS, and "lower case;" *ITALIC CAPITALS*, and "*lower case italics.*" Then there are accented letters, figures, points, and the like, so that the number of punches required to produce a complete set of types such as this book is printed with is not less than 250. When the punches are finished and tempered, the matrices for the face of the type are made. That is done by carefully pressing the punches on pieces of copper, which retain perfect reverse impressions. Each matrix is then "justified," or fitted in all respects for the "mould." When types were made by hand, the mould was enclosed in wood, to enable the workman to handle it; but now that machinery has superseded hand labour, the mould is made entirely of steel. It is composed of a number of parts fitted together with great accuracy, and is so constructed that it may be adapted to the various thicknesses of type; but a mould is required for each size or "body" of type. The matrix having been fixed into the mould, and all the parts adjusted, a series of experimental casts are made, in order to test the accuracy of the work. When the "justifier" is satisfied that all is correct, the mould is fixed into the casting-machine, and the charge of the work passes to the "caster" and his "dresser." Upwards of 100 casting-machines are constantly in operation in the foundry. The machines are adaptations of American and German inventions, which Messrs Miller & Richard purchased, and have brought to great perfection. Their mode of action is exceedingly simple. The type-metal—which is composed of certain proportions of lead, tin, and antimony—is contained in a cylindrical iron vessel, about six inches in diameter, and is kept in a state of fusion by a small fire burned in a compartment beneath. In the front of the metal-holder is a spout or opening. The working part of the machine, which contains the mould, receives an oscillating motion, which throws it alternately forward to the metal-holder, and backward to a tray which receives the types. When the mould is thrown forward the mouth of it is brought close upon the spout of the holder, a piston is raised in the vessel, and a quantity of metal sufficient to form the type is forced over into the mould. As soon as this takes place the mould is thrown backward, at the same time opened, and the type is ejected into the tray. One machine will accomplish as much work in a day as three or four expert hand-casters.

Notwithstanding all the care that is taken in adjusting the mould, and the great exactness with which the machine works, the types have to go through several operations before they are ready for the printer. When they come from the machine each has a taper piece of metal attached which has been formed in the throat of the mould. This is removed by boys, who, seated in rows at long tables, pick the types from a heap before them, and by pressing the superfluous piece smartly against the board break it off. The boys work with both hands simultaneously, and the rapidity of their movements is surprising. There also exist on the sides of the newly-formed types projections which are removed by the "rubbers," another class of boys, who take up the types one by one and pass both sides over a flat piece of sandstone. The next operation is "putting up," or arranging the types in long rows with the faces upward. In that position they are carefully examined, have the roughness left by breaking off the "castable" or "jet" removed, and receive a few finishing touches. These last operations are called "dressing," and great care must be taken with them. When the dressers are done with them the types are arranged in "pages," or oblong square parcels—care being taken to keep the different letters separate. Before the types are finally passed they are "examined." The pages are opened up, and sample letters are taken out, which are subjected to the test of measurement by steel guages of nice construction, and to examination by means of a magnifying glass. One type must not differ from another in height, depth, or width, by even the thousandth part of an inch. To persons not practically acquainted with printing such exactness may appear superfluous; but were the types not thus carefully made, printing would be almost impossible, at least with such machines as Hoe's, which have now come into general use in the offices of daily newspapers.

It is not so advantageous for the printer to have letters above a certain size made of metal, and nearly all types above an inch and a half or two inches in length are formed of wood. In this department of the trade Messrs Miller & Richard have recently introduced various improvements. Instead of the letters being drawn on the blocks of wood by hand, they are printed, and the cutting is done by a machine somewhat similar in principle to that used for carving wood.

Manufacturers of printing types usually devote attention to making brass-rules, cases, and "furniture"—which is the printer's technical name for all the wooden or metal frame-work, &c., used in the trade. The lines which appear in tabular matter are formed by thin slips of brass set in among the types. In the joiner's shop composing-cases are made, all the parts of which are prepared by steam machinery. A number of men are employed in making and repairing the machines used in the establishment.

The division-of-labour principle is largely applied. There is no regular term of apprenticeship in the foundry. Boys are taken in at thirteen years of age, and from "breakers-off," "putters-up," and "rubbers," come to be "casters," and ultimately "dressers." There is a mixture of time and piece-work; and the following may be taken as the general rate of wages:—Boys—breakers-off, 2s. 6d. to 4s. a-week; rubbers, 10s. to 18s.; putters-up, 4s. to 7s. 6d. Men— justifiers, 30s.; machinemakers, 27s.; casters, 22s.; dressers, 28s.; workers in wood, 26s. The time worked is fifty-seven hours a-week.

WOOLLEN MANUFACTURES.

THE history of woollen manufactures in England dates back to the time of the Norman Conquest. The Flemings were so expert in making woollen cloth, that it was said of them that their skill in the art of weaving was a peculiar gift conferred by nature. Large numbers of weavers came over from Flanders in the train of the Conqueror, and in the intervals of turmoil prosecuted their calling with success. In course of time they thoroughly established the trade in the country, and in the reigns of Henry I. and of Stephen had accumulated so much wealth, that some of them rivalled royalty itself in the luxurious style in which they lived. About that time guilds or corporations were established in many of the towns where the manufacture of cloth was carried on for the purpose of its encouragement and improvement. Though the art of the woollen cloth maker was thus early introduced and encouraged in England, the state of society in Scotland was such that no attempt was made to create a home supply of clothing until many years afterwards. In the reign of Alexander III. considerable quan-tities of wool were exported to the Continent in exchange for linen, silks, and broadcloth; but there is no mention in the records of those times, so far as we have seen, which would lead to the belief that such things as spinning-wheels and looms then existed in the country. When the art of weaving was introduced is uncertain; but there can be no doubt, that in the fifteenth century cloth from native wool was made and worn in Scotland. In " Morrison's

Itinerary," giving an account of a visit paid to Scotland in 1598 by an Englishman, there is a description of the fashion in dress at that time:—" The husbandmen in Scotland, the servants, and almost all the country, did wear coarse cloth made at home, of grey or sky-colour, and flat blue caps, very broad. The merchants in cities were attired in English or French cloths, of pale colour, or mingled black and blue. The gentlemen did wear English cloth and silks, or light stuffs, little or nothing adorned with silk lace, much less of lace with silver or gold. And all followed at this time the French fashion, especially in court. Gentlewomen married did wear close upper bodies, after the German manner, with large whalebone sleeves, after the French manner, short cloaks like the Germans, French hoods, and large falling bands about their necks. The unmarried of all sorts did go bareheaded, and wear short cloaks with most close linen sleeves on their arms, like the virgins of Germany. The inferior sort of citizens' wives, and the women of the country, did wear cloths made of a coarse stuff of two or three colours in checker work, vulgarly called pladon [plaiding.]"

Cloth was made in those days in much the same fashion that is still followed in the remoter districts of the Highlands, where the wool is carded and spun by the females of the households as a profitable recreation in the winter evenings, and converted into "plaiding" or blankets by the village weaver. The fishermen and crofters of the Western Highlands and islands are generally clad in "plaiding," which, though rough in appearance, is durable, and to them cheaper than any other kind of stuff. Edinburgh was one of the first places in Scotland in which woollen goods were made, and it had at one time about the most important wool market in Britain. The weavers of the city were incorporated by the Town Council in 1475. In the petition asking for incorporation, it was set forth that the articles of the trade had been framed " for the honour and love of God, of His mother the Virgin, and of St Sovrane;" and it was specially stipulated that " the priest shall get his meat."

About the year 1600 seven Flemings were brought to Edinburgh to instruct the people how to make " seys" and broadcloth at home, so as to be independent of a supply from England. There were many difficulties in the way, however, and no record remains to show that anything came of the scheme. When " the Hospital of our Lady," which had been founded in Leith Wynd, Edinburgh, by Bishop Spens of Aberdeen in 1479, passed into the hands of the Town Council in 1619, it was converted into a workhouse, and named Paul's Work. The Council resolved soon afterwards to try the experiment of giving an industrial education to the boys and girls in the workhouse, and for that purpose brought five men from Holland to give instruction in the manufacture of coarse wool stuffs. Though started under the most hopeful circumstances, and encouraged by numerous donations, the manufactory does not appear to have succeeded. Paul's Work was converted into a house of correction, and subsequently sold to Bailie Macdowal, who, about 1770, had it reconverted into a woollen factory, in which he is said to have made superfine broadcloth equal in quality to any brought from England. Paul's Work is now the printing-office of Messrs Ballantyne & Co.—a firm especially famous forty years ago from its connection with the works of Sir Walter Scott. It is mentioned in the records of the city of Aberdeen, that in the end of the sixteenth century, a Fleming obtained leave to exercise his profession in the manufacture of "gograms, worsets, and stamings" without any hindrance from the weaver corporation, on the condition of taking into his employment an apprentice, and instructing him in weaving and dyeing the kinds of cloth mentioned. In 1636 the magistrates of Aberdeen obtained a patent from Charles I. to establish a House of Correction, in connection with which, and with a view of "reforming their morals, and promoting good order and industry," a certain class of the community was to be instructed in the manufactures of broadcloth, kerseys, seys, and other coarse cloths. A situation for this institution was found in a part of the city now known as Correction Wynd. The factory was carried on for some years, but did not succeed, and in 1711 it was abandoned.

The first really energetic and promising effort to establish in Scotland a manufactory of woollen fabrics was made in 1681, by an English company under the management of Colonel Stanfield. The company acquired, in the vicinity of Haddington, a portion of the grounds which had belonged to the Franciscan Monastery, and erected thereon workshops, fulling-mills, dyeing-houses, &c., on an extensive scale. A number of workmen from England were employed to instruct the natives of the locality in the processes of manufacture. For a number of years the company prospered, and received great encouragement from the Government, several Acts of Parliament having been passed relieving them from payment of taxes, and conferring other favours. The services of Colonel Stanfield were acknowledged by his being made a knight. On the death of the colonel the affairs of the company got into confusion; and after struggling on for a few years, the company was dissolved, and the enterprise abandoned. Colonel Charteris bought the factory and grounds, the name of which he changed from Newmills to Amisfield. A new company with a large capital was organised in 1750, and an effort was made to revive the manufactory, but unsuccessfully. Subsequently a third company gave the thing a trial, but with no better result. It was found that fine cloths could not then be made in Scotland so cheaply as in England, in consequence of the manufacturers of the latter country having attained great perfection in the various appliances and processes, besides enjoying other advantages. An Edinburgh gentleman who wrote on the subject 135 years ago, thus compares the two countries with respect to the manufacture of woollen cloth:—"The English have been long masters of the woollen trade. Their clothiers and piece-buyers are owners of stocks able to carry it on, to keep their goods on hand until a market offers, to sell them at reasonable rates and upon long time. England is sufficiently stocked—nay, one may say over-stocked—with the best of workmen in every branch of the woollen trade; and no country can succeed so as to be great gainers by any manufacture until it is sufficiently stocked with good manufactures, that their wages may be brought low enough to enable them to undersell their neighbours in that commodity at a foreign market. Whereas we have no stocks equal to so great an undertaking; we must also be at the expense to bring from England workmen for several branches, and to pay them higher wages than they get at home; and we cannot expect to get their best workmen. These, and many other difficulties not easily to be conquered, render it absolutely impossible for us to succeed in the woollen trade—at least in broadcloth, druggets, fine kerseys, and the woollen goods of Norwich." The author we are quoting from advocated the adoption of the linen manufacture as the staple of Scotland. Referring to what he considered to be the impossibility of Scotland ever becoming a seat of the woollen trade, he says:— "Nor is this any loss to us, since we have a staple manufacture of our own—at least, we may have the linen, in which the English deal not. They are too wise to encourage any manufacture in the weaving way that might interfere with their great staple, the woollen; and we should learn of them to discourage every trade that may interfere with or hinder the progress of our only staple." However strange these notions may appear to us who see the woollen manufacture holding the position of one of the most extensive branches of industry north of the Tweed, they were reasonable conclusions to draw from the facts on which they were based—namely, that woollen goods could be bought in England from 10 to 15 per cent. cheaper than in Scotland, and that linen cloth made in Scotland could be sold in

England at a profit of from 5 to 10 per cent. Considerable attention was bestowed on the linen trade, and its extension was encouraged in various ways; but such factitious encouragement did not deter some persons from persevering in the manufacture of woollen goods.

Various Acts of Parliament were passed for the encouragement of the woollen trade in Scotland. In the sixteenth century English-made cloth was coming into fashion, and it was feared that the effect would be to ruin the home manufacturers of that material. An Act was accordingly passed in 1597, which denounced "the hame-bringing of English claith, the same claith having only for the maist part an outward show, wanting that substance and strength whilk ofttimes it appears to have." Another serious reason urged against the traffic in English manufactures was, that it was the chief cause of "transporting of all gold and silver furth of this realm, and consequently of the present dearth of the cunyie."

Act 8 of King William's First Parliament, dated January 31, 1701, "strictly prohibits and forbids the importation of all cloths or stuffs of any kind made of wool, or wherein there shall be any wool; as also of hats, caps, stockings, gloves, or any other kind of manufactured wool, or wherein any wool shall be found, excepting flannel allenarly." Heavy penalties were imposed for breach of this statute. On the same day an Act was passed setting forth that, "Considering the great hurt and prejudice arising to this kingdom and manufactories thereof by the exportation of wool, and of skins with wool upon them, His Majesty, with advice and consent of the Estates of Parliament, doth not only ratify and revive all former Acts of Parliament made against exportation of wool, or skins with wool upon them, in so far as they strengthen this present Act, and without derogation thereto in any sort, but also of new again do hereby strictly prohibit and discharge all and every person whatever, native or stranger, to export out of this kingdom any wool whatsomever, or skins with wool upon them, or any worsted or woollen yarn, or any sort of foreign wool, or skins with wool upon them, certifying such as shall contravene this present Act, that the wool or skins shall be confiscated, and two-third-parts thereof applied to the discoverer, and the other third-part to the Procurator-Fiscal of the Court where the confiscation is pursued, and the exporter fined in the sum of one thousand merks *toties quoties*. . . . In case any one concerned in woollen manufactures shall contravene any part of this Act any manner of way, they shall not only amit and lose their share and stock in the manufacture in which they are concerned, to be applied to the discoverer, but also shall be fined in the sum of six thousand pounds."

In 1705 an Act was passed declaring linen and woollen manufactures to be free of duty on exportation.   It had been enacted in 1686 that, in order to encourage home manufactures, all bodies of persons dying within the kingdom should be buried in Scotch linen, and that Act was ratified in 1695, and made more complete for its purpose.   Subsequently, in order to give some encouragement to the woollen manufactures, the Acts referred to were rescinded, and in 1707 it was enacted that "no corpse of any person of what condition or quality soever shall be buried in linen of whatever kind; and that where linen has been made use of about dead bodies formerly, plain woollen cloth or stuff shall only be made use of in all time coming." The penalties imposed in the Acts relating to burial in linen were transferred to this Act.   It may be mentioned that it is to the corresponding Act for England that Pope makes his moribund fine lady allude in the famous lines:—

> " 'Odious! in woollen! 'twould a saint provoke,'
> Were the last words that poor Narcissa spoke ;
> 'No! let a charming chintz and Brussels lace
> Wrap my cold limbs, and shade my lifeless face ;
> One would not, sure, be frightful when one's dead—
> And, Betty, give this cheek a little red.' "

There was published in Edinburgh in the year 1733 a book entitled "The Interest of Scotland considered; or, Reasons for Improving the Fisheries and Linen Manufacture of Scotland." The author was Mr Patrick Lindsay, Lord Provost of the city, who evidently possessed a thorough knowledge of the subjects of which he treated.   His book contains the following account of the woollen trade as it then existed:—"At Kilmarnock are made of our own wool low-priced serges, known by the name of that place where they are made.   These are partly for home consumpt, and partly for the markets of Holland; and, by the help of a little care and encouragement, burying crapes, at least those of a low price, might also be made there for home consumpt.   At Stirling and its neighbourhood large quantities of serges are made, and several other low-priced woollen goods for furniture, all for home consumpt, and rather cheaper than such goods can be purchased in England.   This business, by the care and vigilance of the Justice of Peace in those parts, is much improven of late.   At Aberdeen, and counties adjacent, large quantities of our own coarse tarred wool are manufactured into coarse serges, called fingrams, and knit stockings of all prices. Some of these goods are consumed at home, some of them exported to Holland, and some of them sold at London, and from thence are exported to foreign parts. At Edinburgh fine shalloons are made of our best wool, for home consumpt, and cheaper than they can be had in England. At Musselburgh there is a considerable manufacture of low-priced narrow goods, from thence called Musselburgh stuffs, for home consumpt, and export to the plantations; but these are now fallen so low in the price that the makers can scarce get their bread by them. At Galashiels are made a few coarse kerseys, called 'Galashiels greys,' for home consumpt; and was their wool better scribbled, their goods more milled, and better dressed, they might serve in place of the lowest-priced Yorkshires for country wear, to ordinary people, and day labourers. At Kirkcudbright, Hawick, Monygaff, and other places near the wool countries, several packs of tarred wool have been washed and cleaned, and some of it sorted and combed, spun, and wrought up into blankets, and other coarse goods, by private hands, for their own use; all done by the help of public encouragement, to advance the price of wool in those parts, but as yet to little or no purpose. As for the manufacture of broad-cloth, that consists of so many parts that we cannot carry it on without evident loss. First, we have no such thing as a wool stapler in the country, which lays the clothier under a necessity to buy his wool in the fleece; and unless he work up all the sorts himself (which no clothier can do without great loss) he must lose by those sorts he does not use. The washing, cleaning, and drying of wool, by beating it on the flecks, we understand pretty well; but we neither dye wool so well, nor so cheap, as the English do; and we have but few scribblers who understand the close mixing of wool on the cards for medleys. Our women are all bred to spin linen yarn, and are not so fit to spin woollen, especially carded wool for cloth, which no one can do to purpose who is not constantly employed at it. We understand the picking of cloth, and the thickening of it at the mill, pretty well, but we are not so adroit at the tasselling it on the dubbing boards, and are at a loss that we have no tassels of our own growth fit for this work, but are obliged to bring them from England in large quantities to lie by us, as we have occasion to use them. The most curious and difficult operation of the whole is the cutting on the shear-board, and finishing in the hot press. We have no shearmen of our own that understand their business to perfection, and as few pressmen, and must bring our press-papers from England; and the profits of the whole manufacture depend upon the close and equal cutting from end to end, and upon proper and clean papers for every staple of cloth, and a just degree of heat and pressure in the hot press, neither too much, nor too little of either."

Mr David Loch, General Inspector of Fisheries in Scotland, published in 1778 a series of "Essays on the Trade, Commerce, Manufactures, and Fisheries of Scotland," in which he urged the people of the country to persevere in the establishment of the woollen manufacture then begun. He predicted that if they did they would "shortly see Scotland raised from abject poverty and mean obscurity to the same degree of opulence and dignity as our sister kingdom acquired only by this invaluable branch." He pointed out the importance, as a first step, of increasing the number and improving the breed of sheep, and threw out many valuable suggestions on the subject. The woollen trade appears to have been a "hobby" with Mr Loch, and he advanced no end of arguments in favour of its extension. Here is one of the most curious:—"The woollen manufacture is peculiarly favourable in promoting matrimony, and consequently population. Children from five years of age may begin to be useful, and are even employed in different branches of it which are singularly adapted to their infant state."

In 1776 Mr Loch made a tour through most of the trading towns and villages of Scotland, and his book contains some interesting information as to the state of trade and manufactures in each. The following are a few facts relating to the woollen trade, which may be put in contrast with the subsequent report on the present condition of that branch of industry :— In Edinburgh Mr Archibald Macdowal employed what was considered at that time to be "a great number of hands" in connection with his factory in Paul's Work, already referred to. He manufactured about 4000 ℔ of Scotch wool and 17,000 ℔ of Spanish wool yearly, and his machinery consisted of fulling-mills and a spinning-machine. Respecting the latter, we are told that it had been greatly improved by John Thomson, a person of mechanical genius. Mr John Ballantyne, wool merchant, combed 264 ℔ of Scotch wool, and dyed every colour to perfection, scarlet excepted. He employed the charity boys in the Canongate workhouse for four years in spinning wool yarn on the great wheel. Twopence a-pound was the price paid for combing long wool, and from 1s. to 1s. 3d. a-spindle for spinning, according to the fineness of the "grist." Most of Mr Ballantyne's workmen were paid according to what they produced ; and it is stated that "they could easily gain one shilling a-day if they chose to exert themselves." Mr Jeeves, of Edinburgh, was "among the best blanket makers in Great Britain." Carpet-weaving and stocking-knitting on frames were also carried on in the city. Both Edinburgh and Leith had wool markets; and the annual sale in each was about 20,000 stones of 22 ℔. At Dalkeith 700 stones of wool per annum were manufactured into broadcloths, ranging in price from 4s. to 14s. a-yard. About 200 persons were employed, and the value of the goods turned out was about L.3000 yearly. The Musselburgh factories used 1000 stones of wool, and in addition worked up a large quantity of worsted yarn spun at Selkirk and Peebles. The chief produce was "manco-stuff" for the Edinburgh market. The prices varied from 2s. 6d. to 16s. 6d. a-yard. Haddington was regarded at the date of the record quoted from as a most suitable place for carrying on an extensive manufactory of woollen goods, and even then the value of its products from wool was estimated at L.5000 a-year. Upwards of 800 persons were employed in the trade. Wool was also manufactured into cloth, carpets, and stockings at Dunbar, Linton, Tranent, Linlithgow, Perth, and Inverness. In Glasgow there was only one woollen factory, and that was chiefly employed in making carpets. The prices paid to workers were the same as in Edinburgh. Stirling had long been known as an important seat of the woollen manufactures; and 160 looms, 38 stocking-frames, and 17 carpet-frames were employed in the trade. Shalloons, serges, and Highland plaids were the chief produce. In Alloa there were twenty manufacturers, employing 150 looms and about 500 workers, chiefly engaged in making "camblets." Though the looms of Fifeshire were mostly devoted to linen fabrics, a good deal of woollen cloth for local consumption was also made. Kilmarnock had 66 looms engaged on carpets, and 80 in other branches of the woollen trade. The manufacture is said to have been introduced by Miss Maria Gardiner, who, observing the indolence of the people of the place, brought spinners and weavers of carpets from Dalkeith about the year 1728. Ayr had about 100 looms and 15 stocking-frames for wool work. Dumfries and Sanquhar did a considerable trade in stocking-making. In Moffat 50 looms were engaged in serges, shalloons, blankets, &c. Elgin produced L.15,000 worth of yarn annually, which chiefly went into the London and Glasgow markets. There were two woollen factories at Peterhead, one of which turned out goods to the value of L.50 a-week, and the other L.60. The people of Ellon knitted stockings by hand to the value of L.100 a-week. Aberdeenshire seems to have been largely engaged in the stocking-trade.

The value of those articles made by Aberdeen manufacturers amounted to L.120,000 annually. In the town 240 looms were engaged on woollen and linen fabrics, but chiefly the former. At Montrose there was a woollen factory, in which seventy hands were employed. In Kincardineshire a considerable trade was done in stockings.

Regarding the Border towns, in which the woollen manufactures of Scotland are now chiefly concentrated, Mr Loch's narrative states that the people of Galashiels were very industrious, and that they were all employed in making coarse woollen goods, but principally "Galashiels greys." This cloth was made three-quarters of a yard wide, and sold at from 1s. 6d. to 4s. a-yard. Blankets were also made from Forest wool. There were 30 looms and 3 waulk mills in the village. There were then only 600 persons in the parish. The people made all the yarn required for their own use, and also a quantity for sale. The annual consumption of wool was estimated at 2200 stones. Melrose had 140 looms, most of which were employed in making woollen cloth. There were 65 looms in Hawick, employed on linens and woollens. Fourteen of these were carpet looms, belonging to Messrs Robertson & Co., who commenced operations about eighteen years previously, with a capital of L.400. There were six stocking-frames in the town, four belonging to Mr Hardie and two to Mr James Halden. Jedburgh is described as "a royal burgh where there has been much dispute and dissension about their town politics, so that the people have neglected all business, and paid little or no attention to manufactures." There were 56 looms in the town, but these were all employed in jobbing. In Kelso about 40 looms were employed, chiefly in making blankets and flannels. The weavers usually made seventy yards of flannel in two weeks on two looms worked by a man and a boy. The annual consumption of wool was 2200 stones. In Peebles 40 looms were employed in making coarse woollen goods. In Selkirk a few looms were devoted to jobbing work in wool, but a considerable quantity of yarn was made in the district. It was estimated that L.55 a-week was paid in the town for spinning. Looms might be found in almost every village of Scotland at the time referred to, but only in the cases mentioned did they do any but what was known as "customer work"—that is, the weavers worked up the yarns spun in the households of farmers and others, and the cloth was returned thither for the use of the families.

The progress made in the quality of the woollen goods manufactured in Scotland in the end of last century is indicated by the following paragraph, which appears under the head "Edinburgh," in the " Annual Register " for 1793 :—" An eminent manufacturer in this town has just finished two elegant gown-pieces, manufactured from Shetland wool, the one for Her Majesty, the other for the Duchess of York. He has also just finished a very handsome vest-piece for the Prince of Wales, and a beautiful gown-piece for the Duchess of Gordon, both from common Scotch worsted. Encouraged by such patronage, it may reasonably be hoped soon to see the woollen manufacture attain a degree of perfection hitherto unknown in this country. A gown-piece similar in pattern to that of Her Majesty has been ordered for the Empress of Russia."

The woollen manufactures made considerable progress during the first quarter of this century. Improvements in machinery and the mode of working up the wool had brought about a gradual change for the better in the condition of the manufacturers and their work-people; and the good quality of their productions began to be more widely appreciated in the markets at home and abroad. In 1825 the number of persons employed in the various branches of woollen manufacture in Scotland was 24,800. The value of the raw material consumed was estimated at L.300,000, and the profit of labour at L.150,000; so that the total value of the produce was L.450,000. The superfine broadcloths made in Aberdeenshire competed success-fully in the London markets with the productions of English looms, even although the latter enjoyed a world-wide reputation; while the advanced prosperity of all classes at home led to an increased de-mand for narrow cloths, tartans, checks, flannels, and the like. The power-loom was introduced into the trade about the year 1830, and by its aid the quality of the narrow cloths, or "tweels," was im-proved, while the rate of production was greatly increased. A year or two afterwards a little incident occurred which, while proving that there is something in a name, gave an impetus to the "tweel" trade, and helped to lay the foundation for the extraordinary de-velopment of that branch of manufacture which has taken place during the past thirty years. Messrs William Watson & Son, of Hawick, sent a quantity of "tweels" to one of their customers in London—the late Mr James Locke, who was one of the earliest merchants of that kind of goods in the metropolis. In the invoice the word "tweels" was written indistinctly, and was read "tweeds" by Mr Locke, who, in ordering a further supply, adopted what he conceived to be a new and happy designation. The writings of Sir Walter Scott had made the Border land and the Tweed famous all over the world, and the use of the name of the river to designate a material for dress manufactured on its banks, and those of its tributaries, was shrewdly calculated to extend the popularity of the article. The name, added to the strength, flexibility, and other serviceable qualities of the "tweeds," made them fashionable among English noblemen and gentlemen who came to Scotland to shoot and fish, and they gradually worked their way into popular favour.

With this incident the history of the Scotch "tweed" trade may be said to open. It is consequently embraced in a brief period of time; and if we were to proceed in chronological order, this branch of woollen manufacture would fall to be dealt with after all the others. A slight link which connects it with the earliest products of Scotch looms, added to the fact that it is now one of the most important industries in the country, entitles it to precedence. The Galashiels "greys," "blues," and "drabs" ruled the fashion in male attire for many years; but the manufacture of these received a check by the commercial disasters of 1829; and the sameness of hues having by that time palled upon the public taste, it was found impossible to revive the trade. Something new was demanded by the public; and the manufacturers exercised their ingenuity to meet the demand. The first departure from the conventional "blues" and "drabs" is attributed to various persons. Sir Walter Scott, while Sheriff of Selkirkshire, had a pair of trousers made out of a Scotch checked plaid, and his example was followed by many persons. A new direction was thus given to the woollen trade, and the hopes of the manufacturers revived. The tweed trade, in its fullest development may, however, be said to owe its origin to the simple idea of twisting together two or more yarns of different colours. The author of this idea is not known; but Jedburgh claims the honour of having first produced cloth made of yarn of mixed colours. Granting Jedburgh the honour of the birth of the trade, the chief credit of its perfection and development must be given to Galashiels, which early stepped into the foremost place, and yet creditably maintains it.

In 1829 the tweed-makers could boast of only fifteen sets of carding-engines, but these represented a much greater number of manufacturers. In those days the fortunate possessor of "quarter of a set" was a "maister," and a man of means. In a business note-book belonging to one of the oldest firms in the trade, and containing a list of manufacturers in 1829, there are no fewer than thirty-four names, with a footnote to the effect that there was "a number of smaller ones." The total turnover per annum is put down at L.26,000. It is only by comparing these modest figures with the present overturn of upwards of L.2,000,000, that one can gain anything like an adequate conception of the extent and rapidity of the development of the tweed manufacture.

The trade, no doubt, owes much of its success to the genuineness of the article produced, and the consistent "anti-shoddy" policy of the leading manufacturers. Unlike the much milled, much raised, and much shorn cloths of the Continent or the West of England, a thoroughly good Scotch tweed undergoes no process tending to injure the texture or impoverish the cloth, but comes to the wearer with all the natural strength of the material unimpaired—an honest stuff honestly manufactured. So long as the Scotch makers adhere to this policy, and refuse to be tempted into competition with unscrupulous imitators, they will no doubt continue to hold their own. Their strength lies in persistently sticking to the article which specialises them and most fully presents the character and features of the Scotch tweed. The moment they leave their own ground to compete with others, either in closeness and fineness of fabric or perfection of finish, their productions fail in comparison with the Belgian, French, and West of England cloths. A strong point is the purity and brilliancy of the colours obtained. The amount of attention given to style is another special feature which contributes largely to the general success and appreciation of Scotch tweeds. In this respect they rank second to none; and many makers, both English and foreign, who beat the Scotch in certain niceties of manufacture, notoriously imitate their styles. Imitation is the sincerest of flattery, but unfortunately, in this case, the Scotch trade pays for the compliment. A good style is often no sooner out than it is reproduced by Yorkshire makers in a lower quality; and beyond a doubt those "Yorkshire Scotch tweeds" interfere considerably with the sale of their more costly but, in the end, cheaper and more honest originals. In the jurors' report on the Exhibition of 1862, the following allusion is made to Scotch manufacturers:—
" To the Scotch manufacturers belong the credit of having found out what the public like, and of having led for a considerable period the public taste. So largely have their productions been imitated on the Continent, that many of the choicest fancy trouserings of France and other countries are easily traceable in design and colouring to their Scotch origin."

The extent of the Scotch tweed trade may be learned from the following statistics, which have been drawn up with great care :—Number of firms, 85 ; sets of carding-engines, 340 ; spindles, 255,000 ; looms, 2720 ; horse power employed, 3400 ; weight of wool consumed annually, 10,642,000 ℔. Number of persons employed :—

Males, 5440; females, 8169—total, 13,600; total population depending on the trade, 23,800. Capital employed, L.1,360,000; wages paid annually, L.340,000; value of wool used, L.1,064,200; value of goods manufactured, L2,040,000. To these statistics must be added the fact that a considerable quantity of yarn is spun for the tweed market in various parts of the country—a trade largely taken advantage of by manufacturers in busy times, but not included in the above figures. To show the progress made by the trade during recent years, it may be stated that in 1851 there were seventy-two tweed factories, employing 329 power-looms, and 225 sets of carding-engines, and the value of goods made was estimated at L.900,000. In 1862 there were eighty-two factories, employing 1069 power-looms, and 305 sets of carding-engines, and the value of goods made was L.1,830,000.

Though the tweed trade has extended from its birth-place to various towns between and including Inverness on the north and Dumfries on the south, it maintains its principal seat in the valleys of the Tweed and its tributaries. Most of the factories are of modern construction, a considerable number of them having been built within the past ten or fifteen years, while the old mills have been altered to suit the changes which have recently taken place in the modes and processes of manufacture, as well as in sanitary ideas. Many of the mills are stately edifices of four, five, or six stories, and their spacious floors are laden with machinery of the most ingenious and beautiful construction.

Nearly the whole of the wool used in the trade is imported from the colonies of Australia, New Zealand, and the Cape, and from Buenos Ayres, home-grown wool forming only a small proportion. In order to its being converted into tweed, the wool, after it enters the factory, has to undergo upwards of twenty processes, nearly all of which are performed by machinery. The wool arrives in compact bales, made up under hydraulic pressure, and bound with bands of iron. The "sorters" open the bales and separate the wool into classes according to quality. After being sorted, the wool is placed in a large bath, across which ranges of iron prongs are placed at intervals. This is the scouring-machine; and when it has been charged with wool, water, and a certain proportion of alkali, the prongs are set in motion, and the natural grease or "yolk" is thoroughly washed out of the wool. Connected with the scouring-machine is a wringer, through which the wool is passed, and from which it emerges in a slightly moist condition. It is then spread in a thick layer on the grated top of a chamber connected with a fan. The air in the room on the floor of which is constructed the chamber referred to, is, by means of steam pipes, raised to a high temperature. When the fan is set in motion, it draws the heated air through the wool, which is thus rapidly dried. At this stage the wool is usually dyed, though in some cases it does not undergo that process until it has been converted into yarn. We shall suppose, however, that the dyeing is not to take place until the wool has been spun. After being dried the wool is passed through the "willy," or teasing-machine, which opens it up and extracts the dust and other impurities, and turns it out in loose flakes. When the natural oil has been separated from it, the wool becomes crisp, and in order to get it into a workable condition, it has to be slightly moistened with olive oil. This was done by hand until quite recently, but a machine has been invented which distributes the oil more evenly, and mixes it more thoroughly than it could be by hand. The wool is now ready for carding, which is one of the most important processes, as on the manner in which it is done depends to a great extent the quality and smoothness of the yarn.

Nearly all the machinery used in the woollen manufacture was devised for working cotton, but was readily adapted to the coarser though more valuable fibre. The credit of inventing such machinery lies chiefly with the cotton districts of England, so that the history of the inventions need not be minutely followed here. Hand-cards, for preparing the wool for spinning, were introduced into this country about five hundred years ago. But it was not until the middle of last century that any attempt was made to improve on those primitive tools. The first idea in the way of improvement was the fixing of a large card on a table or stand, and suspending over it two smaller cards, one of which the operator worked with each hand. The cards used in that way were called "stock cards," and as they enabled one person to accomplish more than twice the amount of work that could be got through by the old system, they were considered to be a great step in advance. When Hargreaves, Arkwright, Crompton, and Cartwright came upon the scene, and devoted their ingenious brains to the improvement and perfection of the appliances for working cotton, a revolution took place in the textile manufactures of the country. Machinery driven by steam or water power was applied to carding, spinning, and weaving, while to many of the other operations the assistance of the iron arm was extended.

Carding-machinery has been brought to great perfection by subsequent inventors. As used in first-class woollen mills, each set of "carding-engines"—the productive power of a mill is calculated by the number of sets—consists of four machines. The first of these is called the "scribbler," and has two large cylinders and twenty-five small ones, all closely covered with spikes of wire. The wool is fed on an endless apron ; and in order to ensure regularity, the apron is marked off into sections, and the girl who feeds the machine has to spread a certain weight of wool on each section, a pair of scales being attached to the machine for weighing the wool. In passing through the cylinders the wool is separated from all entanglement, and is drawn off in a continuous rope or "silver," which is fed into the second machine, named the "first carder." There is only one large cylinder, and about half a dozen small ones in the first carder, and the wire with which these are covered is finer and more closely arranged than in the scribbler. The "second carder," through which the wool next passes, is in like manner finer than the first. In the early days of the tweed manufacture, the scribbler and one carding-machine were considered sufficient, but now second carders are invariably employed. Not only so, but another machine, which continues the carding and expedites subsequent operations, is being generally introduced. This machine is the "condenser," respecting which something will be said further on.

Where the condenser is not employed, the wool is delivered from the second carder in detached pieces called "cardings," equal in length to the breadth of the machine. As these are produced, they drop into the "piecing-machine," which joins them together, and winds from ten to a dozen or more of them in the form of continuous threads on large bobbins, or spools. The spools are then placed on the "billy," which does the first part of the spinning process, each line of twisted cardings supplying one spindle of the billy. The piecing-machine was invented by Mr John Melrose, of Hawick, in 1844 ; and, though now being superseded by the condenser, is a most ingenious piece of mechanism, and has done good service in its day, having been adopted not only by the Scotch and English manufacturers, but by those of Russia and other foreign countries. Before it was invented, the carding-machines could not be made above half their present width, and the piecing had to be done by hand, so as to get the wool passed through the "slubbing" on the billy. The inquiry which took place previous to the passing of the Factory Act in 1833 revealed the fact, that great hardships were endured by the children employed in the woollen factories ; and the hardest lot of all was that of the "piecers," or children who joined the cardings on the creeping-cloth of the slubbing billy. The carders turned out the wool in rolls about thirty inches long; and, gathering up a handful of these, the piecer stood behind the billy, and as the cardings were drawn in he kept joining fresh lengths to the end of each. This work, which required constant watchfulness and great activity of the fingers, had to be continued for twelve, fourteen, and even sixteen hours a-day. The "creeping-cloth" was formed of coarse canvas, and, by frequent contact with it, the skin was rasped off the fingers of the children, and in the winter time especially they suffered greatly. Nor were these their only causes of unhappiness. If by any mischance or neglect they failed to piece one of the rolls, and thus caused an interruption in the work of the "slubber," they were severely punished. For eighty-four hours' work the boys received 1s. 8d. a-week, and an annual gift of a suit of "Galashiels grey" and a Kilmarnock bonnet.

The restrictions put upon the employment of women and children by the Factory Act led mechanicians to consider whether the labour of which the manufacturers were thus to some extent deprived could not be supplied by machinery. Several contrivances were tried, but Mr Melrose's piecing-machine was so complete that it was, as already stated, at once introduced into the factories. About the same time a great improvement was effected in the carding process. Formerly the carder was supplied with wool in the same way that the scribbler is now fed; and it was, of course, impossible that the fingers could by mere instinct or knack of feeling adjust the wool on the feeding-table so as to be of a uniform thickness. As the wool went unequally into the carder, so it came unequally from it; and as the rolls were thus unequal in thickness, so were the threads into which they were spun. This was a continual vexation with all yarns intended to be twined together in different colours for making tweed. Hand-piecing, too, necessarily stretched the loosely combined cardings, and careless children sometimes carried this "rackin' the rowin's," as it was called, to such an extent as often to make the yarn quite unfit for tweeds. As Melrose's piecing-machine got rid of hand-piecing with its faults, so the "feeding-machine" got rid of the evil of feeding the carders by hand. The feeding-machine was an adaptation of cotton machinery to woollen cards, and it operated by taking the wool from the scribbler in the form of a sliver, and applying it to the carder, as described above. To illustrate the advantage of the invention, it may be stated, that if the scribbler feeder makes an inequality on the feeding-table, that affects only the sliver coming off the machine at the time. Sixty slivers from the scribbler are fed side by side into the carder, and these sixty are reduced to one. Then sixty slivers from the first carder are put

L

on the second carder, and undergo the same process of reduction ; so that the original inequality is reduced in a ratio represented by 60 × 60 to 1.   Sliver-feeding is of equal importance as regards condensers, and, strange to say, the general introduction of those machines was delayed for many years owing to the want of sliver-feeding, although an apparatus for making the slivers was sent from America by the gentleman to whom Scotland is indebted for the condenser.   The condenser was so called because it abbreviated the processes—taking the place at once of the billy and of the piecing-machine.

A little bit of history is connected with the condenser.   Mr Thomas Roberts, of Galashiels, went to America in 1830, and there saw the condenser at work.   Conceiving that it would be advantageous to the trade of his native town if such a machine were introduced, he set himself to study the condenser, and, having mastered all its parts, he made drawings which, with a minute description, he sent to his brother (Mr George Roberts, lately Provost of Selkirk).   Models of the more important parts of the machine were subsequently sent, from which Mr George Roberts had a condenser constructed; but a trial of the machine at Huddersfield Mill, Galashiels, was not satisfactory, and it was set aside.   Mr Wilson, of Earlston, and Mr Houldsworth, of Glasgow, took up the idea, however, and obtained a patent for what they considered to be a perfect condenser.   Manufacturers in Galashiels, Hawick, and elsewhere, gave this machine a trial, but found that they could not get it to make equal yarns, and in the case of mixed coloured goods the inequality caused shading, or "barring," as it is technically termed.   So the machine was condemned, and the gentleman who had first recommended it, and had made a considerable sacrifice of money and labour to have it introduced, was censured for what he had done, as a large sum had been sunk in machines which it was thought could never be got to work.   Mr Roberts returned from America some years afterwards, and discovered to his surprise that the makers of the condemned condensers had neglected to supply an essential part of the machine—namely, that for feeding by slivers instead of by hand.   The omitted portions of the apparatus, which had been lying for years in a lumber garret, were sought out by Mr Roberts.   The other parts of the sliver-making machine were constructed under his superintendence, and the "feeding-machine," as it was then called, was started in a small mill at Selkirk, which was the nucleus of the present Forest Mills, the property of Messrs George Roberts & Co.   The improvement was so obvious that all the carders in Galashiels and Hawick were fitted with it as speedily as possible. Melrose's piecing-machine was invented about the same time, and the two united marked a new era in the woollen manufactures of Scotland.

The condenser has been much improved by an English machine maker, and has been generally adopted. Essentially, the condenser is a carding-machine, the chief difference being in the delivering apparatus. When the sliver comes off the second carder it is wound up on large spools or bobbins. Fifty or sixty of these are set in a frame behind the condenser, and the slivers led in through rows of pegs, which draw them out a little and lay them flat on the feed-apron. The large and small cylinders are completely covered with the hooked wires which comb the wool, but while in the old carder the "doffer," or delivering cylinder, had a card and a space without card alternating in a longitudinal direction, the doffer of the condenser is covered with a series of belts of card, separated by a narrow space from each other. The old-fashioned doffer caught the wool from the cylinder so long as the card was passing it; but when the blank space reached the cylinder it made a break, and so each card of the doffer went round, and its burden was dropped into the piecing-machine in the form of a "carding." The condenser doffer, being a series of complete rings, catches the wool continuously on its side next the cylinder, and gives it off continuously at the opposite side in the form of a very small sliver from each ring. The slivers are caught by rollers and carried forward in a loose, delicate combination to what are called the "rubbers"—two endless webs of leather having both a forward and transverse motion, the latter rubbing each of the fine fillets of wool into a firm condition, so that they may be wound upon a bobbin. When the wool reaches this point it is ready for the next process—the spinning. The delicacy of the condensing process may be judged of from the fact, that it is not unusual to see the slivers, or "rovings," as they are sometimes called, so light that it would take four of them to be as thick as one of the threads from which Scotch blankets are usually made.

The earliest mode of spinning wool and other fibres was by means of the distaff and spindle, and no improvement was made on those appliances until spinning by a wheel was invented in the fourteenth century. Though the spinning-wheel, even in its simplest form, was a great advance on the distaff and spindle, yet it did not speedily supersede them. Ladies had adopted spinning as an easy and profitable recreation for their leisure hours, and preferred the ancient method, which remained in use until a time within the recollection of many persons yet alive. About the middle of the sixteenth century a great improvement was made on the first form of spinning-wheel, by so constructing it that the operator could be seated, and by means of a treadle keep the spindle in motion, thus admitting of both hands being used in manipulating the wool. Spinning-wheels of this kind are to be found in almost every home in the Highland districts, where one of these articles is still considered a most suitable gift to a bride. The spinning-wheel forms a picturesque and significant accessory in many paintings of domestic scenes in the pastoral regions, and our Queen not only owns one, but knows how to use it. The "muckle wheel" was employed extensively in some parts of the country in preference to the treadle wheel. It consisted of a fly-wheel of wood which set the spindle in motion. The operator gave the wheel a smart shove round with the hand, and then walked backward, as ropemakers do. By holding the roll or carding of wool firmly between the fingers while retreating, it was drawn to the required size of yarn. The impetus of the wheel enabled the spinner to retire five or six yards, and a thread of that length having been produced, it was wound up on a spindle as the operator returned to give a fresh impulse to the wheel. As the textile manufactures of the country extended, a more expeditious mode of spinning was desired, and many unsuccessful attempts were made to supply the want. At length the difficulty was overcome, to an extent never dreamed of, by the invention of the "spinning-jenny." The author of that contrivance was James Hargreaves, a weaver at Standhill, near Blackburn. Too poor to patent his machine, and so make a fortune by it, as he would undoubtedly have done, he employed it secretly in making weft for his own loom ; but the knowledge that he had devised labour-saving apparatus got abroad, and his neighbours broke into his house and destroyed the "jenny," little calculating that in that rudely-constructed piece of mechanism lay the germ of much of the subsequent manufacturing prosperity, not only of England, but of the world. Arkwright, also, succeeded in producing a spinning-machine based on an invention of a foreigner named Paul, who conceived the idea of spinning by rollers, though he did not succeed in carrying it into practice. A third inventor—Samuel Crompton, weaver—united the leading features of the "spinning-jenny" and Arkwright's machine in the "spinning-mule," which is now universally used in spinning wool, linen, cotton, &c. A subsequent inventor has given the "mule" the power of self-action, and now it spins 500, 800, or even 1000 threads at a time without requiring any attention beyond that of a boy or two, whose duty is to mend any threads that may break. The self-acting mules are now coming into general use in the Scotch woollen mills, and it is no uncommon thing to find in one apartment, and under the charge of not more than a dozen persons, machinery capable of producing as much yarn in a day as could be made in the same time by 20,000 of the most expert spinners with the old-fashioned wheel.

The yarn requiring to be dyed is reeled into " cuts," and " hanks " of three, four, or six " cuts," as found most convenient. The oil is then scoured out, and if the yarn is not to be coloured, it is hung up in a close room and exposed for a certain time to the fumes of sulphur, which make it a pure white. All colours of yarn are employed in the tweed trade ; and as a good deal of the beauty of the cloth depends on the quality of the dyes used, the dyeing is one of the most important departments of the manufacture. Some knowledge of chemistry is essential on the part of the foreman, and a thoroughly efficient man never fails to obtain liberal wages. In the early days of the woollen manufacture of England, dyeing was but imperfectly understood. Not that there was any want of a variety of dye stuffs ; but their chemical qualities were not sufficiently known to enable them to be successfully applied. The importance of the art was not lost sight of, however, and foreign dyers were encouraged to settle in the country. In 1552 an Act of Parliament was passed limiting the number of coloured cloths to "scarlet, red, crimson, murray, pink, brown, blue, black, green, yellow, orange, tawny, russet, marble grey, sadnew colour, asemer, watchett, sheep's colour, lion colour, motley, or iron grey." In the reign of William and Mary, the list was extended by the following additions :—" Violet, azure, friar's grey, crane, purple, and old medley." Not only were the colours limited, but the mode of producing them was regulated by statute, the use of certain materials—such as logwood and gall— being prohibited. In the sixteenth century manufacturers began to dye the wool before it was spun, instead of, as formerly, after it had been spun and woven. Notwithstanding the variety of colours which the dyers were capable of producing, only a few could be considered good or permanent ; and up till 1667 the fine broad cloth of England was sent to Holland to be dyed. During the last fifty years a great advance has been made in the art. Chemists have successfully sought to increase the number of dyes, and the man who discovers a new and really good tint now-a-days may be said to have found a fortune. From the most unlikely materials beautiful colours have been extracted. A dye closely allied both in properties and

appearance to the famous Tyrian purple has been extracted from
guano ; and the nasty tar which exudes from coal in the process of
gas-making has recently been found to contain the elements of a
series of dyes of great beauty. Wool is always dyed in boiling liquid,
the heat being necessary in order to fasten the colours. The yarn is
hung upon poles which stretch across the boiler and rest upon its
sides ; and while suspended in that way, the dye has free access
to every part of it. In order to facilitate the process the workmen
keep constantly turning the hanks on the poles. The time usually
required for all the operations in dyeing ordinary colours is three
hours. The tweed manufacturers have bestowed great attention on
dyeing, and with the most gratifying results ; but it is not necessary
to go further into the details of the art in this place, though they
are exceedingly interesting.

On being taken from the boilers, the yarn is rinsed to free it from
superfluous dyestuff, and is then carefully dried. The yarn intended for the warp, or longitudinal threads of the web, is more
firmly twisted than that for the weft ; and on the completion of the
dyeing process, the former is wound on bobbins for the warp-mill,
and the latter on pirns for the shuttle. The warper takes the
bobbins, and, by the use of a winding-machine of peculiar construction, arranges the threads in parallel rows, and finally winds them
on a cylinder which forms part of the loom. The warp threads are
then drawn through the " heddles " and " reed " by boys, and the
whole is ready to be fixed in the loom. Before the invention of the
power-loom, weaving was done by hand, and the loom employed was
of the simplest construction. A few specimens of the hand-loom
still linger in the manufacturing centres, and in the rural districts,
where faith in " home-made " stuffs still survives. Up till twenty
years ago the hand-loom weavers formed a large section of the industrial population of Scotland, and most of them worked in their
own houses, before the factory system was developed. They
were a grave, thoughtful, and exceedingly industrious class ; and
from their ranks went forth many men who took an advanced position in the world of learning, or were noted for their commercial
enterprise. Among those who still preside at "the four stoops of
misery," as the hand-loom is designated in some parts, men are to be
found who possess a knowledge of history, politics, and general
literature that would adorn a much loftier station of life. As a class,
they suffered great hardships through the introduction of the power-
loom. Those among them who had spent their early days and
prime of manhood in throwing the shuttle, could ill adapt them-

selves to other pursuits ; and they clung to their vocation, resolved
to be content with an occasional web which they calculated would
fall to their share when the productive power of their mechanical
competitor was unequal to meet the extra demands made upon the
manufacturers for certain classes of goods. In this way those of
them who remain are still employed, and usually when they obtain
a web, they have to sit at it early and late to get it done within a
limited time. Their life is thus, in many instances, made up of
days and nights of close application to work, followed by disheartening
ing intervals of idleness. The handloom weavers are generally to be
found together in a certain quarter of the town, and in several cases
that quarter is known as the Weavers' Row. In busy times, the
" rickle-tick " of the looms may be heard issuing from every door
and window, and a stranger might have the impression that he
was in the midst of a hive of industry in which the bees could not
fail to have every comfort and happiness. But there are frequent
gloomy, weary days, in which the shuttle lies at rest, and the men
hang about the doors with sad countenances, or saunter to the factory
to ascertain what prospect there is of obtaining another job. It is
a curious fact that in Galashiels hand-loom weaving is still paid by
the scale of rates which ruled before the introduction of the power-
loom.

The invention of the power-loom marked an era in the textile
manufactures of the world. Like most other contrivances used in
making woollen cloth, the power-loom was originally devised for
weaving cotton. The inventor was the Rev. Dr. Cartwright, who,
considering it probable that when Arkwright's patent for spinning
machinery expired, so many mills would be erected, and so much
cotton spun that hands would not be found to weave it, suggested
that Arkwright's next task should be the invention of weaving
machinery. The hint thus thrown out was allowed to pass unheeded,
the doctor's manufacturing friends considering the weaving of
cloth by machinery to be an impossibility. Though not a mechanician, Dr Cartwright regarded the idea to be not only of vast
importance, but of perfect practicability. He devoted close attention
to the subject for a year or two, and patented two machines, neither
of which, however, could be got to work satisfactorily. After years
of anxious labour, and the expenditure of L.40,000 on experiments
and patents, the Doctor so far succeeded that, on application to
Parliament in 1808, he received a grant of L.10,000 as a return for
his losses and exertions. Other hands took up the machine, and it
at length was made perfect, and for about forty years has been

employed in the manufacturing districts of England. Among those who tried to devise a machine to supersede the hand-loom was Mr Bell, of Glasgow, who in 1794 had a power-loom constructed; but he did not succeed in getting it to work. In 1796 Mr Robert Miller, of the same city, patented some improvement on Mr Bell's machine; and in 1810 a factory furnished with 200 improved looms was erected at Pollockshaws. Several years elapsed, however, before the enterprise succeeded. In 1825 there were but 1500 power-looms in Scotland, and these were applied only to the production of coarse linen and cotton goods. A few years afterwards an attempt was made to weave woollen yarn on them; but before that could be done certain improvements had to be effected, and it was not until a few years later that the machine was got to work properly. Now the power-loom is almost exclusively employed in the tweed trade, and as adapted to that particular branch of weaving it produces work of unequalled quality.

The general distinction between a tweed and a cloth is that a cloth is woven loosely and felted firmly, while a tweed is woven firmly and felted in a less degree. On the loom the tweed looks so close and fine that it would be thought impossible to improve it. In passing through the weaving department of one of the large tweed factories, one is struck by the great variety of styles and patterns in progress. Each loom works two widths of cloth, and though the same shuttles cross the warp of both, the colours of the completed fabrics may differ considerably. By using warps of different hues, the cloth, though made with the same weft, will of course be dissimilar in shade. To the inexperienced eye the appearance of many patterns of tweeds differs little from that of plainly woven cloths, but that little distinction is often the result of no small amount of ingenuity in the distribution of the threads and colours in the loom. For fancy patterns the looms commonly used are a modification of the Jacquard, limited to work twenty "leaves" or sets of "heddles." The looms are superintended by young women, who earn large wages, and to whom the work is well suited, as it is easy and healthy. They have simply to look out for and mend broken threads, keep the shuttles supplied with yarn, and remove any knots or imperfections in the work. The handloom weavers had a strong prejudice against the power-looms, and would not relinquish their old-fashioned machines and go to work with the new; hence females were set to do the work. Ultimately the men came to think that they should overcome their prejudices, and many of them would fain take charge of the power-looms; but the women having got possession, determined to keep it, and minding a power-loom is now regarded as a woman's and a womanly occupation.

On being taken from the loom the cloth is examined by "birlers," who pick out all irregular threads, hairs, or dirt; and by "darners," who insert with a needle any portions of the warp or weft that may have been omitted in the weaving.   It is then milled or "fulled." From a peculiarity in their formation, the fibres of wool possess the property of felting, and under the action of the fulling-mill, they become hooked together in a compact layer ; and by continuing the process of fulling for a sufficient length of time, the warp and weft of a piece of cloth would get so united as to be indistinguishable by the eye.   As already pointed out, however, the degree of fulling forms one of the distinctions between superfine cloth and tweed ; for while the former receives four fullings of three hours each, the latter generally receives much less.   To variety of style in tweeds has recently been added variety of finish, and while some cloths are highly felted, others receive no more milling than is necessary to cleanse them thoroughly.   Cloths made of scoured yarns are usually milled with fuller's earth, which, while it possesses great cleansing properties, is less liable to injure delicate colours than soap containing alkali.   Before the invention of fulling-mills, the cloth was "waulked" or shrunk by being tramped by men's feet in tubs; and when that method of conducting the operation was superseded, the attention of the Parliament of Edward IV. was seriously called to the fact, that "hats, bonnets, and caps, as well single as double, were wont to be faithfully made, wrought, fulled, and thicked by men's strength —that is to say, with hands and feet—and thereby the makers of the same have honestly before this time gained their living, and kept many apprentices, servants, and good houses, till now of late that, by subtle imagination to the destruction of labour and sustenance of many men, such articles have been fulled and thicked in fulling-mills, and in the said mills the said hats and caps be broken and deceitfully wrought, and in no wise by the means of any mill may be faithfully made."

After being milled, the cloth is "raised" by being passed over a cylinder covered with teasles—the seed-pods of the *Dipsacus fullonum*, a plant extensively cultivated for the purpose in the cloth-making districts of England.   Numerous attempts have been made to supply the place of the teasles by cards or combs of wire ; but nothing that has yet been tried can raise the fibres so nicely, and with so little injury to the cloth, as the teasle.   The cloth is stretched on "tenters" or rails to dry. Each rail has a row of hooks, and on these the cloth is fastened, a selvage of coarse wool being worked on the web for the purpose of accommodating the hooks without injury to the body of the fabric. In that way the cloth is evenly stretched; and when it dries, the threads lie straight and regular. A tentering and drying machine has recently been introduced, which enables the manufacturers to carry on their operations independently of the weather. The cloth, as it enters the machine, is caught by a series of tenter-hooks attached to a pair of endless chains. It is then passed over a range of heated pipes, which rapidly expel the moisture.

The next operation is "cropping," or cutting off the long fibres raised by the teasles. In an interesting paper on the "Wool Manufactures of Hawick," recently read to the Hawick Archæological Society by Mr D. Watson, the following account is given of the manner in which raising and cropping were done in the early days of the woollen manufacture:—" The cloth, being stretched or hung over a frame in front of the workman, was brushed over with hand-cards, previous to the introduction of teasels, to raise the loose fibres of the wool to the surface, and lay them all in one direction. When the whole web had been gone over in this manner it was handed over to the 'clipper,' whose apparatus consisted of a long narrow stool or bench, the top of which was cushioned, and a pair of large and peculiarly shaped shears, about eighteen inches long in the blades, and curved to fit the top of the cushion. These shears, which had a spring like the common wool-shears to open the blades, smoothed a considerable portion of the surface at every clip, and could be used pretty rapidly by an expert workman. Owing to their large size and peculiar form, there was some difficulty in getting them properly ground, which was at length overcome by the erection, at Galashiels, by subscription, of a large grindstone, to which all the manufacturers in the district sent their shears once a-year to be ground, a professional cutler being brought from Sheffield to do the work, which occupied him from a month to six weeks." The cropping is now done by a machine of American invention. The raised fibres are neatly and evenly shorn by the cloth being passed under a small roller, on which a series of steel blades are arranged in a spiral form. By repeatedly raising and cropping the cloth a very fine surface may be given to it, but that can be done only at the expense of damaging the fabric, and taking real value out of it; and in making tweeds these operations are usually limited to the extent of merely giving the surface a slight dressing, so that the texture of the cloth is left unimpaired. Some years ago, however, it was considered necessary that every thread in a pattern should be seen; and in order to produce that result the cloth had to be much raised, and cut very closely.

The cloth is finished by being pressed between warm mill-boards in a hydraulic press of great power, and, after a final careful examination, is measured and rolled up ready for the market.

A set of carding-engines, with the other necessary machinery, produces annually about 1000 pieces of cloth, or 50,000 yards ; and the total production of the trade, including goods made from bought yarns, may be roughly estimated at about 17,000,000 yards.

There are about 3000 yards of thread—1500 each way, warp and weft—in a yard of cloth of average weight; so that a man in a complete suit of tweed, with tweed overcoat, carries about with him rather more than twenty miles of woollen yarn. The weight of a yard of the thinnest tweed made for summer wear is about seven ounces, while that for winter use ranges as high as eighteen, and even twenty ounces. A fair average weight of cloth for each season, summer and winter respectively, is about ten ounces and fifteen ounces a-yard.

There is no fixed scale of wages in the tweed trade, and the rates vary considerably, not only in the different centres, but in the factories of individual towns. In few if any other branches of industry are female workers paid so liberally—indeed, their wages are little below those of the men. According to one statement, the women employed in the weaving department earn on the average about 12s. a-week, but superior hands occasionally make from 16s. to 18s. Another statement places the average at 11s., and the highest at from 14s. to 15s. Male operatives receive from 16s. to 20s. a-week, according to the department in which they are employed. Men in charge of departments have a shilling or two more. It will thus be seen that a family in which the father, and, say, three sons or daughters are at work—and that is no uncommon case—enjoy an income equal to that of many persons who have to maintain a far higher social position.

A fact worthy of notice, and one that does not fail to strike observant persons visiting the tweed manufactories of Scotland, is their superiority in a sanitary point of view over almost any other works of the kind. The occupation itself is a peculiarly healthy one, particularly in the carding departments, where the equal temperature necessary for the work, and the amount of oil held in the atmosphere, are said to present peculiarly favourable conditions for a certain class of invalids. The fact was first taken notice of by the late Dr M'Dougal, of Galashiels, and fully corroborated by the further investigations of Sir James Y. Simpson. It is by no means unusual in the tweed districts for persons suffering from or threatened with pulmonary disease to seek employment in the manufactories as a means of cure, or of protection against that malady.

In almost every establishment in the trade there is a sick society and a saving society. The sick society is merely a mutual insurance against loss in the event of being laid off work, each member contributing a small weekly sum, thereby earning a title to so much per week in case of sickness. The saving society is not intended for permanent savings, but only a temporary provision for setting aside small periodical sums against rent and other domestic contingencies. Both societies are dissolved yearly and started afresh. This enlightened regard for mutual interests sometimes goes much further. A slight misfortune befalling any of the class is certain to set a subscription afoot ; and in some establishments it is customary to subscribe towards the marriage providing of a girl leaving her employment to undertake housekeeping.

Trade unionism, if not a thing unknown, has seldom or never exercised its power for evil in the trade ; and it says a good deal for the general intelligence of those employed—and perhaps something, too, for the employers—that the trade has suffered less from strikes than almost any other in the country.

It is not known when hand-knitting was invented, but that it was practised in very early times is proved by certain passages in the works of the most ancient writers. For instance, in Homer's "Odyssey," Penelope is represented as weaving a web by day, which she unwove at night; and the process of unweaving is considered to be more applicable to a knitted fabric than to one constructed on the loom, since more time would be required to undo a piece of woven stuff than would suffice to make it. No direct historical mention is, however, made of knitting until the reign of Henry IV. About the middle of the fifteenth century the peasantry of England and Scotland began to wear knitted instead of woven woollen caps; and by an Act of Parliament passed in 1488, the price of a knitted woollen cap was fixed at 2s. 8d., or 8s. of our money. In several subsequent Acts, reference is made to woollen caps ; and from "the statute of servants," passed in 1563, it would appear that a woollen cap was an enforced badge of poverty or of service. The statute enacted that " every person not being possessed of twenty merks (L.13, 6s. 8d.) rental should wear on Sundays and holidays, when not on travel, a woollen knit cap, on pain of forfeiting 3s. 4d. (10s.) a-day." An Act of the Parliament of Edward VI. was passed in 1552, in which " knitte hose, knitte petticoats, knitte gloves, and knitte slieves," are mentioned, so that the art of knitting was not confined to the making of caps. The manufacture of woollen hose and caps by the knitting process is supposed to have been first practised in Scotland; at least it is certain that it was done in Scotland before it was in England.

When knitting became known in this country, it was readily adopted as a pleasant and profitable domestic employment. Upwards of three centuries ago the wives of Scotch peasants knitted all the stockings they and their families required, and used the bark of the alder to dye their yarn—a practice not yet obsolete. Until the introduction of tambouring, crochet, tatting, and other modern styles of working in thread, knitting was recognised as an accomplishment befitting every rank of life; and the young lady of the last century felt as much pride in being considered a good knitter of stockings as her modern sister can do in being pronounced a mistress of the more ornate but perhaps less useful occupations which have supplanted knitting.

The first considerable seat of the hosiery trade in Scotland was Aberdeenshire. In the beginning of last century many persons in that county were engaged in making stockings, which were chiefly exported to Holland, and thence dispersed throughout Germany. The spinning and knitting were done by hand in the homes of the people, and a number of merchants were established in the town of Aberdeen, who gave out the wool and received the stockings ready for the market. The extent of this branch of industry a hundred years ago may be judged of by the following passage in "Pennant's Tour of Scotland:"—"Aberdeen imports annually L.20,800 worth of wool, and L.16,000 worth of oil. Of this wool are made 69,333 dozen pairs of stockings, worth on the average L.1, 10s. per dozen. These are made by country people in almost all parts of the county, who are paid 4s. per dozen for spinning, and 14s. per dozen for knitting, so that L.62,400 is paid annually in the shape of wages. About L.2000 worth of stockings are made annually from wool grown in the county." Other manufactures sprang up in Aberdeen, and at present the hosiery business is comparatively insignificant. Hawick is the headquarters of the trade, and years ago had become famous for the excellence of its stockings. The manufacture of hosiery is also carried on to a considerable extent at Dumfries.

It would appear, that up till the year 1771, all the hosiery made in Scotland was knitted by hand on " wires;" for though the stocking-frame had been invented nearly two centuries before that time, there was such a strong prejudice against it that no one would venture to introduce it.

The story of the inventor of the stocking-frame forms one of the saddest chapters in industrial biography. A meagre version of it, in which the invention is assigned a romantic origin, is familiar to most people, but not so the authentic memoir. A "History of the Machine-wrought Hosiery and Lace Manufactures," written by Mr William Felkin, of Nottingham, and published in 1867, contains the most complete and reliable record of the life of the ingenious but unfortunate William Lee that has yet been produced. Setting aside the incident which is said to have induced Lee to think of devising a knitting-machine, the facts which are placed beyond doubt are briefly these:—William Lee having completed his University course and become curate of Calverton, in Nottinghamshire, his native village, devoted the leisure of three years to working out an idea which he entertained of the possibility of superseding the process of hand-knitting. In prosecuting his invention, he is said to have expended a large portion of his patrimonial means. He believed that if he succeeded in getting the machine to work, he would acquire a large fortune; and, buoyed up by that expectation, he persevered in his task. When he at length completed the mechanical knitter, or "knitting-frame," as he chose to call it, he resigned his position and duties as a clergyman, and in company with a brother began the business of hose-making. Though the machine in its first form enabled one person to do as much work as six of the most expert hand-knitters, there was a strong prejudice in the public mind against any contrivances which were designed to supersede or lessen the demand for hand labour. Lee was satisfied that his invention was not only practicable, but that it was destined to achieve great results ; and he removed his machine to London, taking along with him those of his relatives who had acquired the knowledge of knitting on the frame. Great interest was awakened in London when Lee's arrival became known. He sought the patronage of royalty through Lord Hunsdon, and Queen Elizabeth graciously consented to inspect the knitting-frame. Having done so, Her Majesty expressed her sense of the ingenuity displayed by the invention, but, to Lee's mortification, showed marked disappointment that, instead of fine silk hose, as she expected, the machine was shown at work upon a coarse worsted stocking. Notwithstanding this untoward circumstance, Lord Hunsdon had faith in the importance of the invention, and pressed his conviction on his royal mistress, begging that a patent of monopoly might be issued to the inventor. Elizabeth's answer was in the following terms:— " My Lord, I have too much love for my poor people who obtain their bread by the employment of knitting, to give my money to forward an invention that will tend to their ruin by depriving them of employment, and thus make them beggars. Had Mr Lee made a machine that would have made *silk* stockings, I should, I think, have been somewhat justified in granting him a patent for that monopoly, which would have affected only a small number of my subjects; but to enjoy the exclusive privilege of making stockings for the whole of my subjects is too important to be granted to any individual." Lord Hunsdon marked his own appreciation of the invention by indenturing his son as an apprentice to Lee; and thus Sir William Carey, a knight, the son of a Peer, and of the royal blood, became one of the first stocking-maker's apprentices. Fully conscious of the importance—nay, the necessity—of securing the royal favour, Lee set about adapting his machine to the knitting of silk hose. In 1598 he succeeded in constructing a frame that would accomplish the desired object. There was no difference in principle between this machine and the first made; but, instead of having only eight needles to the inch, it had twenty. When the machine was completed, Lee worked a pair of fine silk hose, which he presented to Elizabeth, and was no doubt full of expectation that he would secure Her Majesty's favour now that he had fulfilled the requirements, the absence of which had prevented him from obtaining a patent for his first machine. The hose were accepted by the Queen, but the only reward the inventor received was a formal expression of satisfaction with the elasticity and beauty of the stockings. These repeated disappointments, added to the indifferent treatment which he received on almost all hands, caused Lee to fall into a deep melancholy. He then showed his machine publicly, and offered it to his countrymen, but they, instead of accepting his offer, despised him and discouraged his invention. Henry IV. came to hear of the stocking-frame, and invited the inventor to France, promising as an inducement certain privileges and honours. Lee accepted the invitation, and soon afterwards established himself at Rouen, where he set up nine frames, and met with a most encouraging reception. Good fortune appeared to be dawning on Lee, and he had begun to forget the ingratitude of his countrymen; but when the way to prosperity seemed to lie open before him, he was again bitterly disappointed. Before he had secured the promised privileges his royal patron was assassinated, and the protection of the court was withdrawn from him. Finding himself unprotected in a foreign country, and left to bear the pangs of a wounded spirit alone, he wrote from Paris asking his brother, who had taken charge of the factory at Rouen, to come to him. It was too late, however, for, before the arrival of his brother, the inventor of the stocking-loom, almost an outcast from his native land, and an alien in France, had died of a broken heart in Paris, and was already buried there.

James Lee returned to Nottingham soon after his brother's death, and, in company with a man named Aston, who made improvements on the knitting-frame, established a manufactory of the machines. The value of the invention now began to be appreciated, and numbers of frames were set up in London, Godalming, and elsewhere. In the course of the seventeenth century the framework-knitters formed themselves into a union, for the purpose of regulating prices, and opposing the employment of persons who had not served a regular apprenticeship. The society was incorporated by a charter granted by Oliver Cromwell. There were only 650 frames in England at that time, three-fifths of which were employed on silk work, stockings, waistcoat pieces in colours, and trouser pieces. During the succeeding twenty years the number of frames greatly increased, and many were exported. In accordance with a clause in the framework-knitters' second charter, granted in 1663, such exportation was illegal, and various measures were resorted to in order to stop it. The society enjoyed a considerable income, and spent large sums in expensive pomp and pageantries. A carriage was provided for the master, and gold lace liveries for beadles and attendants, and among other accessories of the society were a gilded barge, a large band of musicians, flags emblazoned with the arms of the trade, and a splendid hall in which they held their feasts. They overdid the thing, however, got into debt, and made such heavy levies of money to support the extravagant style which had been adopted, that many of the members went to the midland counties to prosecute their calling. In 1727 there were 2500 frames in London, and 5500 in the provinces. Frequent disputes occurred in the trade, as the masters would not submit to certain of the bye-laws of the incorporation. In 1753 a Select Committee of the House of Commons was appointed to investigate and report upon the action of the society. The committee reported, among other things, that, in their opinion, "the bye-laws of the Company of Framework-Knitters were injurious and vexatious to the manufacturers, and tended to the discouragement of industry, and to the decay of the said manufacture." From that time the company ceased to exercise any influence on the trade at large, and existed merely as one of the incorporated trades of London.

Though the manner of forming the loops invented by Lee remains the same as in his original machine, many improvements have been made in the other portions of the frame, and it has been adapted to a great variety of purposes, the most important, with the exception of hosiery, being the making of lace. In Nottingham, which is the chief seat of the framework-knitting trade, about 200,000 persons are employed in making hosiery and lace, and the annual value of the goods produced is over L.8,000,000. A return, made in 1866, showed that the number of persons employed in making hosiery in Britain was 150,000, and that the value of the machine-made lace and hosiery was L.13,000,000 annually. A large proportion of the frames are now driven by steam, which was first applied in the trade about the year 1838.

The stocking-frame was introduced into Scotland about the year 1771. The only distinct mention of the fact occurs in the "Annals of Hawick," where it is recorded that the working of stockings by frame-knitting was begun in that town by Bailie John Hardie in 1771. Arnot, in enumerating the manufactures carried on in Edinburgh and its neighbourhood in 1777, mentions stocking-making on frames; but not until thirty years after that date could the trade be said to be fairly established as a branch of the woollen manufactures of Scotland. Bailie Hardie began operations with four frames, on which only linen and worsted stockings were produced up till 1785, when lamb's-wool yarn was employed; and for about fifteen years after that date, all the yarn used was spun by hand. In 1791 the hosiery trade of Hawick employed fourteen men and fifty-one women, and the frames were twelve in number. The goods produced in that year were 3505 pairs lamb's-wool and 594 pairs cotton and worsted hose. When carding-machinery was introduced at Galashiels, the Hawick hosiery makers sent their wool thither to be spun, the means of transit being a pony with panniers. By 1812 the hosiery trade of Scotland had increased to such an extent that the number of frames employed was 1449, dispersed over thirty-eight different towns and villages. A general census of the trade was taken in 1844, from which it appeared that there were in Scotland 2605 frames, distributed as follows :—Hawick and vicinity, 1200 ; Dumfries and vicinity, 500 ; Edinburgh and vicinity, 150 ; Glasgow and Kilmarnock, 280 ; Selkirk and vicinity, 128; Perth, 108; Langholm, 92; Denholm and vicinity, 87; Jedburgh and vicinity, 60. Of these, 620 were not at work. It does not appear that at any time the number of frames in Scotland was greater than in 1844. The number at present in use does not exceed 1600, of which 900 are in Hawick, and about 400 in Dumfries. In both towns power-frames are at work; and though the aggregate number of frames may be smaller than five and twenty years ago, the productive power is greater, as many of the frames are of the broad kind, and work six pieces at a time.

Messrs William Elliot & Sons, Hawick, are the leading firm in the Scotch hosiery trade. Their chief productions are Cheviot wool stockings, drawers, and undershirts, and they employ 617 men, women, and children; of whom 66 are employed in the manufacture of yarn, 70 are winders, 285 stockingmakers, 95 seamers, and the remainder are employed in the finishing department. The spinning and knitting are conducted in separate factories. On the floors of the knitting-factory, the frames, which are worked by hand, are ranged on either side, each being opposite a window, as a good light is indispensable to the workmen. Though to the uninitiated the knitting-frame appears mysterious and complicated in its working, it is simple in construction, and the process of knitting by it is easily learned. The workman sits on a high stool with his feet resting on a series of treadles, which produce certain of the eleven movements necessary to form each row or "course" of loops. With his hands he places the yarn over the "needles," and works a pair of levers, which complete the operation. The ribbed tops of stockings, bottoms of drawers, and wristbands of shirts, are worked on a frame specially devised for the purpose, and the men who make them earn the highest wages in the trade. A system of dividing labour prevails to some extent—a stocking, for instance, goes through the hands of three persons before the knitting is completed. The occupation of the machine knitter is little different from that of the hand-loom weaver, and the amount of muscular exertion required is about the same. It is considered a healthy trade, as the air in the workrooms is kept at an equal temperature, and no deleterious substance is used in any of the operations. From the knitters and seamers the work passes to the scourers and finishers, and is by them made ready for the market.

The only "fashioned" hosiery factory worked by steam-power in Hawick is that of Mr John Laing. The goods produced are of the finest class, being chiefly underclothing made from merino yarn and spun silk—a branch of the trade which has extended considerably during the last few years, and was never in a more prosperous condition than at present. The merino yarn used is got from Pleasley and Nottingham. The knitting-frames are of the finest construction, and so large that six shirt pieces may be knitted at one time. By the application of steam-power the operatives' labour is reduced to a mere act of supervision, and some of the frames are in charge of women. Mr Laing is exceptional in paying his hands at a weekly rate. Men receive from 14s. to 20s., and boys and girls from 4s. to 10s. Messrs Dicksons & Laings, Hawick, also employ a large number of frames worked by power; but the goods they produce are chiefly of a coarser kind.

The relations between the employers and employed in the hosiery trade have not in general been of a satisfactory kind. During the period from 1810 to 1840 the trade in England was in a critical condition. Too many persons had rushed into it, the markets were overstocked, prices fell, and great destitution prevailed. Matters were complicated and feeling embittered by the action of the Luddites, who thwarted the efforts of those who, by introducing improved machinery, endeavoured to set the trade on a new and better footing. A reliable authority says :—" We do not hesitate to affirm that the actual sufferings and privation experienced during the Lancashire famine of 1863–66 were far less than the distress in the midland hosiery district during the period between 1810 and 1845, when it became a widely spread practice to still the cravings of hunger in the adult by opium taken in a solid form, and in children by Godfrey's cordial." No trade ever endured such a severe and prolonged state of depression; and in groping about for remedies, some most injudicious things were done. The guardians of the poor in certain districts actually induced men to take work at whatever they could get for it, and their wages were made up out of the rates. The full earnings of those who got work ranged from 5s. to 7s. 9d. a-week of sixty, and in some cases of seventy-two hours. As the peculiar circumstances which brought about the crisis in England did not exist in Scotland, the trade north of the Tweed was not seriously interfered with. When better days dawned on the English hosiery makers, and improved machinery was introduced with the most satisfactory results, the Scotch manufacturers wished to maintain their place in the market by adopting the improved frames. In that desire they were strenuously opposed by the workmen, and no attempt was made to bring in the machines until 1855, and even then it was found impossible to get men to work them. The new frames were more costly than the old, and the masters claimed a return on their capital in the shape of a reduction on the price charged by the men for making stockings, &c., which they considered the workmen could well afford to allow, since the rate of production would be so much increased that a man might, even at the reduced price, make something like ten or twelve per cent. more wages than he could earn at the old frame. The men would have the new frames if the old rates were retained, but not otherwise, and no definite arrangement has been come to.

It has been customary in the trade from its earliest days to charge a certain sum weekly in the shape of "frame rent," and this has long been regarded as a grievance, the men considering that they should be put on the footing of workmen in other trades, whose employers provide machinery free of charge. The usual rent of a narrow frame is one shilling a-week, which is deducted from the wages, together with a small charge for oil and gas. There is something antiquated in this system of charges, as well as in other matters connected with the trade; and whether the present generation of masters and workmen be conscious of it or not, they are maintaining a position disadvantageous to both. If they would sink their mutual animosities, and modernise their system of working, they would certainly improve their positions. A Board of Arbitration and Conciliation was formed in Hawick in 1867, on the model of similar bodies in Nottingham and Leicestershire. The Board consists of nine employers and nine workmen, and a neutral gentleman officiates as referee. So far, the only result that has followed the deliberations of the Board has been the abolition of the "Hawick Hosiery Trade Society" for all purposes save those of a charitable nature. This is a step towards an improved relation between parties, and more good is likely to follow.

The Hawick stocking-makers are better paid than their brethren in the midland counties of England, or other Scotch towns where the manufacture is carried on. There are so many different qualities and classes of work, that it is not easy to quote an average of wages; but it may be struck at 18s. a-week for the broad frames, on which underclothing is made, and 13s. to 14s. a-week for the narrow frames. Men employed on the "rib frames," on which the tops of stockings, and wristbands of shirts, as well as ribbed underclothing, are made, can earn from 30s. to 35s. a-week. The average value of work turned out by each worker on the broad frame is L.6 weekly, and on the narrow frame L.3 weekly.

The quantity of wool manufactured into hosiery in Hawick annually is upwards of 1,000,000 ℔., and the value of the goods produce about L.130,000. A considerable quantity of hosiery yarn is spun in Hawick, which is not woven there. It is sold to manufacturers elsewhere, and a large proportion finds its way to Leicester, where an imitation of "real Hawick hosiery" is produced. The stocking-makers are generally an intelligent class of men, and take much interest in public questions. Many of them are keen politicians, and when occasion serves, can give sensible expression to their views. Some, however, take too great advantage of the liberty which working by piece instead of by fixed wages confers, and idle away all their time except what is absolutely necessary they should devote to labour, if they would keep the wolf from the door. The records of the local courts show that others prefer game-trapping, salmon poaching, and the excitement of the public-house, to the "whirr" of the frame. These, it is pleasing to add, are the exceptions, and it may be said of the greater number of those engaged in the trade that no class of workmen live more honourably and respectably, or bring up their families more comfortably and creditably.

The use of tapestry and embroidered cloths as a covering for furniture and the floors of rooms is of great antiquity. The Babylonians, Parthians, and Gauls, were famous for embroidering carpets in different colours, and several cities early acquired a celebrity for the manufacture. The carpets were usually made with a woolly nap on one side, but occasionally the nap was raised on both sides, and the design enriched by the insertion of threads of gold and silk. The names of distinguished makers of carpets have been handed down along with those of the cities in which they plied their vocation. Pathymas, an Egyptian, with Acesas and Helicon, of Cyprus, were among those who obtained eminence in the art, and it is conjectured that the two latter worked under the direction of Phidias, the famous sculptor. Plato mentions that it was customary in Greece to cover couches with carpets, and place others on the floor. The wealthy patricians of Rome used purple carpets, for which they paid fabulous prices. Babylonian covers for couches were sold in the days of Metullus Scipio for L.4600 each, and that price was quintupled in the time of the Emperor Nero. Carpets figured conspicuously in the pageantries of the ancient nations in the East. They were used to deck horses and elephants in triumphal processions, and respect for the dead was marked by placing carpets on tombs and cenotaphs. In Turkey carpets were used many centuries ago in the same manner as at present. The skill of the natives of India in working textile fabrics is universally admitted ; but, perhaps, in none of their productions have they displayed so much ingenuity and taste as in embroidering carpets. The finest work of the kind is executed by the natives of Scinde, but though much in demand among the princes and chiefs of India, those carpets are too gorgeous and expensive for European tastes. In the most costly the design is tamboured in cloth and canvas with threads of gold, silver, and silk. The only Indian carpets used in Britain are a thick soft kind made in Masulipatam.

Carpets were introduced into this country from the East some time during the twelfth century, but as they were very costly they were brought into use only on extraordinary occasions. The Norman practice of spreading rushes on the floors continued to be followed up till the close of the sixteenth century. About that time the working of tapestry had grown to be a fashionable occupation among ladies of the upper ranks, and walls of houses came to be decorated by elaborately worked devices. Clay floors gave place to wood planking, and mats and rugs of home manufacture were spread out on convenient spots. As time wore on the growing taste for soft coverings to the floors of the more important rooms in the houses of the wealthy led to the introduction of the manufacture of carpets.

The merit of originating the manufacture in England is claimed for one of the Earls of Pembroke, who, observing the tendency of public taste, and being desirous to improve the condition of the weavers in Wilton, induced a skilful French carpet-weaver to be smuggled over from France in a sugar cask, in order that he might teach the weavers. It was not until towards the middle of last century that the manufacture was fairly established in this country. The manufacturers of Kidderminster, who had previously become famous for the excellence of their broadcloths, turned their attention to making carpets, and by the year 1735 had succeeded to such an extent that they gave promise of attaining as great celebrity for carpets as for broadcloths. The kind chiefly made at first was that known as "Brussels" carpet, but subsequently a cheaper fabric was invented, which, under the name of "Kidderminster carpet," became very popular, and is yet much in request.

The date at which carpet-weaving was begun in Scotland cannot be stated exactly, but it is certain that the trade was of limited extent until an enterprising firm in Kilmarnock took it up in 1777, and laid the foundation of the celebrity which that town has since enjoyed for its carpets. Kilmarnock was not long in rivalling Kidderminster, and forty years ago had nearly a thousand looms employed in weaving Brussels, Venetian, and Scotch carpets. The average annual value of the carpets made in Kilmarnock during the past forty or fifty years has been over L.100,000. Though Kilmarnock has occupied the foremost place in the trade, a considerable quantity of carpeting is manufactured in other Scotch towns—chiefly Glasgow, Paisley, Bannockburn, Aberdeen, and Ayr. The manufacture had been introduced and had flourished for a time in a number of towns in which it is now unknown. Edinburgh, for instance, early possessed a carpet factory, and one firm in the city is identified with an important improvement in the manufacture. Now, with the exception of a small factory at Canonmills, the trade is extinct in the city itself, having been transplanted to the neighbouring village of Lasswade about thirty years ago.

In 1825 it was estimated that there were between 1000 and 1200 carpet weavers in Scotland, and of these about 800 were in Kilmarnock. Each weaver produced about six yards of carpet a-day, and was paid at the rate of 3½d. to 4¾d. a-yard. The selling prices of the goods ranged from 2s. 9d. to 3s. 9d. a-yard. A considerable quantity was exported to the United States, and the other markets beyond Scotland were London and Dublin. In 1840, 900 looms were employed in the carpet trade, and the wages of the weavers were from 11s. to 18s. a-week.

The Society of Arts did much to encourage the extension and improvement of carpet manufacture in England, and it was under their auspices that carpets in imitation of those of Persia and Turkey were first made by a manufacturer of Axminster. Though this kind of carpet retains the name of Axminster, that town ceased its active connection with the trade many years ago, and Wilton is now the chief seat of the manufacture. In Scotland, the Board of Trustees for the Encouragement of Manufactures granted premiums to carpet makers, which resulted in the introduction of new branches of the trade. Mr Thomas Morton, of Kilmarnock, was one of the principal pioneers of improvement in the manufacture. He devised a number of appliances which tended to better the quality of the goods produced; but his most important invention was the triple carpet fabric which now bears the name of "three-ply Scotch carpeting." The Kidderminster carpet—also called Scotch—consists, so to speak, of two layers or webs, the design being produced on both sides, but with the colours reversed. This carpet is light and cheap ; and Mr Morton sought, by increasing its substance, to make it approximate more closely to the Brussels fabric. He accordingly added a layer, making the carpet at once thicker and softer, and admitting of a third colour being introduced in the design. An honestly made carpet of this kind is almost as durable as one of

Brussels make, and has the advantage of a double presentment.
Usually one side is brought out darker than the other, and it is customary to have the light side uppermost in summer and the dark
side in winter. Before the Jacquard apparatus was applied to carpet
weaving in this country, the "heddles" were moved by a "draw-
boy," who could not always be relied upon for accuracy of work.
Mr Morton superseded the office of the draw-boy by applying to the
heddles a revolving drum studded with pins, a contrivance which
answered its purpose admirably, until it was superseded by the Jacquard machine.

The Board of Manufactures having got the making of Brussels
carpets established as a branch of industry in Scotland, next devoted
attention to the introduction of the art of producing carpets similar
to those made in Turkey. In 1830 the Board offered two prizes, of
L.150 and L.50 respectively, as inducements to the introduction
of the manufacture of carpets of the kinds referred to. At the
same time, other prizes to the amount of L.115 were offered for
improvements or excellence in carpets of the varieties then made in
the country. In 1832 the committee of the Board reported that
great and unexpected success had attended the offering of premiums
for Turkey carpets. The winner of the highest premium—whose
name does not appear in the committee's report—had made a splendid
start, and had written to say that he was grateful to the Board for
having originated a new branch of trade, and stating that he had
orders for as many carpets as he could make in nine months.
Premiums were next offered for Persian and French tapestry carpets,
and those fabrics were soon afterwards introduced. The Scoto-Persian
carpets made by Messrs Richard Whytock & Co., of Edinburgh, and
Messrs Gregory, Thomson, & Co., Kilmarnock, were considered
equal in quality to any produced elsewhere, and became popular
among the wealthier classes of society.

The Brussels carpet possesses peculiar beauty. A large number
of colours may be introduced, and considerable scope is thus given
to the designer; but the special attractions of the fabric are its
softness, elasticity, durability, and the richness of appearance imparted to it by the cord-like arrangement of the texture. Brussels
carpet may be said to consist of two webs woven simultaneously the
one over the other, but both firmly united. The lower web consists
of a strong groundwork of linen and worsted, and the upper of the
solid woollen "pile." The best carpets of this kind are composed
of six layers of worsted warp, each containing 260 threads. Only
one-sixth part of the warp appears on the surface, so that five-sixths

of the substance of the web as well as the linen weft are unseen ; but portions of all the worsted threads are thrown up at intervals, as the colour they bear is required in the design. As all the surface of the carpet is composed of warp—that is, threads running continuously from one end of the web to the other—it will be seen that each colour, no matter how small a part it may play in the figuring, requires a series of threads of the full length of the web. The layers of warp are arranged in " frames," and thus Brussels carpets are spoken of as being " six-frame," " five-frame," or " four-frame," according to the number of layers of yarn, the kind in which most material is used being, of course, the most valuable. It may be useful for purchasers to know that large quantities of " three-frame" and even " two-frame" Brussels carpets are now being made ; and that, consequently, there is quite as great a difference in the quality as in the price of the highest and lowest makes. The cord-like loops of the carpet are formed by wires, which are inserted and withdrawn at the proper time by an ingenious and beautiful piece of mechanism attached to the loom. The Jacquard apparatus is now generally used in producing the design. Large quantities of Brussels and Wilton carpet are made for home use and exportation at Glasgow and Kilmarnock. The difference between the Brussels and Wilton carpets is that, while in the former the raised loops of worsted are left entire, in the latter they are cut, so that the surface has a velvety appearance. The only difference in the manufacturing process is that the wire which holds up the loop while the weft is being thrown in has a knife blade attached to one end, and as the wire is drawn out the blade cuts the loop.

The high price of Brussels carpet of good quality has operated against its obtaining such an extensive demand as its beauty and elasticity would otherwise have secured for it ; and at various times, in different places, attempts have been made to produce a kind of carpet which, while it should require less material for its fabrication, should retain the appearance and softness of " Brussels." None of those attempts were successful until Mr Richard Whytock, of Edinburgh, took up the problem, and invented his patent tapestry and velvet-pile carpet, which more than realised the objects sought. The tapestry carpet, while consisting of only one layer of worsted warp, may be made to show any number of colours and shades ; and its corded surface gives it the richness of texture presented by the best Brussels fabric. Mr Whytock worked out his invention in his factory at Lasswade ; and so highly were his labours appreciated, that his patent was extended for five years beyond its original period.

He granted licenses to Messrs Crossley & Sons, of Halifax, and Messrs Pardoe, of Kidderminster, and thus the manufacture was introduced into England.   The firms named found a ready demand for the new carpet, which yet retains great popularity.   Considering its price, it is perhaps the most beautiful and durable carpet made. Mr Whytock retired from the carpet manufacture a number of years ago, and his invention, having become common property, is now practised in various parts of England and on the Continent.

Instead of working out the pattern by a limited series of coloured threads extending from end to end of the web, Mr Whytock divided each thread into a number of longitudinal sections, corresponding with the number of loops to be formed in the length of the carpet. He then imparted to each section a colour suitable to the part it would occupy in the design.   Thus, if a thread be drawn out of a tapestry carpet, it will be found to possess many hues, and the space occupied by the respective colours will be seen to differ considerably. At one part there is a bit of green to contribute to the formation of a leaf, next a bit of white for the full-blown flower, and again a small spot of crimson to form the tip of the opening bud—longer or shorter sections of the ground colour coming in between.   It is almost impossible to estimate fully the amount of patient thought, minute calculation, and mechanical skill which must have been exercised in bringing this invention to perfection.

Mr Hugh Miller took much interest in Mr Whytock's labours; and in an account of a visit which he paid to the factory at Lasswade, says :—" Every carpet consists of repetitions of certain sets of patterns, and so there must be a recurrence of the same sort of threads. But no two threads barred in exactly the same fashion go together. There is a continual variation, on whose nice adjustment the integrity of the pattern depends—and hence the necessity of much care and correctness in the calculations. In the thread, too, the bars of colour have to be broader in a certain proportion than it is intended they should seem in the cloth, as allowance has to be made for the amount of thread lost in the loop or pile which forms the surface ; and so the pattern in the warp has to be made quite a different sort of thing from the proposed pattern in the web. Hence another set of difficulties.   But this plan appears to have succeeded in overcoming them all, and in producing in the Brussels tissue patterns of even greater beauty than in the many-plied carpets woven in the common style. We were much interested in this establishment at Lasswade to see how simple were the operations performed by each set of the mechanics in the different departments, and yet how regularly the complicated whole grew under their hands. One set of workmen were employed in carefully barring across with colour layers of threads spread on huge cylinders—another set were engaged in fixing the dyes—a third set, in setting up the threads after a given manner for the warp of the projected web—a fourth, in weaving. All of them seemed to be workers in the dark, so far as the pattern was concerned ; they merely measured off certain bars of colours after certain given proportions, or stretched in a particular given fashion a certain number of threads across a frame ; or, when stretched and arranged, weaved them into a web ; and yet the pattern sprung up before them complete in every sprig, leaf, and petal, as if it had been as much a thing of spontaneous growth as the mosses and wild flowers of our woods and moors." By cutting the loops, as in the case of the Wilton carpets already referred to, the tapestry may be converted into a velvet pile carpet. Twelve years ago, 600 power-looms, producing 10,000 yards a-day, were employed in making Mr Whytock's carpet, and since that time the demand has largely increased.

Messrs James Templeton & Co., of Glasgow, are among the most extensive carpet manufacturers in Scotland ; and, like Mr Whytock, Mr Templeton has won distinction by his inventive genius. Turkish carpets used to be made by knotting the worsted to the warp by hand, a clumsy and tedious process, involving much labour and expense. Mr Templeton considered that carpets equally good, and presenting the chief characteristics of the Persian or Turkish fabrics, might be made by machinery at a considerably reduced cost, and resolved to put his idea to a practical test. He succeeded beyond expectation ; and how he accomplished his task will be made plain in the course of a brief account of what came under notice during a visit to the extensive factory of the firm at Greenhead, Glasgow, in which upwards of 500 operatives are employed. The carpets, which are principally made to order, are woven in one piece. When purchasers go to the expense of a carpet of this kind, they like to have something unique, so that often only one carpet is made of a particular design. In many cases the heraldic emblems of a family are worked in, and in others the ownership is marked by cunningly-devised monograms. Then it is necessary, in most instances, to impart to the carpet some special feature in order to bring it into harmony with the architecture of the room and the style of furniture. It will thus be seen that the designing department is one of the most important. Upwards of forty men and boys are employed in it, and the artistic taste and skill displayed by the superiors of the department are highly creditable. The designs are drawn in colours on strong paper, marked off into squares each of which represents a stitch or dot in the carpet. It often happens that the carpets have to be made to fit irregularly shaped rooms, and in such cases great nicety of measurement and calculation have to be followed. Sheets of paper bearing the design are cut up into stripes of two rows of dots each, care first being taken to have them all numbered. These stripes of paper are sent to the yarn room, where a number of girls seek out a certain number of pirns of yarn of the required colours. The yarn is to form the weft of the carpet; but it has to undergo a preliminary weaving process, which converts it into "chenille," or furred cord. Yarn sufficient for twenty or thirty cords is woven in one web, which is afterwards divided longitudinally by a peculiar cutting machine. This preliminary weaving is done on hand-looms, and as a section of the pattern of the carpet has to appear on each cord, the chenille weaver has to work his web in bars of different colours and of various widths, being guided by the stripes of paper which accompany the yarn when delivered to him. In the variety of its colouring, the chenille cord somewhat resembles a thread of Mr Whytock's tapestry carpet. When the web is cut up a number is affixed to each cord, and after being wound into hanks, the cords are ready for the carpet weavers. The carpet looms are of gigantic dimensions—one of them having a width within the frame of thirty-four feet—and from four to six men are employed at each. A copy of the design, with the rows bearing the same numbers as those attached to the chenille cords, is supplied to the weavers, who, after mounting the loom with a strong linen warp, weave in the chenille which is used as weft. Between the rows of chenille a "shot" of stout worsted is put in as a backing. Working those big looms is a tardy process, the shuttle moving slowly from hand to hand of the workmen, and a pause being made to adjust with a wooden comb the "fur" of each row of chenille, so that the dots of colour may assume their proper place in the design. When the carpet leaves the loom it is carefully examined, and defects are made good. It is then shorn by a machine similar to that used in shearing cloth; and when thus reduced to an even velvet-like surface, all its beauty becomes apparent, and not till then can the result of the operations described be fully appreciated. Mats and rugs are made in the same way on smaller looms, and the designs of some of these are remarkable for their chaste elegance.

Messrs Templeton & Co. were successful exhibitors at Paris last year, having obtained one of the two gold medals awarded to carpet manufacturers in Britain. About one-fifth of the workpeople employed by the firm are females, who earn, on an average, about 10s. a-week. The weavers and other male operatives receive from 15s. to 30s. No special provision is made for giving an art training to the boys in the designing department; but those who show special aptitude for the work are encouraged to attend the School of Arts.

An offshoot of this firm is that of Messrs J. & J. S. Templeton, Milend, Glasgow. About 200 operatives are employed, and there is considerable variety in the goods produced. These are chiefly Brussels and Wilton carpets and rugs, silk and rep window hangings, and a variety of chenille carpeting patented by Mr John S. Templeton, in which the warp is formed of chenille cord, and the weft of linen, being the reverse of the mode followed in the parent establishment. A set of damask chair-covers of exquisite beauty were recently made by Messrs J. & J. S. Templeton for Windsor Castle. Upwards of 3000 persons are employed in the manufacture of carpets in Scotland, and the value of the goods of this kind produced in Glasgow alone amounts to L.150,000 a-year.

The manufacture of woollen goods is a prominent feature in the history of several important towns and villages in Scotland, and in order to make the record of that branch of industry as complete as possible, brief historical sketches of the principal seats of the trade are appended:—

GALASHIELS.—The earliest indication of the manufacture of wool in this town occurs in a charter, dated 1622, conveying the barony of Galashiels to the Crown. Among the pertinents of the barony the charter mentions a corn-mill and "waulkmills," rights to hold fairs and markets, and other privileges of a burghal nature. These " waulk" or "fulling" mills, as they are now called, were probably entirely used for blankets and "kerseys" (coarse woollen cloth), made for consumption within the district. The site of the mills was probably on the haugh, where the present town is situated. At the date mentioned the population numbered about 400, and the bulk of the people were " mailin' men," holding "acres" from the baron of Galashiels by a feudal tenure of military service, and probably enough conjoined to their labour on their " crofts" the arts of dyeing and weaving the produce of the cottagers' spinning-wheels. " The Forest" was as famous for its sheep-walks as for its deer-coverts, and abundance of wool would naturally lead in such inland situations to increase of spinning. In 1774 the whole wool used in Galashiels was 792 stones, of 24 ℔ each. It was obtained in the district, and worked into blankets and "Galashiels greys," a coarse and inferior imitation of "Yorkshire medleys." There were 30 looms, 3 waulk mills, and 600 of a population at that date in Galashiels. Eighteen years afterwards—in 1792—the quantity of wool used was 2916 stones, which Dr Douglas, the minister of the parish at that time, calls "manufacturing to a great extent." Operations were then conducted to some extent on the factory system. The wool, after being roughly scribbled on the "dick"—a carding-machine driven by the foot—was carded into rolls and spun by women, who were paid at the rate of 6d. per "slip"—poor pay, indeed, when the spinning of a slip was a fair day's work. 100 slips can now be spun on the mule for 1s. 6d. In 1790 about 300 women were employed in spinning for the Galashiels trade. Ten years before, a "willy" for teazing the wool had been purchased as a joint-stock speculation by some tradesmen. It was driven by hand, and teazed wool for the whole town. The three waulk-mills which were in the town in 1774 were driven directly from the axles of water-wheels, and were either in the open air or under a boarded roof, so that when frost set in milling was suspended. One of the mills was on the site of the Waulk-millhead Mill, another on the site of Messrs J. & W. Cochrane's works, and the third where the works of the late Mr George Bathgate are. The united rental of the three was only L.15 a-year.

Without further allusion to the social condition of the community, it may be mentioned that at the time referred to, and for a number of years afterwards, apprentice dyers took their meals at their employers' tables, and the masters' daughters worked as hard as the other persons employed ; while the heads of establishments either busied themselves at the works, with their coats off, or travelled to Edinburgh on foot to sell their goods. The trade owes much of its prosperity and importance to the indomitable perseverance and prudence of those respected pioneers.

The year 1790 saw the beginning of a new order of things. Carding-machines had been started in Leeds, and Mr John Mercer, of Galashiels, went to examine them. The result was that a scribbler was soon in operation in the town, and it was the first carding-machine started in Scotland. Wilderhaugh Mill was built for the machine, and was the first woollen factory, in the modern sense of the word, in Scotland. The venerable house was removed so recently as 1866 by Messrs Brown & Shaw, the proprietors of large works which now occupy the site. During 1791 the firm with which Mr Mercer was connected got from England a carder, a "billy" with twenty-four spindles, and a spinning-jenny with thirty-six spindles. Mr James Roberts introduced a spinning-jenny with twenty-four spindles. Thus were the old, tedious, imperfect, and expensive hand processes superseded by a " set of machines" which, in all but one particular—perfect equality of yarn—made almost as good work as modern carding-machines. About that time feus were taken on the haugh, the site of the present town, and a cloth hall was started on a principle which enabled the depositor of goods to draw two-thirds of their value, leaving the remaining third to. be lifted when the goods were sold. The parish minister, Dr Douglas, advanced L.1000 to aid the scheme ; and though it was ultimately found to be impracticable, the memory of the good clergyman is much revered on account of the interest he took in the trade of the town ; and his portrait, paid for by subscription, has been hung in the public hall. The new spinning-machines were in most cases erected in the garrets of dwelling-houses, and were driven by hand. They cheapened spinning, and made it possible to give any quantity of yarn the same amount of twine—a material thing in cloth to be much felted.

For many years afterwards the only notable events in the manu-facturing history of Galashiels were the erection of factories on new feus on the haughs, and of dwelling-houses for the increasing popu-lation. A set of machines was a heavy speculation in those days of limited capital, and several firms were usually associated in one mill. In 1792 Mid-Mill (the nucleus of Messrs J. & W. Cochrane's works) was built ; in 1798 Botany Mill was put up ; in 1802, the Waulkmillhead ; in 1803, Rosebank Mill ; in 1804, Nether Mill ; in 1818, Huddersfield Mill ; in 1819, Galabank Mill; and in 1826, Wakefield Mill ; and during those years some great improvements in machinery were also introduced. In 1810 Mr William Johnstone invented the "twiner"—an adaptation of the spinning-jenny—by which two or three threads could be twisted into one. In 1814 spinning "mules" were introduced by Messrs W. & D. Thomson, at what is now called Rosebank Mill, and these cheapened yarns con-siderably. Spinning "jennies" had previously been lengthened to 144 spindles, and were turned by water power ; but the mule enabled one man to work 500 spindles, and earn much larger wages at rates per slip sixty per cent. less than were formerly paid. The "shearing" of the cloth had always been a difficult and expensive operation, being done by hand with a peculiar apparatus, requiring a good deal of dexterity to work it properly. In 1819 Mr James Paterson brought the "Yankee" from America, which at once sim-plified and perfected the shearing process. The principal trade up till 1829 was the making of cloths of the kinds already mentioned from home-grown wool, but knitting-yarns and flannels were also produced.

The commercial disasters of 1829 fell upon the trade of Galashiels with peculiar force, and manufacturers were completely prostrated when they saw that their cloths—greys, drabs, and blues—were not likely to be again required in the home markets. Many experiments were made in that year to develope new branches of trade. Soft tartans—first made in the district by Mr Thomas Roberts, the gentleman who afterwards sent the "condenser" from America, as already related—and trouserings made from twists and mixed colourings, or tweeds, were the only new varieties of goods that were successfully produced. The soft tartan made by Mr Roberts was extensively used by the nobility and gentry for cloaks, dresses, and shawls. Three-fourths of the machines in Galashiels were kept going on tartan for six out of the twelve months for many years, and the goods paid better than tweeds have ever done. The tartans were principally made from foreign wool of different qualities—from the strong fleeces of Van Diemen's Land to the finest Saxony lamb's-wool. While some firms were chiefly devoted to this branch, others prosecuted the manufacture of mixed coloured trouserings; and in a few years the demand for both classes of goods was far beyond the productive power of the town. In 1828 the number of looms employed was 175, and in 1838 it was 265. The population in 1831 was 2100, and by that time a considerable portion of the modern town was built; banks, churches, and schools were founded; and since 1838 the trade has been continually increasing, while many mills elsewhere have been raised by gentlemen going from Galashiels. Hand labour has given place to machines in almost every department of the trade. The population is now close on 10,000, yet factory hands are always scarce, except in periods of temporary depression. The small carding-machines of 1780 have been put out of the way for the best and largest carding-engines that can be got anywhere; steam-engines are at every factory, some of them individually powerful enough to turn all the machinery that was in the valley in 1810; and the productive power of the town is represented by 76 sets of carding-engines, 58 of which are condensers; 25,508 spindles of self-acting mules, 36,982 spindles of common hand-mules, and 4336 spindles of throstles—in all, nearly 70,000 spindles. Assuming that the annual produce of one set of carding-engines is about L.7500, the annual production of the 76 sets will be L.570,000. Little or no home grown wool is used.

It is impossible to calculate the number of hands employed in the factories ; but with the exception of the building trades, which are partly supported by work in other places, almost the entire population is maintained directly and indirectly by the staple manufacture of the town.

The general social condition seems superior in several respects to what prevails in some larger towns. Such a thing as squalid poverty is totally unknown. The rents paid for dwelling-houses by the operatives are from L.5, 10s. to L.7, 10s., and the scale of dietary is more costly than that of Scotch operatives generally. Building societies, co-operative provision stores, and annual sick and benefit societies, have long existed among the workpeople.

HAWICK.—The burgh records of Hawick bear evidence that manufacturing was conducted there so far back as 1640—how much earlier it is impossible to say, as there are no records prior to that date. But, as the earliest references to the "wobstairs" lead to the impression that they were, in proportion to the population, a numerous class, and show that more than two hundred years ago they were an incorporated body, it may be assumed that the shuttle was plied at the confluence of the Teviot and Slitrig by a peaceable population at the time the mosstroopers of the neighbouring peels occupied themselves in plundering brother-marauders south of the Border, and in trying to hold their own when the compliment was returned by the freebooters of Cumberland and Northumberland. If, however, the Scotts, the Elliots, and the Armstrongs, who followed their chiefs to the foray, recognised no authority but that of the doughty men who led them to plunder and to battle, the Town Council of Hawick appears from its records to have enacted and administered laws for the protection of the property of, and even the recovery of debts due to, the "wobstairs ;" while, on the other hand, care was taken that the weavers should do justly to the public.

Linen was the principal material woven in the middle of the seventeenth century, but woollen plaidings were also made. Many years after 1640, the weavers worked only on their own account to the orders of private customers, manufacturers owning looms and paying men to work them being then unknown. The yarns were prepared by hand cards and spun on the domestic wheel. There is, however, early reference to a fulling-mill; but it was not till about the middle of the eighteenth century that a regular weaving shop was established. It was the property of a company, and linen

N

checks were made in it.   At the same period wool was spun and
sold for manufacture in other towns.   A carpet manufactory was
commenced in 1752.   Weaving of linen tapes was begun by a
company a few years afterwards, and the factory, now a stocking weaving shop, is still known as the Incle House, while the
Under Common Haugh, where the tapes were bleached, is still
termed "the bleachfield" by the older residents in its neighbourhood.   Neither carpets nor tapes have been made for more than half
a century.

As already stated, the manufacture of hosiery was commenced in
1771, and continues to be an important branch of trade, though the
number of frames is not so large as it was some years ago.   Originally
the work was all of the "custom" kind.   The first hosier who made
goods for the general market was Mr John Nixon.   That was in 1780.
Mr Nixon did his own spinning as well as stocking-making ; and as
carding-machines were introduced to Galashiels before they were
worked in Hawick, the wool was sent thither to be carded, and returned to be spun and woven.   In 1804 Mr Nixon built Lynnwood
Mill, where carding and spinning were carried on.   The firm which
he founded ceased to exist only four years ago by the retirement from
business of his sole surviving son and partner, Mr William Nixon.
Twenty-six years ago the younger Mr Nixon, having been left the
only member of his firm, restricted his business to the preparation of
yarns at Lynnwood Mill, and, though he remained owner of the premises, the hosiery department passed into the hands of Messrs Nixon
and M'Kie—a house now represented by the junior partner, and
conducting a large and high-class business.   The premises were sold
in 1868 to another firm in the hosiery trade—Messrs Robert Pringle
& Sons.

Mr William Elliot, late head of the firm now represented by two
of his sons, and of whose establishment, as occupying the leading
place in the Scotch hosiery trade, some particulars have already been
given, began business, as most of the hosiers did, on a limited scale,
about the year 1820.   He subsequently entered into partnership
with the late Mr Thomas Wilson, and continued in that relation for
seven years.   The separation took place in 1830 ; and, after resuming
on his own account, Mr Elliot gradually extended his business till,
many years before his death in 1864, it was the most important in
the country.   It was a common saying that, keen politicians as the
stockingmakers of Hawick always were, they looked forward with
more interest to Mr Elliot's comments on the trade prospects for the
spring, given at his annual factory festival, than to the speech

from the Throne at the opening of Parliament. At first he bought his yarns, and manufactured them, then he went into the wool market and secured the Cheviot clips best suited to his trade, got them spun elsewhere, and made into hosiery in his own works. In 1850 he acquired Stonefield Mill and the Waulk Mill. The latter, understood to be the oldest building of the kind in the place, he pulled down, and placed a large spinning-mill on its site; and since that time all the processes in converting the raw material into hosiery have been conducted by the firm. Like other mill-owners in town, he began the manufacture of piece goods; but the hosiery department always received most attention from him. He preferred it to the other, he used to say, because, "though it was not so profitable, it gave most employment in proportion to the capital invested"—a reason which certainly did honour to his heart.

The first proprietor of carding-engines in the town was Mr William Wilson, father of the late Chancellor of India, of the Ex-Provost of the burgh, and of two other sons who are now at the head of extensive manufacturing firms. Mr Wilson commenced business as a hosier in 1788, and nine years afterwards acquired a lease of part of the Incle Company's property, where he conducted carding and spinning. In 1806 Mr Wilson entered into partnership with Mr William Watson, and the firm considerably extended their mill property. The partnership was dissolved in 1819 and the property divided. Mr Wilson assumed his sons as partners, and the new firm, William Wilson & Sons, was dissolved in 1851, the three sons of the original head continuing in business on their own account. The eldest, Mr Walter Wilson, built a large mill, in which he conducts both the hosiery and tweed trades. The second, Mr John Wilson, who obtained part of the property of the firm, and has since built extensive additions, also conducts both branches. Provost Wilson, the youngest son, who obtained the remainder of the firm's property, is with his partner, Mr Armstrong, engaged in the tweed trade only. The original mill property belonging to Messrs Wilson & Armstrong has been greatly enlarged of late years. Messrs William Wilson & Sons were the first to employ steam-power at their works. Mr Watson also assumed his sons as partners at his separation from Mr Wilson, and the firm of William Watson & Sons is now represented by his grandsons, who lately relinquished the hosiery branch, and devote their entire attention to tweeds. Their property has been largely added to.

The original partners of the firm of Dicksons & Laings, whose works at Wilton Mills are well known to all acquainted with the

Hawick trade, were also among the pioneers of the manufacturing prosperity of the burgh. The Brothers Laing carried on business as hosiers, spinning their yarns on the hand-jenny, before becoming mill proprietors. They entered into partnership with Messrs Dickson, and the first part of Wilton Mills was built in 1809-10. On two occasions the factory was enlarged. The parent building was burned down in December 1867, being then the centre of an extensive range of erections which the increasing business of the firm had called into existence. On the site of the old mill new buildings have been erected. It was at the Wilton Mills that the first spinning-jennies in Hawick driven by water-power were worked, and at the same place that the power-loom made a first appearance in 1830. There are other two thriving businesses, offshoots of the Wilton Mills establishment—that of Mr John Laing, hosiery manufacturer, whose works are particularly mentioned in the account of that trade; and that of Messrs Laing & Irvine, manufacturers and merchants, whose head-quarters are in Hawick, and their mill property at Peebles.

The youngest, but by no means the least important, of the leading Hawick manufactories, is that of Messrs William Laidlaw & Sons. The late founder of this firm conducted a small but thriving hosiery trade for a number of years before he built his mill in 1834, and as the tweed trade assumed importance, he, like the majority of his brother-manufacturers, combined that branch with the making of stockings. Both branches were successfully carried on by himself and his sons, until a few years ago one of the latter, now sole partner of the firm, discontinued the hosiery depart-ment, and made great improvements and additions to his premises, which were thenceforth exclusively devoted to the tweed manufacture. Mr Laidlaw subsequently purchased Lynnwood Spinning Mill from Messrs John Nixon & Sons, where hosiery yarns are still made, so that he is again indirectly connected with the original trade of his firm. His mill property at Teviot Crescent and Lynnwood is among the largest in the woollen trade in Scotland, and certainly the most extensive owned and managed by a single proprietor.

The first kind of woollen cloth made for the market in Hawick was a coarse blue, which was sent to Leeds to be finished. Duffle for petticoats, plaidings, blankets, and flannels were also manufactured during the first three decades of the century; and in 1826 Messrs William Wilson & Sons first used foreign wool in the manufacture of fine flannels.

The number of sets of machines in Hawick is 68. The carding-engines are almost all 60 inches wide. There are 52,864 spinning-spindles, of which 12,564 are self-acting, including spinning-frames; 5894 twining-spindles, half of which are self-acting. There are 270 power and from 100 to 150 hand looms. The weight of wool carded in the sixty-eight sets of machines is 1,801,796 ℔ annually, assuming that the mills work only during the day; but as in busy times the majority of the factories run all night, the quantity is in such seasons much greater. The extent of the hosiery trade has already been stated.

SELKIRK.—The number of manufacturing establishments in Selkirk is seven—two spinning, and five tweed. The tweed firms are Messrs J. & H. Brown & Co., Ettrick Mills; Messrs Waddel & Turnbull, Dunsdale Mill; Messrs George Roberts & Co., Forest Mill; Messrs Dobie and Richardson, Bridgehaugh Mill; and Messrs James Bathgate & Sons—the latter, however, having no carding and spinning machinery. Philiphaugh and Yarrow Mills are carding and spinning establishments only, the former containing four sets of carding machines. Yarrow Mill, which belongs to Messrs Brydone and Brown, was started about the beginning of 1868, with four sets of condensers, self-acting mules, and scouring, burring, teazing, and drying machinery of the newest construction. The three first-mentioned firms have seen the beginning of the tweed trade, properly so called, and have all done much towards bringing it up to the important position it now occupies as a branch of national industry. Messrs Waddel & Turnbull, who had previously rented their premises, became proprietors in 1863, and have since greatly enlarged and improved them. The firm have, in addition to their reputation as general tweed-makers, a well-established name as first-rate producers of heavily milled tartans. Messrs Roberts have the reputation of using superior wools, which of course gives a character to their goods. Messrs J. & H. Brown & Co. turn out, perhaps, more Scotch tweeds than any other firm in the trade. They were the first in Selkirk to introduce, about four years since, the new and improved condensers, a step in which they were closely followed by other firms, till now there are only one "piecing-machine" and two "roving cards," or "perpetuals," in the place. Messrs Roberts were the first to introduce the self-acting mule, an example also largely imitated. A great many other machines for saving labour, or improving processes, have been introduced lately, and large additions made to premises. To a water power of sixty horse, Messrs J. & H. Brown & Co. have recently, by the introduction of a pair of 40-horse power condensing engines, provided a motive force which, in case of necessity, can be raised to 400 horse power. These alterations and additions have also necessitated, on the part of Messrs Roberts and Messrs Waddel & Turnbull, the application of steam power, both having erected powerful engines within the last two years, their water supply having become quite inadequate to the demands made upon it. Except in the case of Philiphaugh Mill, which is driven by water alone, the motive power of the more recently started mills is wholly steam. The following statistics of machines and persons employed in Selkirk have been drawn up with care, and may be relied upon:— Carding-machines, 32 sets; spindles in self-acting mules, 15,612; do. in hand-mules, 12,260; do. in throstles for twisting, 1726; number of power-looms, 181; do. hand-looms, 97; persons employed, 1032. The number of workers appears large compared with the number of machines, but that is explained by the fact, that Messrs Roberts & Co. have a mill at Innerleithen for spinning yarn which is worked up into tweeds at Selkirk; and that Messrs J. & H. Brown & Co. buy all the yarn required for one class of goods, which forms about one-third of their entire production. The weight of wool used will be something like 860,000 or 870,000 ℔ per annum, costing, at present rates, about L.86,000 or L.87,000; and wages paid, between L.28,000 and L.29,000 per annum. The total overturn may be set down at L.220,000.

INNERLEITHEN.—The manufacture of woollen cloth was introduced into Innerleithen in 1790, when a mill was built, at an expense of L.3000, by Mr Alexander Brodie, who was a native of the neighbouring parish of Traquair, and who, having made a fortune as a smith in London, wished to benefit his native district by increasing the field for labour. The project did not succeed to the wishes and expectations of the philanthropic originator, and for many years little progress was made. It was not till between 1830 and 1840 that the woollen manufacture began to take root. The rate of advance may perhaps be best shown by giving the increase in the population of the parish, the whole of which increase is due to manufacturing—the rural part, if it has not decreased, remaining stationary. In 1801 the population of the parish was 609; in 1841, 937; in 1851, 1236; in 1861, 1823; and at present it is computed at between 2400 and 2500. Brodie's Mill, which is now in the possession of Messrs Walker, Gill, & Co., has been much extended; and including two at Walkerburn, which is about one and a half mile distant from Innerleithen, six other mills have been added within the last twenty-five years.   At first water was the only motive power; but from the extent to which hill drainage has been carried on, the supply of water in summer is now very precarious, and to all the factories except two, steam-power has been added. All the mills—except that of Messrs Wilson, who manufacture blankets and plaiding only, and those of Messrs J. & A. Dobson, and Dobson Brothers, who produce a quantity of woollen shirting— are solely engaged in making tweeds.   The following statistics relate to the seven factories of the Innerleithen district:—Sets of machines, 29; number of hand and power looms, 264; spindles, 18,708; workpeople, 700; quantity of wool used, 959,604 ℔ per annum; value of goods produced L.210,900 per annum; wages paid L.24,200 per annum.   The rise of the manufacturing village of Walkerburn has been very remarkable.   Fifteen years ago it had no existence, and now it has a population of upwards of 700, and contains the largest factory on the banks of the Tweed, viz., that of Messrs H. Ballantyne & Sons, the first part of which was erected in 1855.   This mill now contains 10 sets of carding-machines, 80 power and hand looms, 7260 spindles, employs 270 workers, and the quantity of wool annually used in it is 450,000 ℔.   In the same village there is another large factory, which belongs to Messrs James Dalziel & Co.   As it was erected within the last three years in place of another, which was burned down, the machinery is all of the newest and most approved kind.   It gives employment to about 130 persons.

JEDBURGH.—So early as the year 1728 a woollen factory existed in Jedburgh, and it is believed to have been the first in that district. It was erected by the magistrates under the patronage of the Trustees for the Improvement of Manufactures and Fisheries.   The mill soon passed into the hands of a joint-stock company, which received a charter from the magistrates, giving them powers—" 1st, To import and export the several subjects of their manufactories without trouble or molestation.   2d, Forasmuch as, by His Majesty's letters patent, it is appointed that L.700 sterling should be annually employed in carrying on woollen manufactures within the several shires which produced tarred wool, whereof the shire of Roxburgh was one, and that the trustees named in the said patent had selected Jedburgh as one of the stations, they enacted and agreed that the company should have full liberty and allowance to apply to the trustees who had the distribution of the said L.700, to the end that they might receive such a share thereof as might be by them allotted for carrying on the woollen manufactory at Jedburgh.   3d, Power to make bye-laws consistent with the public law of the realm and Acts of Council."   This company did not succeed.   In 1745 a merchant in the burgh named Robert Boswell, took a lease of the mill, at an annual rent of L.7; and so anxious were the magistrates to encourage him, that no dyed or waulked cloth was allowed to be hung on the Canongate Bridge, unless it was waulked or scoured by Boswell. The first manufacturers who were successful in carrying on anything like an extensive trade were Messrs J. Hilson & Sons, into whose hands the mill passed in 1786.   Strong woollen cloth, blankets, and plaiding were manufactured to a considerable extent, and afterwards a large quantity of tartan cloth was made by the same firm, who also began the manufacture of carpets, a branch which was given up many years ago.   Although Jedburgh was one of the first towns in Scotland in which the woollen manufacture was carried on, it has long since been eclipsed by the neighbouring towns of Hawick and Galashiels.   Within the last two years considerable additions have been made to its manufacturing power.   There are five factories, having in all eleven sets of machines.   Rather more than 200 persons are employed, and the value of the annual produce may be set down at something like L.66,000.   Tweeds are the principal goods produced.

DUMFRIES.—The hosiery trade of Dumfries is believed to have originated about eighty years ago, but up till about 1810 it was carried on upon a small scale.   Messrs Haining, Hogg, & Dickson are believed to have introduced the business.   In 1810 the late Mr Robert Scott (who may be called the founder of the woollen trade of Dumfries) entered into partnership with Mr W. Dinwiddie, and to the first-mentioned gentleman is due the credit of having concentrated the hosiery trade in the town, taking it out of the hands of the smaller makers, and opening up an extensive and profitable connection with the London and other English markets.   From that time the trade took a great start, and in making hosiery Dumfries soon rivalled, if it did not surpass, Hawick.   About forty years ago the stocking-making of Dumfries gradually developed into the manufacture of woollen underclothing of all kinds.   In 1832 so many as 300 persons were known to be employed in Dumfries and the neighbouring villages in that branch of business; and 400 dozen stockings, socks, drawers, and shirts were made weekly.   From that time till now the trade in all its branches has been prosecuted with, on the whole, remarkable success.   The leading firms in Dumfries are Messrs R. Scott & Sons; Messrs Milligan & Co.; Messrs James Dinwiddie & Company, Greenbrae; Mr William Halliday, Maxwelltown; Mr Robert Macgeorge, Maxwelltown. Mr Paterson carried on the hosiery business on a considerable scale at Lochfoot, and there are one or two independent makers in a small way in some of the neighbouring villages. There are fully 500 frames or stocking-looms in Dumfries and the district, of which at present about 300 will be in full work, giving employment to at least 500 persons, including weavers, winders, seamers, trimmers, finishers, and warehousemen. The consumption of hosiery yarns by the whole trade of the district is estimated at 120,000 ℔ per annum, in addi tion to a large quantity of tweed yarns. The yarns are obtained chiefly from Hawick, Peebles, Alloa, and Kinross. The capital invested in the trade is about L.40,000, and the annual turnover is, as nearly as possible, represented by the capital. Nearly all the work is paid for by the piece or dozen, a different scale applying to every gauge. Hence, it is difficult to estimate the average weekly earnings; but on the narrow frames they will range from 10s. to 13s.; on the broad frames, from 15s. to 18s.; and on the improved frames, from 20s. to 25s. Altogether, the wages paid in connection with the hosiery trade in Dumfries and district exceed L.11,000 a-year. In addition to plain hosiery there is a considerable manufacture of what are known as fancy goods, including tweed hose, knickerbocker hose, shooting socks, &c., besides a variety of articles in which bright colours tastefully arranged are employed. Messrs Milligan & Co. are the only firm in Dumfries who employ power-looms for the manufacture of the hosiery fabric. Steam-power is used in nearly all the mills for the finishing processes.

The tweed trade of Dumfries dates from the year 1846. Two years previously Messrs Robert Scott & Sons purchased premises at Kingholm, about a mile below Dumfries, for the purpose of spinning hosiery yarns, and afterwards turned their attention to the manufacture of tweeds, which they were soon engaged in making in considerable quantities for the London and Glasgow markets. In 1853 the trade greatly improved, and a much larger quantity of goods was manufactured. Prior to that date the senior partner of the firm (the late Mr Robert Scott already referred to) had retired from the business, which was carried on till 1856 by his sons, Messrs Robert and J. L. Scott. In that year the partnership was dissolved, and Mr J. L. Scott carried on the business at Kingholm, while Mr Robert Scott, who had entered into partnership with his brother, Mr Walter Scott, of Manchester, commenced the manufacture of tweeds in a large and handsome mill erected on ground overlooking the Dumfries Dock Park, and called the Nithsdale Mills. In 1866 the partnership between Messrs Robert and Walter Scott terminated, the Nithsdale Mills becoming the property of the former, who let them on lease to his nephew, Mr Robert Scott, jun., and his partner, Mr Nixon, of London. In 1866 Mr Walter Scott erected an extensive factory on the Stewartry side of the Nith, which is known as the Troqueer Mill. Upwards of 300 hands are employed in it. In 1865 the Kingholm Mills were purchased by a limited liability company, under the designation of J. Lindsay Scott & Co., with a capital of L.80,000. A new factory was recently completed for the manufacture of tweeds by Mr J. M'Ewen Henderson, and is known as St Michael's Mill. There are also tweed factories at Newabbey (Mr Robert Laing); at Cample, near Thornhill (Messrs Arrol & Peace); and at Sandbed, near Lockerbie. Mr Thomas Shortridge, Dumfries, does a large business in finishing for country makers.

The number of sets of carding-machines employed by the three principal firms in or near Dumfries is about thirty. The quantity of wool spun by them is about 800,000 ℔ per annum, and a large quantity of yarns is brought from other towns. The capital invested in the whole tweed trade of the town and district is about L.260,000, which may also represent the annual turnover. About 200 power-looms and 100 hand-looms are in use. The number of persons employed in the tweed manufactures of Dumfries and neighbourhood is upwards of 1000. To the ingenuity and shrewdness of a pattern designer at Kingholm Mills—Mr John M'Keachie—a great deal of the early success of the trade is due. Mr M'Keachie was originally a damask weaver, and secured an engagement at Kingholm when the tweed trade was begun there.

LANGHOLM.—Previous to 1832 the manufacturers of Langholm were extensively employed in the cotton trade. The work was supplied by firms in Glasgow and Carlisle, who had agents in Langholm. Mr David Reid (the father of Mr Reid of Messrs Reid & Taylor) and Mr Andrew Byers (the father of the present partners of Messrs Andrew Byers & Son) were the originators of the woollen trade of Langholm. They began the trade by making shepherds' plaids or "mauds," and "shepherd check" trouserings, which they disposed of in the towns within a circuit of thirty or forty miles. They always made their journeys on foot. This branch of trade was limited at the commencement—one or two hand-loom weavers being sufficient to supply the demand. Langholm is now justly celebrated for the beauty and finish of its "shepherd checks." Messrs T. & A. Renwick had the Langholm Mills for many years, and supplied yarns for the tweed manufacturers. Mr Alexander Renwick, the last surviving member of the firm, was an enterprising and active man of business, and gave a great stimulus to the trade of the town. He had Whitshiels as well as Langholm Mills. At his death in 1851 the latter passed into the hands of Messrs Reid & Taylor, who were the first to begin the tweed trade on a large scale in Langholm. From comparatively small premises, they have now, by almost yearly extensions, one of the most extensive and perfect tweed manufactories in the trade. Messrs James Bowman & Son began tweed manufacturing a little later than Mr Reid and Mr Byers, and by like energetic spirit and good taste, they also have converted a comparatively small business into an extensive one. Mr Byers, senior, built the mill in Buccleuch Square. His sons succeeded him, and added to their business the spinning of yarn, by acquiring Whitshiels Mill. The trade of Langholm cannot be correctly estimated by the number of machines for spinning yarn, as the manufacturers purchase at least as much from other towns as is spun in Langholm. There are seventeen sets of carding-machines, and the yarn procured from other towns for the trade will employ at least other seventeen sets. The quantity of wool used is about 530,000 ℔ annually, and the yarn bought from other towns will represent a similar quantity of wool. The amount of money turned over annually in the tweed trade in Langholm is estimated at over L.200,000. The capital employed is about L.130,000.

AYR.—The principal woollen manufactory in Ayrshire is that of Mr James Templeton, situated in Fort Street, Ayr. This mill, from a very small beginning, has grown to be one of the largest in Scotland of its kind. About the beginning of the present century it was a small cotton-mill carried on by a company. A few years afterwards it fell into the hands of the late Dr Charles (who was for some time Provost of Ayr), who turned it into a woollen mill. It was managed by the late Mr James Templeton, and the principal work carried on in it then was the carding and spinning of wool for the country people of the district. About the year 1821 Mr Templeton bought the premises, which he afterwards greatly extended; and in 1832 began the manufacture of carpets. At that time, in addition to the numerous persons engaged in the different processes of preparing the wool, about sixty carpet weavers were employed. At Mr Templeton's death in 1844, his nephew, Mr James Templeton, the present proprietor, came into possession of the mill, and since that time the works have been gradually extending, till they are now more than twice as large as they were then. The buildings cover a large space of ground, and the machinery, which is of the newest and best construction that could be obtained, is also very extensive. Only two kinds of carpets are produced at the works—superfine and three-ply Scotch. Of these immense quantities are annually turned out, which are readily disposed of in the home market. Besides producing all the yarns for his own manufacture, Mr Templeton does a large trade in carding and spinning yarns for the manufacture of Brussels carpets, which he disposes of to the manufacturers of Kidderminster, Glasgow, and other places. He employs nearly 500 persons. Of these 150 are carpet-weavers, whose average earnings will be about 20s. a-week. This is the only public work in the burgh of Ayr in which a large number of hands are employed. In Mr Reid's manufactory, situated in Russel Street, built seven or eight years ago, from thirty to forty persons are employed in weaving winceys and flannels for Glasgow houses. There are a few other woollen manufactories in Ayr, but on a smaller scale. Messrs Jamieson, whose factory is in Lymond's Wynd, do a considerable trade in making blankets, flannels, plaidings, and various kinds of woollen wearing apparel. A few miles from Ayr, in the parish of Dalrymple, there is also a woollen mill of considerable extent—viz., Skeldon Mill, in the possession of Mr Hammond. About sixty persons are employed there in the manufacture of blankets, plaidings, &c.

KILMARNOCK.—Carpet-weaving is the principal and oldest manuture in Kilmarnock. The first firm engaged in it began in 1777; and the value of the manufacture in 1791 was estimated at L.21,000. After that date the trade took rapid strides; and between 1820 and 1827, a dozen firms were engaged in it. The value of the carpets manufactured in 1837 was L.150,000. Previous to 1848 the number of firms had dwindled down to five (the present number), and the annual value of the produce has since been about L.100,000. Steam-power was first introduced into the works of Messrs Gregory, Thompson, & Co., for the manufacture of Brussels carpets in 1857. This firm are the most extensive manufacturers of Brussels carpets in Scotland. The other firms confine themselves to Scotch carpets. The average wages of the men engaged in the carpet factories range from 15s. to 18s. a-week.

Bonnet-making is next in importance of the woollen manufactures of Kilmarnock. There are six firms engaged in that trade, and the annual value of the goods made is estimated at L.55,000. One firm —that of Messrs Douglas, Reyburn, & Co.—sent out goods in 1867 to the value of L.37,000. By the various firms there are employed 1100 women and girls as knitters, 300 as liners, and 90 men and boys as finishers. In Kilmaurs, two miles off, there are employed 100 knitters, 50 liners, and 20 finishers, and the annual value of the goods made is L.1500. In Stewarton, five miles off, there are employed 1800 knitters, 500 liners, and 200 finishers, and the annual value of goods made is L.90,000. In the three places about L.48,000 is annually paid in wages.

The five carpet firms have spinning-mills of their own (three of them in Kilmarnock, and two in neighbouring villages), and spin entirely for their own looms. Besides, there are in town two other small mills—one for supplying the carpet works chiefly, and the other for the supply of tweed manufacturers. In the latter there are twenty-two men employed, and five sets of carding-machines. In each of the other works there are four sets of carding-machines. A more extensive spinning-mill is situated at Crookedholm, two miles off. Crookedholm Mill employs about sixty men, women, and girls, whose wages are about L.1200 a-year. Their labour produces 187,200 ℔ of worsted annually, the value of which is about L.15,000. The machinery is driven by steam.

Tweeds and blankets are manufactured in Kilmarnock to a small extent only. There is no large factory devoted to these branches; only two or three dozens of weavers scattered and unconnected. One or two of the calico-printing firms, on account of the depression of their own trade, have begun in a small way to manufacture blankets and tweeds.

Plaids and shawls, tartans, and other woollen goods, are manufactured to some extent by one firm, whose principal trade, however, is in fine winceys. At present thirty-two looms are in operation, and the value of the annual production is between L.6000 and L.10,000. There is, besides, an extensive wincey manufactory, which sends out goods annually to the value of not less than L.100,000. 500 looms are in operation in the factory, and about L.9000 a-year is paid in wages.

There is a large wool-combing and spinning factory at Dalry, Ayrshire. It belongs to Messrs Thomas Biggort and Co., and contains 36 carding-engines, which employ from 300 to 400 persons. About 2,000,000 ℔ of wool are used annually.

PAISLEY.—The manufacture of woollen goods was introduced into Paisley about thirty years ago. The trade consisted of clan and fancy checks made into long and square shawls, and piece goods for dresses. It has continued to increase steadily. Scarfs and shirting cloth have also come to form an important part of the woollen manufactures of the town. The Crimean war gave a great impetus to the shirting manufacture, and since then it has been an important branch of local trade. Almost all the harness-shawl houses in Paisley are now engaged in the manufacture of woollen fabrics. Since the falling off in the harness trade, manufacturers have gone more and more into the making of woollen goods. Attempts have been made by one or two Paisley houses to introduce the manufacture of tweeds, but without success. The woollen trade is confined to the branches above enumerated, with the exception of winceys and some other kinds of mixed fabrics. The woollen manufacture is chiefly conducted by hand-looms. It is computed that from 800 to 1000 hand-loom weavers are employed on woollen fabrics in Paisley. The various houses also employ weavers in other districts. Kilbarchan contains about 800 hand-loom weavers, who are entirely employed on woollen fabrics for Paisley manufacturers. About 100 weavers more are similarly employed in adjacent villages—making a total of nearly 2000 employed by Paisley houses in the weaving of woollen fabrics during the busy seasons. The wages of the operatives are low, and have somewhat declined of late years. They average from 8s. to 9s. a-week.

STIRLING AND NEIGHBOURHOOD.—In Stirling two woollen manufactories are in operation. The most extensive is Forthvale Mill, belonging to Messrs John Todd & Sons. The principal branch of the woollen trade for which these works are used is spinning yarns for the manufacture of tweeds, shawls, and fancy stuffs. There are six sets of carding-machines and 6284 spindles employed in the spinning department. The machinery is propelled by an engine of fifty horse power. The quantity of wool (all foreign) used annually is 376,000 ℔, and the value of the annual production is L.30,000. There are sixty-five persons employed.

The Parkvale and Hayford Mills, situated near the village of Cambusbarron, about two miles from Stirling, belong to Messrs Robert Smith & Son, and comprise dyeing, spinning, and weaving by power. The goods manufactured are a superior quality of winceys and other materials for ladies' dresses. Wincey has been brought to the greatest perfection by Messrs Smith. The warps are composed of cotton yarn, which is chiefly spun in Lancashire; the wefts are of wool, the produce of the spinning department of the works. In the weaving factory there are 530 power-looms, and in the spinning department 13 sets of carding-engines. The whole machinery is driven by six steam-engines of 300 horse power in the aggregate. The wools manufactured are English, German, and colonial, and the quantity used annually is 610,000 ℔. The goods made amount in value to from L.170,000 to L.200,000 per annum, according to the price of raw material. There are in all 950 persons employed. The wages paid annually amount to L.19,000.

At Bannockburn there are two extensive works—one owned by Messrs Wm. Wilson & Sons, the other by Messrs J. & W. Wilson. That of Wm. Wilson & Sons embraces spinning, dyeing, and the weaving of carpets, tweeds, and tartans. Fourteen carding-machines are employed. The quantity of wool used annually, including 50,000 ℔ purchased from other spinners, is 680,000 ℔. The value of the annual production is L.80,000; and the number of persons employed is from 500 to 600. Messrs J. & W. Wilson manufacture carpets only. The wool used annually is 500,000 ℔, and the value of their annual production is about L.25,000. They employ 180 hands, including weavers, dyers, and wool-sorters. There are, besides, two small manufactories, in which about 50 persons are employed in weaving tartans and kiltings. The value of their annual production is about L.45,000.

ALLOA.—At Alloa there are four wool-spinning factories, and an extensive business is carried on. At Kilncraigs Manufactory, which belongs to Messrs John Paton, Son, & Co., seventeen sets of carding-engines are in use, and 300 persons are employed. The work done comprises the spinning of knitting, hosiery, and tweed yarns. The yarns known throughout Scotland as " Alloa yarns" are also manufactured at this establishment. At Springfield Mill, belonging to Messrs Thomson Brothers, the work is of a finer description than that done at Kilncraigs, the yarns produced being for making fine shawls, winceys, and tweeds; and there are about 220 persons employed. At Gaberston Mill, belonging to Messrs F. Lambert & Co., both spinning and weaving are carried on. There are five sets of carding-engines, and about 300 persons are employed. The articles manufactured are woollen shawls and shirtings. There are two other spinning-mills in Alloa—the Keiler's Brae New

Mill, belonging to Messrs John Paton, Son, & Co., and Lambert & Co., at which a good business is done in making hosiery and shawl yarns.   Seven sets of carding-engines, and nearly sixty persons, are employed.   Keiler's Brae Old Mill is occupied by Mr Henderson, and contains two sets of carding-engines.   The quantity of wool used in the spinning-mills is about 3,458,000 ℔, and the value of the produce is nearly L.230,000 annually.

TILLICOULTRY.—In 1755 the population of Tillicoultry was 787; in 1793, it was 909.   During that period the woollen trade was in an almost lifeless state; but a reaction soon followed, and in the succeeding thirty-eight years the population was increased by 563.   In twenty years more the census showed an addition of 3210; and in 1861 the population was 5054—giving a total increase of 4145 over that of the year 1793.   In point of antiquity, the Tillicoultry woollen trade ranks among the first, if it was not the first, in Scotland.   Mention is made of its woollen goods in the Cartularies of Cambuskenneth so early as the reign of Mary Queen of Scots.   At that period, and for about two centuries afterwards, Tillicoultry was famous for weaving a coarse woollen cloth called "serge," which is described as a species of shalloon, having a worsted warp and woollen weft. It sold at about 1s. a-yard, and was long known in the Lawnmarket of Edinburgh as "Tillicoultry serge"—indeed, that name appears ultimately to have been applied to all serges made in the district.   Towards the end of the eighteenth century the current of manufacturing enterprise in Tillicoultry seems to have become stagnant, and the making of serges was transferred to Alva; though it would appear from the old "Statistical Account" that a market for Alva goods was not easily obtained, it being a common saying that "a serge web from Alva would not sell in the market while one from Tillicoultry remained unsold."   Notwithstanding such preference, Alva ultimately carried the trade in that class of goods.   In 1792–5 the woollen trade of Tillicoultry appears to have been at its lowest ebb. There were then but twenty-one weavers in the parish; and the stamp-master (who kept no note of the goods) supposed that 7000 ells of serge, and an equal quantity of plaiding, would cover the produce that passed through his hands annually from Tillicoultry. The manufacture of muslins was introduced about that time, but apparently met with small success.   In 1798 or 1800 John Christie, "an ingenious and energetic native of the village," erected the first woollen factory in Tillicoultry.   At a later period he introduced carding-machines with improvements of his own. In 1817 the present firm of Messrs R. Archibald & Sons began business, and soon other woollen factories were erected. The trade at the time was almost solely confined to the production of blankets and plaidings. It is worthy of note that the first " self-acting mule-jenny" and "slubbing billy" made in the kingdom were purchased from the inventor and maker, Mr Smith, of Deanston, by Messrs R. Archibald & Sons, in 1839. The machines are still in the possession of the firm, and in operation.

In these days, when strikes and intimidation are common, it may be interesting to narrate an incident which happened in Tillicoultry in the transition period of its manufactures. When spinning machinery was first introduced to do the work which the wives of the village had formerly done with their "muckle wheels," Mr William Archibald (who possessed the mill now occupied by Messrs Wm. Gibson & Co.) endeavoured to introduce water as a driving power ; and for that purpose erected a dam on the Mill Glen Burn to divert the water to his mill. The wives considered that such a scheme was neither more nor less than "a new way of playing off an old-fashioned trick—taking the bread out of their mouths by taking the work out of their hands ;" and a council of matrons was convened. What transpired at the meeting is not reported ; but the result was that they mustered in a body, and, armed with spades, hoes, pick-axes, pokers, and tongs, proceeded, " without let or hindrance," to the mill-dam, which they speedily demolished.

About 1824 the manufacture of tartans was introduced into Tillicoultry, and such were the enterprise, energy, and taste brought to bear upon it, and the success by which it was attended, that general prosperity prevailed, and that to such an extent as to add in twenty years over 3000 to the population. The tartan trade has undergone a considerable change since then—" clan " patterns, which for many years were paramount, being now almost entirely discarded for "fancy" patterns. Messrs Paton, of Tillicoultry, have long held a high place in the market for these goods ; and their manufactures in tartan have decorated the person of Her Majesty the Queen, and serve as hangings in the Royal Palace at Balmoral. The woollen productions of Messrs Paton are of a varied description, and consist of shawls, tartans for dress, and cloakings of various kinds. The firm employ from 900 to 1000 operatives, and pay annually in wages about L.20,000. They have sixteen sets of carding-engines in operation, each of which puts through wool to the value of about L.2600 —the goods when finished representing about three times that amount. Both water and steam are used to drive the machinery —the former of twenty-five, and the latter of seventy-five nominal horse power.

Messrs R. Archibald & Sons carry on an extensive business in shirtings, shawl goods, tartans, and thin tweeds; and Messrs J. & R. Archibald, Devondale, have long been famous for the excellence of their Scotch tweeds.

In the parish there are twelve woollen factories, containing forty-six sets of carding-machines, and employing upwards of 2000 persons. Besides these, there are nine establishments where hand-loom weaving is carried on, containing in all about 180 looms, and employing nearly an equal number of weavers, whose chief productions are shawls and napkins. In connection with the factories there are 340 hand-looms, and about 230 power-looms. Australian and Cape wools are those principally used in Tillicoultry; neither that grown on the Ochils nor on the Cheviots being suitable for the class of goods manufactured.

The sanitary condition of the factory operatives is highly satisfactory; and the healthy character of the woollen trade is now generally admitted by the Factory Inspectors, as well as by the medical profession. Dr Thomson, late of Tillicoultry, and now of the General Prison, Perth, was the first to write upon the subject, in 1840; and concurrent testimony seems to have emanated from Hawick, Galashiels, and elsewhere, from professional observers, and led Sir James (then Dr) Simpson to institute an investigation. A number of valuable experiments were made by Sir James through Dr Thomson and others, which resulted in demonstrating that the vocation of the operatives in woollen manufactories is a healthy one.

ALVA.—The inhabitants of Alva are almost entirely dependent on the woollen manufactures. The population was 1150 in 1821; 2216 in 1841; and 3282 in 1861; and the trade of the place has increased at a like rate. Blankets and serges were the only goods produced up till 1829, when the manufacture of shawls was introduced. There are nine spinning-mills in the village, employed on yarns for making shawls, tartan dress goods, tweeds, &c. The mills contain thirty-seven sets of carding-engines, driven by steam and water power. The number of persons employed is about 220. The amount of raw material put through in the course of a year is valued at about L.123,000. Some of the yarn is used in the place, but the greater part of it goes to agents in Glasgow.

The weaving of shawls, handkerchiefs, plaids, and shirtings is the staple trade of the village, and gives employment to about 693 journeymen and 99 apprentices in the busy season, besides from 500 to 600 women who do the winding, twisting, and finishing. A number of young boys are also employed as drawers and twisters. Since shawls and tartans ceased to be fashionable articles of female attire, and since the closing of the ports of the United States to our manufactured goods, trade has been limited to a few months of the year; and that circumstance presses hard on those employed in the weaving business, who generally seek work during the winter months in Hawick, Selkirk, Galashiels, &c.

The value of the manufactured goods runs from about L.200,000 to L.250,000 annually. The chief market is Glasgow, but a considerable quantity also goes to Manchester, London, and some of the principal Irish towns.

At Menstrie—a small village two miles distant from Alva—there is a large factory known as Elmbank Mill, owned by Messrs Drummond & Johnstone. It contains eight sets of carding-engines, driven by a steam-engine of ninety horse-power. The machinery is of the most recent construction. The raw material used annually is valued at about L.33,000, and the goods manufactured at double that amount.

KINROSS.—This town has changed its staple trade several times. The manufacture of linen attained some importance in the middle of last century, and in 1790 employed nearly 200 looms, while the value of the produce was about L.5000 annually. Sixty years ago, cotton weaving was introduced and flourished for a time, until machinery driven by steam or water power was set up and superseded the labour of the hand-loom weaver. Previous to the introduction of these manufactures, Kinross was famous for its cutlery; but that trade is now extinct. It is upwards of thirty years since the manufacture of woollen goods was begun in the locality. From about the year 1836 up till 1845, weaving was plentifully supplied from Tillicoultry and by local manufacturers. In 1846 a wool-spinning mill was erected at the south end of the town for spinning yarn for the manufacture of shawls, &c.; and, by dint of good management, it has been kept in active operation. Another mill, at Bellfield, had been started several years previously, and is still at work. About that time, too, the greatest activity prevailed in the manufacture of shawls and plaids, which branch was carried on till 1848, when it received a material check. A gradual decline then set in, and now all the factories, as such, have ceased to exist, the largest having been recently converted into a printing-office. As a consequence of these changes, woollen weavers have had very irregular employment; and most of the original manufacturers, from various causes, do comparatively little in the manufacture of woollen goods. In the spinning of woollen yarns, again, the case is different, for not only have the two mills referred to been regularly employed, but another large mill (under the Limited Liability Act) has been recently erected opposite the old one on the South Queich; and a fourth, superior to all the others in size, was erected in Milnathort the other year.

ABERDEEN.—Machinery was first introduced into the woollen manufactures in Aberdeen about 1789 by Mr Charles Baird, silk dyer, who brought from England two carding-engines and four spinning-jennies. He erected a mill at Stoneywood, a few miles from Aberdeen. Previous to that time the carding and spinning of wool were performed by hand. During the eighteenth century an extensive trade was done in the manufacture of stockings for the home and foreign markets—that being, in fact, the staple trade of the city. In 1703 a company was formed for the purpose of carrying on the trade on an enlarged scale, and during the greater part of the century stocking-making was prosecuted most successfully—most of those engaged in it retiring upon competent fortunes. Perhaps the best known, because the oldest, house in the woollen trade in Aberdeen is that of Messrs Alexander Hadden & Sons, Green. Established about the year 1748, this firm have long held a foremost place in the trade. Their works are situated in the centre of the city, and are of considerable extent. The mills at Garlogie and Don also belong to them. They consume about 2,000,000 ℔ of wool annually, and give employment to upwards of 1400 hands. The tweed trade is principally in the hands of Messrs J. & J. Crombie, Grandholm Mills, and Messrs Hadden. The manufacture of tweeds is now carried on extensively. Messrs Crombie have long been engaged in the tweed business, and their works at Grandholm are well known. They give employment to nearly 600 persons. The value of the tweeds manufactured in and around Aberdeen is upwards of L.120,000 annually. Messrs Hadden are the largest manufacturers of carpets in Aberdeen. The trade has been a growing one, and now the annual value of the productions is about L.50,000. The manufacture of wincey, for which the city is widely known, is in the hands of seven or eight firms, some of which do a large business in the home and foreign markets. The annual value of the wincey goods produced is at least L.250,000. The hosiery trade is rather on the decline in Aberdeenshire, and has been so for some years. The value of the hosiery made annually is over L.20,000. The manufacture of shawls is but in its infancy in Aberdeen, and does not call for more particular notice. There are at least 230 power and 600 hand looms at work in the woollen trade. It is estimated that upwards of 3000 persons find employment by this industry, and that upwards of 3,000,000 yards of tweeds, winceys, &c., are produced annually.

INVERNESS AND NEIGHBOURHOOD.—There are three woollen manufactories near Inverness. They are situated at Holm, Avoch (in Ross-shire), and Culcabock. The first-named—carried on by Messrs Nicol & Co.—has been in existence for seventy years, and is the oldest in the north. The goods produced are tweeds, mauds, plaiding, and blanketing, the greater part of which is made from wools grown in the northern counties. A considerable quantity of colonial wool is also used in the manufacture of tweeds. The number of operatives employed at Holm Mills is 100 ; at Avoch, from 50 to 60; and at Culcabock (formerly a meal mill, and recently converted into a woollen factory), 15. At Holm both water and steam power are used. There are three sets of carding-engines, and spinning and twisting machines. At Avoch two sets of carding-engines are in operation. At Culcabock there is only one set of carding-machines, and the weaving is done with hand-looms. The quantity of both home and foreign wool used yearly at the three factories is about 212,000 ℔.

The woollen manufactures indigenous to the Highlands are home-made cloths and tweeds, and hand-knitted hosiery and shawls. For these there is always a ready market. The shawls are valued on account of their lightness and warmth, and the home-spun tweeds are prized for their peculiarly comfortable and durable qualities, which render them more suitable than any other fabric for the use of sportsmen and tourists. To appearance these tweeds are somewhat coarse, but in reality they are softer and in some cases of finer texture than machine-made goods. Messrs Macdougall & Co., of the Royal Tartan Warehouses, Inverness and London, have for many years given special attention to the improvement and development of these tweeds and tartans, and have brought them into notice with the higher classes of society, thus obtaining for them their present celebrity.

LINEN AND JUTE MANUFACTURES.

FROM the frequent mention of linen in the history of Scotland, it is
evident that the inhabitants were acquainted with the processes of
making cloth from flax six hundred years ago at least. It is related
that, at the battle of Bannockburn (fought in the year 1314), "the
carters, wainmen, lackeys, and women put on shirts, smocks, and
other white linens, aloft upon their usual garments, and bound
towels and napkins on their spears, staves, &c. Then placing themselves in battle array, and making a great show, they came down the
hillside in face of the enemy with much noise and clamour. The
English, supposing them to be a reinforcement coming to the Scots,
turned and fled." There is good reason for concluding that the
linen so successfully displayed on this memorable occasion was
home-made. At first the flax was grown, dressed, spun, and woven
by the people for their own use ; but towards the close of the
sixteenth century linen goods formed the chief part of the exports from
Scotland to foreign countries. About the same time a considerable
quantity of Scotch linen found its way into England. Several
attempts were made to establish linen manufactories, so that the
trade might be extended and carried on more profitably; but the
promoters, though encouraged by royal favours and the concession of

certain privileges, did not succeed. Efforts were made to improve and extend the woollen manufactures of Scotland by various legislative enactments, one of which prohibited the importation of woollen cloths from England. The English people retaliated for this interference with their trade by treating the men who sold Scotch linen in their territory as malefactors, whipping them, and making them give bonds that they would discontinue the traffic. This told seriously on the working population of Scotland, for it was calculated that from 10,000 to 12,000 persons were employed in making linen goods for the English market. An appeal to the king had the effect of removing the restrictions on the trade. In 1686 the first Parliament of James VII. passed an "Act for Burying in Scots Linen," the object of which was to encourage the linen manufacturers in the kingdom, and prevent the exportation of the monies thereof by importing linen. It was enacted that "hereafter no corpse of any persons whatsoever shall be buried in any shirt, sheet, or anything else except in plain linen, or cloth of hards, made and spun within the kingdom, without lace or point." Heavy penalties were attached to breaches of the Act, and it was made the duty of the parish minister to receive and record certificates of the fact that all bodies were buried as directed.

It would appear that the weavers, in order to increase their gains, had, towards the end of the seventeenth century, begun to make linen cloth of inferior quality, and Parliament interposed to put a stop to that practice. In 1693 an Act was passed "anent the right making and measuring of linen cloth." It set forth that "the King and Queen's Majesties considering how much the execution of the good laws for the right making of linen cloth hath been hitherto neglected, to the prejudice of the lieges, and the loss of trade within this kingdom, do therefore, with advice and consent of the Estates of Parliament, ratifie, approve, and confirm all Acts of Parliament made for the right making and improving of linen cloth." The Act then proceeds to describe minutely how yarn is to be made up and sold, and how the cloth is to be woven and measured; and this in consideration of "how much the uniform working and measuring of linen cloth may raise the value thereof with natives and strangers, and render the trade more easy and acceptable to merchants." In order to afford protection against dishonest work, the Act required "that the owner of all linen cloth made for export, before it be exposed to the first sale, shall be obliged to bring the same to a royal burgh where linen is in use to be sold, there to receive the public seal and stamp of the burgh, bearing the coat-of-arms of the burgh upon both the ends of ilk piece or half piece thereof, which shall be a sufficient proof of the just length and breadth, evenness of working, and the due and sufficient thickness and closeness thereof ; and for that effect there shall be in each royal burgh where linen is in use to be sold, an honest man well seen in the trade of linen cloth appointed to keep the said seal for marking linen therewith." The fees to be charged by the stampmaster were also fixed by the Act, and he was subject to penalties if he neglected his duty. For the encouragement of all persons who should establish manufactories of linen cloth it was further " statute and ordained, that all lint, flax, and linen yarn imported for the use of companies or manufactories, and all linen cloth exported by them, shall be free of custom duties or excise."

The linen manufacturers of Scotland derived great advantage from the union with England. The duties charged on goods exported to the sister kingdom were removed, and at the same time the colonies were opened to Scottish enterprise. A period of great industrial activity set in, and the quantity of linen goods produced was much increased. In 1710 upwards of 1,500,000 yards of linen cloth were produced. Ten years afterwards England alone took L.200,000 worth of Scotch linen annually. A great stimulus was given to the trade by the establishment of the Board of Manufactures in 1727. The fifteenth section of the Treaty of Union with England, signed in July 1706, stipulated that " L.2000 per annum for the space of seven years shall be applied towards encouraging and promoting the manufacture of coarse wool within those shires which produce the wool," and that " afterwards the same shall be wholly applied towards the encouraging and promoting of the fisheries and such other manufactures and improvements in Scotland as may most conduce to the general good of the United Kingdom ; and it is agreed that Her Majesty (Queen Anne) be empowered to appoint committees who shall be accountable to the Parliament of Great Britain for disposing the said sum." No action was taken to fulfil the conditions of this clause of the Treaty until 1727, when an Act was passed for the appointment of twenty-one commissioners to take charge of the revenues and annuities allotted to the encouragement of manufactures and fisheries. By that time the money which was to be devoted to the improvement of the woollen manufactures had accumulated to the sum of L.14,000 ; while L.6000 in addition was due for the other purposes referred to in the section of the Treaty under notice. The interest on those sums, added to an annuity of L.2000, placed a considerable amount at the disposal of the " Board of

Trustees for Manufactures," as the commissioners were designated, who laid before the King in council a triennial plan for the apportionment of the revenues.

The first plan prepared was for the three years from Christmas 1727, and provided for the expenditure of L.6000 yearly in the following proportions :—For the herring fisheries, L.2650 ; for the linen trade, L.2650 ; and for spinning and manufacturing coarse tarred wool, L.700. The money allotted to the linen trade was divided as follows :—Premiums for growing lint and hemp seed at 15s. per acre, L.1500 ; encouraging spinning schools for teaching children to spin lint and hemp, L.150 ; prizes for housewives who shall make the best piece of linen cloth, L.200 ; salaries to the general riding officers at L.125 each, L.250 ; salaries to forty lappers and stampmasters at L.10 each, L.400 ; expenses of prosecutions, L.100 ; procuring models of the best looms and other instruments, L.50. It would appear from this that technical education is not such a new thing in this country as some persons suppose—the spinning schools referred to being places in which a technical knowledge of a certain branch of industry was imparted to young persons. The sum of L.10 a-year was allotted to the endowment of each seminary, of which sum the teacher or mistress received L.5 as salary ; L.4, 1s. 8d. was devoted to the purchase of fourteen spinning wheels, at 5s. 10d. each ; 5s. to maintaining pirns, bands, &c. ; and the balance, 13s. 4d., went to provide coal and candles for the session, which lasted from the 13th October to the 15th April. The spinning schools were situated chiefly in the Highlands, as the trustees considered it highly desirable to create habits of industry in those regions where indolence and poverty reigned supreme.

The Board lost no time in taking steps for improving the quality of the linen made in Scotland ; and their records show, that one of their first acts was to propose to Nicholas d'Assaville, cambric weaver, of St Quentin, France, to bring over ten experienced weavers of cambric, with their families, to settle in this country, and teach their art to others. The offer was accepted, and the Board purchased from the Governors of Heriot's Hospital five acres of ground in Broughton Loan, a suburb of Edinburgh, on which they built houses for the French weavers. The colony was named Little Picardy, and its site is now occupied by Picardy Place, which, with York Place, forms the eastward continuation of Queen Street. The Frenchmen were Protestants, and they began operations in 1729—the men to teach weaving, and their wives and daughters the spinning of cambric yarn. A man skilled in all the branches of the

linen trade was at the same time brought from Ireland, and appointed
to travel through the country and instruct the weavers, and others,
in the best modes of making cloth.

It may be interesting to note a few facts contained in the minutes
and annual reports of the Board. In 1728 premiums were offered
to persons who should construct bleachfields. Several persons offered
to make fields; and it was agreed that they should receive L.50 for
each acre of ground so laid out. Considerable sums were also paid
for the introduction of improved modes or appliances for dressing
flax. A dispute arose at Irvine in 1732 as to the adjudication of
the housewives' prize for making linen cloth. On reference to the
Board, the prize was given to the wife of the minister of Dreghorn.
At that time the linen manufacture was reported to be in a flourishing
condition, and it went on steadily increasing till 1740. During that
year the manufacture of coarse linens met with a serious check from
the severe frost which prevailed in the winter season. The weavers
were badly provided with houses, and were unable to work during
the frosty weather. That circumstance, coupled with the high price
of provisions, led to many of the men leaving their employment and
enlisting in the army. In 1745, L.50 was awarded to John Johnston
for the invention of an ingenious method of throwing the shuttle in
broad looms. In 1750 premiums for sowing flax-seed were discontinued, owing to want of funds. An Act of Parliament was passed
in 1753 giving L.3000 per annum for nine years (in addition to
the L.2000 formerly granted) to the trustees, to be applied by them
for encouraging and improving the manufacture of linen in the
Highlands. No part of the said sum was to be given for any other
use than instructing and inciting the inhabitants of that part of
Scotland to raise, prepare, and spin flax and hemp, and to weave
the same into coarse linens. This was regarded as a judicious act,
calculated to wean the turbulent Highlanders from their feudatory
propensities, and to impart a spirit of industry to them. With a
view to the proper administration of the fund, the surveyor of the
Board made a tour of inspection to several districts of the Highlands,
and the report he wrote on the condition and manners of the people
excited much attention, as it revealed the existence of a state of
matters little removed from barbarism. In 1755 the Trustees reported that the cambric manufacture established in Edinburgh by
foreign weavers had not succeeded, the prohibition against importing and wearing French cambric having increased smuggling,
and thrown great quantities of French cambric into the country
duty free. The Trustees opened a Linen Hall in Edinburgh in 1766

for the reception and sale of goods; and for nearly five and twenty years the hall served its purpose of accommodating the trade. In 1790 a representation was made to the Board to the effect that the manufacturers did not then consider the hall to be any advantage, and therefore it was closed. There were 252 lint mills in Scotland in 1772, distributed as follows :—Aberdeen, 7 ; Ayr, 22 ; Banff, 8 ; Caithness, 1 ; Dumfries, 1 ; Dumbarton, 16 ; Edinburgh, 2 ; Elgin, 3 ; Fife, 11 ; Forfar, 31 ; Haddington, 1 ; Kincardine, 2 ; Kinross, 5 ; Lanark, 31 ; Linlithgow, 4 ; Perth, 73 ; Renfrew, 3 ; Ross, 3 ; Stirling, 28. It was reported in 1773 that several new kinds of manufacture had been introduced—such as the making of gauzes and thread at Paisley ; while the spinning of silk, wool, and cotton, had been considerably extended. In 1787 a premium of L.100 was awarded to Mr Patrick Taylor, Edinburgh, for introducing a mode of figuring linen floorcloth. Many improvements in machinery, &c., are noted, and frequent mention is made of the introduction of the modes or appliances used in other countries. In 1790 a great step in advance was made by Messrs James Ivory & Co., who erected at Brigton, Kinnettles, Forfarshire, a mill for spinning yarn by machinery driven by water power. The trustees resolved to reward the enterprise of the firm by awarding to them a premium of L.300 ; but, in consequence of some matter affecting the patent, they subsequently withdrew the award. In the same year the trustees purchased the vested rights of the foreign weavers in Little Picardy. The weavers had found it necessary, on account of the cambric trade not succeeding, to apply themselves to other occupations. By the close of the century the spinning schools would appear to have accomplished the purpose for which they had been originated, as in the year 1800 the Board refused an application from Sir John Sinclair to have spinning schools established in Caithness, the grounds of refusal being that spinning was then so generally known and so easily acquired as to render schools for teaching it no longer necessary. The awards of the Board were not always made in the strict spirit of the constitution, as in 1802 they gave ten guineas to a man in Alyth " for his ingenuity and industry in weaving with a wooden arm and hand." The premiums offered to the linen trade in 1807 were for the best and second best ravens-duck, shirting, diaper, huckaback, plain linen, &c. There were eleven prizes in all, and five of the successful competitors belonged to East Wemyss, three to Dunfermline, two to Edinburgh, and one to Culross. It was reported in 1821 that the crop of flax had decreased very much, owing to the low price current. In 1822 the king approved of L.15,000 being expended on building offices for the Trustees at the north end of the Mound, Edinburgh. The building, now called the Royal Institution, was completed in 1828, at a cost of L.20,424. The abolition in 1823 of the law relating to the stamping of linen in Scotland curtailed the functions of the Trustees. The manufacturers had frequently urged the injurious effect of the operations of the Board, and were ultimately successful, as stated, in having all legislative interferences with the trade abolished.

A record is preserved of the quantity and value of linen cloth stamped in each year during which the jurisdiction of the Board of Manufactures extended to the trade. The figures relating to every tenth year up till 1818, as well as those for the last four years in which the stamp-laws were in force, are subjoined :—

Years.	Number of Yards.	Estimated Value.		
1728	2,183,978	£103,312	9	3
1738	4,666,011	185,026	11	9
1748	7,353,098	293,864	12	11
1758	10,624,435	424,141	10	7
1768	11,795,437	599,669	4	2
1778	13,264,410¾	592,023	5	4½
1788	20,506,310⅛	854,900	16	2¼
1798	21,297,059	850,403	9	9
1808	19,390,497	1,014,629	18	4
1818	31,283,100¼	1,253,528	8	0½
1819	29,334,428¼	1,157,923	4	11
1820	26,259,011¼	1,038,708	18	5
1821	30,473,461½	1,232,038	15	4¾
1822	36,268,530½	1,396,295	19	11½

The above figures show the development of the trade under the encouraging influences extended to it; but the credit does not lie altogether with the Board of Manufactures. The British Linen Company, incorporated at Edinburgh in 1746, did much good by advancing money to the manufacturers, and helping them to dispose of their goods. The company originated with the Duke of Argyll, and other noblemen and gentlemen, who, finding that the linen manufacturers were frequently placed in a position of difficulty by the fluctuations of the market for their goods, and that sales had sometimes to be made under value in order to raise money to meet pressing engagements, resolved to form a company for trading in all branches of the manufacture. With a capital of L.100,000, the subscribers of which were actuated solely by patriotic motives, the company imported flax, linseed, and potashes, which they sold on credit to suitable persons, afterwards buying at a fair price the yarns and linens made from the material supplied in that way. The company had warehouses in Edinburgh and London, in which they stored their purchases, and thence disposed of them by exportation and otherwise. After a time, the company came to think that they could best promote the branch of industry to which their attention was specially devoted, by advancing money to manufacturers, and allowing them to prosecute the trade on their own account, free from the competition of an incorporated body. The company accordingly suspended operations as dealers in linen, and adopted banking as their sole business. In the latter connection the incorporation still exists, retaining its original designation of "The British Linen Company."

For several years after the repeal of the Act anent stamping linen, a system of inspection was in operation; but it was entirely voluntary. The inspectors, in most cases, were the same persons who had acted as stampers under the Act, and so were generally well qualified for the work. Manufacturers either took their cloth to the inspector, or, as was more commonly the case, got the inspector to go to their factories. If the inspector was satisfied with the quality of the cloth he stamped it with his own name. Such a system was liable to abuse, however, and the stamps of the inspectors soon lost whatever value they had. Merchants became better acquainted with the quality of the goods they bought, and were content to deal according to their own judgment, without the intervention of inspectors.

It would appear that linen was an article available in making payments of rent in kind; for the rental lists of the Marquis of Huntly show that in May 1600 the payments of that description included 990 ells of linen. Dressing and spinning lint formed an important part of the domestic duties of the wives of farmers and cottars in those days. In an account of a tour in the Highlands of Scotland made by an Englishman in 1618, it is stated that "the houses of the gentry are like castles, and the master of the house's beaver is his blue bonnet; he will wear no shirts but of the flax that grows on his own ground, or of his wife's, daughters', or servants' spinning; his hose, stockings, and jerkins are made of his own sheep's wool." Sixty years later another visitor wrote :—"But that which employs great part of their land is hemp, of which they have mighty burdens, and on which they bestow much care and pains to dress and prepare it for making linen, the most noted and beneficial manufacture of the kingdom." A third visitor, who came in 1725, wrote as follows:—

"Many of the Scotch ladies are good housewives, and many gentlemen of good estate are not ashamed to wear the clothes of their wives and servants' spinning." Among some notes of the manners and customs of the people of Scotland, written by a lady who was born in 1714, is the following:—"Linens being everywhere made at home, the spinning executed by the servants during the long winter evenings, and the weaving by the village webster, there was a general abundance of napery and underclothing. Every woman made her web and bleached it herself, and the price never rose higher than 2s. a-yard, and with this cloth almost every one was clothed. The young men, who were at this time growing more nice, got theirs from Holland for shirts; but the old ones were satisfied with necks and sleeves of the fine, which were put on loose above the country cloth. Table linens were renewed every day in gentlemen's families, and table napkins were always used. A few years after this, weavers were brought from Holland, and manufactories for linen established in the west. Holland, being about 6s. an ell, was worn only by men of refinement. I remember, in the year '30 or '31, of a ball where it was agreed that the company should be dressed in nothing but what was manufactured in the country. My sisters were as well dressed as any, and their gowns were striped linen at 2s. 6d. per yard. Their head-dresses and ruffles were of Paisley muslins, at 4s. 6d., with fourpenny edging from Hamilton—all of them the finest that could be had. At this time hoops were constantly worn four and a half yards wide."

With reference to the linen trade, Mr Patrick Lindsay, in his book on the "Interest of Scotland," already quoted, expresses himself strongly on the "woeful neglect" with which it was treated at the time he wrote, in 1733; and he makes suggestions for remedying a state of matters so undesirable. The following extract contains some of his views :—

"If all our spare and idle hands were employed in the linen, and thereby enabled to live comfortably by their own labour, and to bring in a little wealth to the country, the improvement of our other manufactures might be safely left to themselves, for it is more our interest to be served with several kinds of goods from England, so long as they are bought cheaper in England and our linen sells to advantage there, than to be overstocked in any branch of business which we cannot export; and in this our greatest danger lies. Many of our young joiners and other young tradesmen go now and then to the plantations for want of suitable encouragement at home. Were all these supernumerary tradesmen bred to be linen weavers, how much might this valuable manufacture be increased, by employing in it so many more hands. As manufacture was in no esteem, men of fortune thought it beneath them to breed their children to any business of that sort; and, therefore, since war ceased to be our chief trade, the professions of law, physic, the business of a foreign merchant and shopkeeper, reckoned the only suitable employments for persons of birth and fortune, have been greatly overstocked. Several young men, bred to no business, pretend to turn merchants, and follow trade in the smuggling way, and thereby do great hurt to the fair trader, and to their country, and in the event ruin (for the most part) themselves. After the Revolution many churches continued vacant for several years, and young men were no sooner qualified for the ministry, than they were sure of a settlement; and even too many were admitted (to the discredit of the profession) before they were so well qualified for it as the dignity of the office requires. Our Church livings are but small, and therefore few people of rank or any condition educate their sons for clergymen; whereby these many vacancies were a great temptation, and an encouragement to people of low rank to follow that profession. One bad effect of this way of supplying vacant churches to the public is, that as these clergymen have nothing but their stipends to depend upon, unless they are frugal beyond measure, and parsimonious to a fault; if they have wives and children, these must be left indigent, as burdens upon the public. The case is now much altered as to vacancies, for at present we are so overstocked with young clergymen, that one-half of the probationers who are now candidates for the supplying of churches as they fall vacant can never in reason hope to be provided for. The public suffers greatly under this heavy burden of so many idle and useless hands; and of all professions, an unemployed clergyman is the most helpless and useless member of society. . . . . . Thus it is evident that every profession, and every trade (except the linen) is, and is very liable to be, overstocked in numbers; but the linen trade, if duly improved, is sufficient to employ our supernumerary hands, and can never be overstocked. . . . . . The linen manufacture may be brought to as great an extent in value as any other business now carried on in Britain, except the woollen; and it may employ near as many hands as the woollen does. And the linen trade of the north is of as great consequence to the nation in general as the woollen in the south, and equally deserves the same care, countenance, and encouragement from the public."

When the Board of Trustees for Manufactures began operations in 1727, the manufacture of linen was carried on in twenty-five counties of Scotland, the quantities produced in each varying from 65 yards (valued at L.3, 7s.) in Wigtownshire, to 595,821½ yards (valued at L.13,989, 10s.) in Forfarshire. Perth, Fife, and Lanark came next in order. Subsequently, linen was made in all the counties except Peebles. Forfarshire kept the lead all through, and still occupies the foremost place. In 1822—the last year in which the stamp-laws were in force, and, consequently, the last respecting which any accurate statistics exist—the chief seats of the trade and the quantities of linen stamped were as follow:—Forfarshire, 22,629,553½ yards; Fifeshire, 7,923,388¼ yards; Aberdeenshire, 2,500,403¾ yards; Perthshire, 1,605,321 yards; Kincardineshire, 632,896 yards; Inverness-shire, 318,465 yards; Cromarty, 297,754 yards; Edinburgh, 129,709 yards. During the past thirty or forty years, the manufacture of linen has died out in many towns and villages in which it at one time formed the chief branch of industry, and has been drawn together into the counties of Forfar, Fife, and Perth.

The following notes, drawn chiefly from the "Statistical Account of Scotland," will show how the manufacture was dispersed over the country seventy or eighty years ago; how the people of some districts failed to take it up; and again how it grew and flourished for many years in certain towns in which it is now unknown. Beginning with the "far north," it is recorded that about the year 1790 an attempt was made to introduce the linen manufacture into Shetland, but without success. As the people could purchase linen cheaper than they could make it, they did not take kindly to the new industry; and, besides, their habits and constitutions would appear to have been ill-suited to the vocation, for it is said that "the fair sex were so accustomed to roam about the rocks, that they could not apply themselves with diligence to the manufacturing business; and the constant sitting was said to have brought on hysterical disorders." In Orkney the case was different. The making of linen yarn from home-grown flax was introduced in 1747, and in course of time the trade spread over nearly all the islands. The yarn made acquired a good name in the southern markets, and from 1750 till 1785 about 250,000 spindles were exported annually. After that time the trade gradually declined, and it was abandoned about the close of the century. Weaving was introduced at the same time as spinning, but it never attained much importance. The greatest quantity stamped in any year was under 30,000 yards. The cloth was sold in Edinburgh, Glasgow, and Newcastle, at an average price of eleven pence a-yard. The chief cause of the decline of the trade was the low price paid by the country agents for spinning and weaving. It is said that latterly the most expert spinners could not earn more than twopence a-day. The substitution of linen underclothing for home-made woollen shirts and vests was alleged to have seriously affected the health of the people, colds and rheumatism having become much more common among them. Before the sea fishing received much attention from the inhabitants of Caithness, and before the now famous pavement quarries of that county were opened up, the making of linen cloth, and other domestic industries, were carried on by the people, but chiefly to supply their own wants. The farmers grew small patches of flax, which supplied the raw material, and in course of time quantities of dressed flax were imported. At Thurso a large number of persons were employed towards the end of last century in spinning flax for the south-country merchants. The Custom-House books show that, in 1794 and the two following years, 253,749 ℔. of dressed flax were brought to Thurso, which would produce 162,342 spindles of yarn. The spinners were paid at the rate of 1s. and the agent 2d. a-spindle. From this it would appear that, in the three years mentioned, the total sum paid for spinning, &c., was L.9294, no inconsiderable amount to be set loose in those days in a poor district of the country. In 1851 an attempt was made by Mr Peter Reid, proprietor of the *John o' Groat Journal*, to revive the cultivation of flax in Caithness. Mr Reid erected at Wick a mill for dressing flax by Schenck's process, and with the aid of the Caithness Agricultural Society, got a number of farmers to devote an acre or two of ground to raising flax. He furnished the seed, paid a rent for the ground, and gave prizes to those who produced the heaviest crop per acre. At the time Mr Reid was induced to take up the trade, agricultural affairs in Scotland were not in a prosperous state. Oats could be purchased at 13s. per quarter, and oatmeal at 11s. per boll; and the proposal to cultivate flax was hailed as likely to improve the prospects of farmers. In Aberdeenshire, Fifeshire, and Lanarkshire, mills similar to Mr Reid's were built, and inducements were offered to farmers to undertake the cultivation of flax, and for a time it was thought that matters would turn out advantageously for all concerned. The hopes which had been raised were not realised, however, for in the course of a year or two the price of grain rose, and the farmers refused to have anything more to do with flax, which they considered a troublesome crop, and not so remunerative as they had expected. The failure of the enterprise was a sad blow to the mill-owners, and

P

in Caithness especially was the subject of much regret, for in that county there is a scarcity of occupation during greater part of the year, and the flax mill was expected to relieve in some measure the overstocked labour market. During the few years that Mr Reid's mill was in operation, the flax scutched at it fetched from L.50 to L.60 a-ton, a higher price than was obtained for flax prepared at mills situated in more favourable localities. About the close of the eighteenth century the spinning of linen yarn from flax imported from the Baltic was carried on in Sutherlandshire, but on a very small scale. Some linen cloth was also woven for home use, and occasionally, when the supply exceeded the demand, a few hundred yards were stamped for sale. In Ross-shire flax and hemp were at one time cultivated. The flax was dressed, spun, and woven to an extent which sufficed for local requirements, and about L.500 worth of cloth was exported annually. The hemp was converted into canvas and cordage for the fishing boats of Avoch and the neighbourhood. Though the trade is now extinct in Cromarty, as well as in the counties mentioned above, it would appear from the stampmaster's returns that the inhabitants were at one time pretty extensively engaged in making linen goods. In 1758 about 7950 yards were stamped; but during the thirty years following there was a considerable falling off, followed, however, by a somewhat sudden and extensive revival. Thus, while the number of yards stamped in 1788 was 4656½, valued at L.186, 8s., in 1822 the figures were—yards stamped, 297,754; value, L.13,461, 17s. At Inverness—a town which possesses great natural facilities for carrying on manufactures, though these have been very little taken advantage of—a large hemp factory was established in 1765, and for some time was so prosperous as to employ 1000 hands. The hemp was brought from the Baltic, and was chiefly converted into sacking and tarpauling cloth, a considerable portion of which was sent to the West Indies to be used in covering bales of cotton. The factory is still in existence, though the business done is not so extensive as it once was. About the year 1780 an enterprising company began the manufacture of linen thread, and for a number of years remarkable success rewarded their efforts. They gave employment to 10,000 persons throughout the county, most of whom worked in their own homes, their labours being superintended by district agents, of whom there were nineteen. The earnings ranged from 1s. to 12s. a-week. The flax was obtained from the Baltic ports; and when the thread was finished, it was forwarded to London, and thence dispersed over the world. The trade was taken up in some other towns, the social and commercial circumstances of which were more favourable to its prosecution; and many years ago, Inverness retired from competition with them. In the first year following the passing of the Stamp Act, 10,696 yards of linen were stamped for sale in Inverness-shire, and the quantity made increased gradually, until in 1822 it reached 318,465 yards. From that time the trade declined steadily, until it left the county altogether. Nairnshire also figured in the stampers' returns, but to a limited extent, and for a brief period. In the parishes of Elgin and Forres, in Morayshire, flax was grown at a very early date, and in the former it appears that teind was paid on lint in the twelfth century. About 1790 a large number of persons were engaged in spinning flax for the southern markets, and about 50,000 yards of linen cloth were produced annually. A like quantity of cloth was made in Banffshire; but, in addition, nearly 5000 persons were employed in the parish of Banff in making linen thread, in which about 3500 bales of Dutch flax were used every year, and the value of the manufactured article was L.30,000. The thread was sold in Nottingham and Leicester, where it was used in making lace, &c. Competition in various quarters spoiled the trade, and the people took to other kinds of work. In 1748 the Earl of Findlater introduced the linen manufacture into the parish of Cullen. At that time the Earl was President of the Board of Manufactures, and the mode in which he carried out his object is thus recorded: —"The Earl took to Cullen two or three young men, sons of gentlemen in Edinburgh, who had been regularly bred to the business, and who had some patrimony of their own. To encourage them to settle so far north, he gave them L.600 for seven years, the money to be then repaid by yearly instalments, free of interest during the whole period of the loan. He also built weaving shops and furnished every accommodation at reasonable rates. From his position at the Linen Board, he obtained for the young manufacturers premiums of looms, heckles, reels, and spinning-wheels, with a small salary for a spinning-mistress. So good a scheme and so great encouragement could not fail of success, and in a few years the manufacture was established to the extent desired. All the young people were engaged in the business, and even the old found employment in various ways in the manufacture, which prospered for half a century." But Cullen could not escape the influences at work in other quarters, and the trade drooped and became extinct in the early years of this century. The parishes of Keith and Fordyce shared in the prosperity which attended the linen trade, but were as unable to retain it as their neighbours.

The next district to be noticed is that in which the trade has survived to some extent the circumstances which led to its extinction in the counties farther north. Aberdeenshire was early engaged in the manufacture of linen yarn and cloth. About the year 1745 the Board of Manufactures sent a spinning-mistress to Aberdeen, at the request of some persons who desired to provide employment for the working population. Pupils were readily found among the wives and daughters of mechanics and labourers, who soon turned out yarn at the rate of 100,000 spindles a-year, for which they were paid in the aggregate about L.5000. The manufacture of white and coloured linen thread was subsequently begun, and was carried to great perfection. In 1795 the thread manufacture employed 600 men, who earned from 5s. to 12s. a-week; 2000 women, 5s. to 6s.; and 100 boys, 1s. 8d. to 2s. 6d. At the same time upwards of 10,000 women were employed in other parts of the county in spinning yarn for making the thread. Several large manufactories for spinning flax by machinery were established on the Don, near Old Aberdeen, about seventy years ago, and for many years the trade continued in a flourishing condition. At Huntly the trades of dressing flax and making linen cloth were carried on for many years, during the best of which the value of the goods produced was about L.25,000 annually. Peterhead did considerable business in the manufacture of thread. In 1794 there were fifty-two twist mills in the town, at which the yarn spun by women in their own homes was made into thread. About 1200 persons were employed in the various departments of the manufacture. Linen yarn and cloth were made in several other parts of Aberdeenshire. The quantity of linen cloth stamped in the county in 1758 was 103,109 yards. The quantity in any year prior to 1790 did not rise much above these figures; but subsequently a great advance was made, and the quantity stamped in 1822 was 2,500,403 yards. Kincardineshire claims to be the first county in Scotland in which flax-spinning by machinery was established. In 1787 a mill for spinning linen yarn was erected on the Haughs of Bervie by Messrs Sim & Thom, who obtained a license to do so from the inventors of the machinery at Darlington. The mill is still in operation, but the original machinery has given place to more modern contrivances. At Benholm and Auchinblae, in the same county, the linen manufacture still survives, but on a much smaller scale than formerly. 632,896 yards of linen were stamped in the county in 1822.

The early history of the linen trade in the counties of Forfar, Fife, and Perth, in which that branch of industry is now almost entirely concentrated, will be dealt with when the present condition of the linen manufactures of the country comes to be noticed.

The average quantity of linen cloth made in Kinross-shire, from 1780 till 1790, was 118,434 yards, worth about L.4500. This does not include what was made for home consumption. Between 300 and 400 looms were employed in the trade. The yarn was spun by women, chiefly from flax raised in the county. In 1811 a period of depression of trade was experienced throughout the country ; and the gentlemen of Kinross-shire, with the view of ameliorating the condition of the working population, subscribed L.4000, and began to purchase on their own account and risk, cotton and linen yarn, which they gave out to weavers to be made into cloth. The result did not come up to the expectations that had been formed of the scheme, as the market was overstocked, and the goods could be got rid of only at a losing price. The trade was then abandoned, and has not been tried again. In 1756 the weavers of Kinross formed a trade union, the members of which made themselves subject to stringent laws as to work and recreation. In order to induce all in the trade to become members, it was enacted that "none of the weavers already incorporated, or that may hereafter be incorporated, shall, without the consent of the whole or greater part of the subscribers, have any correspondence with non-subscribing weavers in the way of borrowing or lending any of the utensils of their craft, under pain of incurring such penalties as the incorporated members shall inflict." Small annual payments were made by the members, and breaches of the rules were punished by the infliction of heavy fines. It was usual, on occasions of public rejoicings or fairs, for the president of the society to issue an order enjoining the members to conduct themselves with decorum and sobriety, and to go to their homes at an early hour, under pain of dismissal from the society. The co-operative principle was adopted by the members for maintaining the funds of their union. The records of the society show that sums were advanced to members for the purchase of yarn, which, when made into cloth, was sold ; and whatever profit remained after the cost of the yarn and the labour of the weaver were paid, was returned to the treasurer of the society. The linen trade was several times started in Clackmannanshire, but it does not appear to have attained a sound footing at any period. About the year 1748 the Duke of Argyll introduced the manufacture at Inverary, but it prospered only for a short time. The people of Buteshire also gave it a trial, but without success. In Stirlingshire, from 30,000 to 40,000 yards of linen were made annually about the beginning of this century, but for many years past no one in the county has engaged in the manufacture. Dumbartonshire produced 310,827 yards of linen cloth in 1758, but after that year the trade declined, until, in 1822, only 11,331 yards were made ; and the industry is now extinct. A small quantity of linen was produced in Linlithgow. Mid-Lothian long stood high in the trade, which was chiefly concentrated in Edinburgh, as many as 1500 looms being employed on linens in the city. The manufacturers were famous for making the finest damask table-linen, and linen in the Dutch manner equal to any that came from Holland. So early as 1698 there is mention of a bleachwork having been established at Corstorphine. The following figures show the quantity and value of the linen cloth stamped in the county in the years named :—1728—747 yards, valued at L.198, 17s. ; 1738—18,988 yards, L.2986, 11s. 9d. ; 1748—236,954 yards, L.9616, 18s. 10d. ; 1758—712,719 yards, L.36,132, 16s. 10d. ; 1768—389,962 yards, L.32,191, 17s. 6d. ; 1778—178,290 yards, L.22,674, 16s. 2d. ; 1788—244,710 yards, L.36,338, 1s. 2d. ; 1822—129,709 yards, L.22,287, 18s. The price of the cloth made in Edinburgh was always high on account of the fineness of the quality. While the average price over Scotland was about 10d. a-yard, the price of the Edinburgh linen ranged from 2s. 6d. to 2s. 11½d. The manufacture of linen goods has long ceased to rank among the industries of the city. Salton, in Haddingtonshire, is noted as having been the first place in Britain in which weaving of the linen cloth known as "hollands" was established, and the first in which a bleachfield of the British Linen Company was formed. In the beginning of last century the lady of Fletcher of Salton, animated by a desire to increase the manufactures of the country, travelled in Holland with two expert mechanics in the habit of lackeys. Her rank procured her access, with her supposed servants, to the manufactories ; and by frequent visits, the secrets of operations were discovered, and models of the various works were made by the disguised artisans. The parish in that way became acquainted with two valuable processes of manufacturing—the making of pot barley, and the weaving of "hollands ;" and for several years it supplied the whole of Scotland with those articles. In Lanarkshire linen was manufactured on an extensive scale at Glasgow and East and West Monkland. The trade was established at Glasgow in 1725, and for a long period formed the staple industry of the city. Nearly 3000 looms were in 1780 employed in linen fabrics in the Barony parish alone. Ten years later, however, cotton had almost entirely superseded flax, and the weavers were mostly occupied in making muslins. At present about a dozen firms are engaged in the manufacture of flax. In 1728 upwards of 272,000 yards of linen were stamped in Lanarkshire ; twenty years later the quantity was 1,191,982 yards ; in 1768 it was 1,994,906 yards ; but in 1822 only 228,692 yards were submitted to the stampmaster. The cotton trade had become the staple of the west, and linen was neglected. Large quantities of linen cloth were made in Renfrewshire. The highest figures are those for the year 1778, when 1,467,935 yards were stamped in the county—being chiefly made in Paisley, where also a large number of persons were engaged in making white sewing thread. The art of making this thread was introduced into the neighbourhood from Holland in 1725, and was carried on for a long time in the family of a lady, who first learned the secret and began the trade. The linen manufacture of Paisley gave way before the introduction of cotton, and was long ago abandoned. Ayr shared in the profits of the linen trade in its early days, but many years since the people took to other pursuits. Kirkcudbright and Wigtown made a small show in the returns, and Roxburgh and Berwick produced from 30,000 to 60,000 yards annually. Melrose was famous for its " land linens " from an early date, and the weavers received many orders from London and the Continent ; but the trade began to decline about the year 1770, and never rallied. Linen was a commodity in which a great business was done at St Boswell's fair, held in the parish of that name ; but for a number of years past none has been offered, as the trade has died out in the district.

Of 197 flax, hemp, and jute factories ascertained to be in existence in Scotland in September 1867, 176 were situated in the counties of Forfar, Fife, and Perth. This concentration of the trade has, as already shown, taken place in comparatively recent years, and the causes of it are not difficult to discover. The human hand, aided only by the rude appliances of ancient times, can ill compete with modern machinery propelled by steam ; and manufacturers in places where circumstances were adverse to the introduction of the tireless agent, naturally found it impossible to succeed in a competition with people more advantageously situated. Hence the spinners and weavers of linen in the outlying districts had to relinquish their wheels and looms, and follow the trade to the absorbing centres, or seek new kinds of employment. The change caused much hardship, and broke up many homes. Not a few of the weavers had been able, in the more prosperous days of the trade in the rural districts, to acquire little freeholds, on which they lived with their families in the midst of happiness and contentment; and it was a sad day when the failing of occupation compelled the sons and daughters to leave the parental roof and go, it might be, many miles away to find a market for their labour. In the long run, the change has been advantageous to a much greater number of persons than those who suffered by it, and now its effects are almost entirely obliterated, if not forgotten.

The linen and jute manufactures are almost the only branches of Scotch industry dealt with in this book which have previously had their histories written. A few years ago Mr Alexander J. Warden, merchant, Dundee, published under the title of "The Linen Trade, Ancient and Modern," an exhaustive and thoroughly trustworthy treatise on every department of the manufactures referred to, and from a second edition of the work, issued in 1868, some valuable information here embodied has been drawn. Mr Warden gives the following statistics relating to the flax, jute, and hemp factories of Scotland, as existing in September 1867 :—

Districts.	Number of Works.	Nominal Horse Power.	Number of Spindles.	Power Looms.	Persons Employed.
FORFARSHIRE.					
Dundee, ...	72	5,822	202,466	7,992	35,310
Arbroath, &c.,......	18	892	36,732	830	4,941
Montrose, &c.,......	6	495	33,966	122	2,483
Forfar, ...	6	232	...	1,401	1,865
Brechin, ...	4	190	5,400	539	1,322
Carnoustie,..........	2	84	...	445	650
Total,............	108	7,715	278,564	11,329	46,571
FIFESHIRE.					
Kirkcaldy, &c.,......	18	909	28,670	1,612	3,887
Dunfermline,........	5	410	1,100	1,858	2,410
Leven District,......	9	856	32,350	252	3,044
Eden do.,............	16	444	10,478	1,271	2,038
Tayport, ...	3	72	2,000	45	200
Total,............	51	2,691	74,658	5,038	11,579
PERTHSHIRE.					
Blairgowrie, ...	9	562	18,296	393	2,050
Coupar-Angus, ...	3	62	1,268	224	467
Alyth,..............	2	42	...	178	315
Perth, &c.,..........	3	181	1,500	553	908
Total,............	17	847	21,064	1,348	3,740

Districts.	Number of Works.	Nominal Horse Power.	Number of Spindles.	Power Looms.	Persons Employed.
GENERAL ABSTRACT					
Forfarshire,..........	108	7,715	278,564	11,329	46,571
Fifeshire,.............	51	2,691	74,658	5,038	11,579
Perthshire,..........	17	847	21,064	1,348	3,740
Kincardineshire,....	5	74	2,818	...	120
Aberdeen, ...	1	785	16,814	428	2,175
Total,............	182	12,112	393,918	18,143	64,185
Other Parts of Scotland, }	15	2,840	93,661	1,774	13,010
Grand Total,.........	197	14,952	487,579	19,917	77,195

It will be seen from the above figures that Forfarshire has considerably more than half of the entire linen trade of Scotland. Putting Dundee aside for more special notice afterwards, Arbroath claims first attention. Though the conversion of flax into cloth was practised on the banks of the Brothock for a number of years previously, it was not until about 1738 that the trade began to assume importance. The first impulse to the manufacture arose from one of the Arbroath weavers accidentally discovering the mode of making the variety of linen cloth called " Osnaburg," after the place in Germany where it was first made and from which it was imported. The man had worked up a quantity of flax which was unsuited for the kind of cloth then in demand in the home market, and on taking his web to a merchant offered to give him a bargain of it. The merchant recognised the similarity between the web the weaver was disposed to look upon as almost unsaleable and the Osnaburg cloth; and not only purchased the piece, but gave an order for some similar webs. The weaver reluctantly accepted the order, little dreaming what a fortunate discovery he had made. Before many months had elapsed, a large number of weavers in the town and neighbourhood were engaged in the production of Osnaburgs; and thus was laid the foundation of the almost uninterrupted prosperity which the linen manufactures of Arbroath have enjoyed. Soon after the discovery was made, a number of gentlemen of property.in the town formed themselves into a company for the manufacture of Osnaburgs and other brown linens. They obtained the best machinery that was then known in the trade; and, by devoting great care to the manufacture, succeeded in producing a better quality of goods of the kind than was made elsewhere, and the brown linens of Arbroath became

famous in the markets at home and abroad. So great did the demand for them become, that most of the weavers in the county, and
many beyond it, devoted themselves to making brown goods. In
the year 1792 the quantity of Osnaburgs and brown linen stamped
in Arbroath was 1,055,303 yards, valued at L.39,660. At that time
nearly 500 weavers were engaged in making sail-cloth. Their productions were nearly equal in value to the other linens. In 1740
the manufacture of linen thread was introduced into the district;
and, after a run of prosperity extending over nearly half a century,
rapidly declined, and became extinct about seventy years ago. Machines for spinning flax had been invented about 1790, but were not
brought to any degree of perfection until a number of years afterwards. In 1807 or 1808 a portion of the Inch Flour Mill at
Arbroath was devoted to giving the spinning machinery a careful
trial. The efficiency of the machines having been established, the
flour grinding gear was cleared out, and the entire mill devoted to
flax-spinning. Subsequently additions were made, and the mill is
still in operation. The experiments at the Inch Mill were watched
with much interest ; and when their entire success was demonstrated,
a change came over the trade, and the erection of factories was
proceeded with rapidly. A period of extraordinary prosperity set
in about 1820, and continued for five years. What followed is thus
recorded by Mr Warden:—" During this halcyon era were erected
many spinning and other works of an extent greatly beyond the
means of the proprietors, and very much beyond the legitimate
requirements of the trade. There was then a plethora of banks in
the town, and in their competition for business unwarrantable facilities were afforded to men without capital, and many of them without
experience or judgment. The natural consequence followed when,
in the beginning of the year 1826 (a year memorable in the annals
of the trade for the dire calamity which then burst upon the commercial world), the manufactures of the place were all but unsaleable,
money became scarce, credit failed, and almost the whole manufacturing community, adventurer and honourable merchant alike, were
engulphed in one common ruin. Almost every mill and factory was
silent, distress prevailed throughout the town, and it was some time
before Arbroath became its former self again." In 1832 there were
sixteen spinning mills in Arbroath and its immediate neighbourhood;
but these were not so extensive as those at present in operation.
The rates of wages then current were:—Men from 10s. to 15s.
a-week; women, 4s. 6d. to 5s. 3d.; boys and girls, 3s. 3d. to 3s. 6d.
Ten years later the quantity of flax spun annually in Arbroath was

about 7000 tons, and the value of the yarn, L.300,000. There were then employed 732 linen weavers, of whom a third were women, and 450 canvas weavers, of whom about a fifth were women. In 1851 eighteen firms were engaged in the staple trade of the town. The horse power of their engines was 530; the number of spindles, 30,342 ; power-looms, 806; persons employed, 4620. These figures show, by comparison with the preceding table, that during the seventeen years ending with 1867, the trade had not increased much. It has, however, been maintained in a healthy state, and the quantity of canvas made in Arbroath annually is about 500,000 pieces.

In the early years of last century an annual market for linen yarn was held at Montrose, and thither manufacturers from the adjoining counties repaired to dispose of their goods. The making of sail-cloth was the first manufacture of any consequence established in the town. It was begun in 1745 by a company, whose success induced others to embark in the trade. The result was that it was overdone, and canvas-weaving became almost extinct. Pennant states that, when he visited the town in 1776, considerable business was being done in the manufacture of sail-cloth, fine linen, lawns, and cambric. He adds that "the men pride themselves in the beauty of their linen, both wearing and household, and with great reason, as it is the effect of the skill and industry of their spouses, who fully emulate the character of the good wife so admirably described by the wisest man." The flax manufacturers of the district readily adopted the machinery which had been invented for spinning, and the first factory was built in 1805. In 1834 there were four large factories in the town, all of which were worked by steam power. Besides these, there were three factories on the North Esk, owned by Montrose firms, and propelled by water. The aggregate spinning power was equal to the production of 1,157,093 spindles of yarn annually. Some of the yarn was woven in the town and district, but the greater part was sold to manufacturers in other towns, or exported. The quantity of canvas and other fabrics made in the town and neighbourhood was then about 50,000 pieces. Though the factories have not increased in number since 1834, the productive power of all has been extended. About 50,000 tons of flax, tow, &c., are used annually. In addition to the persons engaged in the factories, a large number of hand-loom weavers are employed. The chief characteristic of the trade in Montrose is its steadiness, resulting from the caution of the manufacturers.

When the people of Forfar took up the linen trade, they devoted their chief attention to weaving, and to that they have adhered throughout, obtaining their yarns from other towns which, like Montrose, are for the most part engaged in spinning. In 1792 the linen weaving trade was in a flourishing state in Forfar. The principal kind of cloth made was Osnaburg, and from 15s. to 20s. were paid for weaving a piece of 120 yards in length, which occupied a man eight or ten days, according to his ability and industry. In the early years of this century the quantity of linen stamped in Forfar annually was about 1,800,000 yards; and from 1816 till the abolition of the stamp laws in 1822, it was over 2,600,000 yards. Five and twenty years ago 3000 hand-loom weavers were engaged in weaving coarse linens, of which about 2000 pieces were produced weekly; and the value of the yearly produce would not be less than L.250,000. Since that time the trade has increased considerably. The manufacturers, having found that they could not compete successfully in the markets unless they followed the example of other places and adopted the power-loom, have introduced that machine; and the hand-looms, of which nearly 5000 were in use a few years ago, are being gradually discarded. Upwards of L.100,000 is spent annually in wages among the linen workers of the Forfar district. The linens made are chiefly of the brown kind; and the manufacturers have long been celebrated for the uniform and sterling quality of their goods.

In the parish of Brechin flax was cultivated at an early date; and after the manufacture of Osnaburgs was established in the country, the people paid increased attention to the cultivation of the fibre, and also to working it up into cloth. The quantity of linen stamped at Brechin in the beginning of last century was upwards of 500,000 yards a-year, and in 1818 it reached 750,000 yards. The number of persons employed in the trade at present is less than it was thirty years ago; but the production is much greater, owing to the extensive introduction of improved machinery. The premises of the East Mill Company are very extensive. Though the original building was considered to be a large concern in its day, its bulk is insignificant in comparison with the additions that have from time to time been made. Up till a few years ago all the weaving in Brechin was done by hand, but now there are three power-loom factories in operation. There are two extensive bleach-fields in the town, capable of bleaching about 4000 tons of yarn a-year. The principal fabrics made are bleached shirtings, dowlas, and similar goods.

Kirriemuir, another Forfarshire town which has retained its connection with the linen trade through all its changes, is in the singular position of doing a large and prosperous weaving trade by means of the hand-loom alone. In 1805, and the two years following, the quantity of linen stamped in Kirriemuir averaged 2,226,200 yards a-year. In 1833 it was calculated that the rate of production had increased to 6,760,000 yards annually; and at present it cannot be less than 9,000,000 yards. About 4000 persons are employed, of whom more than one-half are weavers. The manufacturers are disposed to erect power-loom factories; but hitherto they have been unable to obtain suitable sites. In the local history of the town, the name of David Sands, a weaver of extraordinary ingenuity, who lived about the year 1760, is mentioned. He invented a mode of weaving double cloth for the use of staymakers, and subsequently succeeded in weaving and finishing in the loom three shirts without seam. One of these he sent to the king, one to the Duke of Athole, and the third to the Board of Manufactures.

In various other quarters in Forfarshire, spinning, bleaching, and weaving linen are carried on, but chiefly for manufacturers in the towns mentioned above.

The history of the linen trade in Perthshire differs little from what has been recorded respecting other counties. Blairgowrie is the chief seat of the manufacture—nine of the seventeen firms in the county having their works on the banks of the Ericht at that place. In the end of last century the linen trade was carried on in no fewer than twenty-seven parishes of the county. 477,743 yards of linen cloth were stamped in Perthshire in 1728; 793,228 yards in 1758; 2,651,674 yards in 1778; and 1,605,321 in 1822, the last year in which the stamp-laws were in force. In the New Statistical Account of Scotland a curious remark, emanating from this county, is made respecting the effect of spinning mills on a rural population. The reporter from Caputh (writing so recently as 1839 be it remembered) says:—"Happily for the peace and purity of our quiet rural population, no spinning mills have yet been erected, neither is any great public work going on at present in this parish."

The figures relating to Fifeshire show that, while the number of factories in that county is close upon half the number in Forfarshire, the persons employed show a marked difference in proportion, indicating that the factories of Forfar are, on the average, more extensive than those of Fife. It will also be observed that, in proportion to the spinning power, the number of power-looms at work in Fife is greater than in Forfar—the number of spindles to each loom in the former being about fifteen, while in the latter it is nearly twenty-five. Kirkcaldy is the chief seat of the trade in Fife, and possesses some fine mills. Though the art of making linen was known and prac-

tised in the district about 200 years ago, the quantity produced was
insignificant until about 1743, when upwards of 300,000 yards were
stamped in the town annually. Kirkcaldy did not make the whole,
however, as the figures include the cloth brought in from Abbots-
hall, Dysart, Leslie, &c., to be stamped. An annual market for the
sale of linen cloth was established in 1739, and various other steps
were taken by the magistrates to extend the trade. Handkerchiefs,
checks, and coarse ticks were the kinds of goods first made ; but the
market for these having been spoiled by the war of 1755, which interrupted communication with America and the West Indies, trade
became so bad that nearly all the looms were standing idle, and the
manufacturers were considering how to employ their capital more
profitably. Before abandoning the linen trade, however, Mr James
Fergus resolved to try to produce something that would sell in the
home market. He studied the making of "ticking," and succeeded
in producing a fabric of first-rate quality. This new branch of the
trade was readily adopted by the desponding manufacturers, and
since then ticking has been one of the principal articles made in the
town. Towards the close of last century it was calculated that upwards of 1,000,000 yards of linen, worth about L.50,000, were made
annually in Kirkcaldy. In 1818 the quantity stamped in the town
(including the produce of the neighbouring towns and villages) was
over 2,000,000 yards. About one-seventh of the linen made was
from home-grown flax, the remainder being made from flax imported
chiefly from Riga. In 1793 three flax-spinning mills were erected
at Kinghorn, and two large spinning mills belonging to a Kirkcaldy
firm have long been in operation in that town. The number of
persons employed in the linen manufacture in Kirkcaldy about
seventy years ago was nearly 5000, and their average earnings did
not exceed L.7 a-year. The next statement of wages applies to
the year 1838, when the net weekly earnings of linen weavers
averaged 7s. 3d. for ticks ; 5s. 11d. for fine sheeting ; 3s. to 6s. 6d.
for dowlas ; and 9s. 3d. for sail-cloth. In the year 1821 a power-
loom factory was built in the town, and is supposed to have been
the first establishment of the kind. The late Mr James Aytoun,
of Kirkcaldy, made some important improvements in the machinery
used for spinning flax, and adapted it to the production of yarn
from tow. During the past six or seven years the trade of the
district, which had remained almost stationary for twenty years,
has been considerably extended, the additions made to the spindles
and looms being equal to nearly 100 per cent. There is an extensive
linen factory at Dysart, owned by Messrs James Normand & Son.

At Leslie several extensive mills, beautifully situated on the banks of the Leven, give employment to a large number of persons. These mills are owned chiefly by Messrs John Fergus & Co. and Messrs D. Dewar, Son, & Sons, of London. Power-loom factories have recently been erected at Tayport, Auchtermuchty, Falkland, Kingskettle, Ladybank, Strathmiglo, and elsewhere in Fife, all indicating that the trade of the county is in a healthy state.

Dunfermline is the chief seat of the manufacture of table linen in Britain—indeed, it may be said, in the world. When the linen trade was established throughout Scotland in the beginning of last century, the people of Dunfermline shared in its profits, and always aimed at the production of a high class of goods. They were most successful in making table linen, and to that branch they have mainly adhered. Long ago they had outstripped all competitors in their staple industry, and the produce of their looms has for many years graced the tables of royalty at home and abroad. At the Exhibitions of 1851 and 1862, the goods shown by Dunfermline manufacturers attracted much attention, and helped to extend their fame.

In the early days of the linen manufacture only coarse goods were made in Dunfermline—first the variety known as " huckaback," and subsequently " diapers." The weavers appear to have been rather fond of trying the more difficult kinds of work, and some of them adapted their looms to producing novel patterns of cloth. Great ingenuity was also expended in weaving articles of dress without a seam. In 1702 a weaver in the town made a seamless shirt in the loom, and a like feat was afterwards successfully accomplished by others. Two of those novel productions are worthy of mention. In 1821 Mr David Anderson completed in the loom a gentleman's shirt elaborately ornamented. It was of very fine linen, and bore on the breast the British arms, worked in heraldic colours and gold. For the accomplishment of the work he received L.10 from a fund which had been formed in Glasgow for the encouragement of inventions and improvements in manufacturing. The shirt was presented to His Majesty George IV., who was graciously pleased to accept it, and to order L.50 to be sent to the maker. Mr Anderson subsequently wove a chemise for Her Majesty Queen Victoria. It was composed of Chinese tram silk and net-warp yarn, and had no seams. The breast bore a portrait of Her Majesty, with the dates of her birth, ascension, and coronation, underneath which were the British arms and a garland of national flowers. The flag of the Weavers' Incorporation is also a remarkable piece of work. It consists of a solid body of silk damask, bearing a different design on each side, and yet both are interwoven.

Damask weaving was introduced into the town in 1718, and the story of its introduction is somewhat curious. Mr James Blake, a man of ingenuity and enterprise, went from Dunfermline to Drumsheugh, near Edinburgh, where damask weaving was carried on. The process of weaving was kept a close secret; but Blake was determined not to be frustrated in his mission, which was to find out the secret and work it for his own advantage. Feigning to be of weak intellect, he lounged about the workshop in which the damask looms were employed, and ultimately ventured in. The expression on his countenance when he saw the looms was so full of puzzled wonder that the weavers allowed him to gratify his curiosity by minutely examining the machines. He asked to be allowed to creep under one of them that he might more closely watch its mysterious working. This odd fancy of an idiot, as the workmen believed him to be, caused some amusement; but no one objected to him going under the loom. While the weavers were smiling at his bewildered stare, Blake was carefully noting in his mind the manner in which the mechanism was arranged and how it operated. He appeared to be fascinated by the looms, and was in no haste to go away. When he did leave, however, he was in full possession of the secret. Returning to Dunfermline, he at once began to construct a loom from memory, and soon had the gratification of possessing a perfect machine. He had a workshop in the old tower of the Abbey, and there, in company with one or two faithful assistants, he devoted his whole time to making damask goods, keeping well the secret which he had become possessed of in such a singular way. It would appear that he was more successful in maintaining the secret than the weavers of Drumsheugh, for his loom was the only one of the kind in the town for many years. After the principle of the damask loom became generally known, however, it was not readily adopted, the machine being costly and difficult to work. Fifty years after Blake had set up his machine, there were only ten or twelve damask looms in Dunfermline; and ten years later, in 1778, the number did not exceed twenty. Three persons were required to work the loom at first—two weavers, one at each side, to throw the shuttle and move the "lay," and a boy to work a series of cords which raised the warp threads necessary to produce the design. Sometimes one man undertook to work a web two yards wide without an assistant, and in that case he had to rush from one side of the loom to the other continuously in order to keep the shuttle going. That was a laborious mode of working; but it was more profitable than the other system, as, though a smaller quantity of cloth was produced in a given time, that shortcoming was more than compensated for by saving the wages of an assistant.

An important improvement was made on the damask loom by Mr John Wilson, of Dunfermline, who devised a mechanical arrangement which dispensed with the services of the draw-boy. The value of Mr Wilson's invention was publicly acknowledged by his being made a burgess of the town in 1780; and a further reward was conferred on him by the Board of Manufactures, who presented him with L.20. As damask was then woven, it was necessary that the weavers should commit to memory the details of the patterns; and when a loom was changed from one design to another, the workmen had to devote four or five days to getting the new pattern by rote. An error of memory was unfailingly registered in the cloth; and as the value of the piece was thereby deteriorated, only persons who had good memories, and who took great pains to learn the patterns, could pass as efficient workmen. Subsequently an invention was made which rendered it unnecessary to trust to memory for the proper working of the design. The new apparatus was known as the "holey-board." In 1803 Mr David Bonnar obtained a patent for what he called a "comb draw-loom," which had the effect of still further simplifying the operations of the damask weaver. The trade gradually increased under these various improvements in the weaving machinery; but it received its greatest impulse from the introduction of the Jacquard machine in 1825. By the year 1830 that machine had come into general use. The advantages derived from the Jacquard machine are numerous; but the most important are that the facility of production enables the damask manufacturer to sell his goods at a lower price per yard on the average than was formerly paid for weaving alone, taking into account also the reduced price of the raw materials, and that there is no limit to the variety of designs that may be produced. The designs of the damask made by the old process were crude and indistinct, but by means of the Jacquard machine the greatest distinctness of outline and delicacy of detail have been attained. The Jacquard machine makes every thread of warp and weft play its part in the design; but by the "draw" system the pattern was brought out by moving four or five threads at a time; the result is that the old damask looks as if the design were worked in mosaic, each spot being a square equal to the thickness of four or five threads, and in some cases even more. A change of design was a serious matter for the weaver before the Jacquard machine was introduced, as the mounting of a fresh pattern occupied five or six weeks, and during that time he received no remuneration.

When the damask trade had become fairly established in Dunfermline, the manufacturers received orders from noblemen, bishops, and private gentlemen, for sets of table linen bearing their coats of arms, &c. His Majesty William IV. was their first royal customer, and Queen Victoria had some linen made for her household in 1840. Since the latter date many orders have been received from royal personages at home and abroad. Great attention has been paid to the designing department of the trade, which has more than kept pace with the mechanical improvements. In 1826 a drawing academy was established in the town, with the view of teaching young men the principles of drawing, and fitting them to fill the office of designers. The institution did not succeed, and was given up in 1833. Some designers of eminence were trained, however, and good resulted to the trade generally. The academy was supported at the joint cost of the Board of Manufactures and the manufacturers of the town, who expended L.126 on it annually. The Board also gave premiums for excellence of design in damask goods, and one firm received L.516, 10s. in premiums of that kind in eighteen years. Thirty years ago an export trade to America was opened up by the manufacturers, and about L.150,000 worth of damask goods found a market in the United States every year. The Americans have ever since been good customers.

The largest factory in Dunfermline, and the most extensive of the kind in Britain, is the St Leonard's Power-Loom Factory, which belongs to Messrs Erskine Beveridge & Co. The factory is beautifully situated on the south side of the town, and is in every respect a model establishment. The main building is but one storey high, and the roof consists of a series of ridges. Externally, the place is unpretending enough, but there is an air of tidiness and cleanliness in all its accessories which impresses one favourably. The coarser sorts of yarn used in the factory are brought from Dundee, Kirkcaldy, &c., and the finer sorts from Yorkshire and Ireland. Some of the yarn is received in a brown state, some bleached, and some dyed. Many tons are kept in stock in rooms set apart for the purpose. The yarn is given out in hanks to the winders, who, by the use of simple but ingenious machines, wind it on bobbins for the use of the warpers, or on pirns for the weavers. The warpers take a certain number of bobbins and arrange them in a frame. The threads of the bobbins are then led to the warping-machine, in which they are arranged side by side, and wound with equal strain upon a roller. One of the great difficulties that had to be overcome by the inventors of the power-loom was the tendency which the rapid and somewhat violent motion had to soften and break the warp. That difficulty was removed by "dressing" the warp with paste, and for performing that operation a machine was devised. After the yarn has been wound upon it, the roller is taken to the dressing-machine. The yarn is led between a pair of cylinders, one of which revolves in a bath of paste made either of flour or of Irish moss; it then passes over a number of cylindrical brushes, which smooth down the fibres and remove the superfluous paste; next it is brought into contact with a large cylinder heated by steam; and having been thus thoroughly dried, is wound upon the beam of the loom. Before the warp is placed into the loom, however, a preliminary operation is necessary. The threads have to be drawn through the "heddles" and the "reed," and then the beam is placed into the loom.

In following the warp to the loom, it is necessary to enter a workshop, the floor of which is covered by machinery, while overhead the eye gets lost in a maze of belts, shafts, and other mechanism, the motion of which makes one giddy, while the noise closely resembles the roar of a great waterfall with a metallic tinkle superadded. In this workshop, which is nearly 700 feet long by 160 feet wide, 900 power-looms are at work. Each large loom and each pair of small looms is attended by a young woman. The looms are prepared and kept in order by mechanics and tenters, each of whom has charge of a certain number of looms. The tenters take off the finished webs and put on fresh warps, and the mechanics look after the working portions of the apparatus. The looms employed in weaving damask are each fitted with a Jacquard machine, that beautiful contrivance which, next to the loom itself, is perhaps the most important invention ever made in connection with textile manufactures. The Jacquard machine was invented in the year 1800, and was at once adopted by the silk and muslin weavers of France, as it enabled them to introduce unlimited variety into the figuring of their goods.

In the St Leonard's factory the capabilities of the Jacquard machine are admirably illustrated. Each loom seems to be producing a pattern different from all the others, and yet the beauty and elegance of the designs are nearly equal. Some of the table-covers made bear the arms and insignia of the persons for whom they are intended, while others display designs of exquisite beauty,

composed of flowers, shields, &c. A somewhat coarse variety of
table linen worked in brown and white yarn is made in large quantities for the American market; but the chief produce of the establishment consists of the finest quality of white table linen.

Most of the designs are made by Mr Joseph Paton, the father of
Sir Noel Paton, the celebrated Scottish artist; and that gentleman
has done more, perhaps, than any other to maintain the fame of the
local trade. His fine artistic taste and thorough knowledge of the
capabilities of the material that he has to work upon have enabled
him to produce designs which for elegance and appropriateness
are unsurpassed. It would appear that taste in the matter of table
linen changes as frequently as taste in matters of dress, and that the
favourite design of to-day may be a drug in the market next month.
At one time a stately classical style is in vogue, at another nothing
but florid Italian will sell, and with the next change perhaps the
public taste may be met by a bit of modern device. Sometimes the
centre of the cloth is filled with elaborate work, and the border
treated in a simple way. Again, the centre is plain, or dotted over
with leaves, and the border is composed of a broad band of flowers,
&c.

The designs are drawn on paper, the surface of which is divided
by lines into minute squares, each square representing a loop of the
fabric. The paper bearing the design is taken to the card-maker,
who, by means of a curious little machine, punches in a piece of
card-board a series of holes corresponding with the design. The
cards are attached to the loom, and serve to guide the apparatus
through the pattern. The holes in each card relate to one throw of
the shuttle, so that a card is required for every thread of weft that
goes to make up a table-cover. As many as 50,000 cards have been
used in making one piece of damask. When the design is transferred to the cards all trace of it is lost until it reappears in the web
on the loom; for the holes in the card convey no idea whatever of
the drawing.

Most of the white goods are sent to Perth to be bleached, there
being no convenience for conducting that operation at the factory.
The bleaching is followed by hot-pressing, which gives a beautiful
"finish" to the cloth. Goods that have not to be bleached are calendered
on the premises. The "lappers" receive the goods from the finishers,
fold them up, and prepare them for the market. These operations
are carried on in the warehouse, a stately detached building in the
Italian style of architecture, and three storeys in height. The main
entrance of the warehouse opens on a spacious hall paved with

ornamental tiles, and otherwise decorated. On the ground floor are the counting-room, manager's offices, and the packing-room, while the floors above are occupied by a series of large halls, fitted with enclosed shelving for storing goods, and tables for exhibiting them upon. The whole place is handsomely fitted up.

In addition to the 900 power-looms in St Leonard's factory, Messrs Beveridge & Co. employ 180 hand-looms in a separate workshop. Altogether they give employment to about 1500 persons, of whom about ninety per cent. are females. The quantity of linen made by the firm averages about 200,000 square yards a-week, so that the yearly produce, supposing the average width of the web to be one yard, amounts to 10,400,000 yards, or upwards of 5900 miles, which would be sufficient to cover a board at which the entire population of Scotland and Ireland might dine at one time. The value of the cloth made is about L.360,000 a-year, and the price per yard ranges from 5d. to 5s. The women and girls employed in the factory earn from 4s. to 15s. a-week, and the men from 10s. to 40s. Many of the women and a few of the men live at a considerable distance, and, when they go to work in the morning, take their day's provisions with them. Two large dining-halls are provided for their accommodation at meal time. These are comfortably fitted up, and adjoining them is a large stove for warming food. Provision is also made for the education of the children of the workpeople. A schoolhouse has for many years been in existence in connection with the factory. The school is open to the public generally, but there is this difference—while the children of persons connected with the factory are charged only half the established fees, other children have to pay full. There are usually about 300 children in attendance.

Next in extent to St Leonard's is the Bothwell Power-Loom Factory, built about five years ago by Messrs D. Dewar, Son, & Sons, of London. It has accommodation for 580 power-looms, but only 470 have yet been set up. Upwards of 500 persons are employed. The goods made are similar to those produced at St Leonard's. Two other factories have recently been erected—one for Mr Alexander, fitted with between 400 and 500 power-looms ; and the other for Messrs Inglis & Co., with about 300 power-looms. Messrs Andrew Reid & Co., and Messrs Henry Reid & Son, who have long been engaged in the trade, have made considerable additions to their factories, which now contain about 300 power-looms each. There is another power-loom factory occupied by Messrs Hay & Robertson, but it is of small extent compared with those mentioned. In Mr Darling's factory there are 180 hand-looms.

It is calculated that there are scattered throughout the town and suburbs from 600 to 700 hand-looms, which, with those in the factories of Messrs Beveridge and Mr Darling, give a total of, say, 1000 hand-looms at present in operation. The total number of power-looms is 2670, and the quantity of cloth made annually by hand and power is over 30,000,000 square yards, which, formed into a web of the uniform width of one yard, would measure the distance between Great Britain and New Zealand, with a thousand miles or so to spare. There is more linen cloth manufactured in Dunfermline than was made in all Scotland in any year preceding 1822, and the value of the goods produced cannot be much under L.2,000,000 a-year.

Mr Balfour, the designer at the Bothwell factory, produced a piece of work about ten years ago which excited much attention. It was styled "the Crimean Hero Tablecloth," and many copies of it were supplied by Messrs Dewar to royal and other orders. The cloth was pronounced to be the greatest achievement in damask work ever accomplished. It is thus described in Chalmers's "History of Dunfermline:"—

"The designing and executing of the work occupied about eight months, and occasioned an outlay of nearly L.600. The cloth was inspected and greatly admired by the Queen and Prince Albert at Balmoral, as also by the Emperor and Empress of the French at Paris, who gave an audience to the proprietor, introduced to their majesties by the Earl of Clarendon. Orders were given for the imperial as well as royal tables. The cloth is composed of the finest linen warp and white silk weft, six and a-half yards in length, and three in breadth; but when wrought for sale, it will consist of linen only. The pattern consists of a beautifully elaborate leafy scroll-work for border, in which, at proper intervals, are inserted twenty-four faithful portraits. In one end-border are Her Majesty Queen Victoria in the centre, and on either side the Prince Consort and the Duke of Cambridge. In the other end-border are the Emperor Napoleon in the centre, and on either side the Empress Eugenie and Prince Napoleon. In the centre of one of the side-borders is placed the King of Sardinia, and on either side Bosquet, Brown, F. Nightingale, La Marmora, St Arnaud, Cardigan, Raglan, and Bruat. In the other side-border, the Sultan in centre, with Omer Pasha, Williams, Canrobert, Evans, Campbell, Pelissier, Lyons, and Simpson, on either side. Each portrait of the sovereigns is surmounted with their respective armorial bearings, placed towards the middle of the cloth; and alternately with these are trophies containing the names of the chief battles, with their dates—Alma, 20th September 1854; Balaclava, 25th October 1854; Inkermann, 5th November 1854; Tchernaya, 16th August 1855; and in the centre of the cloth there are magnificent trophies illustrative of the fall of Sebastopol, with the motto, *Deus proteget justitiam*, and the date 8th September 1855—the ground around all of these being interspersed with the stars of the orders of the different sovereigns. In the corners of the border are the standards of the four powers rising from behind a shield containing their insignia united—the Rose, the Fleur-de-lis, the Crescent, and the Cross. An idea may be formed of the extent of the design by persons acquainted with the nature of the work, when it is mentioned that there were 50,000 cards, and seven 600-cord Jacquard machines employed in forming the pattern on each loom. These machines required to be kept in operation at the same instant, and the whole was put in motion by a single move- ment of the foot. The web was 1600 threes in the reed, equal to 4800 threads upon the yard, and which, again, multiplied by three, the number of yards in the breadth, gives the total number of threads in the breadth to be 14,400."

Dundee is the metropolis of the linen trade, and till recently had a monopoly of the manufacture of jute. The perfection to which the latter branch has been brought by the enterprise and ingenuity of those engaged in it is remarkable, considering the comparatively brief period which has elapsed since the fibre was introduced. Capitalists in Dundee had for many years shown a disposition to make the place a seat of manufactures; but though they tried to establish a permanent trade in various articles, they had little success until they turned attention to working in flax. The date at which the manu- facture of linen cloth was begun in Dundee is not known; but it is recorded that at the time of the union of England and Scotland 1,500,000 yards of linen were made in the town annually. Then, as now, the chief fabrics were of the coarser kind. A writer in the "Gentleman's Magazine" in 1742, described the linens of Dundee as being the "poorest and meanest;" but whatever truth may have been in that remark at the time it was written, it certainly could not hold good in recent years. In the account which Pennant gives of his tour through Scotland in 1776, it is stated that Dundee used to be celebrated for the manufacture of "plaiding," which was ex- ported undressed and undyed to Sweden and Germany for clothing the troops of those countries; but that trade was superseded in 1747 by the manufacture of Osnaburgs, which became the staple trade of the county of Forfar. In 1789 there were made for sale and stamped in the parish of Dundee 3,181,990 yards of coarse linen, valued at L.80,587. Besides a large share of the above, there were made in the town of Dundee upwards of 700,000 yards of sail-cloth, valued at L.32,000, and esteemed to be superior in quality to any made elsewhere in Britain. The manufacture of cotton was introduced about the year 1790, and several companies engaged in it; but the trade survived for a few years only. During the latter half of last century a considerable trade in the manufacture of coloured sewing thread was carried on in the town. There were seven companies or masters engaged in it, who owned sixty-six twisting-mills, and employed upwards of 1700 persons. 269,568 ℔. of thread, valued at L.33,696, were made annually. The writer of the report on Dundee in the Statistical Account of Scotland (1792) says:—"The particular cause of the increase and prosperity of Dundee is undoubtedly the bounty allowed by Parliament on linen manufactured for exportation. By that the industry of the inhabitants was first set in motion and encouraged; and their consequent prosperity, if it be not an evidence in favour of bounties in general, is at least a decisive one that in some cases they are wise and judicious, and may be productive of the greatest benefit."

Whatever effect bounties may have had on the trade in its early days, the present extent and prosperity of the staple industry of Dundee is principally owing to the perseverance of the manufacturers in adopting and improving machinery for superseding hand labour, cheapening production, and improving the quality of the work. Up till the beginning of the present century, all the yarn used was spun by hand chiefly by persons residing in the country districts; and on the market-days the housewives brought the produce of their spinning wheels into Dundee for sale. The manufacturers went to market and bought what yarn they required. Great difficulty was experienced in obtaining any considerable quantity of yarn of similar size and quality, and that defect considerably interfered with the operations of the manufacturers, and caused them to lose much time. The first step taken to remedy that state of matters was the appointment of agents, who travelled through the district and purchased the yarns from the people, carefully selecting and separating the various sizes and qualities. Those agents also got flax from the manufacturers in the town, and employed persons to spin it. The system continued in existence until about forty years ago, when the spinning machinery, which had been introduced, kept the looms fully supplied with yarn.

Flax-spinning by machinery was first tried in Dundee, in a small mill built at Chapelside, by Messrs Fairweather & Marr, about the year 1793. The machinery was propelled by a ten horse power steam-engine. A second mill, of about the same extent, was built soon afterwards; but though both were kept going for some time, the element of success was wanting, and the enterprise was abandoned for a time. In 1798 five spinning mills, having an aggregate of sixty horse power and 2000 spindles, were erected in various parts of the town. One of these—the Bell Mill—was considered to be a gigantic concern at the time, the building and machinery having cost about L.17,000. The early years of this century were disastrous to the trade, owing to the foreign markets being in a state of stagnation, arising from political complications and war. In 1811 only two spinning mills continued in operation, and the Bell Mill, which had come to a stand, was offered for sale. Between that time and the year 1822 a great change took place in the district. In 1822 seventeen flax-spinning mills were in operation in Dundee, all of which were driven by steam-engines, representing in the aggregate 178 horse power. About 2000 persons were employed, and the number of spindles going was 7944. In the neighbourhood of Dundee there were thirty-two spinning mills, containing 6978 spindles. The mills in operation in Dundee and neighbourhood in 1832 were driven by engines of 800 horse power, and the yearly consumption of flax was 15,600 tons, which produced 7,488,000 spindles of yarn. 3000 persons were employed, and the capital invested in machinery was estimated at L.240,000. In 1846 there were thirty-six spinning mills in Dundee, with a motive power equal to 1242 horses, while the number of spindles was 71,670. Five years later the number of factories was forty, but some of these were devoted to power-loom weaving.

Experiments were made with the power-loom in Dundee so early as 1821, but the result does not appear to have been favourable to the introduction of that machine. Messrs W. Baxter & Son built a factory in Lower Dens, into which they proposed to introduce ninety power-looms; but they appear to have had misgivings as to the practicability of weaving by power, and did not carry out their intention. In an account of Dundee written in 1833, it is stated that "power-looms have not been employed here, or at least not to any advantage, and they are understood to be entirely laid aside." In 1836 Messrs Baxter built a power-loom factory at their Upper Dens Works, and that was the first establishment of the kind in Dundee. Three other power-loom factories were erected soon after, but for a considerable time there was no addition to the number.

The linen trade of Dundee has passed through a series of crises which threatened its destruction; but it has survived them all, and is at present in a healthy state. In 1810 the price of flax suddenly fell from L.150 to L.80 a-ton, and the effect on the manufacturers was most disastrous. Many of them were ruined; and during the succession of violent fluctuations which occurred in the six following years, few of those who withstood the first convulsion were fortunate enough to escape bankruptcy. The stopping of the machinery threw many workpeople idle, and great distress prevailed in consequence. Had this state of matters continued much longer, it is not improbable that the trade would have been abandoned; but, fortunately, a brighter day dawned for the manufacturers, and for seven or eight years preceding 1825 they enjoyed a period of prosperity, which helped to repair their shattered fortunes, and gave them hopes of better things to come. In the autumn of 1825, however, the trade was completely paralysed by the commercial panic which broke out in London, and rapidly spread over the country. The tide of prosperity suddenly turned in Dundee, and again many firms had to suspend payment. The stagnation of business became so serious, that the Government were induced to lend a helping hand to the merchants of the town, to whom they granted Exchequer bills for goods deposited. That aid was timely, and some of the most extensive merchants availed themselves of it. In 1827 a revival of the trade with America took place, and things began to mend. Great quantities of bagging for packing cotton, &c., were made for the United States, and from that article handsome profits were realised. In 1832, the last year in which bounties were paid on goods exported, the value of the linen sent out from Dundee amounted to close upon L.600,000, on which the manufacturers received L.46,854 of bounty. A fire which occurred in New York in 1835 was the cause of the next check which the trade received. A great quantity of bagging had been consumed in the conflagration, and the manufacturers of Dundee were hopeful that an opportunity had occurred for profitable exertion. Most of the machinery was set to work to produce bagging, and the result was that the supply far exceeded the demand, and the market was spoiled. Some of the goods sent out lay in store for years, and the price ultimately obtained entailed serious loss. Again a number of the merchants and manufacturers became insolvent. In order to save the trade from ruin, the banks opened warehouses, and received goods in deposit, on which they advanced money. After a little time the trade rallied, and continued in a fair state of prosperity until 1847, when it was seriously affected by the crisis brought about by over-speculation in railways. The experience of the linen manufacturers of Dundee goes to prove that the calamity of war may directly pro-mote the arts of peace, for they profited largely by the demand created for their goods, first by the Crimean and subsequently by the American war. In the former case, however, some of them did not act judiciously, for instead of regarding the demand created by the war in the East to be a temporary one, they would appear to have looked upon it as permanent, since they sunk a great deal of capital in extending their factories. When peace was declared, those who had acted thus found out their mistake, and were unable to keep the whole of their machinery going. The American war was the most fortunate event that ever occurred for the linen manufac-turers of Dundee. Both armies became extensive customers, and for three years the factories were kept fully employed. Great wealth was realised, and the stability of most of the firms was so well secured by the accumulation of capital, that they are not now likely to sink under fluctuations of trade which would otherwise have ruined them.

Before proceeding further, mention must be made of the introduc-tion of jute, which has had a most beneficial effect on the trade of Dun-dee. Jute is the fibre of plants of the *cochorus* order, which are com-mon in almost every part of India. In the end of last century the East India Company caused inquiry to be made throughout their vast territory with the view of discovering a substitute for hemp. Among the specimens sent to this country was a quantity of jute, but no particular notice appears to have been taken of the material. Small parcels were sent on several subsequent occasions, and at length some of it fell into the hands of manufacturers at Abingdon, Oxfordshire, a town famous for its sacking, twines, &c., by whom it was spun into yarn, and used in making carpeting. Subsequently, about the year 1824, a bale or two of jute was sent to Dundee, to Mr Ander-son, a linen manufacturer. He got his mother, who was an adept at the spinning wheel, to make a trial of spinning it, but she did not succeed to her satisfaction. Mr Anderson seemed to recognise the value of the fibre, and made numerous experiments with it, but without much success, beyond producing a coarse yarn suited only for sacking. The new material was regarded with suspicion by the public, and goods suspected to contain jute were difficult to dispose of. In 1822 Mr Thomas Neish, merchant in Dundee, got a small consignment of jute from London, and tried to get some of the manufacturers to spin it, but none of them would make the attempt ; and, after lying aside for four or five years, the jute was sold for the purpose of being made into door-mats. Ten years after receiving this parcel of jute, Mr Neish got another consignment, which was again offered to the manufacturers in vain. After being much pressed by Mr Neish, Messrs Balfour & Meldrum reluctantly resolved to make experiments with the fibre. Success attended their efforts, and the foundation of the jute trade in Dundee was laid. Mr James Watt, another merchant in the town, rendered great service in bringing jute into favourable notice. For the first year or two after the possibility of spinning jute had been demonstrated, the manufacturers did not spin it pure, but mixed it with flax and tow. In 1835, however, pure jute yarn was made and regularly sold in the market. The raw material could be bought in 1833 for L.12 a-ton ; but four years afterwards, when the value of the fibre had to some extent been recognised, the price was L.22 to L.23. The growth of jute in the popular favour will be best shown by the increase in the quantities imported into Dundee in successive years since 1838 ; and for the sake of comparison, the quantity of flax, tow, and hemp imported in various years since 1815 is also given—

JUTE.		FLAX, TOW, AND HEMP.			
Year.	Tons.	Year.	Tons.	Year.	Tons.
1838	1,136	1815	2,187	1848	30,585
1848	8,905	1820	4,958	1853	47,113
1853	15,400	1825	13,902	1858	25,842
1858	30,086	1830	20,496	1863	28,988
1863	46,983	1835	27,130	1865·	44,821
1865	71,702	1837	15,237	1867	41,409
1867	63,674	1838	30,850	1868	36,712
1868	58,474	1843	26,268		

The present annual consumption of flax in Dundee is estimated to be about 28,000 tons ; of hemp, about 1500 tons ; and of jute, about 60,000 tons—in all, 90,500 tons ; so that in half a century the quantity of raw material used has increased fully fortyfold. It may be mentioned that all the jute is imported from Calcutta. Formerly it was sent through London and Liverpool, but considerable quantities are now brought direct into Dundee as well as into Greenock.

Mr Warden thus describes some of the qualities of jute :—" It is one of the most easily dyed fabrics known, and the colours it takes on are bright and beautiful. The common dyes are quickly applied, but they are very fugitive, and when exposed to the sun's rays soon become faint and dull. By the common process the colouring matter strikes little more than the outside of the fibre, and, as it were, paints it; and this mode of dyeing requires little material, and is done at small cost. The fibres of jute do not subdivide so minutely as those of flax, and they are of a hard, dry nature, and to a considerable extent impervious to moisture. It therefore requires a more complex process to make the colouring materials thoroughly penetrate the fibres so as to make the dye lasting. This can, however, be accomplished, and the better class of goods made of dyed jute undergo this process, which makes the colours both brighter and faster. It is hardly possible to make every colour perfectly fast, although some of them are as durable as those upon other materials. Jute is very readily brought to a rich cream colour either in the fibre, in yarn, or in cloth. It is, however, very difficult to bring it to a full white without injuring the strength of the fibre. Many experiments have from time to time been made to bleach jute, but at best they have been only partially successful, and it may be said that a perfect white has never yet been attained without impairing strength. Fresh sound jute of fine quality can without danger be brought to a moderate degree of whiteness; but as the fibre gets older, exposure to the atmosphere changes it to a browner tinge, and it then becomes more difficult to bleach. The sightly nature of jute, the regular even thread which by the improved machinery is formed of it, and the smooth, tidy, and clean appearance of jute cloth, are all pleasing to the eye, and therefore attractive. These qualities, combined with its cheapness, have served to recommend it to consumers, and bring it into general use. Now, instead of being used stealthily by spinners, as of old, it is the only material spun in a large proportion of the factories, and to a greater or less extent it is used in every establishment in the town."

The effect of the introduction of jute on the linen trade of Dundee is shown in the following passage from a paper read before the Social Science Association at Edinburgh in 1863, by Mr Robert Sturrock, Secretary of the Dundee Chamber of Commerce :—" By the introduction of jute into the linen trade great changes have been brought about. In place of sackcloth, bagging, and other coarse fabrics being made from hemp, hemp codilla, flax codilla, and coarse tows, they are all now entirely made from jute, and some of these raw materials are not now known in the trade. Though much the same quantity of flax and tow is now imported as many years ago, the real linen trade is in this way supplemented, the quantity formerly required in the coarser branches being now available for other purposes. On the first introduction of jute, it was only used for fabrics of the coarsest description—in fact, it was then considered that it never could be used otherwise; but from the improvements in machinery, and from gradually increasing experience, this has been found to be erroneous. The more common descriptions of Osnaburgs, sheetings, and many other fabrics, are now manufactured solely from it; or these goods, in place of being made of flax or tow as formerly, are now composed partly of tow and partly of jute. Fine goods are also manufactured from a combination of jute and cotton. In this manner has the linen trade again been most largely supplemented. The jute trade has increased so rapidly, and the goods made from the fibre are now so highly appreciated over the whole world, that, looking to the future, one is entitled to say that in extent it will probably only be rivalled by the cotton manufacture. The packsheet, baggings, sackings, sacks, and woolpacks of Dundee, are used in almost every quarter of the globe. When I state that they are by far the cheapest manufactures of this description that can be made from any raw material, it will be no matter of surprise though this trade still continues to advance with great strides. There is still one fabric worthy of particular notice, which owes its existence solely to jute. It is the manufacture of jute carpets. These have nearly the appearance of carpets made from wool; and though they are neither so durable nor retain their colour so well, still, when I state that the cost varies from 6d. to 1s. 4d. a-yard, it is not remarkable that they should be greatly used. Rugs, in imitation of wool, are also manufactured from the same material. The reporters appointed by the jury on jute goods at the International Exhibition last year, remarked, 'It is in Scotland exclusively where goods made from jute represent a large branch of industry. This very cheap raw material is employed there—either pure or mixed—to make ordinary brown cloth, but more especially sacking, packing-cloth, and carpets. The jute yarns used for carpets are of the richest and most varied colours, and are sometimes used with cocoa fibre. Even the Brussels or velvet carpet is imitated with success in appearance, if not in durability.'"

The flax and jute factories of Dundee are substantial edifices. They are fitted with every appliance that has been devised for promoting the health and comfort of the operatives, and facilitating their work. As a rule the proprietors are possessed of a spirit of considerate liberality towards those who toil for them—or, rather, with them, for the life of even a prosperous manufacturer is anything but a sinecure. Many of the employers are men who had but a humble start in life, and have created their own fortunes by close application to business. Some of those who were leaders in the earlier and more trying days of the trade have retired to spend their remaining time in the enjoyment of fortunes accumulated during years of anxious labour, leaving sons and successors to carry on the work which they brought to such a successful issue. Others who have attained an age and position which would entitle them to retire, continue to work on, as if determined to die in harness. Indications of the prosperity prevailing among the class are abundant in the stately mansions which they have reared for themselves in the outskirts of the town and in quieter localities adjacent.

The most extensive factory in Dundee is that of Messrs Baxter Brothers & Co., which is entirely devoted to the manufacture of flax and hemp. In 1822 the late Mr William Baxter—father of the present head of the firm, Sir David Baxter, Bart. of Kilmaron—who owned a small mill in Glamis, entered into partnership with his eldest son Edward, and built a flax-spinning mill, with an engine of fifteen horse power, on the Dens Burn, in the north-east quarter of Dundee. That was the germ of the present vast establishment. The mill proved a successful speculation, and three years after it was set going, a similar work of double the power was built farther up the "burn." At that time other members of the family had been taken into the partnership, and the firm was known by the designation, which it at present bears, of Baxter Brothers & Co. Success continued to follow the extended firm, and from time to time the mills were enlarged, until it became necessary to introduce an engine of ninety horse power. In 1833 they built another mill still farther up the stream, and to that they added, in 1836, the first power-loom factory ever erected in Dundee, and in that department alone provided employment for upwards of 300 persons. In 1846 the firm had in operation in Lower Dens Mills one engine of ninety horse power, driving 3028 spindles ; and in the Upper Dens Mills two engines, equal together to 105 horse power, and driving 8000 spindles. In the power-loom department they had two engines of thirty horse power each, and 256 looms, with accommodation for nearly double that number. They had also a calendering shop with a ten horse power engine. The site chosen for the works originally was ill adapted for convenient extension, being on the banks of a natural gorge or "den," as the name of the place implies. The difficulties of the situation have been completely overcome, and the existence of the valley becomes apparent only when one enters the establishment, and sees how the gorge has been dammed up to form a series of deep ponds, which intercept and retain for the use of the boilers the whole water of the Dens Burn.   The ground at Dens belonging to the firm extends to twenty-one acres, of which ten are occupied by buildings, courtyards, and ponds.   The extent of the buildings may be judged from the fact, that the superficial area of the floors is not less than twelve acres, the greater part of which is covered by machinery of the finest description.   In the spinning department there are 22,000 spindles, with the requisite preparing-machines ; and in the weaving-rooms are 1200 power-looms.   In such a large place it is, of course, necessary to distribute the steam-power, and no fewer than twenty-two engines are employed, the combined nominal force of which is 750 horse power.   There are thirty-two steam-boilers, which consume nearly 300 tons of coal weekly. The largest chimney has twenty-two of these boilers connected with it. By using properly constructed furnaces, and with good management on the part of the firemen, scarcely any smoke is to be seen issuing from the chimneys.   No recent census of the establishment has been taken; but the number of persons employed is stated to be from 4000 to 4500, of whom a large percentage are women or girls.   7000 tons of flax are used annually, a quantity far exceeding what is worked up in the same time by any other firm in the world.   A considerable quantity of hemp is also manufactured.

The factory occupies a commanding site; and its elegant belfry and obelisk-shaped chimneys are conspicuous objects in the view of the town obtained from the east.   A wide public street separates the upper division of the establishment from the lower, but there is direct connection between them by means of a tunnel.   Owing to the nature of the site, and the way in which extensions were made, the mills have an irregular appearance, which somewhat masks their extent.   The front presented to Princes' Street by the Upper Mills is, however, an imposing piece of masonry.   It is 250 feet in length, and consists of five lofty storeys, with attics.   Over the centre of the front is a statue of James Watt similar to that which stands in Adam Square, Edinburgh.   This building forms the largest division of the spinning department.   The flax is imported in bags, which are deposited in a range of extensive warehouses.   In order to trace it through the various stages of manufacture, the visitor must follow the flax from the stores to the heckling shops.   Heckling is an operation whereby the fibres of the flax, as they come from the scutchers, are subdivided longitudinally into filaments of a fineness suited to the quality of cloth to be made.   In order to produce a fibre of sufficient fineness for cambrics and lawns, only the best quality of flax is used, and the heckling has to be done with great care on fine heckles. Before the invention of heckling-machines the operation was performed by hand, and the persons employed in that occupation formed a large proportion of those engaged in the linen manufacture. The hecklers were generally a rough lot of men, who were continually making unreasonable demands, and "striking" when these were not complied with. The personal annoyance and interruption to trade caused in that way led the manufacturers to devise means which would enable them to dispense with hand heckling. Machines were invented which performed the work as well as, and more expeditiously than, the hecklers, and the result was that the hand hecklers were thrown out of employment. The heckling machines used in the Dens Works are most ingeniously constructed. The flax is taken from the bales in small bunches, and each bunch has its ends presented to the "ending machine," which draws the fibres into a parallel position and removes any entanglement from the extremities. The bunches then pass to a heckling-machine, where each is spread out and fixed between two pieces of wood, leaving the ends free. Thus held, the flax is placed on the machine, the chief part of which consists of a revolving apron of stout leather studded with spikes arranged in five or six bands, the spikes increasing in fineness from the feeding side of the machine. The wood clamps, with their dependent flakes of flax, slide along a rail placed above, and running transversely to the apron of the machine. When a fresh flake is laid on all the others move one space to the left, and are brought into contact with the various bands of heckles in succession, and finally emerge with all the fibres nicely dressed, and bearing a gloss which makes the flax look almost like silk. The clamps are then unscrewed, the flax fixed by the end which has passed over the heckles, and again put into the machine, which completes the process by bringing the fresh end of the fibres to the same degree of fineness as the other. Both sides, as well as both ends of each flake, are brought into contact with the heckles, a self-acting motion in the machine turning the clamps over each set of heckles. When the flakes of flax come from the machine the second time they are twisted in the central part, by which means each is kept separate for convenience of handling. The occupation of the hecklers is not a pleasant one, and, to those not acquainted with the trade, it seems wonderful how people can live for many days in an atmosphere so laden with dust as that of the heckling-rooms of a flax-mill. So

R

dense is the air that it is almost impossible to distinguish persons at the remote end of a room thirty yards in length; but, despite that fact, the workers do not suffer so much in their health as would be supposed. Some of them wear respirators, extemporised from a bunch of flax; but few of them take the trouble to use that simple preventive. The rooms are fitted with "dust extractors"—openings in the floor, through which a strong current of air is drawn from the room—and by these the more deleterious particles of dust are removed.

The next process is spinning, which involves several operations—such as spreading and drawing—the object of which is to increase the fineness of the fibres, give them a parallel arrangement, and unite them in a continuous line or sliver. The flax is sent through the drawing-machine again and again, until it comes forth in a smooth even band, about an inch in width and a quarter of an inch in thickness. As the flax in this state has no twist to keep the fibres in place, it is caught off the machine in tall tin cans, and is not subjected to any handling. From the cans the sliver is fed into the roving-machine, by which it is still further drawn out, and slightly twisted. The roving is wound upon large bobbins, which are next placed in the spinning frames, whereby the roving is drawn out to the required degree of fineness, and firmly twisted. Some of the flax is spun by the wet process, which better adapts it to certain purposes. In wet spinning, the roving, in passing over the spinning frame, is made to dip into a receptacle filled with water heated by a steam-pipe. The hot water softens and separates the fibres, and admits of their being drawn out into a finer thread than if spun dry, while at the same time it causes the loose fibres to combine better with the body of the thread. The machinery used in these processes has been much improved in recent years, and is as great an advance upon that employed fifty years ago as the first spinning-jenny was upon the rock and spindle. The full bobbins from the spinning frames are passed to the reelers, who make the yarn up into cuts, heers, hanks, and spindles, each spindle containing 14,400 yards. In some cases the yarn is prepared for the loom directly from the spinning frame; but generally it is bleached first. Messrs Baxter are the principal proprietors of a large bleachfield, and are also the chief employers of several works of the same kind in the district. It would be impossible to conduct the bleaching at their mills in town in consequence of the smoky atmosphere and scarcity of water.

Leaving the bleaching process to be described afterwards, we shall follow the yarn through the other departments. In the winding and warping lofts the hanks of yarn are placed on a frame and wound upon bobbins for the warping-machines, or upon pirns for the shuttles. After the yarn is warped it is dressed by being coated with paste; and when the threads have been drawn through the heddles and reed, the yarn, together with the beam on which it is wound, is placed in the loom. These processes have already been described. At the Dens Works they are all carried on in lofty, well-ventilated, and well-lighted rooms.

The weaving department, in which, as already stated, 1200 power-looms are employed, is broken up by the peculiar construction of the factory; but the principal section of it presents an interesting sight. In a noble apartment most conveniently situated 754 looms are congregated, and a walk along the avenues between the lines in which these are arranged affords an opportunity for seeing the various kinds of fabrics produced by the firm. The principal is navy sail-cloth, of which an immense quantity is made, the chief supply for the British navy being furnished at present, as for many years past, by Messrs Baxter & Co. Bleached and brown sheeting, ducks, paddings, towellings, hammocking, Osnaburgs, and Hessians, are among the other goods manufactured. The total quantity turned out yearly is about 20,000,000 yards. When the webs are taken from the looms they are passed through rubbing-machines, which, by a peculiar action, draw the warp threads closer to each other, and give a more solid body to the cloth. A web, which in the loom measures forty-two inches in width, will, after being rubbed, measure two inches less, and its length is at the same time increased. The cloth is next picked and cropped. The cropping-machine, which is similar to that used by woollen manufacturers, removes loose fibres and any roughness of surface. Calendering is the next operation, and on it the appearance of the cloth greatly depends. There are in Dundee a number of establishments solely devoted to calendering and press-packing linen and jute goods; but in the Dens and other extensive factories the work is done on the premises in a special department. The calender generally consists of five massive rollers, from five to six feet in length, set in an upright frame. Two of the rollers are composed of paper and the others of iron, one of the latter being hollow to admit of its being heated by steam. The rollers may be raised or lowered by hydraulic power, according to the degree of pressure desired. The treatment to which the goods are subjected in the calender varies according to the nature of the fabrics. Thus the cloth may be either beetled, sarceneted, cylindered, chested, or mangled, as may be desired, the different style of finish given being the effect of putting the cloth through the rollers in particular ways, and continuing the operation for a longer or shorter period. The goods are next measured and folded by machinery, and the pieces pressed separately in a hydraulic press worked by steam. When the goods are made up into bales they are again put into the press and reduced to the smallest possible bulk, the amount of pressure put on being upwards of 1000 tons. Previous to the adoption of this mode of treating the goods, the bales were so light in proportion to their bulk that the vessels laden with them had to carry a large quantity of ballast. As the bales are made up a consecutive number is painted upon each, along with the trade-mark of the firm, &c. For the removal of the raw material and goods to and from the factory a dozen horses are constantly employed. In addition to the persons directly engaged in the working of flax, Messrs Baxter & Co. employ a large staff of mechanics, who make and repair all the machinery required about the works. The machine shop and foundry occupy an extensive building, fitted with the finest and most improved machines and tools for working in iron and wood, and, taken apart from the great establishment of which it forms a component part, would be reckoned a considerable place of its kind.

The great care taken to preserve the health and promote the comfort of the bodies of the workpeople has already been mentioned; and now some notice may be taken of what is done for the improvement of their minds. Adjoining the works is a handsome and commodious school-house, to which all employed about the establishment have free access. Every expense connected with the school is defrayed by the proprietors, who take great interest in the education of their operatives, and, by a liberal distribution of prizes, encourage them to persevere in acquiring knowledge. There has been a school in connection with the works for upwards of thirty-eight years, but the present building is only ten years old. The thirty-eighth anniversary festival was held in May 1868, when upwards of 100 prizes were distributed to the pupils. The chair was occupied by Sir David Baxter, the much respected head of the firm, a gentleman who is known far and wide for the liberal and substantial aid which he has given to every good cause that commended itself to him. The annual report of the teacher showed that the average attendance was 570 day scholars and 356 evening scholars, making a total of 926. For the instruction of these there are, in addition to the master and mistress, thirty paid monitors for the evening school and twenty-four for the day school. The branches taught are reading, writing, and arithmetic, in addition to which the girls are instructed in sewing, knitting, and fancy work. There is a library in connection with the

school, to which the elder pupils have access.   The workpeople
generally appear to appreciate the kindness of their employers, to
whom they have never given any trouble by combinations or strikes.
They are a steady, well-conducted class; but this remark, it is but
justice to say, applies equally to the other factory operatives in the
town.   In order to meet in some measure the rapidly-increasing
demand for house accommodation by the working classes, Messrs
Baxter & Co. recently built a large number of houses in the neighbourhood of their factory.

To complete the description given above of the various operations
in manufacturing flax, some account of the bleaching process is necessary.   About a hundred and fifty years ago the Dutch were esteemed
the best bleachers in Europe.   Their method was to steep the cloth
for about eight days in ley made from vegetable ashes.   It was then
washed out with black soap, and placed to steep for about a week
in a vessel filled with butter milk.   After another washing with
black soap, the cloth was spread on the grass for two or three weeks,
during which time it was sprinkled at regular intervals with clear
water.   All these operations had to be repeated several times before
the cloth was brought to the required degree of purity, so that the
material was for six or seven months in the hands of the bleachers.
All the fine linen made in Scotland was at one time sent to Holland
to be bleached.   The Board of Manufactures paid great attention
to this department of the linen trade, and, as already stated,
granted liberal rewards to persons who established bleach-fields.
The Board paid the following sums for experiments in bleaching:—
To James Spalding, L.180; to Dr Wm. Cullen, Glasgow, L.21; and
to Dr Francis Home, L.100.   The first important improvement in
bleaching in this country was made by Dr Home, who for butter
milk substituted water acidulated by sulphuric acid.   This greatly
facilitated operations, as it enabled the bleachers to do in twelve
hours what formerly required nearly as many days.   In 1785 chlorine
was discovered and successfully applied to bleaching by Berthollet,
a French chemist.   An establishment for bleaching by chlorine was
erected at Aberdeen in 1787, and was the first of the kind in this
country.   Chloride of lime, a substance of more convenient application, was discovered in 1798 by Mr Tennant of Glasgow, and is
now the principal chemical stuff used in bleaching.   Mr Alexander
Drimmie, in 1820, substituted soda ash for potash ley in bleaching,
thereby reducing the cost of the operation, while linen cloth might
be bleached in a few days by the use of the soda ash alone, almost
without exposure on the grass.   In 1825 Mr Drimmie effected a

further and important improvement by inventing a machine for washing the cloth. The substances which require to be got rid of by bleaching are—first, the organic colouring matter naturally present in the fibre; second, resinous and fatty bodies, also inherent in the fibre; third, weavers' dressing and perspiration taken up during the process of weaving; and fourth, certain saline or earthy substances. To separate these from the cloth, it is subjected to a series of operations such as washing, boiling in lime-water, steeping in a solution of sulphuric acid, and so forth. The cloth is then sent to the calender and finished. Cotton loses about one-twentieth of its weight by bleaching, and linen about one-third. There are a number of extensive bleach-fields in Forfarshire, Perthshire, Fifeshire, &c., some having a direct connection with linen factories, and others being carried on as separate undertakings.

The importance to Dundee of the introduction of jute has already been pointed out. In the course of a few years the Indian fibre has come so extensively into use, that the manufacture of linen on which it was grafted has been deposed from the position which it long occupied as the staple trade of the town. In dealing with the trade of Dundee, however, it would be difficult to dissociate flax and jute ; for, though one or two firms confine their attention solely to the manufacture of flax, and a few to jute alone, all the other manufacturers work both fibres, sometimes mixing them in certain proportions, and at others keeping them distinct. The establishment of Messrs Baxter Brothers & Co. was chosen to illustrate the manufacture of pure flax, and now some account will be given of an equally extensive factory in which jute alone is used—namely, the Camperdown Linen Works at Lochee, belonging to Messrs Cox Brothers.

The village of Lochee, which lies within the municipal boundary of Dundee, had an early connection with the linen trade. Towards the close of last century there were nearly 300 looms in the village, and these were chiefly employed on coarse linens, of which 4860 pieces, valued at L.12,520, were produced annually. The cloth was bought by several merchant weavers, who disposed of it in Dundee and Perth, or sent it into the English market. The first of those merchant weavers was a Mr Cox, who died in 1741, and whose family are mentioned in the statistical account of the parish, published in 1793. It is stated that the family were then engaged in the same line, much to the credit and advantage of themselves, and that to their industry and example the district was principally indebted for its flourishing condition. They bought cloth from the weavers, as their ancestors had done, and after bleaching it at their bleach-work near Lochee, sent it into the market. The trade they carried on would appear to have always been in a healthy state, and a gradual extension of the bleaching department took place, until the fields in connection with the work measured not less than twenty-five acres. At the close of the bleaching season in 1819, when the warehouses were filled with finished cloth, a fire broke out and consumed the whole, entailing enormous loss on the representative of the fourth generation of the family who then owned the premises. Instead of rebuilding the bleach-work in a permanent way, Mr Cox ran up a few temporary buildings to serve till the expiry of his lease. Meantime he had turned attention to the manufacturing department of the trade, and established a weaving factory in Lochee. Mr Cox was succeeded in 1827 by his eldest son, who in 1841 took three of his brothers into partnership with him, and these gentlemen constitute the present firm of Cox Brothers, whose factory is one of the most extensive and complete of the kind in Britain. Messrs Cox were among the first who made experiments with jute, and such was their success therein, that they gradually discarded flax, and now their vast establishment is entirely devoted to the manufacture of jute. They have their own buyers at Calcutta, and import the raw material direct; while they are, perhaps, the only firm in the district who complete within their works all the operations of spinning, bleaching, dyeing, weaving, printing, calendering, and packing.

The Camperdown Linen Works, the chief portion of which was built between the years 1845 and 1850, occupy eighteen acres of level ground on the north side of the village of Lochee. The works have been constructed on a regular and well-considered plan, so as to admit of almost unlimited extensions without interfering with the convenience of arrangements whereby the various processes are conducted without waste of time or labour in shifting the material about. The design of the buildings is characterised by much neatness; and an elegance and airiness pervade the place which show an extraordinary advance on the notions as to what a factory should be. It is not many years since the ideal of a factory was a hideously plain building of many low storeys, into which the light struggled through windows about two feet square, the dust and dirt on which it would have been considered something like sacrilege to have removed. Anything approaching to ventilation was not thought of, and, consequently, no provision was made for the admission of air to the sickly operatives. Now, and particularly in the case of the factory under notice, the storeys are from fourteen to seventeen feet in height, and every room is thoroughly ventilated ; indeed, no class of workers are better cared for in the matters of light and ventilation than those in the more extensive factories in the linen manufacturing districts. It may be that the Factory Acts are to some extent to be credited with this ; but it is due to the owners of many of the factories to say that in the matters referred to they have far exceeded the requirements of the law.

A branch line of railway connects the Camperdown Works with the main system, and by it some of the raw material is brought in and the finished goods are sent out. The jute is deposited in two large stores, detached from the main body of the factory, whence it is withdrawn as required. The first operation in manufacturing the fibre is "batching." One of the great obstacles which the early workers in jute had to contend with was the hard and dry nature of the fibre. They could neither get it to spin nor weave satisfactorily. Old machines were altered and new ones devised with the view of overcoming this peculiarity of the jute ; but none of these were successful until the idea occurred to some one that the jute might be softened by being moistened with oil. This was tried and found successful to a degree beyond expectation. The oil is applied in a special apartment called the "batching-room," in which the jute is spread in layers, each layer receiving an abundant sprinkling of oil and water. In that condition the material is allowed to lie a certain time, according to the season and temperature. The fibres of jute are from five to eight feet in length, and sometimes even more, and in order to bring them to a spinning condition, they used to be cut ; but as a square end was not favourable to complete hackling nor correct spinning, the fibres are now torn asunder by being fastened by the ends to iron bars placed on either side of a wheel having a number of stout spikes on its rim. After a handful of jute is fastened to the bars, the latter are thrown forward, the spikes strike the jute in the centre, the fibres are dissevered, and a fine pointed end appears on each side. From this stage the processes which the jute goes through in being converted into cloth are so similar to those to which flax is subjected that it is unnecessary to describe them in detail. The machinery used in the Camperdown Works is of the latest and most improved construction, and is all made on the premises.

In the weaving-shed 700 power-looms are employed in making plain and twilled sackings, and all the other fabrics usually made of jute ; and in another part of the establishment 300 hand-looms are engaged on carpeting. The firm have paid much attention to the last-named branch, and have brought the manufacture of jute carpets to great perfection. For certain kinds of carpeting Messrs Cox hold a patent, and some of their productions are characterised by considerable beauty of design. There is an extensive dyeing shop at the works in which all the yarns required for coloured goods are dyed. The colours used are of the most brilliant hues, and the jute takes them on more readily than any other fibre known. Jute carpeting is so cheap that it is within the reach of the humblest householder. Some of it is sold so low as 8d. a-yard, and, considering its appearance and durability, is a wonderful bargain.

A few statistics relating to the factory will show its extent and importance in a more forcible way than any minute description of the departments. As already stated, the works occupy 18 acres of ground, a considerable portion of which is covered by buildings. The area of the floors is 50,000 square yards, or about 10½ acres. The machinery is propelled by steam-engines, varying from 3 to 120 horse power ; the aggregate nominal horse power is 580 ; and the indicated horse power 1850. The steam for the engines is generated in 22 boilers ranged side by side in one line. The smoke from the furnaces is carried off by an ornamental chimney 300 feet in height and 35 feet in diameter at the base. The chimney alone cost over L.3000. The quantity of coal consumed is about 15,000 tons annually. There are 4300 persons employed within the works, and, in addition to these, the firm employ 400 sack-sewers, who work in their own houses. The wages paid are the same as those current in the trade. 14,000,000 yards of sacking are turned out annually, and about half that quantity of other fabrics. There is a free school for the workers, at which there is a regular attendance of about 400 pupils. The factory operatives are informed of the flight of time by a splendid turret clock, which chimes the quarters.

In order to complete this record of the linen and jute trade, it is essential that some mention should be made of a factory devoted to the production of mixed fabrics. Messrs James Smieton & Son, Dundee, have for many years taken the lead in introducing new fabrics made of combinations of various materials. In 1857 they selected a piece of ground to the west of the village of Carnoustie, and thereon built a power-loom factory, which bears the name of Panmure Works, and is regarded by the trade as a model establishment. It is small in size compared with the great factories which have been previously described, but in convenient arrangement of departments and completeness of organisation it is unsurpassed. The ground occupied by and pertaining to the factory is about 10 acres in extent, and has a frontage to the Caledonian Railway of 700 feet. The central part of that frontage is occupied by a fine two-storey building, 325 feet in length. There is a siding on the railway for the service of the factory. The waggons containing the yarn are passed on to this siding, and brought into the works one by one. The yarns used are respectively composed of flax, tow, hemp, cotton, and jute. Some of these are made in Forfarshire and Glasgow, and some are imported from France and Ireland. The weaving shed occupies a central position in the works, being bounded in front by the main building, in rear by the calendering and finishing department ; on one side by the warping and dressing room, and on the other by the engines and packing warehouse. The weaving shed is 180 feet in length by 150 in width, and contains 400 power-looms. The departments in which the successive operations are carried on adjoin each other ; and the yarn, which passes in at one side of the main entrance, makes a circuit of the place, and emerges at the other side packed up ready for transportation. This is an admirable arrangement, which at once facilitates operations and saves expense. The variety of fabrics that may be produced by using different qualities and combinations of yarn is immense. Usually there are no fewer than 80 different kinds of cloth in the looms at one time, and the list of fabrics made embraces upwards of 500 varieties. Mixtures of tow with jute and of flax with jute are the principal; but a great variety of fabrics are produced by mixing cotton with jute and cotton with flax. The quantity of cloth made is about 5,000,000 yards annually, a large proportion of which consists of "drills," " paddings," and "Russian sheetings " for the United States, West Indian, and Mexican markets. Checks and stripes in endless variety are also made for the same countries. The machinery, which is of the most improved kind, includes 4 cropping-machines, 3 calenders, and 1 mangle, the latter working under a pressure of 60 tons. The combined force of the 3 steam-engines in use is upwards of 200 horse power. The number of persons employed is about 600.

About five years ago Messrs Smieton spent L.2000 in the erection of an institute for the use of the workpeople. The building, which is two storeys in height, is one of the finest in or near Carnoustie, and is a handsome monument of the liberality of the founders. The ground-floor of the institute is occupied by a house for the keeper, a class-room, library, reading-room, and cloak-room. The reading-room is liberally supplied with newspapers and magazines, and furnished with a bagatelle board, draught boards, &c. The upper portion of the building consists of a fine hall, provided with a piano and a harmonium. The day scholars are taught in the class-room, and in the evening such of the female workers as choose attend in the hall to receive instruction in reading, writing, and needlework. Two teachers, and all the books, stationery, newspapers, magazines, &c., are provided free of charge. Indeed, the whole expenses of the institute are defrayed by Messrs Smieton, to whom the yearly cost is about L.300. Finding that their workpeople had difficulty in obtaining suitable accommodation in Carnoustie, the firm recently built eighty houses in the neighbourhood of the factory.

At the close of 1867 there were in Scotland sixteen firms engaged in the manufacture of flax, hemp, and jute, who employed 1000 persons or upwards, the aggregate number of operatives being 31,162, or an average of about 1948. Four of those firms were spinners, but not weavers, and all the others were spinners and weavers. The number of spindles employed by them was 205,454; of power-looms, 5177; and the nominal horse power of their engines was 6057. The works of eight of the firms are in Dundee, two in Glasgow, two in Greenock, and one each in Aberdeen, Johnstone, Markinch, and Arbroath.

Of the great firms in Dundee, two have already been noticed. The third in order is that of Messrs A. & D. Edward & Co., established in 1828. Their factory is situated in the Scouringburn, and the main portion of it consists of a fine building 300 feet in length, and five storeys in height, in which the spinning operations are carried on. The weaving department occupies an extensive range of buildings behind the main block. In the spinning mill there are 18,476 spindles; and in the weaving factory, 600 power-looms. All the goods are finished and packed on the premises. The works, like nearly all those recently built, are fire-proof throughout. The goods manufactured embrace all varieties of flax and jute fabrics; and the establishment is exceptional, as being the only one in the district in which linen damasks are made on an extensive scale. The number of persons employed is 3300. Messrs Gilroy Brothers employ upwards of 2000 persons in their Tay Works in the Lochee Road. The establishment has a frontage of 1000 feet, of which 400 feet are occupied by a splendid block of building recently erected. The latter is by far the largest and most imposing structure in Dundee. It is five storeys in height in the centre, and four in the wings.

The central part terminates in a pediment, in the tympanum of which the Dundee arms are boldly sculptured in stone, and on the apex there is a colossal figure of Minerva with her spindle and distaff. The firm have greatly extended their works since 1851. Then they employed engines of 80 horse power; now more than four times that power is required to move their machinery. The material manufactured consists chiefly of jute. Messrs Gilroy are extensive shipowners, and the jute used in the factory is brought from India in their own vessels. A spinning mill, built by Messrs J. & A. D. Grimond at Bow Bridge, Dundee, in 1857, was, and is still considered to be, the finest structure of the kind in existence. The machinery is of the best construction, and the building throughout is elegantly fitted up. There are 3600 spindles in the mill, and 136 power-looms in the weaving department. Messrs Grimond have hand-loom and power-loom factories at Maxwelltown, and altogether employ about 2000 persons. Among the goods manufactured are carpeting, matting, and hearth-rugs. Mr O. G. Miller owns five mills in Dundee. They stand contiguous to each other, and all are devoted to spinning. There are ten steam-engines of 260 horse power, 16,970 spindles, and nearly 2000 workpeople. The St Roque Spinning Mill, and the Wallace Power-Loom Factory, owned by Messrs W. R. Morrison & Co., are extensive concerns. They contain 5000 spindles, 510 power-looms, and the motive power is supplied by engines of 220 horse power. The workpeople number about 2100. The Seafield Works, belonging to Messrs Thomson, Shepherd, & Briggs, were started about fourteen years ago, and have had a rapid growth. They now contain 6000 spindles, 120 power-looms, and give employment to upwards of 1000 persons. The firms out of Dundee who, at the period to which these remarks refer, employed 1000 persons and upwards were:—Messrs Richards & Co., Aberdeen; City of Glasgow Flax-Spinning Company; Glasgow Jute Company; Messrs Finlayson, Bousefield, & Co., Johnstone; Gourock Rope-Work Company, Greenock; Messrs John Fergus & Co., Prinlaws; Mr Andrew Lowson, Arbroath; and the Greenock Sacking Company.

The following figures compiled towards the close of 1867 represent pretty exactly the extent of the linen and jute trades of Dundee:—The estimated quantity of yarn spun annually in the town is 31,000,000 spindles, valued at L.3,487,500; in the surrounding district, 29,000,000 spindles, valued at L.3,262,500—making a total of 60,000,000 spindles, valued at L.6,750,000. Taking the power-looms at 8000, and the quantity of cloth produced by each at 200

yards a-week, the cloth turned out in a year will amount to the prodigious quantity of 83,200,000 yards, or 47,272 miles. The value of
the yarn and linen together is estimated at L.8,000,000. The capital invested in the factories in Dundee is stated to be L.2,500,000
in the district of which that town is the centre, L.2,200,000; in
other parts of Scotland, L.1,000,000—total, L.5,700,000, to which
has to be added the value of the bleach-works, calenders, &c., in the
trade, which cannot be put down at less than L.1,300,000. It
takes about six months from the purchase of the raw material
before the goods can be manufactured and the proceeds drawn, so
that the stock-in-trade of manufacturers and merchants will amount
to L.5,000,000. It would thus appear that a capital of L.12,000,000
is required for carrying on the linen trade of Scotland.

COTTON MANUFACTURES.

THE EARLY DAYS OF THE BRITISH COTTON TRADE—THE INVENTION OF SPIN-
NING AND WEAVING MACHINERY, AND ITS EFFECT IN EXTENDING THE
MANUFACTURING INDUSTRIES OF THE COUNTRY—INTRODUCTION OF THE
COTTON TRADE INTO SCOTLAND—NOTES ON THE FIRST FACTORIES—THE
MANUFACTURE OF MUSLIN—TRADE-UNIONS, STRIKES, AND RIOTS—PRO-
GRESS OF THE COTTON MANUFACTURES IN SCOTLAND—EFFECTS OF THE
AMERICAN WAR ON THE TRADE—THE COTTON FAMINE—DESCRIPTION OF A
GLASGOW COTTON-MILL.

COTTON-WOOL has been known in Britain for at least six hundred
years; but for four centuries the only use to which it was put was
the formation of candle-wicks. At least no mention occurs of the
manufacture of the fibre into cloth until 1641. Mr Baines, the
historian of the cotton trade, has recorded his belief that the art of
spinning and weaving cotton was brought into England by the
Flemish refugees, in the end of the sixteenth century. Others are
of opinion that the art slowly spread from its birthplace in India,
where it has been practised from time immemorial—first into Arabia,
then westward through Spain to Britain. The first authentic reference to the manufacture of cotton in this country occurs in "The
Treasure of Traffic," a small work published in 1641, and is as follows:—"The town of Manchester, in Lancashire, must be also herein
remembered, and worthily, for their encouragement, commended,
who buy the yarne of the Irish in great quantity; and weaving it,
return the same again into Ireland to sell. Neither doth their industry rest here, for they buy cotton-wool in London, that comes
first from Cyprus and Smyrna, and at home worke the same, and
perfect it into fustians, vermilions, dimities, and other such stuffes,
and then return it to London, where the same is vented and sold,
and not seldom sent into forrain parts." No attempt appears to
have been made at that time to imitate the fine cotton fabrics of
India, which were imported in large quantities by the English and

Dutch East India Companies. The muslins, chintzes, and calicoes of the East became extremely fashionable for ladies and children's dresses. In 1678 this import trade had become so extensive that an outcry was raised against it, on account of the prejudicial effect it had on the woollen and silk manufactures already established in England. Public opinion found vent in numerous pamphlets, some of which presented extraordinary views of political economy. The home cotton trade was too insignificant then to be recognised as at all interfering with the other textile manufactures, and no mention was made of it by the pamphleteers. The author of a *brochure* entitled " The Naked Truth, in an Essay upon Trade," published in 1696, says :—" The commodities that we chiefly receive from the East Indies are calicoes, muslins, Indian wrought silks, pepper, salt-petre, indigo, &c. The advantage of the company is chiefly in their muslin and Indian silks (a great value in these commodities being comprehended in a small bulk), and these are becoming the general wear in England. Fashion is truly termed a witch—the dearer and scarcer any commodity, the more the mode. Thirty shillings a-yard for muslins, and only the shadow of a commodity when procured !" The Government were at length induced to interfere to prohibit the use of Indian goods. An Act of Parliament was passed in 1700, which forbade the introduction of Indian silk and printed calicoes for domestic use, either in apparel or furniture, under a penalty of L.200 on the weaver or seller. So strong was the desire to possess the forbidden goods, that an extensive system of smuggling sprang up, and further measures were necessary in order to accomplish the purposes of the first Act.

The manufacture of cotton was carried on in every quarter of the globe—its chief seats in Europe being Italy, Spain, Turkey, Bavaria, Saxony, Prussia, and the Low Countries—before it was introduced into England. But though thus almost the last to take up the trade, no country has done more than England to perfect the manufacture, nor has any one profited more by it. The cotton manufacture is the staple trade of Britain, employing about a million of the inhabitants, and yielding in the form of wages and profits the immense sum of L.60,000,000 a-year. Cotton cloth is the cheapest article of clothing manufactured, and Britain is the chief source from which the markets of most countries are supplied. Such being the case, a brief sketch of the general history of the cotton manufacture may be properly given here, as a prelude to a notice of the introduction and develop-ment of the trade in Scotland.

The first mention of cotton as an article of import occurs in the

Customs' books for 1697. In that year 1,976,359 ℔. of cotton-wool were imported into Britain; and the cotton goods exported were officially valued at L.5915. The quantity of cotton-wool imported showed rather a decrease in succeeding years until 1746, when it suddenly leaped up to 2,264,868 ℔. The following table of the imports and exports for a number of years will show how the trade has increased:—

Years.	Imports. ℔.	Exports. ℔.
1781, . .	5,198,778	96,788
1790, . .	31,447,605	844,154
1800, . .	56,010,732	4,416,610
1810, . .	132,488,935	8,787,109
1820, . .	151,672,655	6,024,038
1830, . .	263,961,452	8,534,976
1850, . .	694,996,000	108,294,800
1860, . .	1,345,597,600	243,600,000
1867, . .	1,400,308,400	406,016,000

The cotton trade owes its marvellous development to a variety of mechanical contrivances, the history of which forms one of the most interesting chapters in the records of inventions. The machines used in the textile manufactures generally were of a primitive kind till past the middle of last century. In 1760 the Society of Arts offered a premium for the greatest improvement in the common spinning-wheel, which, excepting the distaff and spindle, was the only apparatus then known by which a thread could be formed. The society afterwards offered a prize of L.100 for the invention of a machine that would spin six threads of wool, cotton, flax, or silk at the same time. This roused to action the minds of many mechanicians, and the result was a triumphant success. When the demand for cotton cloth increased beyond the powers of production, the price rose considerably, and weaving became a favourite occupation; but the spinners were not equal to the task of keeping all the weavers employed. The trade was then a domestic one, the husband generally weaving the yarn spun by his wife, and both being assisted by their children. Up till 1773 cotton was never used alone in the formation of cloth. The yarn was not considered to be strong enough for warp, and accordingly linen yarns, procured chiefly from Ireland and Germany, were used for that portion of the fabric. The merchant supplied the linen yarn, with a certain proportion of raw cotton, to the weaver; and if the latter had not a family who could spin the cotton, he employed other persons to do so. It is stated to have been no uncommon thing for a weaver to walk three or four miles in a morning, and call on five or six spinners, before he could collect weft sufficient to serve him for the remainder of the day; and when he wished to weave a piece in a shorter time than usual, a new ribbon or a gown was necessary to quicken the exertions of the spinner.

The productive power of the loom was doubled by the invention of the "fly shuttle," and as that happened at the time it was found most difficult to obtain yarn, Mr Kay, the inventor, was subjected to great persecution by the weavers, who feared that their occupation was endangered by the invention, and the result was that he had to leave the country. A dozen years before the Society of Arts moved in the matter, a machine for spinning by rollers was invented by Mr John Wyatt, and its practicability successfully demonstrated; but Mr Wyatt shared the too common lot of inventors, and failed to reap the fruits of his ingenuity. His memory will be preserved, however, as the inventor of the primary principle of a most important part of the spinning-machinery now in use throughout the world. But the chief honours in machine spinning belong to Hargreaves, the inventor of the spinning-jenny; to Arkwright, the inventor of the spinning-frame; and to Crompton, who, in 1779, combined the action of both machines in the "mule jenny." Though the increase of spinning power, so necessary and so much desired, was thus provided, the workpeople offered great opposition to the introduction of the new machines.

In 1779 a mob rose and scoured the country for several miles round Blackburn, demolishing the jennies, and with them the carding-engines, and every machine set in motion by horses or by water. It would appear that the rioters admitted the jennies containing twenty spindles to be useful, and they spared all such; but those which contained more than twenty spindles were either destroyed or cut down to the standard which the mob had fixed. The sentiments and actions of the rioters were sympathised with and participated in by the middle and upper classes, who, failing to perceive the tendency of inventions for improving and cheapening the manufacture to cause an extended demand, and thereby give employment to more hands than were in the first instance superseded, became alarmed lest they should be subjected to increased taxation for the support of the workmen who would be thrown idle by the use of machinery. Spinners and other capitalists were driven from the locality in which the riots took place, the trade became almost extinct, and it was many years before cotton-spinning was resumed at Blackburn. Mr Peel, the great-grandfather of the present Sir Robert Peel, was among the sufferers at the hands of the rioters, the machinery of his cotton-spinning mill having been thrown into a river, and his personal safety threatened. A large mill built by Arkwright at Birkacre, near Morley, was destroyed by a mob in the presence of a powerful body of military and police, who failed to act in consequence of not being called upon to do so by the civil authorities.

Contrary to the absurd notions and desire of the workpeople, the spinning machines were generally adopted by manufacturers, and results were achieved which convinced everybody of the value of the inventions. Having succeeded in overcoming the prejudices of their operatives, the manufacturers began to entertain feelings of jealousy towards each other. The spirit which prevailed is described in the following extract from "Baines' History of the Cotton Manufacture:"—"This period of high intellectual excitement and successful effort would be contemplated with more pleasure, if there had not at the same time been displayed the workings of an insatiable cupidity and sordid jealousy, which remorselessly snatched from genius the fruit of its creations, and even proscribed the men to whom the manufacture was most deeply indebted. Ignorance on the one hand, and cupidity on the other, combined to rob inventors of their reward. Arkwright, though the most successful of his class, had to encounter the animosity of his fellow-manufacturers in various forms. Those in Lancashire refused to buy his yarns, though superior to all others, and actually combined to discountenance a new branch of their own manufacture, because he was the first to introduce it. He has related the difficulties with which he had to contend in his 'Case.' 'It was not,' he said, 'till upwards of five years had elapsed after obtaining his first patent, and more than L.12,000 had been expended in machinery and buildings, that any profit accrued to himself and partners.' 'The most excellent yarn or twist was produced; notwithstanding which, the proprietors found great difficulty to introduce it into public use. A very heavy and valuable stock, in consequence of these difficulties, lay upon their hands; inconveniences and disadvantages of no small consideration followed. Whatever were the motives which induced the rejection of it, they were thereby necessarily driven to attempt, by their own strength and ability, the manufacture of the yarn. Their first trial was in weaving it into stockings, which succeeded; and soon established the

manufacture of calicoes, which promises to be one of the first manufactures in this kingdom. Another still more formidable difficulty
arose: the orders for goods which they had received, being considerable, were unexpectedly countermanded, the officers of excise refusing
to let them pass at the usual duty of 3d. per yard, insisting on the
additional duty of 3d. per yard, as being calicoes, though manufactured in England; besides, these calicoes, when printed, were
prohibited. By this unforeseen obstruction, a very considerable and
very valuable stock of calicoes accumulated. An application to the
Commissioners of Excise was attended with no success; the proprietors, therefore, had no resource but to ask relief of the Legislature,
which, after much money expended, and against a strong opposition
of the manufacturers in Lancashire, they obtained.'"

While the spinning machinery was being brought to perfection, it
became evident that something would require to be done to improve
the preliminary process of carding the cotton. Lewis Paul, who was
the patentee of Wyatt's invention for spinning by rollers, had so
early as 1748 taken out a patent for a cylinder carding-machine, the
various parts of which bore a close resemblance to the carding-
engines at present in use. Paul's machine was defective in many
ways, however, and little progress was made towards producing a
really efficient mechanical carder until Arkwright devoted his
ingenious mind to the subject. On the 16th December 1775 the
great mechanician took out a patent for a series of apparatus, comprising the carding, drawing, and roving machines. These were so
decidedly advantageous that they were at once adopted, and the
effect on the trade was almost magical. The factory system, which,
except to a small extent in the silk manufacture, was then unknown
in England, became established throughout the country; and the
mechanism devised for spinning cotton was applied to the spinning
of wool, flax, &c., so that all the textile manufactures of the country
received a gigantic impulse from the introduction of machinery to
supersede hand labour. Arkwright prospered in business, but he
was not allowed to enjoy an undisputed title to his inventions. His
success stimulated the jealousy of his fellow-manufacturers; and as
there was a belief prevalent in Manchester that he was not really the
author of the inventions for which he claimed patents, several
persons ventured to set up machines similar to his without obtaining
a license. An association of Lancashire spinners was formed to
defend the actions raised by him against the persons who infringed
his patents. On various pretexts his patent for preparing-machines
was set aside in 1785. He resented the treatment to which he had

been subjected by his competitors in the trade, and exerted himself to raise up in Scotland a successful rivalry to Lancashire. With that view he favoured the Scotch spinners as much as possible, and formed a partnership with Mr David Dale of Lanark Mills.

Though the invention of the spinning-frame, spinning-jenny, and carding-engines did much to advance the manufacture of cotton, something was left to be desired. The water-frame produced suitable yarn for warp, and the jenny made excellent weft; but they were not capable of making the finer qualities of yarn. In 1779 a weaver named Samuel Crompton succeeded in producing machine which combined the chief features of the water-frame and spinning-jenny, and was capable of producing yarns suitable for muslins. Owing to its hybrid origin, the new machine was called the "spinning mule." Crompton toiled at his loom in an old mansion-house, and spent all his spare time and money in working out his invention. He was of a retiring disposition; and when the machine was completed he wished, in his own words, "to enjoy his little invention to himself." The yarn he produced was so superior in quality that persons from all quarters sought him out in order to ascertain how he spun it. He found that he could not retain the secret of his invention, nor was he rich enough to patent it, so he gave it to the public on condition that a petty sum (L.60) should be raised by subcription. He subsequently got a grant of L.5000 from Parliament. The next step in advance was made by Mr Kelly of Lanark Mills, who in 1790 applied water power to work the mill, and two years later communicated a self-acting motion to the mule. Mr Buchanan, of the Catrine Cotton Works, also invented a self-acting mule. Many minor improvements have since been made, and as showing the degree of perfection that has been attained, it may be mentioned that the spinning-mule has been found capable of forming, from one pound weight of cotton, a thread 950 miles in length, whereas the water-frame of Arkwright made only forty hanks to the pound, or a length of nineteen miles. A pound of the finest cotton yarn may be worked into lace worth L.250.

Prior to the year 1790 the only motive power applied to the machinery of the cotton mills was water, and mills could be erected only where an abundant supply of that element was available. Lancashire owes its early and extensive connection with the cotton trade mainly to the fact that the county is intersected by a large number of streams which descend rapidly from the hills in the hundreds of Blackburn and Salford. Thirty years ago there were on the river Irwell alone 300 mills propelled by water. When all the available water power was taken up in any locality, there was a bar to any increase or extension of the mills; and were no other motive power available, the cotton trade would be distributed more generally over the country. But when the manufacturers were beginning to realise the disadvantages of depending on the water power, James Watt was completing his improvements on the steam-engine. Watt's engine found favour in the eyes of the cotton manufacturers, and came into general use. Supplied with a motive agent of unlimited and inexhaustible power, which could be made available in almost any locality, the manufacturers felt their position to be much improved. It was no longer necessary for one proposing to build a mill to range the country in search of a waterfall of sufficient strength to keep his machinery going, and, having found such—it might be far away from any centre of population—to convey thither not only the appliances and material necessary to carry on the work, but to induce an adequate number of workpeople to take up their abode in the neighbourhood of the mill. The steam-engine enabled him to set up his mill in the midst of the people.

When the spinning machinery was brought to a degree of perfection, it was found that a good deal more yarn was produced than the weavers could use up, and a large quantity was exported. Attempts to construct a machine for weaving had been made first in 1678 and again in 1765; but they were unsuccessful, and the fact of their having been made was all but forgotten. In the year 1784 a company of gentlemen had met at Matlock, and, some of them being manufacturers from Manchester, conversation naturally turned on the inventions of Arkwright. The Rev. Dr Cartwright, of Kent, who was present, remarked that Arkwright, having completed his spinning machines, would next require to invent machinery for weaving. The other gentlemen expressed a unanimous opinion that such a thing was impracticable. What followed is related by Dr Cartwright in a letter which he wrote to Mr Bannatyne of Glasgow. "In defence of this opinion," he says, "they adduced arguments which I certainly was incompetent to answer, or even to comprehend, being totally ignorant of the subject, having never at that time seen a person weave. I controverted, however, the impracticability of the thing by remarking, that there had lately been exhibited in London an automaton figure which played at chess. Now you will not assert, gentlemen, said I, that it is more difficult to construct a machine that shall weave than one which shall make all the variety of moves which are required in that complicated game. Some little time afterwards I employed a carpenter and smith to carry my ideas into effect. As soon as the machine was finished, I got a weaver to put in the warp, which was of such materials as sail-cloth is usually made. To my great delight a piece of cloth, such as it was, was the produce. As I had never before turned my thoughts to anything mechanical, either in theory or practice, nor had ever seen a loom at work, or knew anything of its construction, you will readily suppose that my first loom was a most rude piece of machinery. The warp was placed perpendicularly, the reed fell with the weight of at least half a hundredweight, and the springs which threw the shuttle were strong enough to have thrown a Congreve rocket. In short, it required the strength of two powerful men to work the machine at a slow rate, and only for a short time. Conceiving, in my great simplicity, that I had accomplished all that was required, I then secured what I thought a most valuable property by patent, 4th April 1785. This being done, I then condescended to see how other people wove; and you will guess my astonishment when I compared their easy mode of operation with mine. Availing myself, however, of what I then saw, I made a loom in its general principles nearly as they are now made." Dr Cartwright in 1809 received a grant of L.10,000 for his ingenious invention. Other inventors entered the field, and, while some devoted their attention to improving on Dr Cartwright's loom, others set themselves to construct an entirely novel machine. Reference has been made to some of those inventors in dealing with the woollen manufactures.

It had been the custom in hand-loom weaving to "dress" the warp in the loom, for which purpose frequent stoppages had to be made. In order to keep the power-looms going steadily, a man was required to dress the warp while another attended to the weaving. The extra hand consumed all the profits of the improved loom ; and the next thing demanded was an apparatus that would dispense with his services. Mr William Radcliffe, cotton manufacturer, of Stockton, set to himself the task of overcoming the difficulty. On the 2d January 1802 he shut himself up in his mill along with a number of weavers and mechanics, resolved to produce some improvement. After two years of experiments the dressing-machine was produced, and by its use the power-loom was rendered fully efficient. The power-loom most commonly employed at present was invented by Mr Horrocks, of Stockport, between the years 1803 and 1810. It is constructed entirely of iron, and is a neat, compact, and simple machine, moving with great rapidity, and occupying little space.

A few facts will illustrate the effect which the inventions men-

tioned have had on the cotton trade. In 1786 the yarn known in
the trade as No. 100 sold at L.1, 18s. a pound. Seven years afterwards the price had fallen to 15s. 1d. In 1800 the price was 9s. 5d.,
and in 1832, 2s. 11d. The cost price of a piece of calico was L.1,
3s. 10½d. in 1814. In 1822 the same could be made for 8s. 11d.,
and in 1832 for 5s. 10¾d. The official value of the cotton goods
exported from Britain in 1720 was L.16,200 ; in 1780, L.355,060 ;
in 1795, L.2,433,331 ; in 1810, L.17,898,519 ; and in 1830,
L.35,395,400. The value of the raw and manufactured cotton
exported from Britain at present is about L.80,000,000 a-year.

The wonderful inventions of Arkwright and others, and the
impulse that these gave to the cotton trade, did not escape the
notice of Scotch manufacturers ; but it would appear that, though
they entered into the manufacture of cotton with great spirit and
enterprise after it had been carried north of the Tweed, the credit of
introducing it belongs to Englishmen. The first cotton mill in
Scotland was built at Rothesay in the year 1778 by an English
company, but was not long in being acquired by Mr David Dale,
who became one of the most extensive cotton manufacturers in
the country. The mill was of small extent, and in the present
day would be regarded as an almost insignificant concern ; but
it was the nucleus of one of the most important branches of
industry that has ever been carried on in Scotland. The germ
planted by English hands had a rapid growth ; and, before the
Rothesay mill was sixty years old, there were nearly two hundred
cotton factories in Scotland. Lanarkshire and Renfrewshire were
chosen as the chief seats of the trade, partly on account of the
abundant supply of water power available, and partly because persons with the capital and enterprise required to carry on the new
trade were more numerous in the west, while many of them had
previously been engaged in manufacturing soft goods, and so were
most likely to appreciate the value of cotton. Glasgow has all along
been the centre of the trade ; and nearly the whole of the cotton
goods manufactured in Scotland are made by or for firms having
their headquarters in that city.
In the year 1787 there were nineteen mills in Scotland, all driven
by water. They were distributed as follows :—Lanarkshire, four ;
Renfrewshire, four ; Perthshire, three ; Mid-Lothian, two ; other places,
six. The second mill in Scotland was built at Dovecothall, on the
banks of the Leven in Renfrewshire. It consisted of three storeys,
measuring fifty-four feet in length, twenty-four feet in breadth, and

eight feet in height.   This mill proved so remunerative that it was soon enlarged, and five similar establishments of considerable size were erected in the same locality.   A cotton mill of what was then considered to be an immense size was built at Johnstone in 1782, and the locality being favourable for the trade, others followed, until in 1837 there were eleven mills in the town.   Elsewhere in Renfrewshire numerous spinning mills were set up about the beginning of the present century, and a large number of the inhabitants were engaged in weaving the yarn produced.   About thirty years ago one of the finest cotton mills in the country was built on Shaws Water, near Greenock ; but the only thing that remains remarkable about it is a water-wheel of 120 horse power.   The wheel is said to be the largest in the world, being seventy feet in diameter.   It is composed of iron, and weighs 180 tons.

Renfrewshire had cotton mills before Lanarkshire ; but once the industrial race had fairly begun, the latter county shot far ahead of the former, and in less than fifty years from the building of the first factory in the county, Glasgow had become the centre of a hundred cotton mills.   Of the earlier factories in Lanarkshire, particular mention may be made of that erected at New Lanark in 1785 by Mr David Dale, one of the pioneers of the Scotch cotton trade.   The only recommendation the site possessed was the prime one, in the eyes of a manufacturer of those days, of an unlimited supply of water from the Clyde ; otherwise, it was a mere morass.   Mr Dale knew the capabilities of the spot, and set to work accordingly; and simultaneously with the mill he laid out and built a range of houses for his workpeople.   Spinning operations were begun in 1786; and so well did matters turn out, that a second mill was put up in 1788. The second mill was destroyed by fire before it was completed, but was rebuilt in the following year.   Subsequently two other mills were erected.   Each mill was 160 feet in length by 40 feet in width, and seven storeys in height.   Ranges of stores, offices, mechanics' workshop, foundry, &c., were also provided. The population of the village, which had been built in the neighbourhood of the mills, was 2400 in 1820, and of these 1700 were employed in the works.   The surrounding country being unable to supply so many workpeople, Mr Dale had found it necessary to invite families from a distance to take up their abode in the village.   He also obtained a number of children from the charitable institutions in Edinburgh.   The mills were placed under the management of Mr Dale's son-in-law, Mr Robert Owen, who subsequently became notorious on account of his visionary projects for the regeneration of our social system. Notwithstanding his peculiar notions, Mr Owen did much to improve the condition of the workpeople under his charge. An educational institution of considerable size was built in the village for their sole use. It embraced rooms for the various classes of pupils, a lecture-hall, and a chapel. The course of education included a higher range of subjects than is usual now-a-days in similar institutions, but at the same time attention was given to imparting a knowledge of practical matters. The boys were instructed in gardening and agriculture, and the girls attended in rotation at the public kitchen to receive lessons in domestic economy. Judging from an account of the village and its inhabitants written about forty years ago, the community must have been an exceedingly happy one. The establishment has changed hands since Mr Owen's day, and is at present a thriving concern.

Before proceeding to trace the growth of the cotton trade generally, it is necessary to notice a circumstance which tended to increase and establish it in the country. The manufacturers of Glasgow and Paisley had acquired celebrity in making the finer kinds of linen fabrics before cotton was introduced ; and they had not been long engaged in working the new fibre, when they attempted to imitate the products of Indian looms. Mr James Monteith, of Glasgow, was the first to make the experiment ; but as the yarn then made in Scotland was not fine enough for the purpose, he obtained some "bird-nest" Indian yarn, and from it produced the first web of muslin woven in Scotland. He was so successful that he wove a second web, from which he had a dress made and embroidered with gold for presentation to Her Majesty Queen Charlotte. That was about the time the spinning-mule was invented, and as that machine produced yarn sufficiently fine for making muslins, many manufacturers turned their attention to the production of that class of goods. There is evidence of their early success in a report by the directors of the East India Company, made in the year 1793, on the subject of the cotton manufacture in this country. The report states that "every shop offers British muslins for sale equal in appearance and of more elegant patterns than those of India, for one-fourth, or, perhaps, more than one-third less in price." Glasgow came to have an extensive trade in plain and printed muslins, while Paisley acquired celebrity for fancy fabrics. The joint productions met the public taste, and the trade being found to be remunerative, was extensively entered into. The competition with India was made easy by the cheapness of production at home and the heavy duties which were imposed on goods imported from the East. The duty in 1787 on Indian muslins and nankeens was L.18 per cent. ; in 1802 it was raised to L.30, 15s. 9d. per cent. ; and a gradual increase took place, until, in 1813, the maximum rate of L.44, 6s. 8d. was reached. The duty on white calicoes was even heavier than on muslins, and in 1813 amounted to L.85, 2s. 1d. per cent.

The change that took place in the dress of the people consequent on the introduction of home-made calicoes and muslins is thus described in Macpherson's Annals of Commerce under the year 1785 : —" The manufacture of calicoes, which was begun in Lanarkshire in the year 1772, was now pretty generally established in several parts of England and Scotland. The manufacture of muslins was begun in England in the year 1781, and was rapidly increased. In the year 1783 there were above a thousand looms set up in Glasgow for the most beneficial article, in which the skill and labour of the mechanic raised the raw material to twenty times the value it was when imported. Bengal, which for some thousands of years stood unequalled in the fabric of muslins, figured calicoes, and other fine cotton goods, is rivalled in several parts of Great Britain. . . . . A handsome cotton gown was not attainable by women in humble circumstances, and thence the cottons were mixed with linen yarns to reduce their price. But now cotton yarn is cheaper than linen yarn, and cotton goods are very much used in place of cambrics, lawns, and other expensive fabrics of flax ; and they have almost totally superseded the silks. Women of all ranks, from the highest to the lowest, are clothed in British manufactures of cotton, from the muslin cap on the crown of the head to cotton stockings under the the sole of the foot. The ingenuity of the calico-printers has kept pace with the ingenuity of the weavers and others concerned in the preceding stages of the manufacture, and produced patterns of printed goods which, for elegance of drawing, far exceed anything that ever was imported ; and for durability of colour, generally stand the washing so well as to appear fresh and new every time they are washed, and give an air of neatness and cleanliness to the wearer beyond the elegance of silk in the first freshness of its transitory lustre. But even the most elegant prints are excelled by the superior beauty and virgin purity of the muslins, the growth and manufacture of the British dominions. With the gentlemen, cotton stuffs for waistcoats have almost superseded woollen cloths, and silk stuffs, I believe, entirely ; and they have the advantage, like the ladies' gowns, of having a new and fresh appearance every time they are washed."

The rapid extension of the cotton trade in Scotland was owing, among other things, to the facility with which workpeople could be obtained. There was no regulation requiring special qualifications in those who desired to be employed in the cotton mills, and, generally, a few lessons sufficed to make boys, girls, or grown-up persons, quite conversant with the simple duty of attending to the spinning machines or working at the loom. The wages paid in the factories were considerably higher than those given to agricultural labourers, and the result was that many relinquished the plough and became spinners or weavers. In course of time a redundancy of hands had entered the trade, and the natural result followed, that wages were reduced. That led to a succession of ruptures between the masters and workmen, the first of which occurred in 1787, when the masters combined to reduce the prices paid for certain kinds of work. A scale of prices was drawn up and presented to the workmen, and a conference was held with the view of arriving at an understanding on the matter. The result of the conference was not satisfactory to the operatives, and they formed a combination to resist the action of their employers. They held meetings in Glasgow, at which resolutions were adopted to expel from the trade those masters who had become most obnoxious. The expulsion was to be brought about by the workmen refusing to enter the service of said masters. It was further resolved that no man should work under a price fixed by the Union. The contest continued for some time, until a number of the workmen, unwilling or unable to remain unemployed, took work at the masters' prices ; but they were compelled by their brethren to return the cotton, and in many instances it was burned. The men continued to assemble in large bodies, parading the streets ; and on the magistrates attempting to apprehend the ringleaders, they were resisted. The Riot Act was read, the military called out to the assistance of the civil power, and the workmen not dispersing, several were killed and others mortally wounded. Prosecutions followed, which ultimately broke up the combination, and the operatives were obliged to submit to any terms the masters chose to impose. Subsequently the workpeople made several attempts to carry their purpose. In 1809 the weavers of Scotland, in conjunction with those of Lancashire, applied to Parliament for a bill to limit the number of apprentices and fix a minimum for the price of labour. Deputies were sent up to support the application, and the whole circumstances of the trade were investigated by a committee of the House of Commons ; but the conclusion arrived at was that such a measure would be injudicious, and, consequently, the House declined to interfere. A similar result followed an application of the same kind made two years afterwards by the Scotch weavers alone. After being thus thwarted, the operatives endeavoured to have their wages fixed by two committees, one of masters and the other of workmen ; but though committees were appointed, the principle of fixing wages was not established. Not discouraged by these rebuffs, the weavers next had recourse to proceedings under some old Acts of Parliament, the relevancy of which, though disputed by the masters, was affirmed by the Court of Session. An action was begun in 1812, which lasted for ten months, but ultimately failed in consequence of the masters refusing to bring in counter-evidence, and eventually refusing to recognise the decision of the judges. A week after the close of the action, 30,000 weavers struck work in one day, and 10,000 followed soon after. The authorities interfered, and prosecuted the leaders of the strike. That practically broke up the Weavers' Union, and the men returned to work after being idle for six weeks. The dispute between the weavers and their employers was unattended by serious acts of violence.

The circumstances which led to combinations among the weavers prompted the cotton-spinners to unite for the protection of what they conceived to be their interests. Their first union was formed in 1806 ; but it did not become conspicuous until 1810, when, in consequence of its operation, the masters stopped all their mills, and would not re-admit any of the operatives unless they signed a declaration that they would not be concerned in any illegal combination, and would not interfere with their employers as to whom they should employ. For about six years little was heard of the union ; but fresh misunderstandings arose between the employers and employed in 1816, and between that year and 1824 serious outrages were perpetrated. Several obnoxious employers were shot at, and their mills were set on fire; while some of the men who disregarded the dictates of the union were shot, and others were shockingly injured by having vitriol thrown upon them. In December 1820 an attempt was made to shoot Mr Orr, manager of the Underwood cotton mill at Paisley, on the night before his marriage. In August of the same year a workman named Fisher, who had a large family dependent on him, was shot at when in bed. He was again shot at next month, and in November he was waylaid while going to his work, and a quantity of vitriol was thrown on his face and breast, which burned him dreadfully. As soon as he recovered he resumed work, but the unionists seemed determined to stop his career, and he was shot at a third time, fortunately without receiving injury. Several men were wounded by pistol shots. In 1823 a conspiracy to assassinate one mill-owner and five spinning-masters was discovered in Glasgow. Threatening letters of the most diabolical kind were sent out in great numbers.

The oath by which the combined cotton-spinners bound themselves was in the following terms :—" I, A. B., do voluntarily swear, in the awful presence of Almighty God, and before these witnesses, that I will execute, with zeal and alacrity, as far as in me lies, every task or injunction which the majority of my brethren shall impose upon me, in furtherance of our common welfare, as the chastisement of nobs, the assassination of oppressive and tyrannical masters, or demolition of the shops that shall be incorrigible; and also that I will cheerfully contribute to the support of such of my brethren as shall lose their work in consequence of their exertions against tyranny, or renounce it in resistance to a reduction of wages. And I do further swear that I will never divulge the above obligation, unless I shall have been duly authorised and appointed to administer the same to persons making application for admission, or to persons constrained to become members of our fraternity."

A report on trade combinations was made to Parliament in 1838, in which it was stated that the Glasgow master cotton-spinners had a combination of a somewhat mysterious character ; they had no written rules, no fixed times or places of meeting, no regular subscriptions or expenses, except a charge on each master, at so much for every thousand spindles worked in his factory, for the support of a secretary. The union of operative cotton-spinners had by that time been established on a more civilised basis than formerly; and of 1000 spinners in Glasgow 750 were members of the union. Between 1826 and 1836 a series of partial strikes occurred, the result of which was to equalise the wages in the various mills and districts. In the autumn of 1836 the operatives applied for an advance of sixteen per cent. on their wages, which was granted by the Glasgow manufacturers, but not by those in the surrounding districts. The Glasgow unionists were then, as they said, compelled by the threats of their own masters to strike against a wealthy country manufacturer who had refused to comply with the demands of his men. The strike lasted sixteen weeks, and cost the union L.3000, but without improving the position of the men. In the spring of 1837 trade grew dull, and the masters who had given an advance of wages proposed to return to the previous rate. That step the men resolved to resist, and went out on strike; but at the end of fifteen weeks they gave in, and returned to work at a lower rate of wages than had been previously offered. During the strike several outrages were perpetrated. One man was murdered, a woman had vitriol thrown upon her, and there were two attempts at incendiarism. The direct and indirect loss occasioned by this strike was estimated to be upwards of L.160,000. The wages of the spinners were always much higher than those of the weavers, and notwithstanding the reduction to which they had been subjected, they were earning after the strike from 20s. to 40s. a-week of sixty-nine hours. No serious strike has occurred in the trade since 1837, though frequent disputes have arisen.

Unlike the woollen and linen trades, the cotton manufacture in Scotland shows little increase in recent years. The quantity of cotton manufactured weekly in Scotland averaged 1652 bales in 1831, 2035 bales in 1835, and 2364 bales in 1840. From that year up till 1861, the quantity consumed rose and fell according to the state of the markets. In 1866, when the trade in England had almost recovered from the depression caused by the American war, the quantity of raw cotton used in Scotland was 2500 bales; but in the following year only 1700 bales were consumed. The trade is being gradually concentrated into fewer hands, the latest returns showing a falling off in the number of factories. In 1838 there were 198 cotton-mills in the country, distributed over the following counties: —Aberdeen, 4; Ayr, 4; Bute, 2; Dumbarton, 4; Dumfries, 1; Kirkcudbright, 1; Lanark, 111; Linlithgow, 1; Perth, 7; Renfrew, 60; Stirling, 3. The total number of steam-engines employed was 193, with an aggregate of 5612 horse power, and 73 water-wheels, with an aggregate of 2728 horse power. The number of power-looms was upwards of 15,000, and of workpeople 35,576. A return made to Parliament in 1862, gives the following statistics of the trade in the years 1850, 1856, and 1861:—

Counties.	Factories.			Spindles.			Power-Looms.			Persons Employed.		
	1850.	1856.	1861.	1850.	1856.	1861.	1850.	1856.	1861.	1850.	1856.	1861.
Aberdeen, . .	2	2	2	49,536	60,000	66,276	60	60	70	730	726	770
Ayr, . . .	4	4	3	25,816	29,396	30,240	787	680	968	1,067	947	1,089
Bute, . . .	4	4	4	32,780	51,632	52,148	948	995	977	1,011	1,104	976
Dumbarton, .	4	4	4	56,952	75,584	75,296	233	244	246	749	862	758
Dumfries, . .	...	1	1	...	7,464	16,308	...	...	...	...	61	112
Kirkcudbright,	1	...	...	5,808	...	...	156	...	...	173	...	...
Lanark, .	94	83	96	875,310	1,111,352	1,138,602	18,811	16,774	24,149	22,759	21,650	27,065
Linlithgow, .	1	1	1	15,399	15,624	19,800	...	...	...	118	128	121
Perth, . .	3	3	3	60,364	63,244	57,796	528	488	552	1,278	1,104	1,069
Renfrew, . .	51	46	44	508,928	556,423	408,742	1,977	2,270	2,968	7,884	7,580	8,749
Stirling, . .	4	4	5	52,200	70,410	50,190	64	113	180	556	536	528
Total, . .	168	152	163	1,683,093	2,041,129	1,915,398	23,564	21,624	30,110	36,325	34,698	41,237

An attempt was made by the author to ascertain the present extent of the trade by sending schedules to all the manufacturers; but as several declined to give any information as to the extent of their factories, though it was explained that aggregate results only would be published, it is impossible to make a reliable statement on the subject. From the fact that a number of the schedules returned blank were accompanied by notes explaining that certain establishments have recently been converted to other purposes than manufacturing cotton, while some have been closed, there is reason to believe that the number of cotton factories in Scotland has undergone a considerable decrease since 1861, though at the same time it will be seen from the following figures that the production, as represented by the quantities of goods and yarn exported, shows an increase:—

IMPORTS INTO SCOTLAND.			EXPORTS FROM SCOTLAND.			
Years.	Raw Cotton.	Cotton Manufactures.	Cotton Manufactures.		Cotton Yarn.	
	Cwts.	Value.	Yards.	Value.	Lbs.	Value.
1861	172,055	L.19,009	150,754,031	L.2,644,419	6,550,401	L.467,612
1862	10,794	63,105	120,119,627	2,538,076	5,516,094	504,045
1864	7,216	30,027	94,766,371	2,693,731	5,827,611	831,395
1865	10,063	18,411	126,912,955	3,154,183	5,787,075	653,155
1866	19,736	24,711	164,194,915	4,346,157	7,733,268	847,833
1867	26,320	12,745	206,394,756	5,002,158	9,495,469	842,326

Owing to various circumstances affecting the sources from which the supply of raw material is obtained and the markets for manufactured goods, the cotton trade of Britain has been liable to considerable fluctuations; but at no time was it so seriously disturbed as during the period between the years 1861 and 1866. The American war almost completely disorganised the trade. The manufacturers virtually depended on the United States as the one great source of supply, and to that fact is to be attributed the disastrous crisis which overtook them on the outbreak of hostilities between the Northern and Southern States. Amid the turmoil of war the cultivation of cotton was neglected, stocks in this country rapidly diminished, and the fibre attained a value which checked consumption—in short, the manufacturing districts became involved in a cotton famine, attended by an amount of suffering unparalleled in the industrial history of the nation. When the American war broke out, a general impression prevailed that it would be of short duration, and only a slight advance took place in the price of raw cotton.

A few notes on the course of the market during the eventful period referred to will show the effect of the war on the cotton interests. The quotation for "middling Orleans" cotton on the 13th September 1861 was 9¼d. per ℔. From that period onward, however, the advance became decided and rapid, and on the 25th October middling Orleans was quoted at 12d. per ℔. On the 30th November advices were received of the capture of Messrs Slidell and Mason, from a British vessel, which event, with the probability of a war ensuing between the United States and Great Britain, caused a decline in the raw material to the extent of a 1d. to 1½d. per ℔. From that period the market rapidly advanced, and continued in a state of intense excitement for several months, the increase in value being truly prodigious. On the 1st January 1862 middling Orleans was worth 11⅞d. per ℔., but on the 15th August was quoted at 19⅝d. per ℔., the advance in that interval being 7¾d. per ℔. The most extraordinary phenomenon, however, in the history of the trade, occurred in the succeeding week, ending 22d August. The stock of American cotton was computed at 20,080 bales, and of East Indian at 25,150 bales, the total stock amounting to 81,980 bales. Middling Orleans was quoted at 23½d. per lb., being an actual advance in the short space of six days of 3⅞d. In the following week middling Orleans again increased in value to the extent of 3d. per ℔., while the total stock of all kinds was reduced to 62,980 bales. Another advance to the extent of 2½d. took place in the following week, middling Orleans attaining the enormous value of 29d. per ℔., or an increase in three weeks—from the 15th August to 5th September—of 9⅝d. During the next eleven weeks the market was on the declining scale, but partially recovered itself by the end of the year—middling Orleans on the 31st December being quoted at 25d. The following year, 1863, opened inanimately, and prices continued without material variation until after the 4th September, when the several facts of extensive orders, both from India and the Continent, for manufactured goods, decreasing stocks of cotton in Liverpool, and the probability of the war continuing, engendered an upward movement in the raw material, which continued with more or less regularity till the end of the year, middling Orleans on the 31st December being quoted at 27¾d. per ℔. The imports of East India cotton into Great Britain during 1863 were 1,228,900 bales, being an increase, as compared with 1860, of 666,226 bales.

T

The first six or seven months of 1864 were characterised by remarkably high prices, middling Orleans on the 22d July attaining the maximum value of 31¼d. per ℔., and "fair Dhollerah," on the 12th August, 24d. per ℔. The stock of American cotton on the former date was stated to be 3810 bales, but a week previously it was estimated at only 1721 bales; the total stock, however, was computed to be 212,176 bales. The imports of East India cotton into Great Britain were 1,399,514 bales, or an increase, as compared with 1860, of 836,840 bales. In 1865 the market was very irregular, oscillating according to the tenor of advices from America. During the first three or four months rumours of peace negotiations were received from time to time, which, combined with the ultimate fall of Richmond, exercised a depressing influence on the market, resulting in the serious decline of 13¼d. per ℔., the quotation in the first week of January for middling Orleans being 26¾d., but on the 21st April only 13½d. At that time, however, news was received of the assassination of President Lincoln, which caused an immediate reaction, the market continuing to advance until the 13th October, when middling Orleans was quoted at 24¾d. Another period of depression ensued, and middling Orleans, by the close of the year, declined to 21¼d. The following year (1866) was a remarkable one, and will long be remembered in the annals of the commercial world for the extraordinary combination of adverse and desponding influences upon every branch of commerce.

Though the manufacturers suffered severely during the crisis, the sorest burden of distress fell upon the workpeople. Out of 355,000 persons employed in the Lancashire cotton factories, only 40,000 were in full work at the close of 1862, and the loss in wages alone was estimated at L.105,000 a-week; but even then the worst had not been reached. The state of matters that prevailed in Lancashire excited the sympathy of the whole country, and brought out one of the most munificent responses that has ever been made to the cry of distress. The cotton operatives of Scotland, residing chiefly in places where other kinds of labour were abundant, did not suffer to the same extent as those in England, but they did not altogether escape the effects of the calamity. Local efforts were made to supply their wants, and the manufacturing communities in the west contributed liberally to support the poor people. The cotton manufacture has now nearly resumed its normal condition, but the recollection of those terrible years of famine will linger long in the minds of all classes engaged in the trade.

Among the most extensive cotton factories in Scotland are those of Messrs A. & A. Galbraith, situated at Oakbank and St Rollox, Glasgow. The factories comprise two immense ranges of somewhat irregular buildings. The original portions were erected many years ago, and the successive additions may be traced in the different tints of the masonry. The joint establishments cannot be pointed to as models, so far as the buildings are concerned; but they are filled with machinery of the finest and most recent construction, and their internal economy is equal to that of any other mills in the country. In the spinning department there are 95,000 mule and throstle spindles, the produce of which is made into cloth by 1532 power-looms. There are several large steam-engines, the aggregate indicated force of which is 1600 horse power. 1700 persons, of whom only 100 are males, are employed; and the quantity of cloth made is 350,000 yards a-week, or 17,000,000 yards a-year. All the cloth is of the plain kind for printing and dyeing.

The cotton arrives at the mill in compactly pressed bales, each 4 feet in length by 2 feet in breadth and height, and weighing on the average about 428 ℔. The varieties generally used are American, Indian, and Egyptian, mixed in certain proportions. The bales are deposited in large stores, and the first process is performed in that department. A few bales of each kind are opened at a time, and their contents thrown into a heap, the quality of the mixture depending on the proportions of the different varieties of fibre used. The cotton is then conveyed in wicker-work trucks to a room, where it is subjected to the action of the "opener," a machine which loosens the large flocks and separates the grosser impurities. The opener does not complete the process of purification, and the cotton is passed to the "scutcher," in which it is made perfectly clean. The inventor of the latter machine was Mr Snodgrass, of Glasgow, and it was introduced into the trade in 1797. The third machine through which the cotton is passed is the spreading-machine, which prepares the fibres for carding. The cotton is spread evenly on a feed-apron, and after passing through fluted rollers, and being beaten by a series of revolving-arms, is delivered in a continuous web or fleece which is wound upon gigantic bobbins. The cotton is now ready for carding, the object of which process is to disentangle the filaments and lay them parallel to each other; and the more carefully the operation is performed, the more perfect will be the yarn produced. The filaments of cotton, when viewed through a powerful microscope, have the appearance of flattened tubes of glass, twisted on their axis; while those of flax are perfectly cylindrical, and jointed like a cane. The normal form of the fibre in both cases remains unchanged through all the processes of manufacturing, and is not altered by their reduction to pulp and conversion into paper. The form of the filaments makes the cotton bear a pretty close resemblance to wool, which flax never assumes.

The carding-engines consist of a series of cylinders covered with wire spikes of various degrees of fineness. The cotton is fed into the carder from the bobbins of the spreading-machine, and emerges in the form of a ribbon or sliver. As it comes from the carder the sliver is exceedingly tender and loose, and is received from the machine in tin cans. If the cotton be examined at this stage it will be seen that, though the fibres show a general tendency to parallel arrangement, many of them are doubled and twisted in a way which would render it impossible to form them into the finer qualities of yarn. On following the slivers to another set of beautiful machines it will be seen how the filaments are arranged in perfect order. The process is called drawing and doubling. The drawing-frame consists of a combination of rollers, which serve to draw out and elongate the sliver. Their action is exceedingly simple, and as they form a part of all subsequent machines through which the cotton is passed until the spinning is completed, it may be well to explain how they act. Let this represent an end view of the rollers, which are commonly arranged in three pairs, ⊙⊙⊙. The rollers are four inches in circumference, and each pair moves with a different velocity. We shall suppose that the cotton is fed in between the left hand pair of rollers, and led on to the others. The speed of the second pair is so much greater than that of the first, that one inch length of sliver in passing between them is drawn out to one and three quarter inches, while the third pair draws it to a length of five inches. This process is repeated until the requisite degree of parallelism is attained in the fibre, but it will be evident that were a single sliver put through several times it would become so attenuated that it would be impossible to manage it. This difficulty is overcome by what is called "doubling," that is, laying several slivers together at every repetition of the process. The effect of the drawing and doubling may be illustrated by taking a tuft of tangled wool and drawing it asunder several times with the fingers, at each drawing laying the two separated portions together and seizing them by the extremities. After being thus drawn five or six times, the wool will be found to have a perfectly parallel arrangement.

The first spinning process is done on the bobbin-and-fly frame. The sliver is fed into the machine from the cans, and, after being drawn out a little, has a slight twist imparted to it, which makes the fibres cohere, and fits the cotton for bearing further elongation. As a sudden extension to the wished-for fineness is not practicable, the cord or "roving" which is formed by the bobbin-and-fly frame is gradually reduced—first by the fine bobbin-and-fly frame, and finally by the throstle-frame or the mule. The coarse and fine bobbin-and-fly frames are essentially similar in principle, but the latter is more delicately constructed than the former. The bobbins from the fine frame are placed on the throstle-frame, by which the thread is drawn out to the requisite fineness and firmly twisted. As the cotton is subjected to one operation after another, it is elevated from floor to floor, according to the arrangements of the successive machines; so that by the time the spinning processes are completed, the cotton has been elevated to the fifth or sixth floor. Most of the machines are attended by women or girls, whose work is exceedingly light, and much more healthy than it used to be.

It would be impossible to conceive machines more perfectly adapted to their purpose than those which crowd the spacious floors of Messrs Galbraith's factories. Each seems to work with a will and instinct of its own, and no one can witness their operations without admiring the ingenuity that devised their thousands of parts and brought them all into harmonious play. Fingers of iron and wood work more deftly, and with apparently more delicacy of touch, than fingers of flesh and blood could ever do; and the finest productions of the Indian hand-spinners are surpassed by the gossamer-like threads which the self-acting spinning-mule produces by hundreds at a time.

Respecting the weaving department there is not much to say, as all the looms are employed on plain cotton of various qualities, ranging from fine muslin for summer dresses to stout calicoes. The manner in which the warp is prepared, the action of the looms, and the other details of weaving, are so similar to those followed in the manufacture of linen, already described, that it would be superfluous to notice them here.

Some of the early travellers in the East brought home marvellous accounts of the muslins made in India. Tavernier states that in the city of Calicut, whence comes the designation of "calico" which is usually applied to muslin, some cloth was made so fine that it "could scarcely be felt in the hand, and the thread was scarcely discernible." A missionary at Serampore states that "muslins are made by a few families so exceedingly fine that four months are required to weave one piece, which sells at 400 or 500 rupees. When this muslin is laid on the grass and the dew has fallen upon it, it is no longer discernible." Oriental hyperbole goes still further, and describes the muslins of Dacca as "webs of woven wind." It has been proved that, however marvellous these fabrics may have been when the rude appliances by which they were produced were taken into account, they were much coarser than the muslins now spun and woven by steam in Glasgow and elsewhere, and sent in immense quantities to the countries where the fabric had its origin. Messrs Galbraith do not make any of the finer qualities of muslins, but many looms in Glasgow are engaged in producing cloth of exquisite delicacy.

Some of the old hand-loom weavers to be met with speak of the large wages made in the early years of the cotton manufacture. At one kind of work a man could earn 7s. 6d. a-day. The price paid for weaving "causey," or printing-cloth, was 6d. an ell, and a good hand had been known to turn out twenty yards a-day. That was the golden age of the hand-loom. In the rural districts farmers apprenticed their sons to the trade. Weaving was more remunerative than the other mechanical occupations; consequently, those engaged at it were looked upon as the aristocracy of the working-class, and a smart weaver lad did not think himself beneath the honour of seeking and obtaining the hand of the daughter of the farmer or merchant. Now, hand-loom weavers in the cotton trade occupy a humble position, and their earnings range, according to ability and energy, from 12s. to 20s. a-week; but very few reach the latter sum.

According to a return recently issued by the Board of Trade, the wages of the cotton operatives in Glasgow are as follow, for the week of sixty hours:—Men—overlookers, 45s. ; warpers, 22s.; drawers and twisters, 20s.; dressers, 33s.; sizers, 35s. Women—reelers and winders, 9s. to 10s. 6d.; warpers, 14s.; weavers, taking charge of two or three looms, 11s.; of four looms, 15s. 6d. Girls—taking charge of one loom, 6s. The persons employed in the cotton factories have little to distinguish them socially from the great body of the working population, except, perhaps, that there is a large number of Irishwomen among them, whose manners are somewhat coarse.

CALICO-PRINTING AND TURKEY-RED DYEING.

ANTIQUITY OF THE ART OF DYEING AND PAINTING CLOTH—ITS INTRODUCTION
INTO EUROPE—PROGRESS OF THE ART IN BRITAIN—VARIOUS STYLES OF
CALICO-PRINTING—CORDALE PRINTFIELD AND DALQUHURN DYEWORKS—
DESCRIPTION OF THE PROCESSES OF CALICO-PRINTING AND TURKEY-RED
DYEING.

EXCEPT when used for underclothing or linings, cotton cloth is
generally ornamented with colours. The art of dyeing and painting
cotton and linen fabrics was known to some of the Eastern nations
from a very early date. The Egyptians practised dyeing in blue,
purple, and scarlet, at least 1500 years before the Christian era;
and Herodotus mentions that a tribe who lived on the borders of
the Caspian were in the habit of painting with vegetable dye figures
of animals on their garments, the impression being so strong that it
could not be washed out. In Pliny's time a great advance had been
made by the Egyptians in the art of dyeing and staining. Pliny
thus describes the process, which bears some resemblance to the
modern mode of dyeing by mordants:—" Garments are painted in
Egypt in a wonderful manner, the white cloths being first smeared,
not with colours, but with drugs which absorb colour. These
applications do not appear upon the cloths; but when the cloths are
immersed in a cauldron of hot dyeing liquor, they are taken out the
moment after, painted. It is wonderful that, although the dyeing
liquor is only of one colour, the garment is dyed by it of several
colours, according to the different properties of the drugs which have
been applied to different parts. Nor can this dye be washed out.
Thus the vat, which would doubtless have confused all the colours
if the cloth had been immersed in a painted state, produces a diversity of colours out of one, and at the same time fixes them immov-
ably." It is here stated that the cloth was painted or smeared with
the chemical substances, which shows that the production of patterns

by means of engraved blocks was not then known to the Egyptians.

The art of calico-printing—or rather painting—was not introduced into Europe until the seventeenth century. It had been brought from India, and was at first practised according to the rude method adopted by the inhabitants of that country. In Anderson's "History of Commerce," it is stated that calico-printing was begun in London in 1676; but a considerable time elapsed before the trade came prominently into notice. Towards the close of the seventeenth century a demand had sprung up in Britain for the cheap and gaudy prints of India, Persia, and China; and the result was that the woollen and silk manufactures began to suffer. An outcry was raised against the importation of printed calicoes, and at length the attention of Parliament was called to the matter. In 1700 an Act was passed prohibiting the importation and use of Eastern prints, under a penalty of L.200. Meantime the home calico-printing trade went on extending, and as it was not looked upon as interfering with the consumption of woollen and silk goods, it was allowed to be carried on without legislative interference. By the year 1712, however, it had become so important that Parliament recognised its existence by imposing an Excise duty of 3d. on every square yard of calico printed, stained, painted, or dyed; and as it appeared to bear the impost easily enough, the duty was doubled in 1714. Printed linens were subjected to half the rate levied on cotton. Notwithstanding the duty, printed fabrics were extensively used by the public, and again the cry was raised that the woollen and silk manufactures were in danger. In 1720 an Act of Parliament was passed which prohibited the using or wearing of printed or dyed calicoes, whether printed at home or abroad, and even of any printed goods of which cotton formed a part, excepting only calicoes dyed all blue. This law put an end to the printing of calicoes, and the printers were limited to the printing of linens. After the Act had been in force for sixteen years, the portion of it which forbade the use or wear of printed goods of a mixed kind containing cotton was repealed; and thenceforth cloth composed of linen warp and cotton weft was made and printed. It was estimated in 1750 that 50,000 pieces of this mixed fabric were printed annually. The cloth-printing trade was confined almost exclusively to the neighbourhood of London until the year 1738, when it was introduced into Scotland. Twenty-six years later it was begun in Lancashire, and after that time the London trade gradually declined, as the printers there could not maintain competition with those in the districts where the cloth was manufactured. In 1774 the law which prohibited the printing of English made calicoes was repealed, and, by the aid of a series of wonderful inventions and improvements, the art of calico-printing flourished and increased, though the Excise duty was not removed until 1831.

As showing the progress of the trade, it may be mentioned that in 1796 the quantity of British calicoes and muslins which paid the print-duty was 28,621,797 yards; in 1829, it was 128,340,004 yards; and in 1857, the quantity of dyed and printed calicoes exported was 808,308,602 yards, the declared real value of which was L.13,921,428, while it was estimated that in the same year 135,000,000 yards were retained for home consumption. The quantity exported remained about the above figures till the time of the American war, when there was a falling off. In 1866 the trade had recovered to 897,825,547 yards of coloured calicoes, the declared value of which was L.22,095,216. The year 1867 showed a decrease of about 17,000,000 yards. There are no figures to indicate the quantity of cotton cloth printed and dyed in Scotland; but the fact that upwards of 12,000 persons are employed in the print and dye works shows the importance of this branch of industry.

In both the chemical and mechanical departments of calico-printing and Turkey-red dyeing, many important inventions and improvements have been effected in Scotland, one of the most valuable being the invention, in 1785, of cylinder-printing by Mr Bell, of Glasgow, which worked a revolution in the trade. There are five general styles in calico-printing, namely—1. The fast-colour or chintz style, in which the mordants are applied to the white cloth, and the colours of the design are afterwards developed in the dye-bath. 2. Where the whole surface receives a uniform tint from one colouring matter, and figures of other colours are afterwards brought up by chemical discharges and reactions. 3. Where the white surface is impressed with figures in a resist paste, and is afterwards subjected to a general dye. 4. Steam-colours, in which a mixture of the mordants and dye-extracts is applied to the cloth, and the chemical combination is effected by the agency of steam. 5. Spirit-colours, consisting of mixtures of dye-extracts with nitro-muriate of tin. The latter are brilliant but fugitive.

The following account of the Cordale Printfield and Dalquhurn Dyeworks—two extensive establishments situated on the banks of the Leven near the village of Renton, Dumbartonshire, and belonging to Messrs William Stirling & Sons, Glasgow—will convey an idea of the processes of calico-printing and Turkey-red dyeing:—

As already explained, there are various styles of calico-printing, and sometimes two or more of these are carried on in one printfield; but at Cordale (which is one of the most extensive works of the kind in the country) Turkey-red printing only is practised. The cloth, after being dyed red at the Dalquhurn works, is taken to Cordale and figured with other colours by certain chemical processes. In order, then, to trace the successive operations, it is necessary that the Dalquhurn Dyeworks should be first described. In the beginning of last century an extensive bleach-field was formed at Dalquhurn, which in 1791 was acquired by Messrs Stirling, and used for bleaching the cloth printed at Cordale. In 1828 the firm extended the premises, began to dye Turkey-red, and founded their present celebrity in that branch of trade. The grounds pertaining to the works extend to about seventy acres, of which ten are covered by buildings.

The cloth and yarn dyed and printed by Messrs Stirling & Sons are made chiefly in Glasgow and Manchester. The cloth is sent in as it comes from the looms, and the first process to which it is subjected is a partial bleaching. From 2000 to 3000 pieces, averaging about twenty-five yards in length, are formed into a continuous web by being sewed together by a steam sewing-machine. This web is led on to a washing-machine of peculiar construction, which removes the simpler impurities. The washing-machine consists of a trough surmounted by a framework. In the bottom of the trough is a roller extending from end to end, and there is a similar roller in the framework above. A web of cloth is fed in at each end of the rollers, and after winding spirally round the upper and lower rollers the ends are brought out in the centre. When the machine is set in motion the cloth, which before it enters is compressed into the form of a rope, is drawn round by the rollers; and from the time it enters the machine until it comes out, every part of it has been a dozen times immersed in the trough, and as often wrung nearly dry by compression between the upper leading roller and one which bears against it. Each washing-machine disposes of about 800 pieces, or 24,000 yards, in an hour, consuming in that time about 24,000 gallons of water. As the cloth comes from the washing-machine it is deposited in a large iron boiler, technically called a "kier." When the boiler is full the cover is fixed on, and high-pressure steam is admitted. Water impregnated with a certain proportion of caustic soda is then injected in a boiling state, and by a system of pipes and taps is drawn downward through the cloth. This operation is continued for about eight hours, when the cover is removed, and the end of the web attached to the washing-machine, which draws the cloth out of the kier, and washes it.   From the washing-machine the cloth passes to the souring cistern, where it is steeped in a weak solution of sulphuric acid, and afterwards washed in pure water.   The cloth is then dried, when it is ready for the first stages of the dyeing process. Before the cloth reaches the dye-bath it is subjected to no fewer than twenty-six operations after those which have been described.   It is passed through a variety of dung and oil liquors, exposed on. the grass, dried in stoves, and so on, several of these operations being repeated three or four times, and the whole extending over several days.

The art of dyeing Turkey-red was introduced into England in the end of last century by M. Borelle, a Frenchman, who established himself in Manchester, and received a reward from Government for the disclosure of the secret.   A year or two afterwards, another Frenchman—M. Papillon—went to Glasgow, and, in company with Mr George Macintosh, began to practise the art.   The method followed by M. Papillon was more successful than that adopted by M. Borelle at Manchester, and Glasgow became famous for dyeing Turkey-red.   Up till 1810, however, the colour could be imparted only to thread and yarn.   In that year, M. Kœchlin, of Mulhausen, in Alsace, discovered a mode of giving the colour to cloth; and a year afterwards invented one of the most beautiful and interesting processes in calico-printing, namely, the mode of discharging the colour from the dyed cloth according to any pattern desired, and inserting designs in other colours.   This is the system practised at Messrs Stirling's dye and print works.

The art of the dyer and calico-printer is based on the proper understanding and use of " mordants."   The term " mordant " is applied to certain substances with which the cloth to be dyed must be impregnated, otherwise the colouring matters would not adhere to the cloth, but would be removed by washing.   Thus the red colour given to cotton by madder would not be fixed unless the cloth were previously steeped in a solution of salt of alumina.   The cloth has the property of decomposing the salt, and of combining with and retaining a portion of alumina.   The red colouring prin-ciple of the madder has an affinity for the alumina and combines with it.   The consequence is, that the alumina, being firmly re-tained by the cloth, and the colouring matter by the alumina, the dye becomes "fast"—that is, it cannot be removed by water, even when soap is added, though water alone is sufficient to remove the red colouring matter from the cloth if the alum mordant has not been previously applied. After the cloth has been subjected to the thirty preparatory operations referred to above, it is steeped for a night in the alum mordant. It is then washed and wrung, but not dried. Through all the operations up to this point the cloth is retained in the long webs into which it was formed for the first washing, and in that shape is passed through tubes from one part of the premises to the other. The various departments are distinct from each other in order to ensure perfect work in each ; and in its progress the cloth travels many miles through tubes, over pulleys, and round cylinders ; and inexperienced persons are apt to wonder that after so many washings, boilings, and squeezings there is any strength left in it. Preparatory to being placed in the dye-bath, the cloth is separated into lengths of two or three pieces, several of these subdivisions going into each bath. The baths are fitted with automatic reels, or open revolving frames, round which the cloth is loosely wound, so that it hangs in loops down into the dye. Constant motion is necessary in order to ensure equality of colour, and the cloth is kept revolving round the reels so long as it remains in the bath. The cloth is put in when the dye stuff is cold, and the heat is brought up slowly by means of steam-pipes. In two or three hours the liquid is made to boil, and the ebullition is kept up for about fifteen minutes, when the cloth is withdrawn and washed and cleared several times with soap and soda in copper boilers.

Madder, when used with an iron mordant, produces a purple colour, with alum it produces red, and with alum and iron in certain proportions, it produces chocolate or black. In the production of Turkey-red the madder is mixed with bullock's blood, of which about 130,000 gallons are used annually at the Dalquhurn Works. Yarn is dyed by a hand process, and the operatives engaged in that department have a most unhealthy and disagreeable occupation. They have to stand over the cisterns of scalding, steaming liquor, and keep the yarn in constant motion by shifting and turning about the rods on which it is hung. Self-acting machines for superseding hand labour are being tried. The yarn is subjected to preliminary processes similar to those which the cloth undergoes; but as it cannot be formed into a continuous line, special appliances are required for dealing with it. Machines for washing, liquoring, and wringing the yarn have been devised and constructed at the works.

The establishment contains a fine collection of machines, and the organisation of the place throughout exhibits a variety of labour-saving arrangements and appliances which will not fail to arrest the attention and excite the admiration of visitors. There are from 900 to 1000 persons employed in the works, about two-thirds being women, of whom a considerable proportion are Irish. A more healthy-looking class of women than those employed in bleaching is not to be found, though the labour in the winter time is somewhat trying.

The machinery is driven by 26 steam-engines, the aggregate force of which is about 180 horse power nominal. Steam for these is generated in 14 boilers, and the quantity of coal consumed in the works is from 25,000 to 30,000 tons a-year. About 600,000 pieces, or 18,450,000 yards of cloth, and from about 600,000 ℔. to 800,000 ℔. of yarn are dyed annually, and, when extensions at present in progress are completed, that quantity will be much increased. The wages paid annually by the firm amount to about L.40,000. L.50,000 worth of madder and L.20,000 worth of olive oil are used every year. It would be impossible to have such an establishment in a locality where there was not an abundance of pure water; for the quantity consumed at Dalquhurn would be sufficient to supply every man, woman, and child in the city of Edinburgh with ten gallons a-day.

All the yarn and more than one-half of the cloth dyed at Dalquhurn are exported in a plain red state. The remainder of the cloth is taken to the Cordale Printworks to be printed. As already stated, the style of printing practised is that whereby cloth, after being dyed of a uniform tint, has designs in other colours worked into it by a system of chemical discharges and reactions. If a paste composed of certain proportions of oxalic acid, tartaric acid, lime juice, pipeclay, and gum (the two latter being used merely to give consistency to the mixture), be applied to a piece of Turkey-red cloth, and the cloth be afterwards dipped in a solution of chloride of lime, it will be found that all the parts covered by the paste have become white. The discharge paste of itself produces no effect on the colour, and may be removed by washing in pure water, and the chloride of lime in like manner may be applied without affecting the dye; but when both are brought to act together, the colour at once gives way. Again, if it be desired to erase the red ground according to any particular pattern, and insert, say yellow, in the cleared spaces, all that is necessary is to mix in the discharge paste a mordant that will seize yellow dye; and after the discharging is completed, immerse the cloth in the yellow bath. Thus, a design printed on red cloth with a paste composed of lime juice, tartaric acid, nitrate of lead, pipeclay, and gum, comes out white after immersion in the chloride of lime solution; and on being plunged immediately afterwards into a bath of bi-chromate of potash, comes out yellow, the red meantime remaining unaffected, except where the paste was applied. A knowledge of these facts is necessary to enable one to understand what to the uninitiated are most mysterious and wonderful operations.

The tedious process of painting designs on calico by hand was superseded by the use of blocks about the time the art was introduced into Europe. In block-printing a section of the design is cut upon a piece of sycamore, and, after being coated with paste or colouring matter, is laid upon the cloth and struck smartly. As the blocks most commonly used are only ten inches long by five broad, a great number of applications are necessary in order to print a single piece of cloth. The block has been superseded by the cylinder, except in special cases—such as at the establishment under notice, where, owing to the peculiar nature of the work, blocks are still used to some extent. The blocks are made on the premises by a staff of designers and engravers. As most of the goods are for the Indian market, the colours are somewhat "loud" and the designs peculiar. The dress-pieces made for people of the Hindoo religion have a broad border of peacocks round the skirt, the upper part bearing a spotted or diaper pattern. The ground-work of all is Turkey-red, but the birds and other designs are produced in blue, yellow, and green. The Mahometans consider it sinful to try to imitate nature too closely; and though peacocks figure in the designs prepared for ladies of that faith, they are drawn in the rudest fashion and worked out in mosaic. None of the designs of these Indian garments would find admirers in this country; and as the artists are bound down by certain conventional rules, they have no scope for the creation of original patterns. In cloth for turbans there is the same limitation in variety. The dress pieces are short, being only from $1\frac{1}{2}$ to 8 yards in length; and owing to that and other technical causes, it would be unprofitable to print them on a cylinder machine, so they are done by the block method. As has been already explained, what is printed on the cloth is not the complete colour, but a substance to discharge the red and absorb another colour. This substance is applied in the form of paste, which has no resemblance to the ultimate colour.

The cloth having been calendered and wound round a roller, is taken to the printers, who work in pairs, one standing at each end of a table. At one end of the table the cloth passes up from beneath, and as each space the size of the table is printed, it is drawn down at the other end and brought into contact with a drying

:ylinder. The blocks used in this case are of a large size, some of them being nine inches in width, and long enough to extend across the web. The paste is managed by a boy, called a " tearer," who spreads an even coating over a woollen cloth stretched in a frame, and resting on an elastic bed. The printer presses his block on the woollen cloth, and takes up a quantity of paste sufficient for one impression, the " tearer" giving the cloth a fresh coat after each dip. Guided by a series of brass points, the printer lays his block evenly on the calico, and strikes it with a hammer. In the case of small-block printing, each man has a table for himself, and his work is much lighter than that of those who use the large blocks. Sometimes the pastes for three or four different colours are put on at one time, the printers changing the blocks and pastes at each impression. When two colours only are used, the men take one each, and it is astonishing how much cloth they will turn out in a day. The cylinder printing-machines, however, possess great advantages over the hand process; but they can be used only when the cloth is of one design from end to end, which the Indian dress pieces are not. For machine-printing the design is engraved on a copper cylinder, five or six inches in diameter; and machines are sometimes made to print at one operation designs containing eight, ten, and even more colours. In Turkey-red printing, however, only three or four colours are put on, and when that is the case, a machine will turn out about fifteen yards in a minute. After the cloth has been printed and dried, it is taken to another part of the premises, and treated with chloride of lime, &c., as already described. A few further operations—such as washing, drying, and calendering—make the cloth ready for market.

Messrs Stirling devote a part of their establishment to the production of bandana handkerchiefs. The natives of India used to make silk handkerchiefs, which bore white spots on a uniformly dyed ground. They produced the white spots by tying up the parts with thread, and then subjecting the cloth to the dye. In this branch of dyeing, British manufacturers could not, until a comparatively recent period, compete with the Indians, and the latter held the market for bandanas, until M. Kœchlin made his grand discovery. Messrs Monteith & Co. of the Barrowfield Dyeworks, near Glasgow, adopted the principle of M. Kœchlin soon after it was discovered, and succeeded in making bandanas far surpassing in excellence the best productions of India. Other firms followed, and Glasgow has since had almost a monopoly of the trade. The process of making bandanas may be briefly described. A dozen pieces or so of dyed cloth are laid evenly one over the other, wound upon a roller, and taken to the press-room—a large apartment occupied by a range of hydraulic presses. The roller is fixed in a framework behind one of the presses, and the end of the cloth brought forward between the upper and lower plates. Suppose, for instance, the pattern is to consist of a series of circular spots of red on a white ground. In order to produce this two plates of lead have to be prepared. The surface of the plates is cut away, leaving a series of lozenge-like eminences. The lozenges of the upper plate must fall exactly upon those of the lower. A number of channels are cut on the back of the plates, which communicate by holes with the sunk part of the engraved side. The cloth is spread over the lower plate, and the latter is pressed against the upper plate with a force of hundreds of tons. The result is that the cloth is tightly compressed between the raised parts of the plates. A stream of bleaching liquid is then allowed to run along the channels in the plates, and is forced into contact with the cloth except at the points where it is compressed. The liquid discharges and carries off the red dye, and, on opening the press in a few minutes after the liquid has been let in, it is found that a handkerchief with red spots on a white ground has been produced. The process is capable of being applied to an endless variety of patterns. Thus the ground might be made yellow, green, or any other colour, as easily as white. Six presses, worked by as many men, are capable of producing upwards of 4000 handkerchiefs in a day of ten hours. In some cases blank spaces are produced, into which flowers, &c., are printed by another process from an engraved copper plate, and some very pretty work is produced in that way.

The Cordale Printworks cover five acres of ground, and give employment to about 500 persons—men, women, and children—so that in their two establishments Messrs Stirling employ nearly 1500 workpeople. Machine-printers earn from 30s. to 50s. a-week; small-block printers, 25s. to 30s.; large-block printers, 30s. to 40s.; boys, 4s. to 7s. The machinery at Cordale is driven by two water-wheels and an engine of 50 horse power.

MANUFACTURE OF SEWED MUSLIN.

ANTIQUITY OF THE ART OF EMBROIDERING—ITS ADOPTION AS A FASHIONABLE
RECREATION IN THIS COUNTRY—MADE A BRANCH OF MANUFACTURE IN
GLASGOW—EXTENDED TO IRELAND—IMPROVEMENTS IN PRINTING DESIGNS—
HOW THE TRADE IS CONDUCTED—EMBROIDERING BY MACHINERY.

FROM the remotest antiquity ladies have delighted .in ornamenting
articles of dress by means of the needle. The Egyptians acquired
great celebrity in the art of embroidering the linen coverings of their
bodies and of the furniture in their houses. Herodotus speaks in
the highest terms of the delicacy and beauty of the fine linen and
embroidery of Egypt, and describes a linen corselet presented by
Amasis, King of Egypt, to the King of the Lacedæmonians. The
corselet was ornamented with numerous figures of animals worked in
gold and cotton. The Egyptians, indeed, would appear to have had
a passion for embroidering, for it is stated that even the sails of their
pleasure-boats were elaborately decorated with work of that kind.
The vestments used in the Greek Church have from an early period
been gorgeously embroidered, and in the Anglo-Saxon Church the
same fashion prevailed. The talent displayed by the Anglo-Saxon
ladies, and their devotion to the ornamentation of their churches and
ministers, were inherited by the Normans, by whom some remarkable
pieces of work were produced. The Queen of William the Con-
querer sewed the famous Bayeux Tapestry, which is one of the most
interesting historic works of the kind. It became customary for
persons to give proof of their piety by executing needle-work for the
service of the Church; and the quantity of such gifts possessed by
the cathedrals and churches of England prior to the Reformation was
enormous. In Lincoln Cathedral alone there were upwards of 600
vestments of costly cloth, ornamented with divers kinds of needle-
work, jewellery, and gold. Some of these were worth about L.400

U

each. In Eastern countries embroidering in the rich style referred to is still practised; but in Europe, except in the case of the gorgeous vestments still in use for the services of the Roman Catholic Church, only a few simple forms of the art are practised, and these are almost entirely limited to the clothing of ladies and children, and to certain articles of furniture.

In the end of last century the embroidering of muslin was adopted as a fashionable recreation by ladies in this country. The sewed muslin was a cheap and beautiful substitute for lace, and there was practically no limit to the variety of designs. At first the patterns generally were crude and inartistic; but as taste improved the beauty of the stitching was enhanced by the elegance of the designs. In course of time the embroidering of muslin became a favourite occupation for spare hours with all classes. In the early years of this century it was made a branch of manufacture by some enterprising men in Glasgow, and it is in that connection that it claims notice here. The pioneers of the trade began operations at a time when many women who had depended on the spinning wheel for a living were thrown out of work by the introduction of spinning machinery. There was, consequently, an abundance of willing hands ready to accept the new employment. The manufacturers had to proceed with caution, however, and make certain of a market before they embarked to any great extent in the venture ; and it was, therefore, some years before the trade assumed much importance. Up till 1825 only two or three firms were engaged in it, but these gave employment to many hands. The work was given out to the women to be executed in their own homes. The lasses of Ayrshire showed great aptitude for embroidering, and soon made a name for the excellence of their work—indeed, for a long time the embroidered muslins were sold in the home and foreign markets as "Ayrshire needle-work." In various parts of the county schools for teaching muslin-sewing were established. Three months were considered a fair period for training a girl; and when that term expired the pupil usually began to receive payment for her work, remaining in the school, however, for some time until she attained a certain degree of proficiency. When she began to receive payment for her work, a charge of 2d. or 3d. a-week was made for school-rent, or, as it was called, "stool-room."

Among the earliest firms in the trade were Messrs John Mair & Co., of Glasgow, and Messrs Brown, Sharp, & Co., of Paisley, whose representatives still hold a high position for the excellence of their productions. The founders of these firms did much to establish muslin sewing as a permanent branch of industry in the country. It was their custom to visit their workers, who resided chiefly in Ayrshire, once a-year, in order to ascertain how they were getting on, and to encourage the young people, by kind words and gifts, to strive to attain excellence in the use of their needles. This system had a most beneficial effect, and did much to maintain the celebrity of the Ayrshire muslin sewers. When flax-spinning ceased to be a domestic occupation female labour became very cheap in Ireland, and some of the Glasgow sewed muslin manufacturers took advantage of that circumstance, and sent agents into the north of Ireland in order to test the possibility of having the work done as well and at a cheaper rate than was being paid to the Scotch sewers. The experiment proved successful, and about the year 1830 the Irish work began to compete successfully with the Scotch. In order to ameliorate to some extent the condition of the female population in the north of Ireland, philanthropists of all religions aided in establishing training-schools to teach the poor girls how to sew, and a number of Scotch women were employed as teachers. A firm in Donaghadee—Messrs Cochrane & Browns—became known as the best and cheapest makers of sewed muslins in the United Kingdom; and when they removed their head-quarters to Glasgow about thirty-eight years ago, they were the most extensive in the trade. It was a member of this firm who in 1837 adapted the lithographic press to printing the outlines of the designs on the muslin. Formerly the patterns were impressed by blocks worked by hand. The blocks were necessarily of small size, and the process of printing large articles, such as ladies' dresses and babies' robes, was a slow one; and as the engraving of the blocks was expensive, the variety of designs was limited. All that was changed by the lithographic process, and a great impetus was given to the trade. The cost of production being reduced, the goods were sold at a cheaper rate, and an enormous demand was created for them not only at home, but in Europe and America. The United States merchants took large quantities, and have all along been the most extensive purchasers. Those of France, Russia, and Germany have also been buyers for many years past. Between the years 1845 and 1857 the trade extended rapidly, and in the last-named year reached the summit of its prosperity. By that time no town in the north of Ireland, from Derry to Dublin on one hand, and from Belfast to Sligo on the other, was without its sewing-agent, and in some towns five or six agents were required.

The sewed muslin trade went on increasing steadily from the

outset till the year 1857. There were periods of depression, it is true, arising from changes of fashion and other causes, but they were brief, and when they passed the trade went on with renewed vigour. The profits were sufficient to induce a large number of persons to embark in the business, and keen competition prevailed. After the potato blight in Ireland, labour became cheaper than ever in that country, and the Glasgow manufacturers gave employment to many thousands of the people. The demand for the sewed muslins kept pace with the supply for a time; but at length stocks began to accumulate, and it became apparent that a crisis was inevitable. A recent writer on the subject says:—" Public opinion said some of the houses could not afford to stop, and when the old markets were filled to repletion, consignments to new markets became common ; and although it may be doubtful whether those consignments ever brought a profit to the consignees, they nevertheless created a taste for the goods where it did not before exist, and that had a beneficial effect which endures to this day." Messrs D. & J. M'Donald shot far ahead of all competitors in the extent of their business. They built a palatial warehouse in one of the principal streets of Glasgow; and when at the height of their prosperity they employed in that establishment · 1500 men and 500 women, while between 20,000 and 30,000 needlewomen in the west of Scotland and north of Ireland were engaged in sewing for them. It was no unusual thing for the firm to pay L.15,000 a-month to their Irish agents and sewers alone. The value of the sewed muslin sent into the market by Messrs M'Donald was estimated at not less than L.500,000 a-year. The commercial crisis which began in America in the autumn of 1857 extended with crushing effect to the sewed muslin trade in this country, and some of the firms, including Messrs M'Donald, succumbed. The trade continued stagnant for a considerable time after that disastrous year. Manufacturers who withstood the shock found themselves with vast stocks on hand, for which no purchasers could be got, as the retailers were waiting until the stocks of the bankrupt firms were brought into the market. When the sales came, the prices were exceedingly low, and the retailers were enabled to sell the goods at little more than half the price that could have been obtained before the crisis. Sewed muslin was thus placed within the reach of purchasers of the poorest class; and the result was that the article ceased to be fashionable in the upper and middle classes of society. A prosperous trade had thus been ruined by the injudicious operations of some of those engaged in it; and though some improvement has taken place recently, the value of the sewed muslin goods produced in Scotland and Ireland is not one-half of what it was in 1857 and a few preceding years. As might have been expected, the sudden withdrawal of such a large amount of work caused much hardship among the Irish girls who had been engaged in it, many thousands of whom were thrown idle. In Ayrshire, also, the effects of the disaster were keenly felt.

The mode in which the trade is conducted is this:—The muslin is prepared for the sewers at the headquarters of the manufacturer. It is received in an unbleached state, and is cut up into certain lengths, on each of which is printed the design for one or several articles according to their size. The artist who supplies the designs draws them out carefully in full detail. The drawings are then passed to copiers to take off the outlines on transfer paper, and from that an impression is taken on a stone for the lithographic press. Most of the designers have been trained at the School of Arts, and their productions show a great advance on some of the early work. There is considerable scope for variety of designs, and many of those produced are remarkable both for their intricacy and for their exquisite beauty. The largest pieces at present in vogue are baby robes, for which the sewed muslins and cambrics are admirably adapted. The cloth for insertions and trimmings has the design printed on it by means of a small engraved cylinder of wood, fixed in a hand machine bearing a self-inking apparatus. This machine is called a "monkey," and is worked by a girl, who, after stretching on a table a piece of cloth about three yards in length, passes the "monkey" over it from end to end, leaving a certain space between each impression. On every piece of cloth is printed the number of the pattern, the number of days allowed for sewing it, and the price to be paid provided the work be well done. The cloth is divided into portions suitable for distribution among the sewers, and is then made up into parcels, along with the thread required to sew it, for transmission to the agents, who are stationed in convenient localities. At first the Scotch manufacturers sent over men to Ireland to act as agents, but latterly most of the agency work has been done by natives. In some cases the agents are paid fixed salaries, and in others receive a commission of about $7\frac{1}{2}$ per cent. on the amount of money which they pay to the workers. Their duty is to receive the prepared cloth from headquarters and find women to sew it, taking care to select the best hands for the finest work. They are responsible to a certain extent for the quality of the work, and the sewers are punished for faults by being paid less than the promised price when they do not exercise sufficient care. On the other hand, should the work be done in a superior way, something more than the ordinary price is paid. There is a slight difference between the modes in which the Scotch and Irish agents act. The Scotch agent is usually within a convenient distance of headquarters, and before the sewers receive payment their work is sent thither, and the price fixed. In Ireland such a system would be inconvenient, and there the agent judges of the merit of the work, and pays accordingly. When the sewing is completed, the agent returns the cloth to head-quarters. On opening the parcels, it is found that almost every piece of cloth is besmudged with smoke and grease, and that the whole is odorous of a compound of "peat-reek" and bacon. This insanitary condition of the goods is explained by the fact that most of the women who sew are the home workers of their respective households, and have to relinquish the needle whenever there is cooking or cleaning to do, and snatch it up again when the inter-rupting job is completed. Their hands are thus frequently soiled as well as their clothing; and when to that is added the palpable atmosphere of an Irish cabin, the state in which the goods arrive is easily accounted for. In the warehouse the separate pieces of work are stitched together in webs and sent to the bleach-field, whence they return as pure as snow. The various articles are then cut out, finished, and dressed, when they are ready for the market.

About fifty firms in Glasgow are engaged in the trade. The census return for 1861 showed that there were in Scotland at that time 7224 women engaged in embroidering muslin ; and as that return was made at a time when the trade was still suffering from the effects of the crisis of 1857, it is probable that the number at present employed will not be under 10,000. The occupation is a sedentary one ; and in order to make good wages, the sewers have to apply themselves closely to it. About thirty years ago it required from fourteen to sixteen hours' work a day to make 12s. a week ; and the writers of the "Statistical Account of Aryshire" at that time refer to the occupation as being tedious and unhealthy. One describes it as "an employment which, in most instances, unfits women for other occupations, and, besides, it frequently injures their health, and leaves them very helpless, when they get houses of their own, as to the management of their domestic concerns." It was said to have a prejudicial effect upon the chest and the eyes. The scale of payments is lower than it was prior to 1857 ; but many women who are not robust enough for factory work find muslin sewing to be a light and convenient, if not very remunerative occupation.

Many ingenious attempts have been made to supersede hand-sewing in the flowering, sprigging, and pointing of muslin, but as yet no thoroughly efficient piece of mechanism has been produced. The most successful attempt was made by M. Heilmann, of Mulhausen, whose embroidering machine, though limited to a particular kind of work, effects a great saving of labour. It is chiefly adapted to figuring muslin window-curtains, and similar work. The machine is thus described in "Chambers's Encyclopædia:"—" Although the details of the construction of this machine are rather complex, the principle of its action may be easily understood. The needles have their eyes in the middle, and are pointed at each end, so that they may pass through from one side of the work to the other without being turned. Each needle is worked by two pair of artificial fingers or pincers, one on each side of the work ; they grasp and push the needle through from one side to the other. A carriage or frame connected with each series of fingers does the work of the arm, by carrying the fingers to a distance corresponding to the whole length of the thread, as soon as the needle has passed completely through the work. The frame then returns to exactly its original place, and the needles are again passed through to the opposite set of fingers, which act in like manner. If the work were to remain stationary, the needles would thus pass merely backwards and forwards through the same hole, and make no stitch ; but by moving the work as this action proceeds, stitches will be made, their length and direction varying with the velocity and the direction in which the work moves. If 140 needles were working, and the fabric was moved in a straight line, 140 rows of stitching would be made ; if the work made a circular movement, 140 circles would be embroidered ; and so on. In order, then, to produce repetitions of any given design, it is only necessary to move the fabric in directions corresponding to the lines of the design. This is done by connecting the frame on which the work is fixed to an apparatus similar to a common pantagraph, or instrument so constructed that one end repeats on a smaller scale exactly the movements which are given to the other. The free end of this is moved over an enlarged copy of the design, the movement being a succession of steps, made after each set of needles has passed through ; and thus the work is moved into the position required to receive the next stitch of the pattern."

MANUFACTURE OF FISHING-NETS.

NET-MAKING BY HAND—STORY OF THE NET-LOOM AND ITS INVENTOR—THE
MUSSELBURGH NET-FACTORY—HOW NETS ARE MADE BY MACHINERY—EX-
TENSION OF THE TRADE, AND DECLINE OF HAND-NETTING.

KNITTING or weaving fishing-nets is one of the oldest branches of the
textile manufactures of Scotland.   Many centuries ago the dwellers
on the shores of Caledonia knew how to twist the fibres of flax and
hemp into traps for fish ; and for many years past the manufacture
of nets, lines, and other engines for capturing the finny tribes, has
been an important branch of industry in the towns and villages adjoining the sea.   It became customary for the wives and families
of the fishermen to spin and weave the nets required for their mutual
support.   The domestic spinning apparatus has now, however, been
almost entirely discarded in that connection, though hand-knitting
continues to be practised in certain parts.   In Caithness-shire, for
instance, some twine is still sold to be worked into nets by women in
their own homes.   A dozen years ago thousands of nets were made
by hand in that county, and in the villages on the south side of the
Moray Firth, the price paid being from 7s. 6d. to 10s. each.   Such
was the cheapness of labour in the north that, though it requires five
weeks' close application in order to produce one net, there were
always more willing hands than could find employment in that way,
and a real hardship was experienced by many when machines began
to compete with hand-labour.   It is probable that in a few years
net-making by hand will be entirely abolished, for the machine-
made nets are much superior to the others.   The material used is
also undergoing a complete change.   That "there's nothing like
hemp for nets," was a maxim with the fishing fraternity from the
earliest times, but within the past five or six years the virtues of cotton
have come to be understood and appreciated, and now hemp is
rapidly going out of fashion.   It has been discovered that, while a

hemp net gradually loses its strength and becomes almost useless in five or six years, a cotton net remains unimpaired for nine or ten years—so that a cotton net, while quite as cheap as a hemp one, is twice as durable. Another advantage is that a cotton net can be made much finer and lighter than one of hemp. On the west coast, and on the south side of the Moray Firth, some flax nets are employed ; but as they are considered inferior to cotton, they are not likely to remain long in use. When hemp comes to be totally discarded, hand-labour in the making of nets will cease ; for it has been found impossible to produce cotton nets by hand either so cheaply or so satisfactorily as by the machines.

The ease with which hand-labour could be obtained in the coast villages, and the practice of fishermen having their nets made chiefly by their own families, were circumstances which tended to prevent mechanicians from considering the practicability of constructing a machine that would supersede netting by hand. The supply of nets had always been equal to the demand, and there was little prospect that fishermen—who, by the way, are notoriously conservative in matters professional—would patronise any contrivance which was destined to supersede the home employment of their families. Such were the circumstances under which the inventor of the net-loom set himself to work out an idea which he had entertained from boyhood. The history of the net-loom and its inventor has not previously been written, and what follows is the result of personal inquiries made in various quarters.

Towards the close of last century a young man named James Paterson followed the trade of a cooper at Musselburgh, and, being by it connected with fishing, he became conversant with the making and use of nets. As he watched the nimble movements of the net-workers, the idea struck him that it might be possible to devise a machine that would relieve the women from the tedious task of net-making. Not considering himself capable of putting the idea into practice, he did not mention it to any one. Some time afterwards he joined the army, and in the capacity of Deputy-Assistant-Commissary-General served in Egypt, the Peninsula, and at Waterloo. He used to relate how through all the trials of war the idea of working nets by machinery haunted him, and he devoted many spare hours to thinking over the matter. After Waterloo Mr Paterson left the military service, and, having acquired some means, resolved to devote himself to the invention of a net-loom. It would appear that he went first to Queensferry, but before he had completed the machine removed to his native place, Musselburgh. With the aid

of an ingenious mechanic, who was sworn to secrecy, and after many
months of anxious labour and many failures, Mr Paterson saw the
machine take promising shape, and at length things were so far
advanced that a trial was to be made.   Worn out by anxiety and
galled by many unsuccessful experiments, Mr Paterson was disposed
to regard this trial as a decisive one.   He thought he had exhausted
his inventive powers, and felt inclined, if the machine did not then
succeed, to abandon the idea he had cherished so long.   The critical
moment arrived, the last lever and spring had been adjusted, and all
was ready.   With nervous touch the inventor and his assistant
moved the machine.   Its action was easy, and all the parts appeared
to work in harmony to the desired end.   After a few moments, however, the machine came to a stand, and no available power could
move it.   It was carefully examined in every part, but no cause for
the sudden and complete stoppage could be found.   Some hours had
been spent in trying to discover what was wrong, when Mr Paterson
lost his temper, flung the key of the workshop to the mechanic, said
he might "make a kirk or a mill" of the concern, and then went
home.   The man was no less distressed than his master, yet he was
not disposed to abandon the invention.   He went to a smith who
had been doing some work for Mr Paterson, and, accompanied by
the smith, returned about midnight to have another look at the
machine.   The smith was quite puzzled by the apparatus, and held
out no hope that he would be able to make it work; but as he
peered with wonder through the ranges of needles, &c., his eye
lighted on a bolt which appeared to be loose.   He called the
mechanic's attention to the fact, the bolt was adjusted, and a fresh
attempt was made to start the machine.   The loose bolt had been
the cause of the stoppage, and the machine worked beautifully
when the bolt was put right.   After getting the smith to take a
vow of secrecy, the mechanic sent him home, and, notwithstanding
the unseasonable hour, rushed into Mr Paterson's house to announce
the fact that he had got the machine to work.   He met with an
indifferent reception, for the news appeared too good to be true.
Mr Paterson got up, however, and went to the workshop, where he
had the pleasure of seeing that the idea which he had nursed for so
many years was not an impracticable thing after all.   There was
the machine turning off row after row of meshes at a rate which
would leave a dozen hand-workers far behind.   Mr Paterson patented
his invention, and established a net factory in Bridge Street, Musselburgh, about the year 1820.   In order to preserve a secret that had
cost so much anxiety, labour, and money, steps were taken to prevent

the workpeople from carrying away descriptions or sketches of the machine.

It was at first difficult to find a market for the machine-made nets, and many years passed before they found much favour with fishermen. In 1839 Mr Paterson had eighteen looms at work, and these, together with the spinning-machines, gave employment to upwards of fifty persons. About seven years before that date a Mr Robertson from Sunderland went to Musselburgh, and started in opposition to Mr Paterson. It was alleged that Mr Robertson made his machines from drawings which were obtained surreptitiously from some of Mr Paterson's workpeople; and Mr Robertson, in order to avoid an action for infringing the inventor's patent rights, sold off and left the country. The greatest difficulty that Mr Paterson had experienced in working out his invention was the formation of a knot which would not slip. The knot put on by hand is peculiar, and Mr Paterson did not succeed in making one similar to it, though the knot he made was firm enough. About thirty years ago a mode of forming the hand-knot on the machine was discovered by Mr John Low, manager of Mr Paterson's factory, and was adopted in all the machines subsequently made.

The principal manufacturers of nets in Scotland are Messrs J. & W. Stuart, Musselburgh, who in 1849 acquired Mr Paterson's factory and the patent rights to his invention. In their hands the business has grown steadily; and though there are a number of rival manufacturers, Messrs Stuart have been able to maintain the foremost place. Their new factory is one of the finest in the country. Finding that the building in Bridge Street, Musselburgh—which may be called the nursery of the trade—was, notwithstanding several additions, too small and otherwise inconvenient, the firm determined to erect a new factory. They selected a site on the right bank of the Esk, adjoining the railway station, and about fifteen years ago completed the first section of a new factory. It afforded accommodation for 100 looms, with hemp-preparing and spinning-machinery, to the extent of 3500 spindles. During 1867 they made extensions which more than doubled the size of the place. About ten years ago, when cotton began to be used as a substitute for hemp in making nets, Messrs Stuart turned their original factory into a cotton-spinning mill; but as the change of material advanced, the productive power of the cotton machinery was not equal to requirements, and the principal portion of the recent additions to the establishment on the Esk is a fine cotton mill containing 20,000 spindles. The buildings altogether form an oblong square, and cover a space of four acres. One end of the square is occupied by the cotton mill, which is an elegant and imposing structure, consisting of four lofty storeys. The front of the cotton mill is of a highly ornate character, the lower part being rusticated, and the upper decorated with pilasters and finely sculptured emblematic figures. On either side of the square, and at the end opposite the cotton mill, are the hemp, spinning and weaving departments, stores, mechanics' workshops, &c. The inside of the square is occupied by a tastefully laid out ornamental garden, in one end of which stands a detached building containing the counting-room and private offices. The factory is connected by a branch line with the North British Railway.

Cotton-spinning has already been described, and a few sentences on the preparation of hemp will suffice. In tracing the hemp through the various processes of manufacture, the visitor must follow it from the spacious stores to the softening-room, where it is fed through sets of grooved rollers, which make the fibres more pliable, and so facilitate subsequent operations. The heckling comes next in order, but neither in that nor in the subsequent operations of drawing, roving, and spinning is there anything different from what has been described as pertaining to the manufacture of flax and cotton. The machinery used is of the most perfect description, and the fineness and evenness of the yarn produced is wonderful when the coarseness of the hemp fibre is taken into account. In twisting two or more folds or " plies " of yarn together to make twine, the yarn is made to pass through water. The water lays all the loose fibres, so that they get twisted into the body of the twine, thus at once increasing its strength and improving its appearance. After the twisting is completed, the twine, while still wet, is run through a rubbing-machine, which makes it perfectly smooth. The Musselburgh hemp twine has long held a high place in the favour of the fishermen; and though in recent years cotton has become more popular than hemp, Messrs Stuart have never had their hemp-spinning machinery idle. The twine is reeled off the bobbins of the twisting-frames into hanks. The hanks are stretched in frames, and placed in a hot-air chamber, where they are thoroughly dried in a few minutes. The twine intended for sale is made up into bundles of 28 .℔. each, and that which is to be made into nets on the premises is wound upon large spools or bobbins. The machinery is driven by two handsome engines of 100 horse power each.

The weaving shed is an extensive place. It contains 200 looms, with space for 100 more, which are being put in. The shed is lofty, well lighted, and well ventilated. Each loom requires the space of three or four common power-looms. Though called a loom, the apparatus has no resemblance to the ordinary weaving-machine, being more like the knitting-frame—indeed, some portions of the two are identical in shape and name. The part which forms the mesh consists of an arrangement of hooks, needles, and sinkers, one of each being required in the making of every knot. In a loom capable of working a net 400 meshes in depth, there are consequently 1200 moveable parts directly employed in the formation of each row. The looms are from six to eight feet in width, and about six feet in height. The hooks and needles are ranged horizontally at a height of three feet above the floor. The mechanism by which they are put into operation is of a peculiar and complicated kind. Indeed, it is obvious that the machine, however well it may accomplish its work at the slow rate at which it is worked, is not yet perfect, the clumsy looking system of weights, springs, and levers, and the singular movements which the operatives have to make in order to bring them into play, are decidedly primitive, and a long way behind the automatic spinning-machinery situated in another part of the establishment.

Messrs Stuart have made and patented various improvements on the original loom, but there is still scope for the exercise of mechanical skill in the way of effecting further amendments. It is at least possible to adapt the machine to be moved by steam or water power. The mode of working the loom may be briefly described. The operative moves a lever which draws the last completed row of meshes off the sinkers, and transfers them to the hooks. Another lever is moved, and the meshes are caught by the needles. The effect of these changes, and the movement of other parts of the machine, is to twist the lower part of each mesh into a loose knot. The foot of the operative touches another lever, and a steel wire is thrust across the machine through all the knots. There is a hook at the end of this wire—or shuttle, as it is called—into which the end of a piece of twine is fixed. The wire is then withdrawn, and as it goes takes the twine along with it. Now the sinkers play their part. They consist of thin slips of brass having a hook or notch formed on the upper end, and are situated between the needles. When the twine has been drawn across through the loops of the meshes, the sinkers are released in succession, and as they descend each draws down the cross thread into a loop, sufficient to form two sides of a mesh, the other two being formed by the same parts of the previous row. One or two movements more remove the knots off the needles, and draw them firmly, thus completing the operation. In forming each row of meshes, the worker has to press upon half a dozen levers in succession, and pass from one end of the machine to the other. The occupation is consequently an unusually active one. As the work proceeds, the net is wound upon a self-acting cylinder which forms the upper part of the machine. There is an index attached to the cylinder which records the progress of the work. When a sufficient length of netting has been made, it is unwound from the cylinder, and taken to the "guarding-room," where several rows of stout twine are worked on what is to be the upper side of the net. That is called the "guarding," and its purpose is to withstand the friction of the "back-rope." It has not been found possible as yet to put on the guarding in the loom, and, accordingly, girls are employed to work it by hand. The net is next carefully examined, and any defect made good. It is then neatly folded up, and in that condition goes into the market, the subsequent operations necessary to fit it for use being performed by the fishermen.

In addition to their famous herring-nets, Messrs Stuart make nets for the salmon, mackerel, pilchard, and sprat fishing. The pilchard-nets are made of stout hemp twine, while the sprat-nets are composed of fine cotton thread. There is as great diversity in the varieties of the herring-nets. Many of the Dutch fishermen retain their old faith in heavy gearing, and have their nets made of three-ply hemp twine; others, taking a hint from the Scotch fishermen, are beginning to use light nets. The fishermen on some parts of the Scotch coast are now using cotton nets, which look like gossamer webs beside the Dutch "three-ply." An index of the different notions prevailing as to what a net should be is afforded by the fact that Messrs Stuart make for herring-fishing some cotton nets sixty yards in length, and 300 meshes in depth, which weigh only 12 ℔.; and others of the same size which weigh 60 ℔. The hemp-twine used is either two or three ply; while the cotton twine ranges from three to thirty ply. Messrs Stuart have shown their nets and twine at all the recent exhibitions, and received medals and honours at London, Dublin, Paris, Amsterdam, Boulogne, &c.

In their cotton mills and net factories, the firm employ about 800 persons, a large proportion of whom are women and girls. Nearly all are paid by piece. The average wages of men in the spinning department is 21s. a-week; of women, 7s. 6d. to 8s.; and in the weaving department men make 20s., and women 10s. to 18s. Many of the workers live at a distance from the factory, and it was customary with them to carry a supply of provisions for the day when they went to work in the morning. Considering that such a mode of food-taking was neither healthful nor pleasant, Messrs Stuart provided a dining-hall, kitchen, and cook; and now the workpeople can have a freshly cooked hot breakfast and dinner for an almost nominal payment. A breakfast of porridge and milk costs 1d.; of tea and bread, 2d. Dinner of soup and bread costs 1½d., the same with meat 2d. additional.

During the past ten years net-making, as a branch of manufacture, has increased rapidly. In Scotland there are thirteen or fourteen firms engaged in it. Nearly 600 looms are in use, and in making twine and nets upwards of 2000 persons are employed. The hemp and flax twine used is made in Scotland, but all the cotton, except what is made by Messrs Stuart, is brought from Manchester. Singular as it may appear, it is a fact that net-making by machinery was only recently introduced into Wick, the metropolis of the herring fishery. Though the greater portion of the nets used in that quarter were machine made, and though labour during ten months of the year is cheaper than in almost any other part of Scotland, no one had sufficient enterprise to embark in the manufacture until about two years since, when Mr George Manson gave the thing a trial, and was at once eminently successful. The trade has also been begun recently in Elgin, Largo, and Buckie. Messrs N. & N. Lockhart, Kirkcaldy, are next in extent to Messrs Stuart. They have been engaged in making twine and nets since 1854, and three years ago built a fine new factory in which they have sixty-four net-looms, and a large quantity of spinning-machinery for twine and yarns.

MANUFACTURE OF PAPER AND PAPER-HANGINGS.

ORIGIN OF PAPER—THE PAPYRUS OF THE EGYPTIANS—PROGRESS OF THE ART
OF PAPER-MAKING—THE FIRST PAPER MILLS IN BRITAIN—EXTENT AND
DISTRIBUTION OF THE TRADE IN SCOTLAND—THE MATERIALS USED FOR
MAKING PAPER—PAPER-MAKING BY HAND—THE INTRODUCTION OF MACHI-
NERY—INVENTION OF THE PAPER-MAKING MACHINE—THE PAPER DUTY—
COWAN'S PAPER MILLS AT PENICUIK—THE MANUFACTURING PROCESSES DE-
SCRIBED—INVENTION OF PAPER-HANGINGS BY THE CHINESE—SUCCESS OF
THE MANUFACTURE IN FRANCE—DIFFICULTIES OF THE FIRST MANUFAC-
TURERS IN BRITAIN—THE TRADE IN SCOTLAND.

PAPER has come to play such an important part in the social, intellectual, and commercial relations of mankind, that even a temporary
stoppage of the supply would amount to a calamity. It is almost
the only article manufactured for which no convenient substitute
could be obtained. Our woollens, linens, and cottons would stand
substitute for each other; but what substance is there that would
serve so many purposes as paper, or serve them so well? It is the
vehicle of written thought between nations and individuals, and
without it the art of printing could not have been made available,
except in a costly, and consequently limited, way. Viewed in that
light, it may be said that paper has contributed more to the advancement of the human race than any other material employed in the
arts. As an article of manufacture and commerce, it occupies an
important place, giving employment to many thousands of persons,
and utilising materials which would otherwise be useless, or worse.

The history of paper, so far as it is known, is familiar to most
people. Paper had its origin in Egypt, where the bark of the
papyrus—a reed growing on the banks of the Nile—was formed
into sheets for the priests to write upon. The thin slips of bark
were arranged in transverse layers on a table, and subjected to pressure, under which the bark became cemented by its own gum. The
"sheets" of papyrus formed in that way were flexible, and, before

the invention of bookbinding, were usually made up in the form of rolls. It is not known when nor by whom the art of making paper from pulp was invented; but it is considered probable that the credit of the invention belongs to the Chinese, who, it is stated, were familiar with the art about the beginning of the Christian era. The first materials used were the barks of various trees, portions of bamboo stems, and cotton. In the seventh century the Arabians learned from the Chinese how to make paper of cotton. From China, the art was carried into Spain, and the Moors discovered that paper could be made of hemp and flax as well as of cotton. Spain communicated the art to France and Holland, and thence it reached Britain.

The first paper mill mentioned as existing in this country was erected in Hertfordshire by Mr John Tate, who is thus referred to in a book printed by Caxton, about the year 1490:—

> " Which late hathe in England doo make thya paper thynne,
> That now in our Englyssh thys booke is printed inne."

Royal patronage was conferred on Mr Tate for his enterprise, as appears from two entries in the Household Book of Henry VII. These are as follow:—" May 25, 1498—For a rewarde geven at the paper-mylne, 16s. 8d." " 1499—Geven in rewarde to Tate of the mylne, 6s. 8d." In 1588 a German named Spielman established a paper mill at Dartford, for which he was knighted by Queen Elizabeth, who also granted him a license " for the sole gathering for ten years of all rags, &c., necessary for the making of such paper." Mr Tate's paper mill must by that time have stopped, else it is unlikely that a monopoly would be given to Spielman. The latter would appear to have carried on an extensive business, if a poet of the time who wrote the following lines did not exaggerate:—

> " Six hundred men are set to work by him,
> That else might starve or seek abroad their bread ;
> Who now live well, and go full brave and trim,
> And who may boast they are with paper fed."

It is a matter of literary controversy to which of those mills Shakspeare makes Jack Cade allude when he accuses Lord Say of having, " contrary to the King, his crown, and dignity," built a paper mill.

During the fourteenth century the Germans carried the art of paper-making to great perfection; but, realising the value of retaining the trade in their own hands, they kept their processes secret.

x

Even so late as the sixteenth century the Dutch prohibited, under pain of death, the exportation of moulds for making paper. This secretiveness on the part of the people from whom it is supposed we acquired a knowledge of the art, tended to retard its progress in this country; and up till a comparatively recent period only the coarser kinds of paper were made. The first patent for paper-making in Britain was granted in 1665 to Mr Charles Hildeyerd, and was described as being for " the way and art of making blew paper used by sugar-bakers and others." The second was granted in 1675 to Mr Eustace Barneby, for " the art and skill of making all sorts of white paper for the use of writing and printing, being a new manufacture, and never practised in any way in any of our kingdoms or dominions." In 1680 Mr Nathaniel Bladen patented " an engine method and mill whereby hemp, flax, lynnen, cotton, cordage, silk, woollen, and all sorts of materials may be made into paper and pasteboard." Five years later Mr John Briscoe claimed to have effected great improvements in the manufacture of paper, and took out a patent for " the true art and way for making English paper for writing, printing, and other uses, both as good and as serviceable in all respects, and especially as white, as any French or Dutch paper."

In Scotland the manufacture of paper was begun towards the close of the seventeenth century. On the 19th August 1695, a company was formed in Edinburgh " for manufacturing white writing and printing papers." The articles of the company are preserved in the British Museum, and are the earliest documents extant relating to paper-making in Scotland. When or on what scale the company began operations there are no means of ascertaining. In 1709 a paper mill was built at Valleyfield, Penicuik, by Mr Anderson, the Queen's printer in that year. This mill is still in operation, and by successive extensions has become one of the largest in the country. The trade made little progress for many years. In 1763 there were but three mills in the neighbourhood of Edinburgh, and the quantity of paper made was only 6400 reams a-year. Ten years later there were twelve mills in the district, some on a more extensive scale than any in Britain, and the production had risen to 100,000 reams. At that time a large quantity of printing paper was sent to London, from which city it used formerly to be brought. It is related of one of the earliest Scotch paper-makers that he found his trade to be so unremunerative that he tried to eke out a living by exhibiting an elephant. At present, according to the trade directory, there are in Scotland fifty-three firms engaged in the manufacture of paper, and these own or occupy fifty-seven mills, of which twenty-two are in the county of Edinburgh—nine being on the North Esk, and a like number on the Water of Leith. The other Scotch mills are distributed over a dozen counties, Aberdeenshire and Lanarkshire ranking next to Edinburgh. The number of paper-making machines in use is eighty, each of which, at a moderate computation, is capable of turning out seven tons of paper a-week, so that the gross annual production will amount to about 30,000 tons. Nearly 10,000 persons are directly employed in the manufacture. Some of the Scotch firms enjoy a world-wide celebrity for the excellence of their productions, and it is an exceedingly creditable fact that one of them carried off the first prize for paper at the Paris Exhibition. A comparison of the Paper Trade Directory for the year 1860 with that for the present year, shows that while the number of mills in England and Ireland has been decreasing, there has been an increase in the number of mills in Scotland. There were 359 mills in England in 1860, and in 1868, 293; for Ireland the figures are 35 and 21 respectively, and for Scotland 54 and 57. This shows that a reduction of 19 per cent. has taken place in the number of English mills, of 40 per cent. in the number of Irish, while the Scotch have increased about 6 per cent.

No material has yet been discovered to supersede the use of linen and cotton rags in making the finer qualities of paper. Many attempts have been made to find other kinds of fibre that would be equally suitable; and the list of substances which have been subjected to a trial is an exceedingly curious one. Up till the year 1857 upwards of two hundred patents had been taken out in Britain for the protection of inventions of this kind, and they relate to about fifty varieties of fibre. The list includes asbestos, beanstalks, clover, dung, gutta-percha, heather, moss, nettles, peat, sawdust, sea-weed, thistles, and tobacco-stalks. In the year 1772 a book was published at Regensburg, in which there were eighty-one specimens of paper made from as many different substances. The demand for paper has always threatened to exceed the supply of rags, and hence the desire to find a substitute or auxiliary. About fifteen years ago many experiments were made both in this country and on the Continent, with the view of discovering some material for making paper that would give more satisfactory results than any of the auxiliaries to rags previously introduced. In 1854 Mr Alexander Brown patented a mode of making paper from the "bracken" or fern plants of Scotland. About the same time a Prussian inventor produced some specimens of paper made from pine trees. In 1855 the proprietors of the "Times" newspaper offered a prize of L.1000 for the discovery of a new and readily available material, but the conditions attached to the prize were not such as to make it an object of popular desire by persons who were best able to make experiments. The competition resulted, however, in proving that paper of an inferior quality could be made from wood-shavings and bran. In 1856 Mr Edward Grantless, a marble-cutter of Glasgow, obtained a patent for making paper from stone. The material that has come nearest to answering the requirements of anything that could compete with rags is "esparto grass," a plant obtainable in great abundance on both the European and African shores of the Mediterranean. The capabilities of esparto were discovered to some extent in 1839, but the plant did not attract much attention until 1862. Mr Routledge, of the Enysham Paper Mills, near Oxford, had made some experiments with the fibre about the year 1856, and succeeded in discovering a more useful, effectual, and economical mode of treating it than any of those who preceded him in the research. About ten years ago he began to make printing paper from esparto exclusively; but it was only when his productions were shown at the Exhibition of 1862 that the value of his discoveries was fully realised. The report of the jury on paper says:—" Judging from the specimens of paper exhibited by Mr Routledge, manufactured by him at his mills at Enysham, in Oxford, exclusively from esparto, as well as from the other specimens of paper manufactured at various other mills employing his process, in which esparto is used as a blend with the ordinary rag material, the results are very satisfactory, demonstrating that a new material has at length been brought into use, meeting this long-desired requirement both as regards quality and economy." Esparto now holds a permanent and important position in the trade. Used alone, or with a small admixture of rags, esparto supplies the entire newspaper press of this country, and is extensively used in the production of other varieties of paper.

About three years ago a gentleman residing in Norwich, who had given some attention to the subject, became the accidental possessor of a peculiar kind of grass grown in Japan, and exported hither chiefly for the use of gardeners, who employ it in tying plants. The grass attains a length of six or seven feet, and a breadth of something less than an inch, and in its native country flourishes in great luxuriance. Nearly the whole of the blade is composed of a strong silky fibre, running from one end to the other, the proportion of vegetable matter making up the remainder of the leaf being much smaller than in any other grass upon which experiments have been made. Thinking he had discovered what might prove to be a

valuable material for making paper, the gentleman referred to submitted a specimen of the grass to Messrs Magnay & Delane, paper-
makers, Taverham, in Norfolk, and by them it was forwarded to
their chemist in London, whose report upon the material, which was
accompanied by a sample of the pulp obtained from it, was most
satisfactory, and placed it far above esparto grass in all its essentials
—namely, length and quality of fibre, the readiness with which it
might be bleached and reduced to pulp, and its capability of being
converted into the finest quality of paper.   The only objection to
the use of the new grass was its cost; but as it can be produced in
unlimited quantities, both in Japan and elsewhere, there is no doubt
that, if the demand were sufficient to induce the cultivation of the
material, the supply could be multiplied to any required extent.   A
good deal of paper is now being made in Northern Europe from
wood-shavings and sawdust; and a Scotch firm of timber-merchants,
in Gottland, recently completed arrangements for the conversion of
the refuse of their mills into pulp for making paper.

The application of machinery to the manufacture of paper is
of recent date.   Both in preparing the pulp and in making the
paper, the appliances used by the early paper-makers were of the
simplest kind, as may be gleaned from the following account of the
mode of working practised in the mills of Mid-Lothian in the early
years of this century :—The rags, after being thoroughly washed and
bleached, were, while still wet, laid in heaps and covered over with
sacking.   In that way they were allowed to ferment for about a
week, when they were taken out and cut into small portions by
means of a sharp hook.   They were next placed in large mortars
made of oak, and there pounded with iron-shod rods, kept in motion
by either wind or water power.   So slow was this process of pulping, that eighty pairs of stamps produced only one hundredweight
of pulp a-day.   Even then the work was so imperfectly done that
the stuff had to be pressed into boxes, and allowed to " mellow " for
several weeks.   After that it had to be subjected to a series of
beatings in the mortars before it was ready for use.   The pulp was
next placed in a " vat," along with a certain quantity of water.   The
fibrous matter was held in suspension by the liquor being constantly
stirred by a revolving frame or series of wooden arms.   In making
paper by the hand process—now almost extinct—the order of operations is briefly this (and it is merely mentioned here to give some
idea of the immense advantage conferred by the paper-machine
now in use) :—The " dipper," or " vatman," takes a framework
covered with wire-gauze, and dipping it into the vat takes up a

quantity of pulp sufficient for one sheet, the size of the sheet being determined by the size of the "mould," and the thickness by the depth of a moveable ledge placed round the wire. The water soon drains off, leaving the pulp resting in an even layer on the wire. Another workman, designated a "coucher," takes the mould and transfers the pulp to a piece of felt or woollen cloth. Usually two moulds are employed, so that, while the "coucher" is emptying one, the "vatman" is filling the other. The "layer" deposits the successive pieces of felt, with their delicate burden, in a pile; and when a certain number of sheets are thus arranged, they are taken to a press which forces out all the superfluous water, and gives a degree of solidity to the paper. The felts are then removed, and the paper is pressed by itself. The largest size of paper made by hand was "antiquarian," which measured 53 inches by 31, and so great was the weight of the liquid pulp employed in the formation of a single sheet, that no fewer than nine men were required to raise the mould out of the vat, by means of pulleys. The "vatman's" duties used to be of a severe kind. In order to give the paper the proper degree of finish, the liquid pulp was heated, and he had to stand over the steaming vat usually from three o'clock in the morning till one or two in the afternoon. Great skill and strength were required in order to produce the heavier kinds of paper, and the vatmen were regarded as the most important persons connected with a paper mill. The paper, after being pressed, was hung on hair-ropes to dry. The ropes were made of hair because that material did not give off any stain. When dry, the paper was sized by being dipped into baths of size-liquor, after which it was again hung up. It was finished by being pressed between hot iron plates. The productive power of a mill was in those days reckoned by the number of vats. The men who attended each vat were called a "vat's crew;" and so independent were the workmen of fifty years ago, that it is stated to have been no unusual thing for a vat's crew, or even all the vats' crews in a mill, to stop working when they thought proper, adjourn to the nearest public-house, and there enjoy themselves for such time as suited their tastes, no matter how pressing the demand for paper. The vatmen and couchers looked upon their occupation as one that machinery could never affect, and were most inconsiderate in their actions and demands. Their employers were so much pestered by the irregular manner in which the work of their mills went on, that they were ready to try any promising invention that offered, and when the paper-machine was brought to a degree of perfection it was readily adopted.

The first important mechanical contrivance introduced into the trade was the pulping-engine, invented in Holland about the middle of last century. The tedious method of fermenting the rags and bruising them in a mortar was superseded by this machine, which, in its present improved form, reduces the fresh rags to pulp in a few hours, and is capable of comminuting five or six tons a-week. Though the means of producing a vastly increased quantity of pulp were thus provided, many years elapsed before any successful attempt was made to perform the work of the vatmen and coucher by machinery. Among the more remarkable contrivances produced for the purpose were a set of automatic figures made by the ingenious M. Montgolfier of Paris. The figures were designed to perform the operations of the respective workmen ; but they could not be got to work profitably, and were thrown aside. While mechanicians of world-wide fame were racking their brains to find out a system of paper-making machinery, Louis Robert, a humble clerk in a paper mill at Essonne, was quietly studying the subject, and one day excited the wonder of his employers by producing for their inspection a working model of a machine which was capable of producing from the pulp webs of paper of any length. The model was a tiny thing, from which the paper came forth no wider than a piece of tape. A machine of larger dimensions was made and patented, and when the invention came under the notice of the Government, the ingenious author of it was liberally rewarded. M. Robert's employers bought the invention, and transferred it to an English firm— Messrs Fourdrinier—who spent L.60,000 in perfecting it ; but, in consequence of some defect or change in the Patent-laws, they reaped no advantage from their enterprise. The invention became public property, and Messrs Fourdrinier were reduced to bankruptcy. The paper-making machine is one of the most ingenious contrivances employed in the arts, and has been a chief cause in the creation of that cheap literature which has done so much to make the present century a remarkable era in the history of human progress. The finest and most complete paper-making machines in the world have been made in Edinburgh, by Mr George Bertram, Sciennes, and Messrs James Bertram & Son, Leith Walk. A machine exhibited by Mr George Bertram at the Exhibition of 1862 created universal admiration, and was pronounced by papermakers of all nations to be by far the most perfect machine ever produced.

When the manufacture of paper began to obtain a footing among British industries, an Excise duty of 3d. a-pound on first-class papers, and 1½d. on second-class, was levied. This duty produced L.46,868 in 1784, L.315,802 in 1815, L.619,824 in 1830. In 1839 the Excise duty was reduced to an equal rate of 1½d. a-pound on all qualities of paper. Under the new scale the revenue from paper increased in 1857 to L.1,244,652, of which sum Scotland contributed L.263,786. In 1860 the paper duty produced L.1,397,349. Having come to be regarded as a "tax on knowledge," its abolition was mooted on several occasions, but not until 1860 was any movement made for its removal by the Government. In that year Mr Gladstone (then Chancellor of the Exchequer) introduced a bill to repeal the duty. The bill passed its third reading in the Commons on May 8, by the narrow majority of nine. A fortnight afterwards the Lords rejected the bill by a majority of eighty-nine. This action on the part of the Lords was regarded by the Ministry to be contrary to the recognised practice in dealing with money bills, and a political crisis threatened to follow. Mr Gladstone denounced the rejection of the bill by the Lords as "one of the most gigantic and dangerous encroachments on constitutional usage which had been made in modern times," and his resignation was considered imminent in consequence of the adverse vote in the Lords. A Customs duty of 2½d. a-pound was charged on foreign paper imported into Britain, the difference between that sum and the Excise duty charged on paper made in this country being looked upon as an equivalent to the restrictions on the home manufacture necessary for the collection of the Excise duty. Though he failed to get rid of the Excise duty, Mr Gladstone considered himself bound, by a clause in the French treaty, to take immediate action for assimilating the duty on paper imported from France to the Excise duty charged on that manufactured at home. He calculated that the difference amounted to seven-eighths of a penny; and on the 6th August 1860 he moved a resolution in the House of Commons for lowering the duty on French paper by that amount. The resolution was carried by a majority of thirty-three in a very full house. A second resolution, making a similar concession to other countries, was carried at the same time without a division. The paper-makers had risen to a man against the removal of the Customs duty, and considered themselves aggrieved in that, while our ports were open to foreign competitors, they had to pay a heavy export duty on rags before they could obtain them from the only available sources. On the 15th April 1861 Mr Gladstone, in his financial statement, said that he felt himself able to repeal the paper duty as from 1st October following; and on the 6th May the House of Commons passed a resolution giving effect to the abolition of the tax, which for a whole year had been a bone of contention both in Parliament and out of it.

It is impossible to say exactly to what extent the abolition of the duty has affected the consumption of paper, as the only statistics of the trade were those drawn up by the Excise; but the increase during the past seven years may be reckoned at a third of the quantity formerly made.

At no period has the paper made in Britain been equal to the requirements of the country, and consequently the article has had to be imported from the Continent. The home supply of rags has always been short, and our paper-makers have thereby been placed in a very disadvantageous position compared with their continental brethren. For a number of years past from 12,000 to 25,000 tons of rags have been imported annually, and on these a heavy export duty had to be paid. In 1860 that duty amounted, in the case of Prussia, to L.9, 3s. per ton; Holland, L.8, 8s.; Austria, L.7, 5s.; and so on. In France, Belgium, and Spain, the export of rags was prohibited. A favourable change has taken place since 1860, however—a considerable reduction of the export duty having been made in most cases.

Messrs Alexander Cowan & Sons are among the oldest, best known, and most extensive manufacturers of paper in Scotland. They have three mills at Penicuik, in the county of Edinburgh; but as these stand within a few hundred yards of each other, they are worked as one establishment. The central position is occupied by the Valleyfield Mill, which is by far the largest of the three. The nucleus of it was built in 1709, by Mr Anderson, printer to Queen Anne, or by his widow. In 1779 Mr Charles Cowan, grandfather of the present proprietors, bought the mill, and, with the exception of the years 1810 to 1814, when it was used by Government as a place of confinement for French prisoners, the premises have since continued in the family. As time wore on, the accommodation in the Valleyfield Mill became unequal to the requirements of an increasing trade, and a neighbouring corn mill was acquired in 1803 and converted into a paper manufactory. This mill is now known as the Bank Mill, because it was at first devoted to making paper for bank-notes. In 1815 the operations of the firm were further extended by the purchase of a paper mill belonging to Mr Nimmo, of Edinburgh, and now known as the Low Mill. Before the last addition the number of persons employed was about thirty, who could turn out by the hand process from two to three tons of paper a-week.

A few years after the close of the French war the Valleyfield
Mill was repurchased from Government, fitted out with the most
improved appliances, and started afresh in the year 1821.   The late
Mr Alexander Cowan was among the first in Britain to appreciate the
value of the paper-making machine, and to introduce it into the
trade; and both he and his successors have ever shown a readiness
to seek out and adopt whatever appliances or arrangements gave
promise of improving or facilitating the manufacture of paper.   In
addition to their three mills, Messrs Cowan have an establishment
at Musselburgh, in which the esparto they use is reduced to pulp,
and another place at Leith, where rags are sorted and cut.

The buildings are so much detached and scattered that it is difficult to realise the extent of the Valleyfield Mill without making a
tour through its various sections.   To proceed in proper order, the
rag stores are the first places to be visited.   They occupy a substantially constructed two-storey wooden building, which had been erected
at Greenlaw Barracks by Government to extend the accommodation
for the detention of French prisoners.   The building is upwards of
100 yards in length, and both floors are covered with bales of rags.
A sample bale lying open here and there enables one to judge of the
contents of all the others.   The collectors of rags arrange them into
a certain number of classes, each of which has its distinguishing
mark.   Only the highest qualities are used by Messrs Cowan, as all
the paper made by them is of the finest kind.   From the store the
rags are taken to the cutting and sorting rooms, where a large number of women are employed.   Each woman stands in front of a
bench, the upper surface of which is covered with wire netting.
Taking a handful of rags from a bale, she shakes them out on the
bench, examines them, removes pins, buttons, &c., and then cuts
them into small pieces by drawing them across the edge of a huge
knife fixed to the bench in a perpendicular position.   According to
the quality of the various portions, she deposits them in one or other
of the compartments of a box which stands near.   The heavier portion of the dust and dirt set loose by these operations falls through
the wire top of the bench into a receptacle beneath, while the lighter
particles float about, and give the atmosphere a very unhealthy appearance.   The rags are next put through the dusting-machine,
which consists of a large cylinder covered with wire net, and having
a series of pegs or spikes inside.   The cylinder is fitted up in an
inclined position, and as the rags pass through it they are thoroughly
shaken and beaten by the spikes.   In order to get rid of the remaining dirt and some of the colouring matter, the rags are boiled in an

alkaline lye or solution.   For that purpose they are conveyed to the boiling-house, which contains a range of large caldrons.   These have double sides, and the rags are boiled by the injection of steam through the perforated lining.   Cylindrical revolving boilers are now coming into use, the advantages claimed for these being that the rotary motion, by turning over the rags, facilitates the action of the steam and lye.

After being boiled, the rags are ready for conversion into pulp. The "engines," as they are termed, for acccomplishing that part of the operations were, as already stated, invented in Holland about the middle of last century.   The first pulping machine is the washing-engine, which is simply a shallow cast-iron cistern, fitted with a cylinder having a number of steel bars $\frac{1}{8}$th or $\frac{3}{16}$ths of an inch in thickness firmly wedged in it, and projecting about $1\frac{1}{2}$ inch from the circumference of the cylinder.   The cylinder extends half-way across the engine, and the inner end of it rests on a mid-feather.   In the bottom of the cistern, immediately beneath the cylinder, a series of bars, corresponding with those on the cylinder, are inserted ; and when the cylinder is put in rapid motion (revolving about 120 times in a minute), the rags are drawn in by the cylinder, and rubbed or torn into what is termed "half-stuff."   The current caused by the turning of the cylinder sends the rags round the ends of the midfeather, keeps them moving in a continuous stream up one side of the cistern and down the other, and draws them in between the cylinder and fixed bars.   As the chief object in the first instance is to cleanse the rags, they are merely broken into fragments by the first engine into which they are put.   During the operation, which requires two hours for its completion, a stream of pure water flows through the machine, and carries off all impurities.   Though the rag-cutters exercise a degree of vigilance in looking out for and removing buttons, pins, and the like, many of these escape notice; and it is necessary to take precautions for their removal in the washingengine.   Sloping down in front of the fixed cutters in the bottom of the engine is a cast iron grating called a " button-trap."   The ribs of the grating lie parallel to the cutters, and, as small particles of pins, needles, buttons, or other foreign bodies are separated from the rags, they drop into the grating, and there lie until the trough is emptied, when very curious collections of dress-fastening appliances are revealed.   The appearance which the rags—especially the coloured ones—present at this stage is very unpromising to the eyes of casual visitors to a paper-mill.   When the washing and " breaking-in" are completed, the rags are drained and deposited in the bleaching vats, where they are subjected for twenty-four hours to the action of a strong solution of chloride of lime. The colouring-matter is thus destroyed, and the fibres are left perfectly white. By pressure in a hydraulic press, the bleaching liquor is extracted, and the stuff is placed for an hour in another washing-engine, which removes the chloride of lime, and still further separates the fibres. The stuff is then placed in the beating-engine and thoroughly pulped. The beating-engine is similar in construction to the washing-engine; but the rubbing bars are sharper and more closely set, and the motion is much quicker. The beating-engine completes its work in about five hours, by which time the fibres are reduced to about $\frac{1}{80}$th of an inch in length, and float free in the water. The contents of the beating-engines are drawn off into large shoots, from which the papermaking-machines are supplied.

Messrs Cowan have in operation five machines of the most perfect construction, and these of themselves occupy several large buildings. The machine most recently set up is one of the largest and finest in Britain. Including the drying apparatus, it is 250 feet in length, and is capable of turning out 2500 square yards of paper in an hour. In order to describe the operation of this machine, it is necessary to return to the vats and trace the pulp from that point forward. Before the pulp is allowed to enter the machine, it is mixed with an additional quantity of water, so that the fibres may float freely. So great is the proportion of water to fibre, that the fibre contained in a gallon of the liquid as it passes to the machine would not, when dry, weigh more than a grain or two. As the pulp is drawn off the vat, it receives the proper quantity of water, and is then sent to flow along an open wooden trough thirty or forty yards in length. The object of this journey is to ensure the deposition of impurities of a heavy kind. The liquid then passes through a sort of sieve, technically termed a " knotter," which retains all knotted or matted fibres. It then passes into the cistern of the machine, where it is kept constantly in motion by a revolving agitator. From the cistern it flows in a carefully regulated stream, the full width of the machine, on to an endless web or apron of wire gauze, having about 4000 holes in each square inch. Most of the water passes through the apron instantaneously, and leaves the pulp deposited in an even layer on the wire, by which it is carried forward over a pneumatic drainer, which draws off the greater part of the remaining water. The portion of the machine which pertains to the apron receives a constant vibratory motion, which assists the extraction of the water, and causes the fibres to interlace and become felted. In determining the thickness of the paper to be made, both the speed of the machine and the quantity of pulp allowed to flow upon it must be taken into account. Before the web of semi-formed paper leaves the apron, it passes under a wire roller, on which letters or devices in wire are sewed, whereby is produced what is known as the "water-mark." From a very early date in the history of paper-making, it would appear to have been customary to form certain devices in the substance of the paper, either to distinguish the productions of particular makers, or the various sizes and qualities. Water-marks have been the means, on several well-known occasions, of detecting frauds, forgeries, and impositions. Curran once made a capital hit, and won a verdict for a client, by referring the Court to the water-mark in a document in the case. The monks in a monastery at Messina preserved with great care, and exhibited with much reverence, a letter alleged to have been written by the Virgin Mary. The " relic" remained an object of interest and curiosity until the monks had a visit from a gentleman, who, after examining the document, said, with affected solemnity, that the letter was wonderful in more ways than one—in fact, that its production involved a miracle, since the water-mark showed that the paper on which the letter was written had not been made until several centuries after the death of the reputed writer. The water-mark is produced, as stated, by a roller bearing the design. The pressure of the wire on the surface of the pulpy web indents it slightly, so that when the paper is held up to the light, the design may be traced by the transparency of the outline. The web next passes between two felt-covered rollers, which extract some of the moisture, and give the material a degree of consistency. When the paper leaves these rollers, it is received upon an endless web of felt, which conducts it through two pairs of pressing rollers, and thus it becomes consolidated. A set of five large drying cylinders, heated by steam, stand next in order ; and over all these the paper is led by the felt. When it comes forth from that ordeal, it has sufficient consistency to travel through the other parts of the machine, without the support of a felt. From the first set of drying cylinders it passes to the smoothing rollers, and thence over a second set of steam-heated cylinders, which complete the drying.

At this point printing-paper, which is sized in the pulp, would be calendered; but the paper made in the machine under notice being for writing upon, is sized by being passed through a bath of animal size after leaving the first set of drying cylinders. The superfluous size is removed by pressure between rollers, and the web is led on to the drying-machine, which is a most extensive piece of mechanism. It is 200 feet in length, and consists of 200 cylinders, each about a yard in diameter. The cylinders are arranged in three double tiers, which rise to a height of twenty feet. Each cylinder consists of two parts—an outside framing covered with wire-net, inside of which is a fan, the action of which drives a current of air against the paper as it travels round the open periphery of the cylinder. In its passage through the machine the paper is brought into contact with every one of the cylinders; and when it emerges, is found to be thoroughly dried. The object of drying the paper in this way is to render it stronger and harder than it would be if dried on cylinders heated by steam. The paper next passes through burnished rollers, which impart a glaze to the surface; and as it leaves these it is wound in convenient lengths upon movable wooden rollers, which, as they are filled, are shifted to the cutting-machine, by which the paper of six or eight rollers is simultaneously cut into sheets of any required dimensions. The intricate journey which the paper has to travel between the pulp-vat and the cutting-machine is over a mile in length, so that there is always about a mile of paper in the machine. The attendants have to watch the progress of the work very closely, for the slightest disarrangement would injure or break the fabric. There is an index attached to the machine, which shows the number of yards made; and that, taken in connection with the weight of the paper given off in a certain time, supplies data for ascertaining whether the paper is of the proper thickness. The machine can be adjusted to produce paper so thin that a thousand sheets would measure less than an inch in thickness, yet so strong would it be that a stripe four inches in width would bear a weight of 20 ℔.

The paper is taken from the cutting-machines to the finishing and packing departments, where the sheets are carefully examined and placed singly between polished copper plates. The plates and paper are passed between rollers, which impart a pressure of thirty or forty tons, the effect being to give the surface of the paper a highly glazed and ivory-like appearance. The paper is then made up into quires and reams, and removed to the warehouse.

Between 2000 and 3000 tons of paper are made annually—all being, as already stated, the finer kinds of writing and printing papers. The quantity made daily is probably equal to a web twenty miles long and above five feet wide. The water-wheels and steam-engines employed in the mills are equal to over 200 horse power.

About 600 persons are employed in the various departments, and these are treated with great consideration and liberality by Messrs

Cowan. The wives and daughters of the partners have always taken a special interest in the sick, and since 1823 have managed a school for the education of the children of the workpeople. In 1840 a commodious school-house was built near the mills, and at present it is attended by about 120 children. The young persons engaged in the mills are compelled to attend free evening classes during three months in the winter season. Usually about eighty girls and forty boys are in attendance at the classes. The mills have only recently been placed under the Factory Laws; but thirty years ago the following rules were put in force by Messrs Cowan:—"1. No child under thirteen years of age shall be employed. 2. No young persons shall be employed before they are able to read, write, and figure—and, in the case of girls, to sew. 3. Wives shall not be employed, as it is considered that they should be 'keepers at home,' for the sake of their husbands and children." Nearly L.1300 a-month is paid in wages, the following being the rates:—Mechanics, 25s. to 26s. a-week; mill-workers (men and lads), 18s. to 19s.; women, from 8s. to 10s.

The most extensive manufacturers of writing-paper in the kingdom are Messrs Alexander Pirie & Sons, Aberdeen. Messrs Pirie have three paper mills, of which that at Stoneywood, on the river Don, about six miles from Aberdeen, is the largest, the others being known as the Union Paper Works, entirely devoted to the making of envelopes ; and the Woodside Paper Works, where coarse papers are made. Upwards of 2000 persons are employed by the firm, and between sixty and seventy tons of paper, cards, and cardboard, are made every week. At the Union Paper Works about a million of envelopes are turned out every day, and the mechanism used in that department is of the most beautiful construction.

The besetting sin of the paper-makers is the manner in which they pollute the streams on which their mills are erected, by discharging thereinto their waste chemicals and other refuse. These substances are fatal to fish and unpleasant to the nostrils, if not pernicious to the health, of people dwelling on the banks of the polluted waters. Since 1840 the paper-makers on the North Esk have been defenders in two actions raised by the proprietors of grounds bordering the river, for the purpose of restraining them from discharging the refuse of their mills into the stream. An action raised in 1841 was settled by the paper-makers undertaking to adopt measures for remedying the evil complained of. The steps taken were not successful, however, and matters having grown worse than ever, a fresh action was raised in 1866. The case was heard by the Lord Justice-Clerk, and occupied the Court for eleven days. The result was a verdict in favour of the pursuers on all the issues; and fresh efforts have since been made to retain or render innocuous the waste from the mills. The paper-makers have been put to vast expense in the matter; and it is due to them to say that they have tried every plan that promised to be effectual. Filtering machines have been procured and filtering ponds constructed; and the agency of fire even has been introduced, with the view of getting rid of the objectionable stuff. At Valley-field, and elsewhere on the Esk, a large extent of ground is devoted to filtering and settling ponds; but as yet absolute success has not been attained by any of the modes of purification tried there or elsewhere. Four or five years ago Messrs Cowan employed chemists of great skill to endeavour to find a way of removing the vegetable fibres and chemical stuffs from the water, and a costly method of incineration was tried. An immense furnace was built, in which the water was evaporated, and the residue burned to cinders. The cinders had a certain value, as they retained a large proportion of the soda used in treating the rags; but that bore a small proportion to the cost of its extraction. The incinerating process, though still persevered in by some of the manufacturers, is not more successful than the others that have been tried, and it has been abandoned at Valleyfield. Messrs Cowan have, though at great inconvenience, got rid of the most offensive of the polluting elements by having their esparto pulped and a large proportion of their rags cut at subsidiary establishments at Musselburgh and Leith.

The most popular form of decorative art is the covering of the walls of rooms with stained paper—a material capable alike of adorning the abodes of the wealthy, and making more cheerful the humblest dwellings. The Chinese are reputed to be the inventors of paper-hangings. Many centuries ago those ingenious people covered their walls with paper, and painted thereon figures and landscapes in the quaint style which even yet marks the limit of their pictorial art. Some of the early British travellers to the east brought home specimens of the painted paper. These were imitated, and came into use to some extent among the wealthier class, who valued the painted paper as a cheap and tolerable substitute for the costly tapestries and leather hangings which were then fashionable. Notwithstanding our early connection with the art, the credit of its development belongs to France, which has long been the chief seat of the manufacture of paper-hangings. In the middle of the sixteenth century paper-staining was an established branch of industry in France; and in 1620 a manufacturer at Rouen invented blocks for producing the patterns, which had previously been done partly by painting, and partly by the stencilling process. The French set themselves to make paper that would resemble in some degree the tapestry which it was intended to stand substitute for, and consequently their earlier efforts were of a pictorial kind. In England the art made slow progress, and it was not until the year 1712 that it attracted the notice of Government, who recognised its existence by imposing a tax of 1¾d. a-yard on stained paper, in addition to the duty of 3d. a-yard charged on the plain paper. At the same time, each manufacturer of paper-hangings had to procure a license at a cost of L.20 a-year. By these restrictions the trade was nipped in the bud, and became almost extinct, until, in 1786, Messrs George & Frederick Echardts established an extensive manufactory at Chelsea.

The invention of the paper-making machine, about sixty years ago, gave the manufacture of paper-hangings a great impulse. Previously, the manufacturers had to begin operations by pasting together twenty-four sheets of paper to form a web or "piece" twelve yards long. Irrespective of the labour they entailed in the way of pasting and joining, the composite webs looked patchy, as the stout paper that was used showed the joinings very distinctly. The cost of production having been reduced by the invention of the paper-making machine, paper-hangings were placed within the reach of many persons who were unable to purchase them at the old rates. The first English paper-hangings possessed little or no artistic merit; and the manufacturers did not bestir themselves to improve their productions until some time after 1825. In order to encourage home enterprise, the Government had placed a prohibition on the import of foreign papers, and that continued without any good result until the year named, when the prohibition was removed, and a duty of 1s. per square yard was imposed on all foreign papers imported. The stamp-duty on English paper-hangings was at the same time removed. French paper-hangings were so much superior to those produced in this country, that large quantities were imported notwithstanding the heavy duty. The trade in home-made papers was threatened with extinction, and the only alternative left to the English manufacturers was either to make a higher class of goods, or resign the market entirely to the Continental makers. They chose the former course; but the French had had such a long start that it was difficult for the English manufacturers to approach them. To

Y

such an extent had the trade flourished in France, that in 1827 there were in Paris alone seventy-two large manufactories of paper-hangings. In 1846 the import-duty on paper-hangings was reduced to 2d. a-yard, and a great influx of foreign papers followed. About that time cylinder-printing had been introduced into the trade, and the facility of production thus obtained, coupled with the removal or reduction of restrictions, had the effect of bringing down prices. Thenceforward paper-hangings became exceedingly popular, and competition among manufacturers tended to improve the quality and designs.

The manufacture of paper-hangings was begun in Scotland within a comparatively recent period, and, until a few years ago, was conducted on a small scale. There are only about half a dozen firms engaged in the trade, and of these there is but one that does what may be called an extensive business. The number of persons employed altogether will not exceed 500; and were it not that the trade is one that has been carried to a high degree of perfection by those engaged in it in Scotland, and that it promises further development, it might have been omitted from the list of subjects dealt with in this book.

About twelve years ago Messrs Wylie & Lochhead, a well-known firm in Glasgow, added paper-staining to the other departments of their house-furnishing business. They began operations in a small way, and limited themselves to block-printing. The encouragement they received induced them, on the abolition of the paper-duty in 1861, to build a large factory, and introduce cylinder printing-machines. The factory is at Whiteinch, near Partick. The main part of it consists of two large buildings, each upwards of 300 feet in length, by 50 feet in width, and united at one end by a transverse block. About 300 persons are employed, and 80,000 pieces of paper-hangings are made every week. The more costly papers are produced by block-printing, the machines being chiefly devoted to the cheap varieties. The papers supplied to the trade range in price from 10s. to 2½d. per piece of twelve yards; but papers which cost 5s. a-yard have been made to special orders.

The paper arrives at the factory in compact rolls, about eighteen inches in diameter, and the first operation it is subjected to is over-running, in order to see that it is unbroken. It is then "padded" or "grounded," by being passed through a machine which spreads upon it an equal coating of colour, and rubs it in well by a series of brushes moving transversely. As the paper comes off the machine it is dried by being made to traverse a hot-air chamber. The paper is padded in all save the very lowest qualities. The ground of the latter is formed by the tint given to the paper in process of manufacture.   After being grounded the surface is polished by means of friction-rollers and brushes. Fine drawing-room papers are "satined" —that is, made to assume a glossy appearance resembling satin—by having French chalk rubbed into them as they are drawn through a machine.   There are eight cylinder printing-machines in the factory —one capable of putting on twenty colours at a time, and the others twelve, eight, six, and four respectively. The paper is fed into the machine in a continuous web. It passes over a large central cylinder, around which are arranged the engraved printing-rollers, each with a supply of the particular colour which it is to put on. As the paper passes along it receives the colours in succession, and comes out with the design fully developed. The preparation of the engraved rollers is a work of great exactness, as the ultimate appearance of the paper depends much on the nice adjustment of the various colours.   The twenty-colour machine prints 20,000 yards a-day, and the others about 30,000 yards each. On leaving the printing-machine, the paper passes through a hot-air chamber, whence it emerges quite dry, and is folded loosely by an ingenious piece of mechanism peculiar to the establishment. It is then raised to an upper floor, where it is measured, cut into twelve-yard lengths, and made up into rolls, all of which operations are performed by machines attended by young women.   Papers striped with various shades of positive colours, such as blue, crimson, &c., are much in fashion at present, and Messrs Wylie & Lochhead have recently set up machinery for producing them.

The block-printing shop occupies an entire floor nearly 300 feet in length. Along one side of it the printing-tables are arranged. A man and a boy or girl are employed at each table. The man draws the web of paper over the table and impresses the block upon it, while his assistant keeps the block supplied with colour, as in the case of calico-printing already described, and at intervals hangs up the paper in long loops. Steam-pipes are laid along the floor, which dry the paper so rapidly that by the time the end of a web has been printed the first of it is ready to receive another colour, and so on, one colour being put on after the other until the design is completed.

Messrs Wylie & Lochhead have devoted great attention to the production of paper-hangings of the highest class, and have carried the art of making gold-stamped, bronzed, and flocked papers to a degree of perfection unsurpassed in Britain. The firm have obtained the services of workmen from the best factories in France, and have introduced the most improved machines and processes. A specially interesting department of their establishment is that in which papers of the finer kinds are made. The simplest way of introducing gold into a design is, after the colours have been put on and thoroughly dried, to print the portion to be brought out in gold with a strong size. That is done by hand; and as the paper passes from the printing-table, it is thickly dusted with bronze powder, which adheres only to the sized parts. The paper is then sent through a calendering-machine, and comes out with the bronze figuring flattened and burnished, and the appearance of the design enhanced a hundredfold. Gold stamping is a slower and more costly process, and constitutes the highest department in the art of the paper-stainer. In order to attain the full richness of effect which gold imparts, the papers to which it is applied are grounded with a quiet colour, on which little or no pattern save that in gold is introduced. The case is somewhat different when the bronzing process is used, as then the gold is employed only to form slender outlines and sprigs, and a variety of hues may be used with good results. Bronzing is particularly effective on embossed papers. In the gold stamping-room there are nine screw presses and several cylinder machines. The presses are constructed to operate upon a yard of paper at a time. The paper is prepared with a dry gum size, and on the parts on which the figures are to appear leaves of Dutch metal are laid. On the platen of the press a number of metal dies bearing the designs are fixed, and kept hot by jets of gas. When all is ready, the man in charge draws the paper into position, and, by moving a fly-wheel, brings down the dies. While he is thus engaged, his assistants place metal on a fresh portion of the paper. As the paper is stamped it is rolled up with the superfluous metal still attached to it and removed to another room, where the gold that has not been fixed is rubbed off. Papers of this kind are cut into twelve-yard lengths before being stamped, and sometimes before being grounded, when that operation is done by hand. The designs in gold consist chiefly of isolated scrolls, or floriated ornaments with vase-like centres. Nothing is introduced to destroy the flatness of surface essential in a wall-paper, and the quantity of gold is made to harmonise with the tone of the ground. The more sombre the ground is the more gold may be employed up to a certain point, and the figuring may be of a bolder kind than would be suitable for a light foundation. Drawing-room papers are in many cases embossed, which is done by passing the paper between engraved rollers.

There is considerable variety in the style of embossing—the most common forms being waved lines, diaper, and graining in imitation of leather.

There is yet another variety of paper-hangings—that in which the design is worked out in "flock." In 1634 a Frenchman named Lanier took out a patent-in this country for the process of flocking paper, leather, &c. Little came of the invention until the beginning of the present century, when it began to be practised by paper-stainers generally. At the Whiteinch factory there is a special department for flocking. The size is printed on by blocks, and the paper is then introduced into a chest the full length of the piece, where it is covered by a deposit of wool powder, which is caused to adhere evenly by a peculiar vibrating motion given to the bottom of the chest. Flocks are not so much used as they were some years ago, nor are they made so heavy.

The designing of paper-hangings affords wide scope for the exercise of artistic skill; but only in recent years has a high degree of excellence been attained. As already stated, the French have maintained a leading place in the art; but the British manufacturers are pressing closely upon them; and it is a significant fact that, next to Australia, France takes the largest quantity of the paper-hangings exported from this country. The earlier productions of our paper-stainers were remarkable only for poverty of style and feebleness of colour. The manufacturers did not realise the power they had in their hands for improving the artistic taste of the people, and were content so long as they satisfied the moderate demands made upon them by their customers. Not only in the manufacture of paper-hangings is artistic skill requisite, it is also essential in choosing a paper to suit the furnishing and lights of an apartment. There should be a degree of harmony between the colours of the paper and those of the carpet and other furniture; and where that is wanting, the effect is unpleasant, even to an unartistic eye.

A large staff of designers and engravers are employed by Messrs Wylie & Lochhead. The designs are drawn to full dimensions and coloured, and from the drawings the engravers prepare a block or roller for each colour. In the case of the rollers and the finer blocks the design is raised on the surface by driving slips of copper into the wood, and filling up with felt the portions which are to make the impression. In some patterns so many as twenty colours are used, requiring as many separate rollers or blocks. To print a rose, for example, about a dozen blocks are required; for in the simplest way it can be done three or four shades of red have to be used for the flower, three greens for the leaves, two wood colours for the stem, and white for the clear spaces. By changing the colours used in a design great variety of effect may be attained with one set of blocks; but so fickle is the public taste, and so keen the competition among manufacturers, that new blocks are being produced continuously. The papers made at Whiteinch during 1868 embrace no fewer than 240 varieties. Though the establishment has not been many years in existence, an immense stock of blocks and rollers has accumulated, representing in first cost many thousands of pounds. Another important department of the factory is that in which the colours and sizes are prepared. It is a laboratory on a gigantic scale. The colours are ground and mixed by machinery, and for the treatment of those which require to be dissolved or compounded in a hot state there is a range of copper boilers, &c.

The men employed at Whiteinch are paid by piece. The block-printers and machinemen earn from 25s. to 35s., and the women receive from 7s. to 11s. a-week.

MANUFACTURE OF FLOORCLOTH.

THE ORIGIN OF FLOORCLOTH—INTRODUCTION OF THE MANUFACTURE INTO SCOT-
LAND—THE SCOTTISH FLOORCLOTH MANUFACTORY AT KIRKCALDY.

WHEN carpets, from being regarded as luxuries fit only for the upper
ranks of society, came into general use, a great advance took place in
the popular idea of what was necessary in house furnishing, and
many articles and contrivances for enhancing the comfort and convenience of households were devised. One thing that came to be
desiderated was a covering for lobbies and stairs, which, while it
would possess to some extent the richness of effect imparted by carpeting, would be better able to withstand the tread of mud-stained
feet and general wear and tear. In the houses of the upper classes
polished oak was used with good effect, but was too costly to be
adopted in humbler dwellings. Then, paving stones looked poor
and cold, and would not admit of wall decorations; plain deal was
little better, and even painted deal failed to satisfy the eye. Painted
canvas was tried with better success, and in course of time was
generally adopted. At first the canvas was of uniform colour
throughout, then a centre of one hue and a border of another became fashionable. From that to chequered and other designs was
an easy step; and so floorcloth came to be an article of manufacture.
It is little more than half a century since it was introduced; and
it is less than twenty years since it began to be produced by other
than the most primitive means.

Up till three years ago there was only one floorcloth manufactory
in Scotland, but that establishment was and still is the largest of the
kind in the world; and the proprietors of it have done more to perfect
and extend the manufacture than all the other British firms put
together. In 1847 the late Mr Michael Nairn built at the east end
of Kirkcaldy an extensive establishment, in which he began to make

floorcloth according to the most improved methods then practised. He obtained skilled workmen from England, and soon won a reputation for the excellent quality of the goods he turned out. Thus encouraged in his enterprise, Mr Nairn devoted himself to the development of the manufacture, which he looked upon as being then in little more than its infancy. He made many experiments in material and in design, and so successfully that he shot far ahead of all competitors; and the present firm of Michael Nairn & Co. are maintaining the position thus attained. About three years ago Mr James Shepherd, who was taken into copartnery after the death of the late Mr Nairn, withdrew from the concern, and, in company with Mr M. Beveridge, started business on a considerable scale in the neighbourhood of the parent establishment; so that there are now two floorcloth factories in Scotland.

Messrs Nairn & Co.'s manufactory occupies an extensive range of buildings in the vicinity of the ruins of Ravenscraig Castle, at the east end of the "Lang Toon." The original portion of the factory is 160 feet in length, eighty-seven feet in width, and fifty-two feet in height. It stands on the top of a cliff, and occupies all the available ground on that level; so that, when extension became necessary, it could be accomplished only by building on the beach, some fifty feet below. On the latter site a painting house and drying store were erected several years ago. This second block is 129 feet in length, eighty-four feet in width, and eighty-six feet in height; and the space between it and the face of the cliff—sixty feet —is partly occupied by paint mills and a boiler-house, and partly by a covered platform thirty feet in width. The upper and lower buildings are connected by bridges at various heights. The first thing that strikes a visitor to the factory is the almost entire absence of machinery and noise. The cloth is woven, painted, and printed by hand; and the energies of the two steam-engines in the place are almost limited to grinding paint and supplying the stoves with heated air, giving off a little power occasionally to work elevators and move the lathes, &c., in the mechanics' workshop.

The cloth used as a foundation for the paint is made from flax tow yarn of a coarse quality, a rough and fibrous surface being best adapted for taking on and retaining the paint. The cloth is made in immense webs, measuring 150 yards in length and eight yards in breadth, and the looms required to weave it are consequently of gigantic size. Two men are required to work each loom, and the weaving of a web of the dimensions stated occupies them for about fourteen days. When the weaving is completed, the web is cut up into six "cloths," each twenty-five yards in length. The "cloths" are taken to the frame-room and stretched firmly on vertical frameworks of wood. In that position the cloth is sized and painted. The " back," or what is to be the lower side of the fabric, is first operated upon, and after the two coats of size and paint which are bestowed on it have been thoroughly dried by the injection of hot air into the apartment, the cloth is turned and the " face" is subjected to a succession of sizings and paintings.

The paint is prepared in a special department, furnished with a variety of mills for grinding and mixing the colours, tanks for holding oils and prepared paints, and a boiler for boiling oil. The paint is made of the consistency of treacle, and is passed to the various departments either through tubes or by means of buckets. The pigments used are chiefly ochres and leads. About twenty tons of paint are used every week, each " cloth" requiring about half a ton. The mode in which the paint and size are applied is this :—The size is a thin liquid, somewhat resembling soap-suds in appearance, and is put on with a broad flat brush. After all the pieces have been coated, the windows of the chamber are closed, and the hot-air valve is opened. Currents of heated air are made to sweep along the surface of the cloth for a certain time. The place is then allowed to cool, and the workmen, who stand on a series of platforms in front of the stretching frames, shear the surface of the cloth with large knives, which remove all the fibres that have not been " laid" by the size. The paint is then put on. Its consistency does not admit of a brush being used, and accordingly trowels are employed. The workmen rub the paint well in, and then smooth it off with the trowels. This is an operation requiring great care, because any unequal distribution of the paint would be sure to show in the finished goods. Hot air is again admitted, and after the paint is dry the workmen rub the surface with slabs of pumice stone, which remove or reduce any roughness. Another coat of size is then laid on, and so the work proceeds until the required " body" is obtained. The process of manufacture is much retarded by the time required to dry the successive layers of size and paint. After the cloth leaves the loom it cannot be got ready for the market in less than two or three months, and the longer it is allowed to " season" after that time the more durable will it be. An engine of fifty horse power is kept going constantly forcing air through a series of heated tubes into the frame-rooms.

The new block of buildings is devoted to sizing, painting, and drying; and when these operations are completed, the cloth is taken off the frames, rolled up, and hoisted to the printing-room, on the upper floor of the old building. Like other processes in the manufacture of floorcloth, the mode of printing has been much improved in recent years, and much of the credit pertaining to the degree of excellence that has been attained in that department is due to Mr Nairn. The earliest figuring on floorcloth was executed by a common brush, in the hands of a house-painter. As the demand for the material increased, a more expeditious mode of producing variegated patterns was sought, and an attempt was made to supply the want by a modification of the ancient stencilling process. The forms of leaves and other objects were cut out of sheets of pasteboard, the sheets were then laid on the cloth, and the paint applied through the excised parts. Mr Nathan Smith, of London, tried to impress the designs from engraved blocks ; and the experiment was so successful that it is still retained, though in a much improved form. The early designs were very faulty, both in conception and in execution ; and before the Exhibition of 1851 little progress had been made in the direction of improvement. The criticism which the display of floorcloth at the first World's Show called forth, had a wholesome effect on the trade. Then the design was worked out in dots of colour, arranged on a neutral ground, and it was rarely that more than two-thirds of the surface were covered by the colours of the pattern. The surface consequently was uneven, and the raised figuring was soon worn off. Mr Nairn was the first to remedy this defect, and how he did so is explained in the following extract from the "Art Journal's" notice of the floorcloths shown at the Exhibition of 1862 :—" Since 1851 a most important fundamental change has been introduced and matured by the enterprising and able Scottish firm of Michael Nairn & Co., of Kirkcaldy; and now floorcloth, having got over the long-established condition of dot-printing, has demonstrated that it may be produced with all the richness, the minuteness, and the finish of velvet-pile carpet. The Messrs Nairn have devised and adopted a system of printing which enables them to introduce any number of colours, and any variety and combination of tints, and also to impart to their designs a clearness of definition, with a depth of tone, absolutely impossible of attainment by dot-printing. The new floorcloth presents a solid surface of colours, in actual contact, which *entirely covers*, and therefore completely conceals, the ground painting—thus at one and the same time affording facilities for the production of a much higher class of designs, and affording a greatly superior and much more durable surface to the wearer. And the inventors of this real improvement in an important and most useful manufacture, have not been slow to carry out, in the matter of design, the advantages which they themselves had introduced by their novel producing processes. Being enabled to produce far better designs than heretofore had been associated with floorcloth, they have executed examples of several varieties of their designs, and placed them in the Exhibition. . . . . Some few specimens of floorcloths having tile patterns appeared in the Exhibition, in the execution of which there are some laudable attempts to emulate the example set by the Messrs Nairn ; but the Scottish firm is without any real rival whatever ; and more than this, to them belongs the merit of having first projected every important improvement which has been introduced into their manufacture. We must not omit to add that, in the treatment of imitative marbles and woods, and in chintz patterns, the Scottish floorcloth maintains the same supremacy as distinguishes their original designs of a higher order. Altogether, this is one of the most gratifying instances of superior excellence in a manufacture that the Exhibition adduced, in favourable contrast with its predecessor of 1851; and it is with sincere pleasure that we are able, in such decided terms, to record our admiration for a staple article of British industry." Visitors to the Scottish Floorcloth Manufactory will readily confirm the complimentary language in which the productions of the establishment are spoken of by the " Art Journal."

The process by which the designs are conveyed to the cloth may now be described. As in preparing designs for the Jacquard loom and Berlin worsted-work, the figures in most cases are drawn and coloured in dots. The number of colours employed in one pattern ranges from four to fourteen, and each colour requires a separate block. In preparing the blocks great care has to be taken in order to make the successive impressions fall into their respective places. The outline blocks have the figures formed on their surface either in copper or type-metal, and sometimes in a combination of both. The " filling " blocks are faced with boxwood, and before being " cut," the printing surface is sawn into minute squares—eighty-one to the square inch—the saw penetrating to a depth of nearly a quarter of an inch. These squares correspond to the squares in the design, and the block is prepared by all the squares being chipped out, except those that are required for the colour to which the block is to be devoted. The division into squares has another and more important purpose. If a close surface were used, the block would not take up nor lay down the paint properly, and one result would be the squeezing out of the colour, and the formation of a ridge round the edges of the figures. It would appear, then, that "dot printing" has not been abolished after all, persons who have read thus far will be ready to say. As a part of the printing process it still prevails, but a subsequent operation changes, or rather obliterates, its effect.

There are two floors in the printing department of the factory, and the upper is separated from the lower by a space of fifty feet. The printing-room is on the upper floor, and the object of the great space beneath it is to admit of the cloth being suspended as it is printed. Two men and two boys are employed at each printing-table. Having prepared the cloth, the men apply the blocks in succession. The blocks are about eighteen inches square, and one is required for each colour in the design. For each colour there is a skin-covered moveable table, on the surface of which the "tear-boy" spreads the paint in an even layer, and on that the printer presses his block after each impression. The cloth is operated upon in sections equal to the width of the blocks, and extending across the cloth. It is interesting to watch the effect of the successive applications of the colours, and wonderful to observe the want of harmony that generally prevails until the last colour has been filled in. After all the printing-blocks have been applied, the surface of the work has a dotted appearance, but that is dispelled by the application of the "finishing block," which bears no design, but has its surface divided into fine parallel lines. Under the pressure of these lines the colours are blended and more equally distributed, while the fine ribbed marking that remains makes the cloth look soft and rich. The mode of taking an impression off the blocks is much superior to the old plan of striking them with a hammer, as is still the fashion in calico-printing. Over each printing-table is a stout beam, to the under side of which a pair of travelling screws are attached. When the printer lays his block on the cloth, he brings one of the screws over it, and by pulling a lever takes off the impression. In proceeding from side to side of the cloth he slides the screw along with him. As each section of the cloth is finished, it is passed over the side of the table and allowed to descend through an opening in the floor. Both ends of each "cloth" are secured to a beam, the bight only being allowed to go down ; and when the piece is finished, it hangs in a long loop with the printed side outward. Thus suspended, the cloth may be moved from one part of the building to another on a peculiar kind of railway constructed on the ceiling of the lower room. When a certain number of cloths have been got ready, they are run into another compartment of the building, and there dried by means of hot air, a process which occupies several weeks. All that then remains to be done is to cut the margins off the cloth and make it up into rolls. The variety of cloth made as above described is of one pattern throughout ; but other kinds are made with borders, such as the narrow cloth used for stairs, several widths of which are printed on one web of canvas.

Messrs Nairn rank among their customers all the reigning families of Europe, and have received the highest honours at the great Exhibitions in London and Paris. The piece of floorcloth shown at Paris was the most magnificent work of the kind ever produced. It was designed by Mr Owen Jones, and illustrated in an extraordinary degree the capabilities of the art. Only four colours were employed, but there was great variety in the figuring—so much, indeed, that upwards of 100 blocks were required.

The number of persons employed is nearly 200, all men and boys. The printers serve an apprenticeship of five years. They begin as " tear-boys," when they receive from 3s. 6d. to 5s. a-week, and from that are promoted to be assistant printers, with from 8s. to 14s. a-week. As journeymen in charge of a table, they receive from 21s. to 23s. a-week. The " trowellers," or those who put on the ground paint, have pretty heavy work ; but as it requires less skill, they receive only from 18s. to 20s. a-week. The quantity of cloth turned out weekly is 40 pieces, measuring 25 by 8 yards. Each piece weighs from 11 to 12 cwt., the paint being equal to five-sixths of that weight.

MANUFACTURE OF LEATHER.

ANTIQUITY AND IMPORTANCE OF LEATHER—PROGRESS OF THE LEATHER TRADE
IN BRITAIN—CURIOUS LAWS FOR ITS REGULATION IN SCOTLAND—PRESENT
CONDITION OF THE TRADE—DESCRIPTION OF A LEATHER MANUFACTORY.

THE conversion of the skins of animals into leather, and of the latter
into articles of clothing and convenience, has from the earliest
times formed an important branch of industry. The equipment of
the early Briton consisted almost solely of prepared skins ; and notwithstanding the wonderful progress that has been made in the arts
since his day, the material which was indispensable to him is not
much less so to us. Leather answers innumerable useful purposes ;
and though substitutes for it have long been sought, none has yet
been found. True, it has been relieved from doing duty in some
ways by india-rubber, gutta-percha, and the like; but none of those
substances can compete with it in lightness, durability, or beauty, as
a covering for the feet, or as a material for making harness, which
are the chief purposes to which it is applied.

Seventy years ago it was estimated that the value of the articles
manufactured from leather in Great Britain amounted to L.12,000,000.
Mr M'Culloch, in his "Statistical Account of the British Empire,"
thus calculates the extent of the trade about thirty-five years ago :—
" At an average of the years 1833 and 1834, no fewer than 304,279
cwts. of foreign cow, ox, and buffalo hides were entered for home
consumption, exclusive of vast quantities of lamb-skins, goat-skins,
&c. The total quantity of all sorts of leather tawed, tanned, dressed,
and curried in Great Britain may at present (1837) be estimated at
about 65,000,000 ℔., which, at 1s. 6d. per ℔., gives L.4,875,000 as
the value of leather only. Now, supposing, as is sometimes done,
the value of leather to amount to one-third the value of the finished
articles produced from it, that would show the value of the manufac-

ture to be L.14,625,000 or L.14,600,000. We incline, however, to think that the value of the manufactured leather articles does not amount at an average to three times the value of the raw material; and, therefore, we may perhaps estimate the entire value of the manufacture at L.13,000,000 or L.13,500,000." Supposing the estimate made seventy years ago and that of Mr M'Culloch to be correct, it would appear that in thirty-five years the leather manufacture had increased in annual value only by L.1,000,000 or L.1,500,000. It would be difficult to state with certainty the present extent of the trade; but a few figures from the Board of Trade Returns will show the quantity and value of the exports and imports of hides and leather. In 1866 there were imported 1,133,130 cwt. of hides, valued at L.3,360,876; and 9,285,928 sheep, goat, kid, and seal skins, the value of which is not stated separately, but it may be taken at something like L.700,000. Manufactured leather was imported to the following amount:— Boots, shoes, and boot fronts, L.90,707; gloves, L.1,194,665; unenumerated goods, L.67,641—total, L.1,353,013. The exports in the same year were 38,900 cwt. of leather, 3,546,618 pairs of boots and shoes, 14,754 cwt. of leather worked up in various ways, and L.252,484 worth of saddlery—the aggregate value being L.2,030,464. To the hides and skins imported have to be added those produced at home, which must nearly equal in number and value those brought from abroad. There are three hide markets in Scotland— at Edinburgh, Glasgow, and Aberdeen—and the number of hides sold at these in 1867 was as follows:—Edinburgh, 30,512; Glasgow, 54,836; and Aberdeen, 41,600. Though so many hides are brought into the Aberdeen market, there is no important tan-yard in that city, but a large number of cattle are killed in the district for supplying the London markets.

Leather has been manufactured in Scotland from an early date. The Edinburgh corporation of "cordiners" or shoemakers, was founded about the year 1449, and it is probable that the leather they used was home-made. The "skinners," as the makers of leather were designated, were incorporated in 1586; and regulations for their good government and the proper conducting of their manufactures were subsequently made from time to time by the Town Council. Several acts were passed by the Scottish Parliament for the protection of the trade of the skinners. The first of these is dated 1592, and it prohibits "all transporting and carrying foorth of this realme of calve-skinnes, huddrounes, and kid-skinnes, packing and peilling thereof, in time cumming, under the paine of confiscation of the same to His Majestie's use." The act was confirmed next year, and the following addition made to it:—" That His Majestie and Estaites of this Parliament, understanding how necessary and profitable the schurling skinnes ar for lyning cuscheones, making of poikes, lyning, puitches, glooves, and claithing of the puir, and utherwise serving to diverse uther uses to all His Majestie's lieges; quhilkis be the transporting and carrying of the same foorth of this realme, ar become to ane exorbitant dearth, that therethrow, not onlie the skinners are greattumlie hurt and prejudged, be the inlaik of the leather thereof, quhairwith to worke; as alswa His Majestie importis na profite thereof be custome, nor utherwis; bot alswa all uthers His Hienesse lieges ar greattumlie hurt and prejudged thereof, therefore it is statute and ordained, that na merchand, craftesman, or uther person or persones, carie, or transport, onie of the saidis schurling skinnes, nor uther skinnes above mentioned, foorth of this realme, under the paine of confiscation of sa monie as sall happen to be apprehended, and furder punishment of the persones, transporters and contraveenirs of the present acte in their person and guddes, according to His Majestie's pleasure." This act would appear to have become in course of time a dead letter; but in 1661 it was revived by the Parliament of Charles II., who, considering how necessary it was that all former laws for improving native commodities should be re-enacted, and understanding that the skinners had at their own cost brought from abroad perfumers, and makers, and preparers of leather, by whose labour and skill the people might be furnished with gloves at an easier rate, and be able to supply other countries with leather work, ratified the act above quoted, and ordered it to be put into execution. For the further encouragement of the skinner trade, a manufacturing license was given at the same time to export gloves made within the kingdom free of all custom and excise.

Edinburgh has always been the chief seat of the leather manufacture in Scotland. Arnot states that in 1778 a considerable trade was done in leather, that there were several tanneries in the outskirts of the city, and that the skinners were well employed. Shoes were made in great quantities, for not only was the home demand met, but a large export trade in those articles was done with the West Indies. Several British regiments raised after the American war were supplied with shoes from Edinburgh. A special feature of the trade in the city is mentioned. That was the making of leather snuff-boxes, pen-cases, drinking-mugs, and a variety of other articles. By a patented process the leather was brought to assume the appearance and consistency of tortoise-shell, being transparent and susceptible of receiving a high polish. The patent was in the hands of Messrs Thomas Clark & Son, whose productions became famous both at home and abroad. The duty on the leather manufactured in Edinburgh in 1778 amounted to L.1100, 5s., which, at 1½d. a-pound, would represent about 1572 cwt., the value of which at present rates would be nearly L.13,000. From an early period leather had been subject to a duty, and the manufacture was accordingly carried on under the surveillance of the Excise. Up till 1812 the duty was at the rate of 1½d. a-pound; but in that year it was raised to 3d., and was continued at that figure until 1822, when it was reduced to the old rate. The reduced duty amounted to L.360,000 a-year. In 1830 the duty was finally repealed.

There are in Scotland about 120 tanneries, in which nearly 3000 persons are employed. A number of the tanneries are of small extent, and limited to tanning a few hides procured in the locality in which they are situated. Of workers in leather, boot and shoe makers number 27,000, and saddlers and harness-makers 2000, so that in Scotland 32,000 persons are employed in making and working in leather. Since the application of machinery to making boots and shoes, several large manufactories of those articles have been started in Scotland. In one establishment in Glasgow no fewer than 2000 persons are employed.

The leather manufactory of Mr Allan Boak, West Port, Edinburgh, is the largest establishment of the kind in Scotland, having a floor space of 5682 superficial yards. A portion of the buildings has been used as a tannery from time immemorial, and about a century ago was acquired by the great-grandfather of the present representative of the firm; and in his hands, and those of his successors, the place grew under the pressure of an expanding trade, until it became the largest tannery in the city. Two or three years ago the premises were almost totally destroyed by fire, a misfortune attended by one good result only—the opportunity it gave for reconstructing the tannery in a more modern and substantial style, which was done at a cost of L.10,000. The kinds of leather made by Mr Boak form an extensive list, but his specialty is the preparation of pig-skins, in which department he is one of the most extensive manufacturers in Britain. The two principal branches of leather-making were conducted separately until recent years, as, previous to the repeal of the duty on leather, persons were prohibited from carrying on tanning and currying at the same time. The occupation of the tanner and currier are so different that they might well remain apart; and in many cases they are carried on as distinct trades. In Mr Boak's establishment, however, they are united, and joined with them are departments for enamelling and japanning leather.

The tanning department occupies the largest amount of space, and the operations conducted in it have to be carefully managed, as the quality of the leather depends more upon the tanning than upon any or all of the subsequent processes. The object of the tanner is to destroy in the hides and skins the liability to putrefaction common to animal matter, and to render them impervious to the action of agents which would decompose them under ordinary circumstances. This is done by steeping the skins in an astringent liquid prepared from bark. The active principle eliminated from the bark is called *tannin*, or tannic acid, which forms a chemical combination with the skins. The bark of the oak is the most valuable to the tanner, and is most extensively used; and for a long time no other substance was employed in tanning. The demand for oak bark having come to exceed the supply, various substitutes were tried, among these being heath, myrtle leaves, wild laurel leaves, birch bark, and oak sawdust. Varied results attended the experiments, but oak bark has never lost its supremacy. What the tanner has got to do, then, is to treat the hides, &c., with an infusion of bark, so as to produce the desired effect. This is a slow process, and attempts have been made to hasten it. Some years ago an Italian tanner (Signor Cesare Osmani), in the employment of Messrs Moore, Morelitt, & Co., Ancona, discovered that Italian mustard-seed could be successfully employed for tanning purposes, and after many experiments, the following important results are said to have been established:—That the duration of the tanning process is reduced to from one-third to one-fourth of the time hitherto required for the different tannages, whether of oak-bark, valonia, gambier, or other tanning materials. This result is obtained without any additional expenditure, because, although the mustard-seed is an item to be added to the cost of tanning, yet the mere saving of liquors and labour fully compensates this. The quality of the leather produced is unexceptionable, and the weight obtained is fully equal, and often much exceeds, that obtained by the old process. The colour of the leather is clearer, and hence better; and the reason for this is that the mustard-bath, to which the hides are subjected in this process, has the effect of expelling the lime which in certain proportions remains in hides subjected to the lime-pit, despite the most scrupulous purging. The fibre of the hide is not injured by the bath in mustard-seed. Signor Osmani came to Scotland in 1868, and commenced operations in Glasgow, and subsequently in

London. In both places he is said to have fully demonstrated the practicability of his discovery.

But we revert to the establishment more immediately under notice.

The hides arrive at the tannery in one of three states—they are either fresh from the slaughter-house, salted, or dried. Dutch hides are salted, and those which come from America or the East are dried. In the first stage of treatment there is consequently a little variety. The fresh hides give least trouble, but the salted and dried ones require special manipulation to make them soft. The hair in all cases is removed by steeping the hides in a solution of lime and water. In the floor of the workshop in which this operation is performed is a range of pits, about six feet in depth, in which the lime liquor is contained. As the hides, after steeping a certain time, are withdrawn from the pits, they are laid on a sloping bench with a convex top, and subjected to scraping with a huge two-handled knife. The hair having been removed, the flesh side is turned up, and all the fleshy portions are scraped off. Hides or skins intended for boots, harness, coachwork, and other purposes requiring dressed leather, are cleansed from grease and other impurities by being soaked in a decoction of pigeon's dung. They are then softened by being beaten for from fifteen to twenty minutes by a set of beating stocks. That operation completed, the hides are conveyed to the tan-yards, in which are a large number of pits, of various sizes—in fact, the yards may be described as lakes divided by walls into a series of square tanks, about six feet in depth. Each tank is cut off from communication with its neighbours, and is fitted with a waste-pipe, by which it may be emptied. A couple of steam-pumps and a set of hose supply the tanks with liquor, and thus dispense with the services of a large number of pumpers and water-bearers, whom it was necessary to employ before the introduction of steam.

There is a bark mill on the premises, by which the oak bark is ground. The bark is infused with water in a range of large tanks, and from these the other tanks are supplied. The hides are first allowed to steep in a weak solution of bark for from four to six months, during which time they are "handled"—that is, they are taken out of the pits one by one, then put in again, and treated to fresh liquor of slightly increased strength. After being shifted about in that way every day for the period stated, the hides are spread out, one over the other, with a layer of oak bark between, and the pit being filled with strong bark liquor, the whole is allowed to lie for four months if the hides are light, and for six months if they are heavy. Twelve months are required to tan heavy ox-hides to perfection, and none are sent out in less time by Mr Boak. In one old-established tan-yard in Musselburgh—that of Messrs Miller—two years are devoted to tanning heavy hides. As already stated, many plans have been devised for hastening the process; but faith in the old mode is still strong in the trade. Among the inventors who have given attention to the subject are Messrs J. & G. Cox of Gorgie Mills, near Edinburgh, who adopted the plan of attaching the hides to a revolving drum, so that, as the drum went round, the hides should alternately hang suspended in the tan liquor and alternately press upon each other on the upper side of the drum. Messrs Cox also originated the process of sewing the hides in the form of bags, into which they injected the tan liquor. The great object aimed at by all the experimenters is to force the tan into the pores of the skins, and hasten the chemical change which is essential to the production of leather. Mr Boak employs a sort of dash-wheel, which keeps the hides moving about in a trough, and helps them to imbibe the liquor. The hides prepared for sole leather are washed and dried after being taken from the tan-pits, and are then ready for use.

The hides which have to be dressed are removed from the tan-yards to the currying shops. The business of the curriers is to scour the leather, and, by a series of operations, bring it into condition for use. Except splitting the hide, all the work is done by hand, and is very laborious. The hides are first weighed and examined, and, according to weight and quality, are selected for various purposes. The divisions usually made are harness leather, saddle leather, shoe leather, and patent leather, each of which classes has a number of subdivisions. It would be tedious to go into the details of all the processes by which the various kinds of leather are curried and dressed, and a few general notes must suffice to indicate the nature of the work. A great difficulty with which tanners have to contend is the way in which skins are damaged by the carelessness of the butchers in removing the hides from the carcases. In consequence of cuts and flaws thus produced, only a small proportion of the hides can be made available for the highest class of harness work. Following the example of the trade in Glasgow, the tanners of Edinburgh a year or two since instituted a system of inspection and classification of well and ill flayed raw hides, which has caused some improvement in this respect. If great care be not exercised in the selection of the hides, it might turn out that much valuable labour would be lost by dressing a hide for a purpose for which it was not suitable. According to the purpose to which the hide is to be devoted, certain parts of it are cut off. This is called "rounding." Hides which are to be dressed for harness, for instance, have the belly portion cut off, and are then termed "backs." The parts removed are treated separately, and applied to various purposes. The currier begins operations by steeping the hides in clean water. If for harness, they are cut into two longitudinally. Hides intended for boots, and those which are to be enamelled or japanned, are dressed whole. The hides, on coming from the tannery, are found to be of unequal thickness, and the flesh sides are rough. In order to remove those inequalities the hides are "shaved," which is, perhaps, the most important part of the currier's work. Standing in front of a narrow upright board or "beam," which rises to his waist, the workman stretches a portion of the hide across the face of the board, and, by dexterously operating on it with a knife of peculiar construction, reduces it to an equal thickness throughout. As the work proceeds, he pinches the leather between his finger and thumb, and is thus enabled to judge of its thickness. This appears an exceedingly simple operation, but great skill is required to accomplish it properly. In ordinary cases the hides are "shaved" only on the flesh side, but when the surface is to be enamelled the grain side has its natural surface removed, in order that the enamel or varnish may adhere better. After being "shaved," the hides are again soaked in water, from which they are taken and rubbed out well, first on one side and then on the other, on a stone table with a tool called a "slicker." In that way all the wrinkles and superfluous matters are got rid of. This process is called "scouring," and when it is completed the hides are allowed to lie for a day or two in a hot solution of shumac and water, which tends to brighten the colour. They are then stuffed—that is, rubbed on the hair side with oil, and on the flesh side with a mixture of oil and tallow. The effect of this treatment is to make the leather pliable, and prevent it from getting hard. A machine for scouring has been recently introduced.

The leather is finished in a variety of ways. The superfluous "stuffing" is rubbed off, and the surface polished with a smooth stone or piece of glass, when the leather is to be sold plain ; but when it is to be prepared for harness or shoemaking, it is usually blackened on one side. Sometimes the shoe leather is "grained" by doubling the hide and rubbing the fold with a small board. The leather so treated has a rich, crimped appearance, and is now much used by shoemakers. For some portions of carriages and harness, and a variety of other purposes, enamelled and japanned leathers are employed. For the production of those leathers Mr Boak has a special department. Enamelled leather is made from all kinds of hides and skins, which have been previously split into thin layers by the splitting-machine. In some cases one sealskin is split into three or four. The splitting-machine is the only piece of machinery employed in the currying department, and is a most useful contrivance. Formerly, when a hide had to be prepared for enamelling, the currier scraped down the substance of it from the flesh side, and thus half the weight of the hide was wasted, because scrapings of that kind are of no use. Now, one hide can be split with less labour, and with the additional advantage that each slice is for most purposes as valuable as a whole hide prepared in the old way. The splitting takes place both before and after the hides are tanned. The hide to be split is fed into the machine between rollers, and as it passes a certain point comes into contact with a knife, which moves backward and forward with great rapidity and splits the hide into two. When the currying is completed the hides are taken to the japanning shop, where they are stretched out and nailed upon large boards. The enamelling substance is then applied, the number of coats being determined by the purpose for which the leather is intended. Sometimes as many as eight coats are put on, one coat being thoroughly dried before another is applied. The enamel is dried by placing the hide, still attached to the boards, in large stoves.

The curriers', though not one of the most pleasant-looking, is one of the most healthy of occupations, though the work is very heavy. The men are usually paid by piecework, and when engaged on some classes of leather, can earn from L.2 to L.3 a-week. The average wages for Scotland may be stated at 26s. a-week. There is a union in connection with the trade, having for its object the support of members when out of work and the regulation of wages. It is upheld by a General and a Local Fund—the former for the support of members, the latter for management. There is an Emigration Scheme connected with it, by which members, on emigrating, are allowed a certain amount. During sickness members receive for six weeks the highest rate of benefit; and a few of the old and infirm members are granted pensions. The amount of contribution is regulated by a sliding scale, rising or falling according to the state of the funds, but not lower than one shilling a-week. The Edinburgh men have a Sick and Funeral Society, independent of the funds of the union, which has been found most beneficial. Apprentices may become members of the local society, but are not admitted into the

union till they have served seven years' apprenticeship. The number of curriers in Great Britain is reckoned about 3000, half of
whom belong to the union. There are about 250 union men in
Scotland, and about an equal number of non-union. In Edinburgh
tanners earn 20s. a-week, and tannery labourers 16s.

The master tanners and curriers and their workmen have had no
difficulties about wages or modes of working; and this gratifying
state of matters arises from an arrangement by which, when any
question arises between the masters and men, a settlement is effected
by mutual consultation.

MANUFACTURES IN INDIA-RUBBER.

In the year 1735 M. de la Condamine, who had been sent to South
America by the French Government on a scientific mission, communicated to the Academy of Sciences at Paris an account of a
resinous substance collected from certain kinds of trees, which was
used by the natives of Brazil for various purposes, such as making
boots, syringes, bottles, and vessels of different kinds for containing liquors.   M. Condamine stated that he had found the substance useful in forming waterproof coverings, which were made
by simply coating canvas with the liquid as it exuded from the
trees.   That was the first intimation received in Europe of the
existence of caoutchouc, or, as it is more commonly called, india-
rubber, a material now extensively employed in the arts.   It was
first brought to England about a century ago; and a treatise on per-
spective drawing by Dr Priestley, published in London in 1770,
contains the earliest reference to its introduction and application to
a useful purpose in this country.   Dr Priestley says,—" I have
seen a substance excellently adapted to the purpose of wiping from
paper the marks of a black-lead pencil.   It must, therefore, be of
singular use to those who practise drawing.   It is sold by Mr Nairne,
mathematical instrument maker, opposite the Royal Exchange.   He
sells a cubical piece of about half-an-inch for three shillings, and he
says it will last for several years."   It was from the property it
possesses of removing pencil-marks that the name of india-rubber
was given to it.   Mr Thomas Hancock, in his interesting " Personal
Narrative of the Origin and Progress of Caoutchouc or India-rubber

Manufactured in England," says that the substance came first into notice in this country in the shape of bottles and animals, that it was sold at the rate of a guinea an ounce, and was used for rubbing out pencil-marks. Up till about the year 1820 it was applied to no other purpose.

Mr Hancock was the pioneer of the manufacture of india-rubber, and it has been said truly that few departments of manufacture have owed more to the ingenious contrivances of one man than that of india-rubber owes to him. Mr Hancock became impressed with the idea that a substance possessing such peculiar qualities as india-rubber might be made available for other purposes than removing pencil-marks, and in 1819 he began to make experiments. He first tried to dissolve the rubber, expecting that it might become useful in a liquid form; but his attempts were not satisfactory. He then took to cutting the rubber into thin bands, and in 1820 obtained a patent for the application of these to articles of dress in the form of braces, garters, &c. In cutting the rubber into suitable shapes a large proportion had to be cast aside as useless parings. Mr Hancock's next care was to devise means whereby such waste might be avoided, and after several failures he succeeded in constructing a machine which kneaded the scraps into a solid mass. This machine he called a "masticator," apparently out of respect to the process by which schoolboys reduce to the consistency of putty the india-rubber which they assert to be an essential part of their academical equipment. The machine consisted of a cylinder and casing, both furnished with spikes which tore the rubber into shreds. During the operation sufficient heat was generated to cause the shreds to amalgamate, and thus fresh blocks were formed. The masticator was the first machine applied to the manufacture of india-rubber. From 1820 till 1847 Mr Hancock continued his researches with a wonderful degree of success, and in that time obtained no fewer than fourteen patents. Almost simultaneously with Mr Hancock, the late Mr George Macintosh, of Glasgow, began to make experiments with india-rubber, and discovered that naphtha, obtained from coal tar, had the power of dissolving the rubber. The solution thus obtained he applied to cloth, which was thereby rendered waterproof. In 1824 he took out a patent for the manufacture of " waterproof," made by cementing two folds of cloth together by means of the solution. Coats made of that material, and bearing the name of the inventor, soon became famous. Mr Macintosh formed a partnership in Manchester, and began to manufacture waterproof garments, &c., on an extensive scale. The firm thus

created still exists, and their productions are widely known. Mr
Hancock worked some of his inventions in conjunction with Mr
Macintosh, and ultimately entered the partnership.

Encouraged by the success which had attended the researches of
Messrs Hancock and Macintosh, many persons took to experimenting
with india-rubber, and the result was a rapid increase in the variety
of its applications. Mechanicians hailed the rubber as a sort of
missing link in their code of materials for machine-making ; and
such was the rage for introducing it, that it was frequently found in
most unsuitable positions. Now it forms an essential part of many
machines, and even the steam-engine has been rendered more perfect
in its action by the introduction of rubber valves and stuffing. The
manufacture attracted attention in America soon after it had been
brought to a degree of perfection in this country, and many novel
applications of the substance have had their origin in the United
States. Mr Goodyear, an American gentleman, has had a career in
connection with the manufacture of india-rubber somewhat akin to,
though less brilliant, than that of Mr Hancock in England, and
both made independently one of the most wonderful discoveries
bearing on the treatment of caoutchouc. The great obstacles to a
more extended use of india-rubber were the clammy adhesiveness of
the substance, its liability to be affected by changes of temperature,
and its sensitiveness to oil and grease. Mr Hancock had tried to
remove these obstacles, but without success, until the year 1843,
when he discovered the vulcanising process. In 1842 Mr Goodyear
sent an agent to this country offering for sale the secret of a mode
whereby the desired qualities could be imparted to the rubber ; but
as no explanation of the process was allowed to be made until after
purchase, the agent returned without accomplishing his purpose. Mr
Hancock saw some of the specimens which had been sent over, and
became convinced of the practicability of changing the nature of the
rubber. He thereupon renewed his experiments, and in the course of
a year had solved the problem, and protected his process of vulcanising by a patent.

The increase which has taken place in the consumption of india-
rubber during the past forty years may be seen from the following
statement of the quantities imported into Britain in various years :—

1830,	.	.	49,952 ℔.	1852,	.	.	2,195,984 ℔.
1842,	.	.	317,184 „	1862,	.	.	7,405,710 „
1847,	.	.	659,568 „	1867,	.	.	8,932,672 „

The chief supplies of india-rubber are derived from Brazil and New

Granada, and the price ranges from L.120 to L.260 a-ton. The value of the raw and manufactured rubber exported annually now amounts to nearly L.1,000,000. Our best customers are France and Australia.

Two of the largest and finest manufactories of india-rubber in the world are situated in Edinburgh; and a description of these, and the operations conducted in them, will illustrate the nature and capabilities of caoutchouc. The establishments stand near each other on the bank of the Union Canal, on the south-west side of the city, and belong respectively to the North British Rubber Company and the Scottish Vulcanite Company (Limited).

In the year 1855 an enterprising American gentleman brought to Edinburgh the machinery and capital necessary for an india-rubber manufactory, and organised the North British Rubber Company. Possession was acquired of the fine buildings known as the Castle Mills, which had been erected at Fountainbridge as a silk manufactory, and had long stood vacant, owing to the projectors not having succeeded to their expectations. The establishment consists of two large blocks of five floors each, and a number of subsidiary buildings.

The india-rubber arrives at the manufactory in various shapes, according to the mode in which it is collected by the natives of the different countries which produce it. The finest qualities generally come in the shape of curiously formed bottles, and the coarser kinds in roughly kneaded balls about four or five inches in diameter. It is no unusual thing to find stones and other heavy substances mixed with the rubber, for the collectors have learned the art of adulteration. The rubber is carefully examined with a view to the detection of deleterious substances before it is subjected to the processes of manufacture. After being softened by steeping in hot water, the rubber is passed through the breaking and cleaning machines. The first of these consists of two strongly-mounted iron cylinders, one of which is grooved diagonally, while the other has a smooth surface. The balls of rubber are fed in between the cylinders, which crush them out into thin pieces. These pieces are then operated upon by a machine similar to the first, except that both cylinders are smooth. The rubber is sent through again and again until it is thoroughly broken and assumes the form of a web. If it be desired to reduce it still further, the rubber is sent through a third set of rollers. On examining the stuff as it comes from the breaking-machine, it is seen, especially in the case of lower qualities, to contain a mixture of bark, leaves, and other foreign matter; and it is to rid it of these that the washing or cleaning machines are employed. Such is the adhesive nature of the material, that it would be impossible to break or clean it in a dry state, and consequently jets of water are made to flow on the rubber and cylinders when the machines referred to are in operation. The water, besides carrying off the impurities set free by the action of the rollers, causes the rubber to assume a granulated appearance, and, under the pressure of the cylinders, it is formed into a web. Simple though this process appears to be, it is thoroughly effective in purifying the rubber. The webs of washed rubber, which are made only three or four yards long, are taken to a drying-room, where they are hung up in a warm atmosphere for several weeks.

From the drying-room the rubber is taken to "the mill," which occupies two entire floors of the principal block. The floors are covered by machines of the most powerful construction ; for the rubber is stubborn stuff, and submits only to a degree of force that would destroy almost any other non-metallic substance. The grinding-machines, to the operation of which the rubber is next subjected, consist of two cylinders, one of which is slightly heated by steam, and the webs formed by the washing-machines are kept revolving round and round the cylinders until all appearance of granulation disappears and the stuff becomes quite plastic. At this stage the rubber has incorporated with it sulphur, or other chemical substances, which determine its ultimate character, and is then made up into rolls of seven or eight pounds each. There are steam-pipes through all the place, which prevent the rubber from becoming hard again until it receives its final shape.

The further treatment of the rubber depends on the purpose to which it is to be applied. To produce articles of solid rubber, the material is rolled out into sheets of various thicknesses, which are by subsequent operations brought into the desired shape. The company do an extensive trade in shoes, and a considerable portion of the machinery in the mill is devoted to the preparation of the materials used in that department. Large quantities of waterproof fabric are also made, and several machines are employed in spreading the rubber on cloth for that purpose. Some of the cloth used is silk, but more commonly calico constitutes the base. It must be of even texture, and free from knots ; and in order to ensure that it is so, it is carefully examined and picked before being placed on the spreading-machines. These machines consist of a series of metal rollers, one of which takes up a supply of rubber and transfers it to the cloth. No solvent is employed in this process, the rubber being simply softened by the heated cylinders of the machines. Driving-belts and hose are composed of layers of canvas impregnated with rubber forced into the texture under immense pressure. The preparation of the canvas is among the operations conducted in the mill.

There is great variety in the goods produced; and as the appliances required in the production of each kind are special, the manufactory is divided into a number of departments, each with a distinct set of workpeople. One of the upper floors of the main building is occupied by the shoemakers. This department is to the casual visitor, perhaps, the most interesting in the establishment. Boots and shoes of all sizes are made; but the articles most in demand are the galoshes which ladies wear over their boots. Four classes of operatives are employed in completing a shoe from the materials as they are sent in from the mill. The first of these are the cutters, who shape the soles, uppers, &c., with great rapidity. They spread out before them a web of prepared cloth, or a sheet of rubber, and laying thereon a metal pattern, cut round it with a sharp-pointed knife. The linings are shaped in a different way. A number of folds of the cloth are laid one over the other, and cut in a hydraulic press by means of a die. All the work up to this stage is done by men. Ten pieces of cloth and rubber are required to make one shoe, and as the parts are cut out they are transferred to young women, who coat the edges of some of them with a solution of india-rubber, and pass them on to the upmakers, all of whom are young women. The rapidity with which the pieces are put together is astonishing. The women sit at long tables, over the top of each of which is an iron rack for holding the lasts. In order that they may withstand the heat to which they are subjected while the shoes are being vulcanised, the lasts are made of cast iron. Taking one of the lasts from the rack, the operative rests it partly on the table and partly on her knees, and lays the pieces on one after the other, rubbing each smooth with a small roller. No stitching is required, the adhesive power of the rubber and solution being sufficient to bind the parts together. An expert worker has turned out as many as seventy pairs a-day; but at the usual rate of working from thirty to forty pairs a-day may be taken as the average. The productive power of the establishment is equal to making 7000 pairs a-day, or upwards of 2,000,000 pairs a-year. As the work goes on the shoes are collected by men and taken to the varnishing shop, where they are coated with a liquid which gives them a smooth and glossy appearance. They are then arranged in a travelling framework of iron and placed in the vulcanising chamber or stove, from which they are brought forth ten or twelve hours after ready for use.

On another floor are the makers of coats, leggings, cushions, bags, &c. The drab or cream-coloured overcoats for India are the finest articles of clothing made in the establishment, and for lightness, durability, and elegance are unsurpassed. The cloth is cut by men, and the parts are put together by young women, who employ a mode of joining them that is more expeditious than the sewing-machine. All the seams are formed by cementing the edges with "solution," and then overlaying them with a fillet of rubber. When the coats are completed they are placed in the vulcanising chamber, and there undergo a change which prevents heat or cold from having any effect upon them. As made by the old process, india-rubber waterproof coats lost their elasticity in frost, and got so soft under the influence of heat that it was no unusual thing to find that a coat which had been folded away during the summer had actually melted and become useless by the softened surfaces adhering together. No such mishap can befall a coat made by the process adopted by the North British Rubber Company. The mode in which leggings, travelling-bags, and other articles of that kind are made need not be described, as it closely resembles that by which shoes and overcoats are produced.

The mechanical applications of india-rubber are numerous and varied, and the importance of the material in this respect is daily increasing. Its use in the form of carriage-springs was patented so early as 1822; but little farther progress in that direction was made until about ten or fifteen years ago, when rubber began to be employed extensively in the shape of tubes, springs, washers, driving-belts, valves, tires for wheels, &c., the making of which now constitutes an important branch of manufacture. The North British Rubber Company have paid much attention to the development of this section of their trade; and the mechanical department occupies one of the main blocks of their factory. As already stated, the base of hose-pipes and driving-belts is composed of canvas impregnated with rubber. Though known as india-rubber belts, the chief part of these articles consists of canvas, the quantity of rubber used being merely what is sufficient to fill up the texture of the cloth, make the respective folds adhere firmly, and form a shield or wrapper to confine the whole and protect it from moisture. The canvas having been prepared in the mill is cut into stripes of the desired width, and two, three, or four stripes are laid one over the other, and made to adhere by being passed between pressing rollers. The shield or envelope is then put on. It consists of canvas similar to what constitutes the core, but one side of it bears a strong coating of rubber. The wrapping completely surrounds the core, and the edges of it are firmly united by an overlapping fillet fixed with solution. The completed belt looks like a piece of solid rubber, but its strength is infinitely greater—in fact, a belt made of rubber alone would be almost useless in transmitting power on account of its elasticity. The belts are made in lengths of 300 feet, and of various breadths and thicknesses, but there is no practical limit to the dimensions. Hose and other pipes are made in a somewhat similar way. They are formed on mandrels, and have a coating of rubber both inside and out. A pipe one inch in diameter, composed of four folds of canvas with the usual proportion of rubber, will bear a pressure equal to 1000 ℔. on the square inch. Suction pipes, which have to be so constructed as to withstand the atmospheric pressure, have a layer of wire inserted into them. The wire is spun into a spiral form on a machine, and the tubemaker, after covering the mandrel with a rubber lining, puts on the wire and fills up the interstices with soft rubber. The canvas and wrapper are then applied. The vulcanising ingredients having been incorporated with the rubber in the mill, all that now remains to be done in order to complete the work is to place it in an oven, so that the heat and cold resisting powers of the rubber may be developed. Among the uses to which india-rubber has been recently applied may be mentioned the formation of rollers for lithographic and calico-printers and paper-makers, insulators for telegraphs, and cells for galvanic batteries, all of which purposes it suits exceedingly well. The other articles produced in the mechanical department are too numerous and their purposes too varied for enumeration; but one piece of work merits notice on account of its novelty and unprecedented size. Mention has been made of the road steamer invented by Mr R. W. Thomson, of Edinburgh, and the new application of india-rubber embodied therein. The peculiarity of Mr Thomson's carriage is that the tires of the wheels are composed of huge rings of vulcanised rubber. The tires were made by the North British Rubber Company, and are the largest pieces of the material ever manufactured, each tire weighing 750 ℔.

India-rubber is admirably suited for door-mats, which are made by piercing thick sheets or slabs of rubber in geometrical patterns. A new variety of mat has just been produced, into which the name or monogram of the owner is introduced. The forms in which india-rubber is most widely known are those of elastic-cords, ribbons, and webs, and in that department of the manufacture a number of hands are employed at the Castle Mills. The rubber is cut by machinery into threads, which are then, having been deprived of their elasticity by a simple process, either braided singly or woven with cotton and silk yarns in a ribbon-loom. The looms in the weaving shop are each capable of weaving eight ribbons of elastic at a time. The braiding-machines are beautiful pieces of mechanism. The thread of rubber is held in a vertical position, while a series of bobbins move round it, and round each other in an exceedingly curious way. Thus the rubber is enclosed in a casing of silk or cotton, which protects it from abrasion, and renders it applicable to a thousand useful purposes.

The company employ 600 workpeople in their establishment, but in the preparation of the cloth, thread, &c., used in the manufacture, as many more are employed in an indirect way. The health and comfort of the operatives are carefully provided for. All the women are paid by piece. In no department can it be said that the labour is heavy, and the work assigned to the women is peculiarly suited to them.

The Scottish Vulcanite Company (Limited) was formed in 1861 by a number of shareholders in the North British Rubber Company, but the two concerns are quite distinct in every other respect. The company began operations on a small scale under an American patent, and with machinery and instructors brought from America. They built a factory on the bank of the Union Canal near that of the North British Rubber Company, and their machinery was started in 1862. In consequence of the novel nature of the work, many difficulties were encountered at the commencement. A set of workpeople had to be trained, and that was found to be a slow operation, entailing the waste of much material. Under admirable management the company overcame all preliminary difficulties. Their original factory has already had a fourfold increase, and they now employ about 500 persons. The factory consists of a large central block 230 feet in length, and seven smaller detached buildings. The main block has four floors, and the others two floors each. A beautiful engine of 120 horse power, erected in one of the most elegant of engine-rooms, supplies the motive power. Everything required for upholding the establishment is made on the premises by the workmen of the company.

The machines used in breaking, washing, and kneading the rubber are similar to those employed in the North British Company's factory. Only the best quality of rubber is used, and the first process peculiar to the establishment is the conversion of it into "vulcanite" by incorporating with it certain chemical substances, and submitting it to the action of heat in an oven. After the chemicals are put in, the rubber is rolled out into sheets about three yards long, half a yard wide, and of various thicknesses. The sheets are laid on canvas-covered frames or trays, which are piled one above the other until the oven is filled. When the rubber is removed from the oven, it is found to have undergone a complete change. Each sheet is then cut into two, placed between metal plates, and subjected to a greater degree of heat. The effect of this treatment is to convert the rubber—which, when it went into the oven in the first instance, bore a close resemblance to putty—into a hard, black, glistening substance, applicable to a great variety of purposes. The change is a very mysterious one; indeed, in the whole range of chemistry there is scarcely a more wonderful thing than the production of the hard horny substance called "vulcanite" from elements which, in their unmixed state, are so unlike it. The idea of producing such a substance was one that could not have been arrived at by any amount of reasoning on the known properties of caoutchouc and the other ingredients; and unless it had been brought about by accident, it is probable that vulcanite would not yet have been known.

The story of Mr Goodyear, the American manufacturer who invented the process of vulcanisation, is very interesting. After having brought the manufacture of india-rubber to a degree of perfection, he undertook to supply india-rubber mail-bags to the Government. As the substance was then treated, it was not suited for that purpose, and the bags became soft, and failed altogether in a short time. The result was most disastrous to the manufacturer, who was forced to abandon the trade. Mr Goodyear did not despair of discovering a mode of so treating the rubber that it would not be readily affected by heat. He tried to attain his object by mixing certain substances with the rubber. He was in his abandoned factory one day, along with several friends; and after showing them the hopeless product of his experiments, he stood near a stove while he discussed matters with them. He retained in his hand the compound of rubber, &c., which he playfully held against the stove, little dreaming that he was making an experiment that would render his name famous. On removing the rubber, he observed that it had become charred, and was hard and tough like leather. Further experiments completed the discovery, and the fortune of Mr Goodyear

2 A

took a sudden turn. As already stated, Mr Hancock of London discovered Mr Goodyear's secret, and patented it; but there is no doubt that the vulcanising ingredients were suggested to Mr Hancock by discovering traces of them in some specimens of india-rubber which had been vulcanised by Mr Goodyear.

There are three departments in the Vulcanite Company's factory, which produce respectively combs, jewellery, and miscellaneous articles. In the comb department, the first operation is to convert the sheets of vulcanite into pieces of suitable size, which is expeditiously done by a cutting-machine. The pieces intended for the finest quality of dressing-combs are placed in heated moulds, and have a plain or ornamental rib raised on the back part, which at once increases the strength and improves the appearance of the combs. The slips of vulcanite so formed are then taken to the cutting-room—a large apartment, around which are arranged a number of beautiful little machines for forming the teeth of the combs. Each slip of vulcanite makes two combs, the teeth of one being cut out from between the teeth of the other. The machines are fitted with metallic tables, kept hot by branches from a steam-pipe which passes round the room. A pile of slips are deposited on the heated table, and are thus softened, the operative withdrawing the lowest slip, or that which is most pliable each time he supplies the machine. One slip is operated on at a time, and is laid on a travelling plate, which moves forward under a pair of cutters. The cutters rise and fall with great rapidity, and with the assistance of an expert workman each machine will produce from 130 to 200 dozen combs a-day. When the slips are withdrawn from the machine, the operative, by a dexterous pull, separates the two combs, which, in the soft state to which the material has been reduced, appear utterly useless, looking indeed as if they were made of leather, the teeth being twisted in all directions. A moment's pressure on the hot plate makes all right again; and when the combs cool, they are perfectly straight. Such is the minuteness of the division of labour in the establishment, that after the cutting is completed, the combs have to pass through a dozen departments before they are ready to be sent out. It is not necessary to follow them through all these, but one or two of the principal operations to which they are subjected may be mentioned. The cutting-machine gives a wedge-like point to the teeth; but it is necessary that they should also be tapered on the outer surface. For that purpose the combs are sent to the grinders, who reduce them to the desired shape on a stone. On examining a comb, it will be seen that the teeth are sharpened towards the edge, so that they have a diamond shaped section. The operation by which they are thus sharpened is called "grailing," and is performed by hand, the workmen using a broad file, which they apply with astonishing rapidity and certainty. The backs and ends are rounded on the grinding-stone, and then the combs are "buffed," to give them a smooth surface. They are next washed, dried, and polished, after which they are sent into the packing-room to be examined and packed up. A cheap and strong kind of comb is made with a brass or white metal mounting on the back. The metal is shaped by means of a die, and is attached to the comb by compression and by being clenched at the ends. Fine combs—or, as they are vulgarly called in Scotland, "sma'-teeth combs"—are made in a different way. The vulcanite is formed into plates the size of a comb, rounded at the ends, and thinned towards the edges. The plates are then placed singly into a machine, which cuts the teeth. The department devoted to this branch of the manufacture is situated in an upper room, open only to privileged visitors, as there are certain specialties connected with it which the company have introduced at much cost, and consequently desire to retain to themselves. It may be mentioned, however, that the teeth-cutting machines, of which about fifty are in use, are exceedingly beautiful and ingenious. They are arranged on a long table, and each does not occupy more than the space of one square foot. Each machine consists of two parts—a small circular saw and a travelling carriage, in which the plate of vulcanite is fixed. The carriage has three motions—one forward towards the saw, one backward, and the other from left to right. When a plate is inserted and the machine started, the carriage advances, and one interspace is formed by the saw; it then retires, moves the thickness of a tooth to the right, advances again to the saw, and so on. The machine is very rapid in its movements, and can cut the teeth on both sides of a four-inch comb in two minutes. A couple of women keep the machines supplied with plates. The saws, which are little more than two inches in diameter, are sharpened by a self-acting machine peculiar to the establishment. All the machines are driven by steam. Besides dressing and fine combs, a variety of others are made, much taste and ingenuity being expended on ladies' back combs, which are mounted with metal, glass, porcelain, or ornamented with carving, &c., in vulcanite. The company was created chiefly for the purpose of making combs, and that department is the most important in their establishment. No fewer than 24,000 combs are made every day, or about 7,500,000 a-year.

Vulcanite is the only material that has successfully competed with jet for making black jewellery. In appearance it closely resembles jet, and has the advantage of being stronger and cheaper. During the past four or five years vulcanite jewellery has attained immense popularity, and the demand for it is rapidly increasing. Owing to the brittle nature of jet it is difficult to work, and articles made of it will always be costly and delicate. Vulcanite, on the other hand, may be readily moulded, carved, or stamped into almost any form. Among the articles of jewellery made of it are ladies' long and Albert chains, necklets, bracelets, gauntlets, buckles, and coronets. The chains are composed of variously shaped links, but the mode in which they are made is alike in all cases. The vulcanite is first cut into slips about eighteen inches in ength and one inch in width. It is then taken to a room in which are a number of punching machines worked by girls. The links are punched out at two operations, the first making the opening in the centre, and the second cutting out the circumference. The punched edges are rough; and in order to smooth and polish them the links, after being fixed on an iron rod, are ground down to a standard size, and polished on the "buff" wheel. They are then ready for being put together, for which purpose they are transferred to women who sit at benches fitted with hot plates. After lying for a few minutes on the plate, the alternate links are cut open at one end with a knife, and these are readily opened and slipped into their neighbours. Many earrings are made by combining links in various ways; links are also introduced into some kinds of back combs, bracelets, &c.

In the miscellaneous goods' department a great variety of articles are made. Owing to the power which vulcanite possesses of resisting the action of acids, it is of much value in the construction of surgical and chemical instruments, and is now being extensively applied in the manufacture of tubes, syringes, flasks, stoppers, &c. A large trade is also done in making vulcanite cells for galvanic batteries. Knife-handles, card-trays, neckties, girdles, and gauntlets are among the other products of the factory. The vulcanite knife-handles are exceedingly pretty, and are superior to ivory, bone, horn, or whalebone handles, in that they cannot be detached from the blade unless they be smashed off, and that they are neither split nor discoloured by immersion in hot water. The card-trays are chiefly made up of thin sheets of vulcanite ornamented with designs in fretwork. Among the greatest novelties are the neckties, which look exactly like silk, the texture being closely imitated by some peculiar process. They are made up in a variety of styles. There appears to be no limit to the uses to which this wonderful substance may be put; and were the raw material more abundant, and consequently cheaper, it would be employed as a substitute for wood, papier-mâché, and like materials, in the construction of many articles of ornament and use in the shape of furniture.

In a manufactory of this kind the making of packing-boxes constitutes an important department. In the shop in which the paper boxes are made seventy young women are employed, and they are aided by a number of beautiful cutting and moulding machines. The boxes and goods are brought together in the warehouse, where they are made ready for sending out.

All the departments of the factory are kept thoroughly clean; and the rooms are lofty, well lighted, and well ventilated. For the convenience of the women who reside at a distance, there is a large dining-hall, comfortably furnished and heated by steam-pipes. Nearly all the operatives are paid by piece, the women earning from 10s. to 14s. a-week for $57\frac{1}{2}$ hours' work. Some of the men earn a high rate of wages, a journeyman comb-cutter making about L.2 a-week when employed on certain kinds of work. It is a fact worthy of mention, that the men who are employed in mixing the chemicals with the rubber, and in conducting other operations in vulcanising, are peculiarly healthy, and never suffer from diseases of an epidemic type.

MANUFACTURES IN GLASS.

" By some fortuitous liquefaction was mankind taught to produce a body at once in a high degree solid and transparent, which might admit the light of the sun and exclude the violence of the wind ; which might extend the sight of the philosopher to new ranges of existence, and charm him at one time with the unbounded extent of the material creation, and at another with the endless subordination of animal life; and, what is yet of more importance, might supply the decays of nature and succour old age with subsidiary sight. Thus was the first artificer in glass employed, though without his own knowledge or expectation. He was facilitating and prolonging the enjoyment of light, enlarging the means of science, and conferring the highest and most lasting pleasures; he was enabling the student to contemplate nature, and the beauty to behold herself." In these few resounding sentences Dr Johnson sets forth the origin and utility of glass. The date of its discovery is lost in antiquity; but its history, so far as known, is exceedingly interesting. There are indisputable proofs that the art of glass-working was known in Egypt before the exodus of the Children of Israel from that land, more than 3500 years ago ; and its spread among the ancient nations has been distinctly traced. In order to treat of glass as an article manufactured in Britain, we must come down to comparatively recent times.

The first mention of glass in English history relates to the glazing of the windows of a church at Wearmouth, in Durham, which was done in the year 674 by workmen brought from abroad. Other ecclesiastical edifices were subsequently furnished with glazed windows; but so late as the sixteenth century glass was a rare thing even in the residences of the nobility. In a record of a survey of Alnwick Castle, made in the year 1567, it is stated that the glass casements were taken down during the absence of the family, so that the heavy expense of damage done to them by wind and otherwise might be saved. In Harrison's "Description of England," written in 1584, the following passage occurs :—" Of old time [meaning, probably, the beginning of the century] our countrie houses, instead of glasse, did use much lattise, and that made either of wicker or fine rifts of oke in checkerwise. I read also that some of the better sort, in and before the time of the Saxons, did make panels of horne instead of glasse, and fix them in wooden calmes [casements]; but as horne in windows is now quite laid downe in everie place, so our lattises are also growne into disuse, because glasse is come to be so plentiful, and within verie little as good, cheape, if not better than the other." A coarse kind of window glass was made in England in the fifteenth century, and the first flint-glass manufactory was established at Crutched Friars, London, in 1557. Several other works of the kind were subsequently set up in the same locality. In 1615 Sir R. Maunsell obtained a patent for making glass by means of coal, instead of wood fuel. About sixty years later the Duke of Buckingham induced some Venetian glass-makers to settle at Lambeth, where they began to make plate-glass mirrors. As the early British manufacturers produced only an inferior class of goods, considerable quantities of glass were for a long time imported from Venice.

The manufacture of glass in Britain was grievously hampered by Excise duties and regulations, which were originally imposed in 1695, and continued—with the exception of an interval of fifty years, between 1698 and 1745—until 1845. At first the duty was comparatively moderate; but in the hands of successive legislators it swelled to an enormous amount, the effect being first to stop any increase in consumption, and ultimately to reduce by one-half the quantity of glass made. In the three years ending with 1779, when the duty on flint and plate-glass was 9s. 4d. a cwt., it produced, on the average, L.64,188 a-year. In 1822 the duty was 98s. a cwt., which produced only L.289,442, showing that the quantity of glass made was less than half what it was in 1777, and that, too, though the population had nearly doubled in numbers during the interval. These facts, as shown by the official returns, could not fail to convince the most obtuse mind that if the manufacture of glass was to be retained as a branch of national industry, some change was necessary. In 1825 the duty was reduced to the following rates:—Flint-glass, 56s. a cwt.; plate, 60s.; broad, 30s.; crown, 73s. 6d.; bottle, 7s. These rates were continued until 1845, when the duty was repealed. The Excise regulations as to the manufacture of glass were numerous and complex, and were enforced under heavy penalties. A few of these regulations may be mentioned. All glass-makers had to take out a license, which entailed a payment of L.20 annually for each glass-house. No pot could be charged with fresh materials without the owner giving twelve hours' notice in writing to the Excise officials, under a penalty of L.50. If, after notice was given, and the pot had been gauged by the Excise officer, any material or preparation was put into any pot, a penalty of L.200 was incurred. Officers could demand access to the workshops at any time to gauge the materials, and mark the pots as they might think fit. Any attempt to obstruct the officers so employed was visited by a penalty of L.200; and the counterfeiting or altering of any marks so made, by a penalty of L.500. There were other regulations which interfered very much with the details of the manufacture. Since the removal of the restrictions the trade has increased very much, and in recent years work of a more artistic kind has engaged the attention of the manufacturers. Britain occupies the foremost place in the flint-glass branch, and in the production of engraved articles of flint-glass Edinburgh holds a high rank.

The first glass-maker in Scotland was Mr George Hay, who obtained from James VI. a patent conferring on him the privilege of manufacturing glass for a period of thirty-one years. Mr Hay took advantage of a peculiarly formed cave at Wemyss, on the Fife coast, and therein set up his furnace; but the concern did not pay, and was soon abandoned, the glasshouse being allowed to 'go to ruin. The place it occupied is still known as the " glass cove." The next localities in which the trade was carried on were Prestonpans and Leith. Operations were begun in the latter town in 1682, bottles being the principal articles produced. The following list of prices, issued from the establishment, is interesting:—"The wine glass, at three shillings two boddles; the beer glass, at two shillings and six-pence; the quart bottle, at eighteen shillings; the pynt bottle, at nine shillings; the choppin bottle, at four shillings and sixpence; the muskin bottle, at two shillings and sixpence—all Scots money; and so forth of all sorts, conform to the proportion of the glasses; better stuff and stronger than is imported." The site of this old glass-work is occupied by an extensive bottle factory, which has been in operation for upwards of a century, and was for a long time the largest in the country. Of seven cones, or furnaces, which were kept going for many years, only two are in use at present, the trade being more profitably carried on in the north of England, where fuel is more abundant and cheaper. In 1777 there were 15,883¼ cwts. of bottles made at Leith, the duty on which amounted to L.2779 odds. There are at present six flint-glass and eight bottle manufactories in Scotland, employing in the aggregate nearly 2000 persons.

The declared real value of the glass of all kinds exported from Great Britain, was in 1835, L.490,493; in 1840, L.417,178; 1845, L.357,421; 1850, L.308,356; 1860, L.653,198; 1865, L.742,639; 1867, L.803,334. The value of the glass exported from Scotland alone was, in 1861, L.62,140; and in 1867, L.106,555.

There are, generally speaking, five varieties of glass, each requiring a peculiar mode of fabrication, and peculiar materials :—1. The coarsest and simplest is bottle glass. 2. Next to it, in cheapness of material, may be ranked broad or spread window glass. 3. Crown glass or window glass, formed in large circular plates or discs. This variety is peculiar to Great Britain. 4. Flint-glass, crystal glass, or glass of lead. 5. Plate or fine mirror glass. Only two of these varieties are made in Scotland—namely, flint-glass and bottle glass.

The principal flint-glass manufactory is that of Mr Ford, which is situated in South Back of Canongate, Edinburgh. The manufactory, which is known as the Holyrood Glassworks, occupies an extensive range of buildings, the nucleus of which was erected by the present proprietor's granduncle, about the beginning of the century. There are two furnaces in the works, which together contain twenty-two pots. The furnaces consist of huge cones of brickwork, pierced with a series of openings corresponding to the number of pots in each. In the centre of each cone a great fire is kindled, and the flames and heat from it are drawn through flues, and brought into contact with the pots, which are arranged round the interior of the wall. The pots are made of a particular kind of clay, which is capable of withstanding intense heat without cracking or giving off any matter that would be injurious to the glass. Usually the pots are made on the premises, and the operation is a tedious and laborious one. The pots are the source of the glass-maker's greatest anxiety, for, notwithstanding the utmost care in making and annealing them, some give way after being in use only for a week or two; others endure for three or four months; but few reach the age of a year. It occasionally happens that a pot splits when full of "metal," as the fused glass is called, and then the accident entails a serious loss. The withdrawing of a broken pot and the insertion of a new one is about the most trying operation that men could be called on to perform. Besides this waste of the pots, the furnace itself undergoes deterioration at a rate which requires it to be entirely reconstructed at the end of ten years, if not sooner.

The constituents of flint-glass—or crystal, as it is more commonly called—and the proportions in which they are used, are as follow:— Carbonate of potash, 1 cwt.; red lead or litharge, 2 cwt.; sand, washed and burned, 3 cwt.; saltpetre, 14 ℔. to 28 ℔.; oxide of manganese, 1 oz. to 4 oz. The sand used is obtained from France, and is of remarkably fine quality, the particles when examined under the microscope appearing to be regularly formed and pure crystals. In order to free the sand from impurities, it is washed with water, then dried in an oven, and afterwards passed through a series of fine sieves. The other compounds having been prepared by various processes, both they and the sand are weighed out in their respective proportions, and thoroughly mixed. These operations are conducted in a special department. The most economical and convenient system of working a glass-house is to make "metal" once a-week, and matters are so arranged that the supply becomes exhausted on Friday, which leaves Saturday and Sunday clear for melting a fresh supply, the operations of the glass-workers being in the meantime suspended. ·

The pots in which bottle glass is made are open, and the flames and smoke come freely into contact with the "metal." In the case of flint-glass that arrangement would not do, as the smoke would spoil the purity of the glass. The flint-glass pots are accordingly made so as to prevent direct contact between the "metal" and the fire. In shape the pots closely resemble the inverted bowl of a tobacco-pipe, with the stem broken off short. The stem part projects through the masonry of the furnace, and contains an opening through which the glass may be withdrawn. The pots are charged every Saturday morning. Each contains about eighteen cwt. of glass, the ingredients for which are put in gradually as the fusion proceeds, from twelve to fifteen hours being required to complete the charging. Though the ingredients become melted in that time, the "metal" is not in a fit state for working owing to the presence of air-bubbles, which can be got rid of only by urging the furnace to its utmost intensity, and maintaining it thereat for from thirty to forty hours, the mouths of the pots being sealed during that time.

The glass is ready for working by an early hour on Monday morning. There are two sets of workmen, who relieve each other every six hours, and the work goes on constantly from Monday morning until Friday. The weekly consumption of coal is about twenty tons to each furnace.

Under the intense heat to which it is subjected in order to get rid of gaseous bubbles, the glass becomes nearly as fluid as water, and in that state could not be worked. Before the blowers begin operations, the temperature of the pots is lowered until the "metal" assumes the consistency of treacle. The tools used by the workmen are exceedingly simple, and are similar to those which the earliest British glass-makers used. Owing to the peculiar nature of the material, the formation of articles in glass depends more upon the skill, expertness, and tact of the manipulator than upon the employment of complicated appliances. The surface of the glass would be spoiled by a free use of metal tools, and almost the only implements employed are composed of charred wood. The operations of the glass blowers are probably the most wonderful in the whole range of the arts, no manipulations of the conjuror being more mysterious to one who witnesses them for the first time. A small quantity of pasty-looking stuff at a white heat is withdrawn from the pot on the end of an iron tube, and in two or three minutes afterwards is passed to the annealing oven in the shape of a decanter, goblet, or wine glass. The operatives work in gangs or "chairs," each "chair" consisting of a chief workman called a "gaffer," together with a servitor, a footmaker, and a boy—four in all. According to the nature of the article being made, the members of each "chair" divide the labour among them, the "gaffer" executing the most difficult parts. One or two examples of manipulation from many that may be seen in the Holyrood Glassworks will suffice to illustrate the art of working in crystal. In making a plain jug, for instance, one of the workmen takes an iron tube, four or five feet in length, and about three-quarters of an inch in external diameter, and dips one end of it into the pot. This is called "gathering," and a chief object of the operative is to take up the exact quantity of metal required for the formation of the article that is to be made—to take more would be to waste, and to take less would be to make the article too thin and light. Having got the proper quantity of glass on the tube, the workman withdraws it, and rolls the pasty mass on an iron table called a "marver." The purpose of this operation is to give the glass a smooth surface. He then blows into the tube, and causes the glass to expand slightly, repeats the "marvering," and so on until the desired dimensions are attained. An inexperienced observer would up to this point have difficulty in guessing what sort of vessel was being made, for the large crystal pear on the end of the blowing tube bears no evidence of its ultimate form. The glass, with the tube still attached, is then handed over to the "gaffer," who is seated on a peculiar kind of chair, fitted with a rail on each side. The rails project forward about a couple of feet, and incline downwards slightly. Laying the blowing-tube across these rails, with the glass projecting on the right-hand side, the "gaffer" completes the jug by a series of dexterous operations. He begins by flattening the outer end of the glass by pressing a piece of wood against it, and thus forms the bottom of the jug. One of his assistants then approaches with a bar of iron called a "punty rod," the tip of which, having been previously dipped into the pot, is applied to the bottom of the jug, and becomes firmly fixed thereto. The vessel is then detached from the blowing-tube by drawing a piece of cold iron across the part at which the separation is desired. The glass contracts suddenly under the touch of the iron, and parts readily. By this time the glass, which now bears some resemblance to a champagne bottle, has cooled considerably, and become hard and brittle; and, in order to restore it to a plastic state, it is held in the opening of the pot for a few seconds. The workman lays the rod across the rail of the chair, and as he rolls it backwards and forwards, inserts a two-pronged iron tool into the neck of the jug, and widens it out to the required degree. This tool is called a "pucellas," and the prongs are united by a spring arch similar to that of a pair of sugar-tongs. With a tool of the same form, but having prongs of charred wood, the workman then brings the body of the jug into shape. He generally copies from a pattern, and measures the work with callipers from time to time; but the eye is his chief guide, and some men of long experience can afford to dispense almost entirely with measurements. After another heating, the mouth of the vessel is cut round with a pair of shears, a projecting portion being left to form the lip, which a simple operation brings into shape. While the "gaffer" is thus employed, one of his assistants is preparing the metal for the handle, the other being engaged in gathering and bringing forward another jug. The handle is formed by gathering a certain quantity of glass on the end of a rod, and allowing it to elongate by its own weight until it becomes reduced to the required dimensions. The end of the piece of glass thus prepared is brought into contact with the jug, to which it becomes permanently attached. A portion sufficient to form the handle is detached from the rod, and the loose end turned down and fixed by simple contact. This operation has to be performed very expeditiously, as the slip of glass soon cools. The jug receives a final heating to bring up the surface and consolidate the work, after which it is removed to the annealing oven.

Dessert-dishes, salts, &c., are made by taking up a ball of glass on the blowing tube, and pressing it into an open mould during the process of blowing. The part above the mould is blown out until it becomes very thin, and may be readily knocked off at a line above the rim of the dish. Cruet bottles, and flasks for chemical purposes, are formed by blowing the glass into a close mould. A cheap kind of tumbler is made by pressing the glass into shape by a mould and die. On the bed of the press used in the operation is a block of cast iron, in the centre of which is a cavity corresponding to the outside of the tumbler, and attached to a lever above is an iron plunger or core. An assistant having gathered the proper quantity of glass, drops it into the mould, the pressman severing the connection between the gathering-rod and the glass by cutting it with shears. The plunger is then brought down on the glass, and in a second or two it is raised, and the tumbler is turned out. Contact with the mould dulls the brilliancy of the glass, and as the tumblers are formed they are attached to a punty rod, and " fire polished" by being reheated at the pot-mouth, and smoothed off with a piece of charred wood.

Owing to its peculiar structure, glass, especially if of unequal substance, is liable to fracture by sudden changes of temperature ; and if the articles made of it be allowed to cool too quickly on leaving the hands of the workmen, they will either fly to pieces immediately, or become so delicate as to be unfit for use. The process of annealing is intended to reduce or dispel this tendency to fracture, and when properly conducted is completely successful. Annealing is simply an arrangement whereby the articles are slowly cooled. The annealing arch or oven consists of a brickwork tunnel about thirty yards in length, with two or more lines of rails in the bottom of it. At one end of the archway is a coke fire, which raises the temperature in its vicinity to a degree equal to that of the glass when it is brought from the workman at the furnace. The articles are arranged on trays, which slide along the rails in the oven. As a fresh tray is put in, all the others are moved forward into a cooler atmosphere, the one at the extreme end being withdrawn, and so on until in the course of ten or twelve hours each tray traverses the whole length of the oven, and the articles become cooled gradually by passing from the heated to the cold end of the tunnel. The time required for annealing depends upon the size of the articles. The ordinary varieties of table glass may be got through in ten or twelve hours ; but heavy vessels which have to withstand much cutting take from thirty to forty hours. Belonging to each " chair" is a boy, whose duty is to take the articles as they are made and place them in the annealing oven. Mr Ford, not having available ground for increasing his annealing accommodation, has constructed a circular oven with a revolving floor, which carries the articles through a gradation of temperature.

When the articles are withdrawn from the annealing ovens they are piled into baskets, and those which are to be cut are taken to the cutting shop, while those which are to remain plain are removed to the warehouse. The cutting shop is a large apartment lighted from the roof. It contains about forty wheels, attended by as many workmen, whose operations are of a very interesting kind. All articles in the finer class of flint-glass goods are more or less cut, and their appearance is much improved thereby. The glass-cutters use wheels of various forms and dimensions according to the nature of the work. The wheels are fixed in a sort of turning-lathe and are driven by steam, and the variety of patterns that may be produced on them is almost unlimited. The workman rarely makes any attempt at drawing the device on the glass before cutting it. He simply divides the circumference of the article into sections by scratching with a file, and guided so far by these marks he trusts to his eye for the rest. The first wheel used is of malleable iron, about twenty inches in diameter and half an inch thick. When this is in operation, the tap of a hopper overhead is opened, and a small stream of sand and water flows on the wheel. The sand grinds down the glass to the required shape, but leaves a rough surface, which is first reduced on a fine stone wheel, and then polished on a wheel of willow wood supplied with putty powder. As all the grinding is done wet, the occupation of the glass-cutter is not injurious to health as it would otherwise undoubtedly be. Among the other operations carried on in the cutting shop are "roughing," or "frosting," and "stoppering." The first is performed by scratching off the natural surface of the glass by means of a wire brush and emery powder, the article operated upon being fixed in a peculiar kind of lathe. "Stoppering" is the name of the operation by which stoppers are fitted to bottles. The stoppers are cast of a size larger than the apertures they are to fill, and are ground down by the application of emery on a lathe. When the stopper is so far reduced as to enter the bottle a short distance, the bottle is held to the stopper, and as the latter spins round it is supplied with emery, which enables it to eat its way in until it fits properly.

The warehouse embraces several large rooms, which are respectively devoted to the various classes of goods, and a visit to them serves to show the multifarious uses to which glass may be put. Mr Ford's chief productions are table crystal and lamp globes and funnels, and the stock of these forms an exhibition of no mean interest. Among the engraved goods may be seen some magnificent specimens of work executed by local artists.

Mr Ford employs nearly 200 persons, who are chiefly paid according to results. The glass-makers earn from 20s. to 38s. a-week ; cutters, from 20s. to 34s. ; engravers, from 20s. to 40s. ; and boys, from 4s. to 5s. The term of apprenticeship is seven years, and boys are usually employed in doing the lighter work about the establishment for three years before beginning to serve their term, so that in most cases it is ten years after a boy enters before he becomes a journeyman.

The flint-glass makers have a union, which is understood to be one of the strictest associations of the kind in the United Kingdom. It is called the "National Flint-Glass Makers' Sick and Friendly Society of Great Britain and Ireland," and its headquarters are at present in Birmingham. The entry-money for each member ranges from 10s. to L.7, according to age and occupation—"footmakers" being admitted at one-third less than either "gaffers" or "servitors." The contributions are fixed at 1s. 3d. per week for the two latter classes, and 10d. for the former, or lowest paid class ; and in return for these payments each "workman" or "servitor," while on the sick-list, receives 12s. per week for thirteen weeks, 10s. per week for a similar period, 8s. for twenty-six weeks, and 5s. for twenty-six weeks more. Members on strike receive 15s. per week for six months, if necessary, and 10s. for other six months ; while members intending to emigrate are assisted to the extent of L.8, 10s. if they have been three years on the roll of the society. There is also a superannuation allowance of from 3s. to 8s. per week, the amount of which depends upon the length of time the recipient may have been in the society ; or the weekly allowance may be cleared off by a single payment mutually agreed upon, which must not exceed L.50 in any case. The union does not permit the employment of more apprentices than one to five journeymen, except under special circumstances, and employers when in want of workmen are obliged to apply through the district secretary instead of engaging the men themselves. The glass-cutters have a union as well as the glass-makers. The contributions range from 1s. to 3s. 6d. per week, according to the number of unemployed members, who receive support at the rate of 10s. per week. No serious difficulty has arisen between the employers and employed in Scotland save one, which occurred about nine years ago, when there was a general "lock-out."

Though by expert manipulation the glass-blower produces an infinite variety of beautiful forms, to some of which the glass-cutter gives additional grace, glass is capable of being treated by other processes which enrich its appearance and add much to its value. A finely modelled goblet or decanter of flint-glass, which has the smoothly rounded surface it received from the glass-blower relieved by a few judiciously placed facets produced on the cutter's wheel, is an object possessing great beauty; but its appearance may be much enhanced either by the application of colour or by engraving. The flint-glass manufacturers of Britain use colours in exceptional cases only. The task they appear to have set to themselves was to produce glass of the highest transparency and brilliancy, and impart to it appropriate forms and surface decorations; and they have been so successful that they have got far ahead of all competitors. On the other hand the continental glass-makers revel in colours. The most valuable qualities of flint-glass are its transparency and brilliancy; and when these are hid by opaque colours, the only characteristic of the substance that remains is the least desirable one of brittleness. There is no evidence that glass-engraving in intaglio, as practised in modern times, was known to the ancients; but it is certain that they were familiar with the more difficult process of cutting designs in relief. Those marvellous gems of art, the Portland and Naples (or Pompeii) vases show to what excellence the early glass-engravers had attained. The vases are similar in material, but differ widely in form and design. A good idea of the appearance of the Portland vase, the original of which is in the British Museum, may be obtained by examining a model of it in the Edinburgh Industrial Museum. The vessel is composed of dark blue glass, bearing a number of figures in white opaque enamel, cut in low relief, after the fashion of a cameo. The Pompeii vase is in a museum at Naples.

In the modern school of glass-engraving Edinburgh stands in the highest class, and it is exceedingly creditable that that position has been gained after only a few years' exertion. At the Art Exhibition held in Edinburgh in 1856 glass-engraving was in its infancy in Scotland, and the specimens then shown were coarse and inartistic.

It was not until the firm of Messrs J. Millar & Co., of Edinburgh, turned attention to the matter that a decided and hopeful start was made. So rapid was the progress, that Messrs Millar were able to show at the Great Exhibition of 1862 a collection of engraved glass which attracted universal attention, and won the favourable notice of art critics. A happy hit was made by the beautiful fern pattern then first produced, and now copied by engravers everywhere. Following up the success thus achieved, the firm have gone on producing novelty after novelty. At the Paris Exhibition they made a magnificent display, and, notwithstanding the severe test of competition with the famous glass-makers of the Continent, held their own in the department of engraved flint-glass. Some of the decanters and wine glasses shown were exquisitely beautiful, and were eagerly bought by art collectors. In order that engraved glass might become popular, it was necessary that it should be cheap as well as beautiful; and the Edinburgh makers were among the first to meet both requirements, the result being that their productions are finding their way to the tables of the middle as well as of the upper classes of society. The nobility are now having their coats of arms engraved on every article of table crystal; and persons who have no heraldic emblems to display are having their glasses inscribed with monograms. Messrs Millar & Co. have just invented a process by which the monograms may be gilded as well as engraved, and by which the coats of arms may be worked in true heraldic colours. The experiments made have been highly successful, and once certain trifling technical difficulties are overcome, an important addition will be made to the modes of decorating glass. Another novelty which the firm are bringing out is the introduction of a small degree of colouring in table crystal. The Bohemian uniformly coloured glass has a heavy appearance, and has not found much favour in this country, the clear brilliant flint-glass being preferred; but it has been thought that a small streak of some brilliant colour applied to the lip and foot of a wine-glass or decanter would enhance its appearance by forming a contrast with the snowy purity of table linen. Of course there will be differences of opinion as to the appropriateness of such an application of colour—here it will suffice to chronicle the fact of its introduction, under the name of Alexandra-Venetian glass. The prevailing fashion in engraved glass is, in the case of decanters—which, by the way, are made very light—delicate Grecian ornamentation round the lip and neck, with a medallion containing a figure-subject of a classical type on one of the sides ; and, in the case of wine glasses, wine coolers, and finger glasses, the enrichment consists of wreaths, festoons, and arabesque work, all on a minute scale. The figure engraving generally displays marvellous skill in execution.

The "Times" reporter at the Paris Exhibition thus referred to some of the articles shown by Messrs Millar & Co. :—"There is a small glass jug round which has been engraved, with amazing minuteness, one of the friezes of the Elgin marbles. The procession of men and horses in miniature round this jug has a most charming effect, and many other glass vessels might be mentioned of the same manufacture which are conceived in the happiest vein, and are wonderfully fine of execution. Full as the case of Messrs Millar is of work which in its artistic merit does no small honour to the Scottish metropolis, there are two large decanters in it that deserve notice, not only for the skill with which they are engraved, but also as curiosities which illustrate the drinking customs of the time. Trust a Scotchman near good liquor. Here are wine jugs which have been especially ordered for Scottish noblemen—the one to contain six, and the other nine bottles of claret. The French and the Scotch have an ancient liking for each other. The French patronise largely everything Scotch in the Exhibition, as Scotch shawls and Scotch jewellery; and they come to gaze with admiration on those prodigious decanters of Messrs Millar, which bear startling testimony to the bibulous capacity of the Scotch for the chief wine of France."

The principal glass-engraving establishment in Edinburgh is that of Mr J. H. B. Millar, who works only for Messrs Millar & Co., and for Mr Ford, of the Holyrood Glasswork. The workshop occupies a large brick building at Norton Place, and in it about forty men and boys are employed. Mr Millar was one of the first engravers engaged by Messrs Millar & Co., and by his artistic and manipulative skill has contributed in no small degree to the success they have achieved in this particular department of their business. The glass-engraver requires a sharp eye and a steady hand. He must be careful not to make blunders, because the transparent body on which he operates cannot conceal a flaw, nor does it admit of one being successfully rectified. Another peculiarity of this art is that, while the engraver in wood or in steel may draw his design on the surface to be cut, the glass-engraver has no such means of guiding his hand or eye. The tools used in glass-engraving are exceedingly simple. They consist of a series of copper discs, varying from four inches to one-sixteenth of an inch in diameter, and a small lathe for making these spin round. There are about a hundred discs in a complete set, and in the execution of some designs the whole of these are called into requisition by turns. It is most convenient when several articles of one pattern are carried on simultaneously, as then the engraver is not interrupted by frequent changes of tools. When the discs are in use they are kept supplied with a mixture of emery and oil. The workman rests his arms on his bench, and seizing the article to be engraved with both hands, applies it to the edge of the disc, which is kept revolving by the action of his foot on a treadle. The glass is moved about with great expertness, and the rapidity with which designs of a simple kind are produced is astonishing. Only a trifling percentage of glass is broken in the process of engraving. Mr Millar, who is a native of Bohemia, employs a number of his countrymen, but most of his workmen are natives of Edinburgh who have served an apprenticeship of six years.

The use of coloured glass in windows is of great antiquity. It is said to have been imitated from the Byzantine-Greeks by the Saracenic races, and it is continued in the cities of the East to the present day. Coloured glass was used in the palaces of the Roman Emperors in the mosaic work with which the walls and floors were decorated, and that is supposed to have been the first form in which glass was used pictorially. The earliest direct reference to coloured glass windows in Europe occurs in a description of the Basilica of St Paul at Rome, written in the end of the fourth century. The next allusion to it is made in an account of a church built at Lyons in the fifth century, the windows of which are described as being composed of coloured glass " arranged in patterns." Those early windows were furnished with " stained glass"—that is, glass coloured in the pot, as distinguished from " painted glass," which is produced by the application of colours to the surface. The earliest existing examples of painted windows are believed to be those in the Abbey of Tergernsee in Bavaria, which were presented to the abbey in the year 999. Figure subjects do not appear to have been introduced until the middle of the eleventh century, when " The Mystery of Paschasius" was illustrated in one of the windows of a church at Dijon. The French King Charles le Chauve was the first great patron of glass-painting; and under the impetus it received from the encouragement of royalty, the art became universal, painted glass windows being introduced into all religious edifices of any pretensions. In the twelfth and thirteenth centuries glass-painting reached its highest point of perfection, the French being the most successful artists. The mosaic and medallion windows of the thirteenth century are considered to be the best specimens of decorative work ever produced. The oldest English examples of that period are in Canterbury and Salisbury Cathedrals; but the finest are in York Minster and Lincoln Cathedral. In France the chief works of the same age embrace some of extraordinary grandeur and beauty —such as the windows of the cathedrals of Chartres, Bourges, Paris, Amiens, and Rouen, and in the magnificent Sainte Chapelle at Paris. In the fourteenth century, with the advent of the second pointed or decorated style of Gothic architecture, came a fashion of making painted windows more vivid in colour, broader in style, and displaying a more studied gradation of light and shade. Mechanically considered, the work of that period was a great advance on that of earlier date, but the designs were less pure in conception, and the intense colouring had a tendency to reverse the proper order of things; for while the windows ought to have been subordinated to the general architectural effect, they in most cases usurped the superior place. The fourteenth century style of glass-painting, or something based thereon, is perhaps better suited to modern tastes and requirments than that which preceded or that which followed it; and our artists have, in bringing about a revival of the art, chiefly striven to work in the spirit of that style. When the third pointed or perpendicular style of Gothic came into use in the fifteenth century, another change came over the art of glass-painting, but it was a change for the worse. The mosaic pattern-work of stained glass was discarded, and the design was painted on large pieces of white glass. The painting, as such, was executed with much artistic skill, but the colouring looked cold and feeble, because the native brilliancy of the glass was obscured. Towards the end of the fifteenth century, when Gothic architecture was dying out, glass-painting reached its lowest point of degradation. Painters on glass tried to rival painters on canvas, and the windows of palaces and mansions were filled with glass covered by work which, if applied to canvas, would have rendered the artists famous, but which never could be made effective as a transparency. This fashion was followed up till thirty or forty years ago, when Gothic architecture began to be revived. Architects, while studying the details of mediæval cathedrals and other edifices, could not fail to be impressed with the idea that no copy of the stone and lime would be satisfactory unless the windows were to be filled with painted glass—in short, that the revival would have to be a double one. In England, France, and Germany, men were found ready and willing to devote themselves to study and endeavour to imitate the works of the ancient masters in the art of glass-painting; and now we are fairly in the way to a great revival. Many sad failures were made at the outset; but as the artists are acquiring a knowledge of the capabilities of their materials, the successful works are multiplying. The Germans are striving to eclipse all competitors, and the establishments under royal patronage at Munich and Berlin have produced some grand things; but though the Germans excel in drawing, the British artists show a more thorough appreciation of the ancient mode of working, and if they persevere in the course they are pursuing, will one day stand above all rivals.

Painted glass is now being used extensively in ecclesiastical edifices, and in the houses of the wealthy; and it is probable that it will go on increasing in popularity. Even in Scotland, where, up till a few years ago, a bit of painted wall or coloured glass in a Presbyterian Church would be regarded as an indication of Romish tendencies, painted windows are being freely introduced into places of worship, and yet there is no apparent decay of Presbyterianism among us. Glasgow Cathedral is adorned by some of the finest specimens of glass-painting by modern German artists, and the hearty and liberal manner in which that work was carried out augurs well for the future of the art in Scotland.

Glass-painting was almost extinct in Scotland until about forty years ago. No doubt it had been extensively known and practised during the period in which our fine old abbeys and churches were erected; but it is probable that the artists, like the architects and masons of those structures, were imported from the Continent. Now, however, glass-staining forms a considerable branch of artistic and mechanical industry. Year after year, as already stated, there is a growing desire to have the windows of mansions, public halls, and churches decorated with stained glass. Scarcely is there an entrance-hall in the mansions of the wealthy but has its chief window blazoned with heraldic transparencies; and many public buildings have their windows similarly adorned.

The processes by which these works are produced are multifarious, and the manipulation requires great skill. The coloured glasses used are termed pot-metal, from their being produced by the fusion of metals with crystal in a pot or crucible, and afterwards moulded into sheets. From these the glass-cutter, who handles his diamond with great dexterity, cuts the separate pieces according to the required design, each colour thereof being in a separate piece of glass. The first part of the work consists in preparing a coloured sketch-design, generally on a small scale ,from which afterwards full-sized cartoons are drawn. On these are marked the lines for the lead-work necessary to join the various pieces of glass together. Any outlining or shading, such as upon foliage, heads, or draperies, is produced by fusing an opaque metallic pigment upon the surface of the pot-metal glass. Metallic pigments fused upon the surface of glass produce various colours, but the effect is feeble and dull as compared with the tone and lustre of pot-metal colours. The lead framing, which is used for joining the pieces of glass together, is grooved upon both sides, and the glass is closely fitted and soldered therein. Elaborate work contains about 100 pieces of glass in each square foot; the casing of these in lead and soldering is, therefore, a process requiring great experience and nicety. The glass-stainer makes use of fluoric acid for etching and embossing. Many of the coloured glasses are composed of sheets having about a sixth of their thickness coloured, and the remainder clear. These are called "flashed" colours. The workman covers the flashed side of the glass with a preparation which resists the action of fluoric acid, and draws the design thereon with a tool which removes the resisting medium. He then applies the acid, which attacks the unprotected surface of the glass, and brings out the design clear upon the coloured ground. This process is chiefly employed for ruby and blue borderings with white ornament, as seen in lobby doors and windows. Plate glass is embossed in a similar manner. The acid eats down the figure below the polished surface of the glass, and the surface that remains raised is obscured by grinding with sand or emery. The figure thus remains clear upon an obscure ground.

Mr Ballantine, the head of the firm which, twenty-five years ago, after a competition of skill from all parts of the kingdom, was selected by the Fine Arts Commission to execute the windows in the new House of Lords, has been the chief pioneer in the promotion and improvement of glass-painting in Scotland; and he and his son now carry on an extensive establishment in Edinburgh, and send their productions to all parts of the world. They employ generally from forty to fifty men and boys, almost all of whom they have themselves trained. There are five or six other glass-painting establishments in Edinburgh and Leith, and two in Glasgow, but the art is not practised elsewhere in Scotland.

MANUFACTURE OF EARTHENWARE.

THE conversion of clay into articles of domestic use is one of the
most ancient of arts, and the potter's wheel is the first mechanical
contrivance mentioned in history. The Babylonians, the Egyptians,
and the Etruscans carried the art to a wonderful degree of perfection. The Etruscans produced many graceful forms; and though
for a long time the dull red of the clay, and a black pigment, were
the only colours employed by the potter, some of the designs on
vases, &c., were brought out with admirable skill. The figures were
subsequently scratched in outline on the surface, and then painted—
those of men in a sort of flesh-colour, and those of women in cream-
colour. From this stage the art was carried to its highest point of
excellence in form by the Greeks. The finest Grecian vessels were
made of a red-coloured clay which was glazed black, the figures of
the design being left clear in the natural colour of the clay. Large
numbers of those ancient vases have been found, and specimens may
be seen in every collection of any pretensions. The Romans appear
to have endeavoured to give strength and durability rather than
graceful forms to their earthenware. They improved the quality
and processes of manufacture, and made the art universal. When
they had established themselves in Britain, they set up many
potteries, traces of which exist in various parts of the country. In
Maitland's "History of Edinburgh" mention is made of the remains
of a Roman pottery having been found at Cramond about the
middle of last century. At Peterborough and on the banks of the
Medway *débris* of Roman earthenware and kilns extends over many

miles. These are not the earliest indications of the existence of the art in this country, however, for in the "barrows" or dwellings of the ancient Britons urns and fragments of earthenware, evidently of native manufacture, have been found. When the Roman Empire fell art went to decay. The Arabs and Moors had, however, acquired some knowledge of pottery, and preserved it from extinction during several centuries; and it was from them that the Italians learned how to make the majolica ware for which they became famous. Towards the middle of the sixteenth century majolica reached its highest point of perfection, and then gradually declined until the art of making it was lost. When the Portuguese merchantmen penetrated into the far Eastern seas, and returned laden with the riches of China and Japan, they brought specimens of porcelain which, by their beauty, attracted much attention in Europe, and purchasers and imitators were not wanting. The potters of Italy, Germany, and France tried to find out how the Chinese porcelain was made. New interest was awakened in the art, and some most encouraging results were attained. In France, Bernard Palissy, in the course of his experiments, which were conducted with extraordinary perseverance, discovered a new kind of porcelain; and subsequently Böttcher, of Magdeburg, succeeded in producing wares similar to those of China. The royal porcelain manufactories of Sèvres and Dresden were established about that time, and a great impetus was given to the manufacture of all kind of earthenware.

In Britain the only earthenware articles made up till the sixteenth and seventeenth centuries—when the potteries of Lambeth, Burslem, and Liverpool devoted some attention to the production of ornamental wares—were of a coarse kind. But it may be said that not until Josiah Wedgwood came upon the scene, about a hundred years ago, did the British potters produce any noteworthy examples of ceramic art. Wedgwood's success arose from no accidental causes. He was a chemist and a mechanician, and he added to his great practical knowledge of his trade indomitable perseverance. The delicacy, beauty, and taste displayed in the works which Wedgwood produced in conjunction with Flaxman, the famous sculptor, attained for them a world-wide celebrity, and they are still eagerly sought after by art collectors. Up till Wedgwood's time the people of this country derived their chief supply of ordinary domestic ware from Holland, and of the superior kinds from Germany and France. That order of things has been reversed, and home-made earthenware and porcelain are now used almost exclusively. while the people of all parts of the world draw a large proportion of their supply from the British potteries. A French traveller thus refers to the widespread use of English ware :—" Its excellent workmanship, its solidity, the advantage which it possesses of sustaining the action of fire, its fine glaze—impenetrable to acids—the beauty and convenience of its form, and the cheapness of its price, have given rise to a commerce so active and so universal that, in travelling from Paris to Peters-burg, from Amsterdam to the furthest part of Sweden, and from Dunkirk to the extremity of the south of France, one is served at every inn upon English ware. Spain, Portugal, and Italy are sup-plied with it; and vessels are loaded with it for the East Indies, the West Indies, and the continent of America."

The value of earthenware and porcelain exported is steadily in-creasing :—In 1834 it amounted, from all the British ports, to L.493,382; in 1864 to L.1,422,014; and in 1867 to L.1,635,216. From Scotch ports about L.117,547 worth was sent out in 1867; but as a considerable quantity is carried into England by land, and thence exported, the above sum does not cover the actual value of the goods of Scotch make which annually find their way into the foreign markets.

The first pottery in Scotland was established at the Broomielaw, Glasgow, in 1748 ; but for a long time only the lowest qualities of goods were made. There are now fourteen potteries in Scotland, which give employment to upwards of 5000 persons. The most extensive is the Glasgow Pottery, belonging to Messrs J. & M. P. Bell & Co. This establishment covers upwards of three acres of ground, and gives employment to 800 persons, whose wages amount to upwards of L.20,000 a-year. The head of the firm has devoted much attention to the higher departments of the art, and has pro-duced some excellent work. Technical difficulties of the most for-midable kind have been overcome ; and in the show-room of the establishment there is a display of ceramic art of which the makers may well be proud. Both in articles of use and of ornament, Messrs Bell have produced elegance of form and beauty of colouring that might be put in competition with the best work of English makers in the same class. In porcelain table, tea, and toilet ware, and in some ornamental branches of their art, such as the making of Etrus-can vases, they have been very successful, and among their chief triumphs are admirably executed copies of the Portland vase.

The Glasgow Pottery is peculiarly interesting, as in it almost every variety of work may be seen going on. The raw materials, which are three in number, are deposited in a series of enclosures, separated by low walls, and occupying a great part of an extensive courtyard. In one enclosure is a pile of light-coloured stone of a coarse crystallised grain—that is " China-stone," a species of soft granite obtained in Cornwall, and composed of quartz and felspar, with little or no mica. The stone, when fused, forms a pearl-white transparent mass. An adjoining compartment contains " China clay," which is -composed of the felspar of the granite in a disintegrated state in combination with a small proportion of other substances. Another compartment contains flints. In order to reduce these materials to working condition, they have to be treated in different ways. The clay is diluted with water until it assumes the consistence of cream, when it is passed through a series of sieves, which remove all except the finest particles. The liquid thus produced is stored in cisterns. Were the clay used alone, it would be impossible to produce a vessel that would stand firing. In order to prevent cracking and distortion, it is necessary to incorporate with the clay some substance having the power to counteract its tendency to shrink. Silex, or ground flint, is the most effective agent, and is prepared by subjecting common flints to the action of fire in a kiln, which makes them purely white, and renders them quite friable. The nodules of flint are broken by hand with a hammer or in a stamping-mill, and the fragments are then placed in a large circular iron tub, where they are ground with a large admixture of water. When the flint is ground sufficiently, it is of the consistence of cream, and, after being washed, is stored in cisterns to wait further operations. The China-stone is prepared by grinding in the same way. The liquids thus got ready are mixed in certain proportions, and produce what is technically termed " slip." In that form it is too fluid to be operated upon, and a large proportion of the water must be got rid of. That used to be done by means of the " slip-kiln," in which the stuff was evaporated until it became like putty; but in the Glasgow Pottery the superfluous water is drawn off by means of Needham & Kite's filtering-machines, which effect a great saving of time and fuel. When the clay comes from the filtering-machines it is passed through a pug-mill, which works it into a homogeneous mass. Before being used the clay is "slapped" or beaten, until on being cut in any direction it exhibits a perfectly smooth and uniformly close appearance. The clay prepared as described is that from which the finer kind of common ware is made. If it be desired to produce porcelain, there must be added a certain proportion of phosphate of lime, which is obtained from burnt bones ground in the same way as the flints.

The clay is brought into shape either by "throwing" or by "pressing." The first of these processes is performed on the potter's wheel, which is simply a vertical lathe, the upper part of which consists of a wooden disc. Each thrower has two female assistants, one of whom divides the clay into suitable pieces, while the other assists generally. Potters' wheels are usually driven by hand, but here they are impelled by steam. Taking up a lump of clay, the thrower dashes it upon the wheel, and by pressing the clay with his fingers as it spins round gives to it the desired shape. Nothing could be simpler than the thrower's operations appear to be, and yet a considerable degree of skill is required in order to produce exact work. As the vessels are formed, they are detached from the wheel by drawing a wire along the surface of the latter. In this state the vessels will not bear much handling, and they are placed in gently heated stoves until they become firm. They are then placed in a turning-lathe, and have their external surface smoothed down, and the whole body reduced to a standard thinness. Shallow vessels— such as plates and saucers—are made by laying a thin piece of clay on a moulded block attached to the wheel, and rubbing it down evenly, first by hand and then by a profile mould. This applies only to moulds of a circular shape. All others are made on suitable moulds, without the use of the wheel. Handles, spouts, and other appendages are attached, when the clay is soft, by means of "slip," with which the parts designed to come together are moistened. Handles of the simpler kind are made by pressing the clay through a die, which gives it the required form. The moulded clay is made in lengths of several feet, which are afterwards cut up into pieces of suitable size, and bent to the desired shape. Spouts, ornamental handles, &c., are made by pressing the clay into moulds.

After the articles are completed, so far as the modelling is concerned, they are set aside until they become sufficiently dry to withstand the firing process. They are then said to be in the "green state." The kilns or ovens in which the wares are baked are of a cylindrical form, and have to be strongly built of the best materials. Before being put into the ovens, the vessels are placed in strong fire-clay boxes called "seggars," the pieces being carefully arranged. The seggars are built in lofty piles, between which space is left for the heat to have free passage. The fire of the oven is gradually raised until the seggars become of a white heat. The furnaceman has to be very watchful to keep up an equal temperature, and, in order to ascertain how the baking is proceeding, he uses certain tests. From forty-eight to fifty hours is the usual time required to bake the ware. When the furnaceman is satisfied that the operation has been satisfactorily completed, he stops firing, and the kiln is allowed to cool gradually. When the vessels are withdrawn from the oven, they are called "biscuit-ware;" and as they are then quite porous and brittle, they would be unfit for use.

Articles that are to be decorated with printed designs are now passed to the printing-room. The designs are engraved on copper plates, from which they are printed upon tissue-paper. While the ink is still wet the printed side of the paper is applied to the dishes, and rubbed firmly until the colour is sent into the pores of the ware. The vessels are then rinsed in water, which removes the paper, but leaves the colouring matter. It is necessary that the oil used in the colour should be got rid of, which is done by subjecting the articles to a gentle heat in a "muffle" or small oven. The chief colour used in pottery-printing up till about twenty years ago was a blue produced from the oxide of cobalt; but now a variety of colours are employed. The printing is followed by the glazing process. For porcelain and the fine class of earthenware, the glaze used is composed of a compound of borax, ground flint, Paris white, and lead. The materials, having been finely pulverised, are mixed with water until the liquid resembles cream. The pieces of ware are dipped singly into the glaze liquid, then they are placed into the seggars on sharply pointed tripods of "biscuit" clay, and subjected to the heat of the kiln until the glazing substances becoming fused cover the surface of the articles with a film of glass, which at once strengthens and beautifies them, and renders them impervious to the action of acids. Sometimes the printed design is enriched by enamel, laid on above the glaze and made permanent by being burned in. The highest department of the pottery is that in which the porcelain goods are painted and gilded.

There is a wide field for the exercise of artistic taste in the decoration of porcelain, and Messrs Bell & Co. have in their employment artists who seem to have a thoroughly appreciative knowledge of their business. Both in form and colouring, some of the dinner, dessert, and tea services are exquisitely beautiful. The enamel colours used by the porcelain-painters consist of metallic oxides incorporated with a fusible flux, such as borax and flint. The enamels are worked in essential oils and turpentine, and in some cases bear no resemblance to the colours they are intended to produce. Owing to this circumstance the painter is a toiler in the dark, so to speak, and is unable to judge of the quality of his work until it has undergone a process which makes rectification of faults impossible. The enamel that produces crimson, for instance, is when applied of a dirty violet or drab hue. During the firing it varies from a brown to a dull reddish hue, and from that progressively to its proper tint; consequently a good deal depends upon the fireman, for should he fail to raise the proper degree of heat, or, on the other hand, exceed it, he ruins the intention of the painter. By over-heating, what was intended for crimson comes out a dull purple. Then there is great risk of the article being broken by the too sudden raising or lowering of the fire in the kilns. Vicissitudes such as these make enamel painting in its higher branches an occupation requiring great patience and perseverance on the part of those engaged in it. Gold is now extensively used for decorating purposes. It is prepared by being mixed with quicksilver and flux, the result being a fine black dust, which is mixed with turpentine and oils like the enamel colours. Firing restores the gold to its proper tint, and fixes it; but when the articles come from the kiln the gilded parts are dull, and require to be well rubbed with a bloodstone burnisher before their full effect becomes apparent.

In the artistic department of the Glasgow Pottery, Parian statuary and copies of ancient vases are the chief objects produced. The most successful attempts to multiply in a cheap form the productions of high art have been made with Parian, which is simply a variety of porcelain. It is cast in moulds formed of plaster of Paris, and, for facility of manipulation, each figure or group is cast in a number of pieces, which are united while the clay is in a soft state. When the figures are put together and touched up so as to remove all traces of joinings, they are set aside until they become sufficiently dry to stand the heat of the kilns. A troublesome quality of the clay is the great extent to which it shrinks. A figure, which measures two feet in height when the clay is first poured into the mould, will measure only eighteen inches after being baked. Great experience and skill are required to produce works of this kind successfully; indeed, the difficulties which surround the manufacture prevent Parian of an artistic form from being sold at a low price.

In the pottery trade both the masters and the workmen have unions for the protection of their respective interests; but no difficulty worth recording has ever occurred in Scotland, all questions relating to work and wages being arranged by conference between the employers and employed. During the past twenty years the wages of the workpeople have, in most departments, been greatly increased, while their condition has been much improved. Piece-work is the almost universal rule of the trade, and the men employed in the higher branches make from 24s. to 30s. a-week and upwards. The wages of the women are about the average of those received by factory hands.   Three or four years ago the trade was put under the Factory laws ; and as the reduction of juvenile labour increased the cost of production, the manufacturers were placed at a disadvantage for a time ; but now machinery is being introduced, which compensates for the change.

The art of kneading common clay into rectangular blocks for building purposes seems to have been known from about the time of the Flood, though it does not appear to have been practised by the Western nations until a comparatively recent period.   Pipes of clay were used by the Romans to carry off the sewage of their cities ; and vases, lamps, statues, and architectural ornaments were formed of the same material.   Like other arts which flourished among the ancients, working in clay became extinct for a time ; but its value has been long fully appreciated, and the conversion of clay into bricks, tiles, pipes, and more artistic objects, constitutes an important branch of industry in most countries where the material exists. In England the scarcity of good building stone is compensated for by the existence of vast beds of clay, from which many millions of bricks are made annually.   Scotland is rich in building stone of the best qualities ; but, nevertheless, many bricks are made and used, and we have also an extensive manufacture of articles of fire-clay and terra cotta.   In Britain bricks did not come into use until the fifteenth century, and what are supposed to have been the first buildings of importance in which bricks were employed are still in existence.   These are the Lollards' Tower of Lambeth Palace, built in 1454 ; and a portion of Hampton Court, built in 1514.

In 1784 an Excise duty of 2s. 6d. a-thousand was imposed upon bricks of all kinds.   A subsequent Act of the same reign raised the duty and varied its amount according to certain specified varieties of bricks.   The duty on common bricks was in 1835 raised from 5s. to 5s. 10d. a-thousand.   Four years afterwards the distinction of size and quality in charging the duty was done away with, and a uniform rate of 5s. 10d. was levied.   The tax was at all times regarded as obnoxious and as an obstruction to the improvement of the dwellings of the poorer classes ; but notwithstanding repeated representations from the building trades, and the almost unanimous voice of the press against the duty, it was not abolished until 1850.   Tiles were also subject to duty from 1784 till 1833. The number of bricks made in Britain in the year 1802 was 714 millions ; in 1840, it was 1725 millions ; and in 1850, the year in which the duty was abolished, it was 1563½ millions. The number of bricks made in Scotland annually was 15¼ millions in 1802 ; and 47¾ millions in 1840. If the great increase in railway and other works, the rapid enlargement of towns, and other recent causes leading to a more extensive use of bricks be considered, the number now made in Scotland cannot be less than 200 millions a-year.

There are in Scotland 122 manufactories of brick, tiles, and articles of a similar nature ; and in connection with these from 4000 to 5000 persons are employed. The manufactories are widely scattered over the country, the farthest north being at Banff and the farthest south at Dalbeattie ; but the greater number are in Lanarkshire and Fifeshire, in which counties valuable beds of fireclay exist. The most extensive is that of the Garnkirk Fire-Clay Company, situated on the Caledonian Railway line about six miles east from Glasgow. The company was originally formed to work coal, but, finding that extensive seams of fire-clay existed on their property, they took to manufacturing that material, which now almost exclusively engages their attention. The principal seam of clay is seven feet in thickness, and lies at an average depth of twenty-eight fathoms. Its quality is considered equal to that of the best Stourbridge clay. The manufactory covers upwards of six acres of ground. Raw material is brought in, and finished goods are sent out by branch railways. 300 men and boys are employed, and 200 tons of clay and about an equal weight of coal are used daily. The clay is of a dark colour, owing to the presence of a small proportion of bituminous matter ; but when that is dispelled by the action of fire, only silica and alumina remain, and it is the presence of these substances in certain proportions that decides the value of the clay. As it comes from the pits the clay is entirely devoid of cohesion or plasticity ; and in order to bring it into working condition, it has to be ground very fine, and then mixed with water. Several powerful mills are used for this purpose. They consist of great iron rollers, which travel round a circular trough, and pass over the clay.

Bricks are the commonest and simplest articles made. Some ingenious machines have been devised with a view to superseding hand labour in this branch of manufacture ; but as yet hand labour has the advantage of greater economy. Indeed, the item of moulding, to which only the machines could be applied, forms a small part of the labour in brick-making. At Garnkirk all the bricks are hand-moulded, which is a very simple process, and is executed with wonderful rapidity. An expert moulder, with the necessary assistants to keep him supplied with clay, and to remove the moulds as they are filled, will make from 4000 to 5000 bricks a-day. The moulder works at a table, on one end of which is a supply of clay, the other being left clear for his operations. The bricks are formed in a deal framework, resembling a small box with the top and bottom removed. A boy dips the mould in water and lays it on the table. The moulder, taking up a lump of clay, dashes it into the mould, presses it with his hands, and then removes the superfluous clay by drawing a piece of wood over the mould. His assistant, who has meantime laid down an empty mould, snatches up the full one and deposits the newly formed brick on the floor of the workshop. Thus the work goes on until the floor is covered. An important matter in the manufacture is to take care that at least 25 per cent. of the water contained in the clay is evaporated before the bricks are subjected to burning. In some places, and in the case of common bricks, it is usual to expose them in the open air before firing ; but that is a precarious practice in a climate like ours, and the best plan is to dry them under cover by artificial heat. The Garnkirk brick-sheds, and the drying-rooms in the other departments, are fitted with pipes through which the waste steam of the engines is made to pass, and by the heat which these give off the bricks are brought into firing condition in the course of twenty-four hours. The bricks are fired or burned in kilns, but another mode of firing is sometimes employed in which the bricks are built in " clamps," or large square heaps with layers of fuel between. Kiln baking is the best. The kilns are built in ranges of three or four together, the smoke from all of which is drawn off by one chimney. Internally, the kilns are about 12 feet in length, breadth, and height, and the bricks are arranged in them so as to allow the fire to act freely on all. About 20,000 bricks are placed in each kiln, and the baking occupies six days and nights. Flooring tiles are made after much the same fashion.

The improvement of agriculture, and the consequent increase of draining, has within the past twenty or thirty years led to a great and increasing demand for clay drain-pipes, and many millions of these are produced in Scotland every year for both home use and exportation. They are made of common red clay—a much softer and less durable substance than fire-clay. The pipes are formed by ingeniously constructed machines, which turn them out at a rapid rate.

The Garnkirk Company do not work in common clay, and make no agricultural drain-pipes ; but they have an extensive trade in the manufacture of glazed fire-clay sewage and water-pipes. As already stated, clay pipes were used by the Romans to carry off the sewage of their towns and villages. The city of Rome had a complete system of sewage. There were main sewers built with bricks, and branch sewers consisting of pipes of wood or clay. With the decline of the Roman Empire draining as well as many other good things went out of use, and modern minds were only awakened to the importance of the matter when thousands of persons were carried off by diseases which could be traced to no other origin than defective drainage. The importance of providing means to carry away filth from centres of population is now generally known and understood, though in some cases action is tardy. For main sewers nothing better than brick has been devised, and for branches, nothing better than clay pipes—so that, in the all-important matter of town drainage, we are no further ahead than were the people who occupied the foremost rank of civilisation two thousand years ago ; and it is not long since equality could be claimed. The making of sewage-pipes is an important branch of the manufactures in clay. The pipes are formed by pressing the clay through a die. They are made in lengths of three feet, and each piece has a " collar " worked on one end. After being dried in the stoves the pipes are baked in large circular furnaces. In the course of the baking a quantity of salt is thrown on the pipes, and that combining with the silica of the clay forms a glaze which covers the entire surface. The pipes are made from two to thirty-six inches in diameter. The heaviest articles made are gas-retorts and blocks for the furnaces of glass-houses. Some of the latter weigh fifteen cwt.

Works in terra cotta are also among the productions of the Garnkirk Company. Terra cotta is an Italian term signifying baked clay, but it is commonly employed to designate such articles formed of clay as are used in architectural embellishment. It is, if properly made, one of the most durable materials that can be used in building. It was so employed by the Egyptians, Greeks, and Romans, and by various European nations in the middle ages. Monumental vases in terra cotta have been recovered in a state of perfect preservation from tombs in which they had been placed upwards of two thousand years before, and examples are not wanting to prove the weather-resisting powers of the material. Sutton House, in Surrey, built about the year 1530, is covered with ornaments in terra cotta, which yet retain the marks of the artist's modelling tools.

2 c

Many buildings erected in Italy between the twelfth and seventeenth centuries bear terra cotta decorations in a perfect state. The lodge in Merrion Square, Dublin, was built in 1786 of granite taken from the Wicklow mountains, and ornaments in terra cotta were provided for it by an English manufacturer. It is a remarkable fact that, while the granite mouldings have yielded to the action of the weather, the terra cottas are as complete as when put up. Among other honours which belong to the name of Josiah Wedgwood is that of having revived the manufacture of terra cotta in England. When he founded his great pottery in Staffordshire, he began to make articles in imitation of the ancient works in terra cotta, and in that branch he was soon followed by a lady named Coade. The chief materials employed by them were the Dorset and Devonshire clays, with fine sand, flint, and potsherds. Most of the coats of arms and other insignia placed over the shops in London were made of this material. Though they could not deny its advantages of durability and cheapness, builders did not regard terra cotta with a favourable eye, and it made little progress until within the past ten or twelve years. Its employment in the South Kensington Museum buildings, and in the Royal Horticultural Society's Gardens, gave the public an opportunity of judging of its suitability for decorating modern edifices, and the general opinion has been favourable to its use. A large quantity of terra cotta has been employed in the construction of the Albert Hall of Science and Art, and in many other important buildings throughout England. In Scotland our beautiful and easily carved freestone does away with the necessity for introducing terra cotta in an architectural fashion except in the form of chimney-pots, for which it is well suited; but statues, vases, and fountains made of it are now much used for the ornamentation of pleasure-grounds and gardens. A recent discussion in the Royal Institute of British Architects shows that considerable difference of opinion prevails as to what are the best materials for making terra cotta. Some eminent men in the architectural world maintain that, in order to endure the severe climate of Britain, terra cotta should be made with a hard vitreous body composed of Cornish clay, ground flint, and Cornish stone, with a glaze added. Others are of opinion that the composition of the Albert Hall and South Kensington terra cottas are the best, and the weight of argument appears to lie on their side. Messrs Alexander Wilson & Son, fire-clay manufacturers, Dunfermline, have made a great part of the terra cotta required at South Kensington, and are now engaged with the columns, capitals, cornices, friezes, and other ornamental parts.   They have been providing all the ashlar work required.   The clay used is of a very fine quality.

The articles made of terra cotta at Garnkirk are chiefly statues, fountains, vases, brackets, pedestals, and chimney-pots.   The clay for these is carefully ground.   In the firing, and subsequently, a number of articles are broken, and the remains of these are carefully preserved, and, when ground, a certain proportion of the produce is added to the fresh clay.   The object of this is to reduce the "shrinkage," or tendency to contract, which the pure clay possesses. The articles are formed either by modelling or casting in moulds of plaster of Paris.   Most of the statues and vases are after classical patterns.

Only a small proportion of skilled workmen are required in brick and tile works, and the great body of the men rank as ordinary labourers.   They are chiefly Irishmen, and their earnings may be stated at from 15s. to 17s. a-week.   Some of the skilled workmen earn as high as 30s. a-week.   The fire-clay is excavated by men who have been bred as coal miners.

GRANITE, FREESTONE, PAVEMENT, AND SLATE QUARRYING.

THE quarries of Scotland contribute largely to the wealth of the
country, and afford employment to a great number of persons. Centuries ago, when the danger and inconvenience of wooden and
thatched houses came to be generally felt, more substantial materials
began to be used in the construction of dwellings. At first loose
stones were collected by the builders, and when the supply of these
became exhausted the unbroken beds of rock were drawn upon.
Quarrying then became a trade, and has since gone on extending and
increasing in importance. In most quarters of Scotland sandstones
and granites well adapted for building purposes exist, and the result
is that no country in the world can show more substantial houses
than those occupied by the great body of the people north of the
Tweed. The produce of the granite quarries of Scotland is famous
both in England and abroad; no finer paving-stones exist than those
obtained in the counties of Caithness and Forfar; and the slates of
Easdale and Ballachulish are equal to any.

Aberdeenshire is the chief seat of the granite trade, and has been
associated with it for about 300 years. When granite was adopted
to supplement the other building material available in that county,
a supply was drawn from the loose blocks which lay about in great
abundance. Granite quarrying, as now understood, was not begun
till about the middle of last century; and not until a number of
years afterwards were the modes of working the material such as to

promise for it a widely extended use. As time wore on, the quarrying and working of granite came to be better understood. In the year 1764 the export trade, now of so much importance, was begun. About that time it was resolved to have the streets of London paved, and the durability of Aberdeen granite having been tested, that material was fixed upon as the most suitable. From the preference given to it in that instance has arisen an export trade not confined to the metropolis or to the supply of paving material, but dealing with every part of the kingdom, and supplying every kind of stone-work. A few years passed, the trial made of the durability of Aberdeenshire granite proved highly satisfactory, and the consequence was a demand which led to the opening up of quarries in the neighbourhood of the city of Aberdeen where the supply of granite is abundant.

It appears, from a reference in "Kennedy's Annals," that the contractors for paving the streets of London began quarrying operations among the rocks on the sea-coast of the lands of Torrie, and transported the stones, when roughly dressed, to London. They soon found, however, that such was a very expensive mode of furnishing the article wanted, and, relinquishing their original plans, they entered into contracts with the masons of Aberdeen for a supply of stones suitable to their purpose at certain specified rates. This system has continued up till the present day; and although one or two London firms have leased quarries and work them, they do so for local as well as for export trade. Machinery was first used in quarrying granite about the year 1795, when stones of large dimensions were ordered by the Admiralty for the docks at Portsmouth, at that time in course of construction. The expense of procuring large blocks of stone, such as those required for the Bell Rock Lighthouse and for the Waterloo and London Bridges, must have been rather heavy; but the name that Aberdeen granite gained by being used for such works had the effect of increasing the demand. And this was enhanced locally by the building of Union Bridge and the opening up of Union Street about the beginning of the present century. From that time the trade may be said to have been fairly established, and year by year it increased in importance, the demand for London alone being productive of great good, not only to the town and county, but to the shipping of the port. During the year 1817 the quantity of stones exported to London was 22,167 tons, the value of which, including freight, was about L.23,275. In 1820 the late Sir John Rennie recommended to the Admiralty the use of stones from Aberdeen for docks being constructed at Sheerness. In

1821 the exports, principally of building and carriage-way paving-stones, amounted to 34,687 tons, and in 1831 to 36,252 tons. During the interval the greatest increase in the export trade was in kerb and carriage-way stones, which in the ten years advanced by some 12,000 tons.

It is calculated, on pretty accurate data, that the quantity of granite quarried annually in Aberdeenshire is upwards of 80,000 tons, and of that nearly 40,000 tons are exported. The following statistics will show the number of tons of granite exported from Aberdeen from 1840 down to the beginning of 1869 :—

1840, ...	25,557	tons.
1850, ...	30,385	„
1860, ...	24,666	„
1865, ...	32,023	„
1867, ...	43,790	„
1868, about	50,000	„

All the quarriers in Aberdeenshire produce kerb and causeway stones, and some devote their attention almost entirely to that branch. Within recent years various improvements have been made in preparing kerb, causeway, and tramway stones for the London and other markets. As in the case of every other art when in its infancy, for several years little progress was made either in keeping the quarries in good working order or in improving the instruments of labour. In fact, it is only recently that machinery and steam-power were introduced to work granite quarries. From the great strength of the stone, it has to be blown from its bed by gunpowder; otherwise, the processes followed are the same as in quarrying freestone; but on account of the extreme hardness of the granite, the cost is five or six times greater. Of late years the bores in the stone have been made considerably larger in diameter, and now the common depth may be said to range from four to twenty feet. After the blocks are thrown out of the " face" of the quarry by blasting, they are cut with wedges into the various sizes required. The blocks are then handed over to masons, who work them into the different " orders." Keeping out of view the large number of skilled artisans working in granite in the city of Aberdeen, the quarries of the county give employment to upwards of 1000 hands.

The quarries at Kemnay, leased by Mr John Fyfe, are the largest in the north, more material being sent out from them than from all others in the county. Situated in close proximity to the Kemnay station, on the Alford Valley Railway, every facility is present to develope the work. On an average Mr Fyfe gives work to 250 men all the year round; and these, with the aid of steam-power, which he was the first to introduce in the quarrying of granite, turn out several thousand tons of stone monthly, which goes partly to the home and partly to the foreign markets. Mr Fyfe, it may be remarked, has the contract for supplying the principal stones for the Thames embankment; and besides that, he does an extensive business in kerb and paving stones for London and elsewhere. The Kemnay granite is of a light colour, and from its close texture has become very popular. Since 1858, when the quarries were first opened, Kemnay, formerly a country hamlet, has become a place of some importance, a number of houses for the accommodation of the workmen and their families having been erected by Mr Fyfe. Rubislaw quarry, leased by Mr Gibb for several years, produced a large quantity of granite of a fine dark blue colour, of which a great part of Union Street, Aberdeen, is built. The other quarries of note in the county are those of Sclattie, Dancing Cairn, Persley, &c. From all these, as well as from Cairngall, leased by Messrs Alexander Macdonald, Field, & Co., a good substantial stone, known as Aberdeen granite, is got. The Cairngall quarries, in particular, produce a fine small grained stone, admirably adapted for polishing and for ornamental work. In fact, for those purposes, no better material has as yet been found. It was from Cairngall that the sarcophagus for the remains of his late Royal Highness the Prince Consort was taken. All over the kingdom monumental work in granite from this and the other Scotch granite quarries may be met with. The red or Peterhead granite is found at Stirlinghill, near Peterhead, about thirty-two miles from Aberdeen. It is from that district that the principal supply for the Aberdeen stone-yards is obtained. The red granite is commonly used for mural tablets, gravestones, &c.

An important branch of the granite trade is the conversion of the stone by carving and polishing into architectural ornaments and monumental work. It had its origin in Aberdeen ; and the largest granite polishing establishment in the country is that of Messrs Alexander Macdonald, Field, & Co. of that city. The first experiments in granite polishing were made by the late Mr Alexander Macdonald, father of Mr Macdonald of the above-named firm, about the year 1818. Previous to the year 1819, the only tools used in dressing granite were small picks, and it was when the County Rooms, Union Street, Aberdeen, were being built in that year that chisels and "puncheons" were first tried by some free-stone masons, who, as work was scarce in their department, undertook to work in granite with the tools used in dressing sandstone. In preparing granite blocks for polishing, such tools, tempered to suit, were found to be extremely serviceable, and they were generally adopted. About the year 1830 Mr Macdonald removed his works from King Street to the extensive premises now occupied by his successors in Constitution Street. There, in 1839, he was joined by Mr Leslie; and under the firm of Macdonald and Leslie, the granite polishing trade of Aberdeen became extensively known. Up till that time, and, in fact, till within the last seventeen years, this was the only manufactory of the kind; and now, although there are in Aberdeen several similar establishments, none come up to it in extent or importance. One of the chief improvements introduced by the late Mr Macdonald in preparing granite blocks for polishing was the use of patent axes. The rough blocks sent from the quarry require to be finely dressed before they are fit for polishing. After being dressed to a somewhat even surface, axes are used; and until the introduction of the six and eight bladed axe by Mr Macdonald, the work had to be done with light picks and single-bladed axes. The "compound axes," as they are called, are steel chisels clamped together so as to present a face to the stone of six or eight edges on a space of two inches. Mr Leslie continued in partnership with Mr Macdonald till 1852, after which the business was carried on by the last-named gentleman until his death. Since then his son, the present senior partner of the firm of Alexander Macdonald, Field, & Company, and the other partners, have still further extended the name and reputation of the original Aberdeen granite works.

At the polishing establishment under notice the granite is brought into the yard in blocks of various sizes, according to the nature of the work to which they are to be devoted. By means of travelling cranes the blocks are moved from place to place, or hoisted on to trucks running on rails which traverse every department of the works. In the preparation of slabs of granite for panels or mural monuments, the first operation is to saw up a block into pieces of the required thickness. This is done by powerful sawing-machines propelled by steam. The cutting part of the machines consists of a number of parallel blades of sheet iron fixed in a strong frame—the space between the blades determining the thickness of the slabs. The blades are drawn backward and forward on the stone, being meanwhile supplied with sand and water, which help the saws to grind down the granite. The process is exceedingly slow, some months being required to cut up the larger blocks from which slabs are made. The granite is polished by machinery, except in the case of intricate carvings, which have to be operated upon by hand. A mixture of sand and water is applied in the first instance, and after the surface has been rubbed smooth by means of these, emery is used. Under this treatment the stone assumes a glassy surface, the brilliancy of which is enhanced by a final polishing with oxide of tin.

Granite exists in almost exhaustless abundance in the stewartry of Kirkcudbright. The three principal districts where it is found—namely, that in the north-west, extending from Loch Dee to Loch Doon; the second, stretching from Loch Ken to Palnure Water; and the third, from Criffel to Craignair—occupy nearly one-fourth of the surface of the stewartry. Though the existence of the granite was well known many years ago, the proper mode of working it was not understood; and no idea being entertained of the important uses for which it could be made available, no attention was until recent years paid to quarrying or working this valuable rock. When a gentleman's residence, or other decorated building, was being erected in the locality, the corner and other principal stones in the structure were formed of the Old Red Sandstone from Dumfriesshire. And when the bridge over the Dee at Tongland was built—in the years 1804-5-6, after a plan by Mr Telford, the celebrated engineer, at a cost of nearly L.7000—it appears never to have entered the mind of any one concerned to utilise the granite boulders which lay scattered at convenient distances from the site of the bridge. On the contrary, the stones employed in the erection of the structure were transported at great expense from Annan and the Isle of Arran; and they were subject to a heavy duty, which greatly augmented the aggregate outlay.

The late Mr Andrew Newall, Dalbeattie, had for some years previously carried on in a small way the trade of granite hewing; but it was not till 1825 that anything like extensive operations in the working of granite were commenced in the district. In that year the Liverpool Dock Trustees leased from the late proprietor, of Munches—Mr John H. Maxwell—a portion of Craignair Hill, in the immediate vicinity of Dalbeattie, with the view of opening a quarry on it to obtain blocks suitable for dock purposes; and having entrusted the supervision of the works to a confidential and skilled manager, that gentleman began operations with so much energy and satisfactory success, that within a few weeks he shipped to Liverpool the first instalment of stones. With these the engineer of the docks—the late Mr Hartley—was so well pleased, that orders were immediately received to open another quarry on the hill, and engage as many more men as could be obtained. These instructions were as speedily carried out as the efficient conduct of the already extensive existing operations at the hill would allow. At those two quarries some hundreds of workmen and labourers were daily employed, and some thousands of tons of the wrought material were annually sent to Liverpool, and applied to dock purposes only. That was the first impetus given to the granite trade in Galloway. Its beautiful product, as inspected on the banks of the Mersey, soon made Craignair famous; and there the works continued to be prosecuted with unflagging energy and satisfactory success till 1832, when the manager having failed in his efforts to obtain blocks of the very large dimensions required at the docks, except by incurring great expense in the removal of the superincumbent mass of *débris,* opened a quarry on the farm of Kirkmabreck, in the parish of that name, in the south-west of the stewartry. Two of the Craignair quarries are worked by Messrs D. H. & J. Newall, who turn out large quantities of dressed granite. In addition to these quarries there are two others in the hill. One is worked by Mr Charles Newall, who employs a considerable number of hands, and is one of the contractors for supplying stones for the Thames embankment.

About four years ago Mr Hugh Shearer of London was successful in obtaining the first contract for supplying granite for the Thames embankment, and opened several new quarries on the Munches estate, reopening at the same time four quarries which had formerly been worked by the Liverpool Dock Trustees. Such is the beauty and durability of the stone that it is said the engineers of the Board of Works prefer it to any other; and Mr Shearer and the partners he assumed about a year ago—the designation of the firm now being Messrs Shearer, Smith, & Co.—have supplied, and are still supplying, many thousands of tons of material for the great work referred to, and have besides sent large quantities of paving stones into Russia, several Mediterranean ports, and South America. They have also furnished a great part of the paving stones recently used in Edinburgh, Glasgow, Leith, and some of the principal towns in England; and are at present applying their forces to the conversion into paving-stones of the hundreds of thousands of tons of material accumulated in their "spoil heaps," which has hitherto been treated as waste. It is largely owing to the enterprise of Mr Shearer in securing the use of the stone in the Thames embankment, that the

Kirkcudbrightshire quarries owe their present celebrity. Messrs Shearer, Smith, & Co. employ from 300 to 600 workmen, according to the demands upon them. Their works have a direct connection with the railway system of the country, and they have a loading pier on the river Urr, so that they have ample facilities for transporting their produce. The stones for the Thames embankment are forwarded by rail, dressed and ready for the builders.

In the parish of Kirkmabreck very extensive operations in granite working have been carried on for a long series of years. Adjoining the public road leading to Gatehouse, about two miles from Creetown, is an extensive quarry, leased by the Liverpool Dock Trustees. All the stones procured there are used for dock purposes only. There are four vessels constantly employed by the Liverpool Company in transporting the stones to their docks, and other vessels are often engaged in the same service. On one occasion, when the quay at this place was transferred from one proprietor's land to that of another, there were nineteen vessels constantly occupied for six months in the removal to the banks of the Mersey of the material that had accumulated at the abandoned wharf. About two miles east from these works, on the farm of Bagbie, on the estate of Kirkdale, belonging to Major Rainford Hannay, a quarry was opened in February 1864 by Messrs Forrest, Wise, & Templeton. The stone of this quarry is of first-class quality, and, like that at Kirkmabreck, can be worked without blasting. In connection with their works the firm have a commodious quay erected for the shipment of such portions of their prepared materials as are transported by sea, and between it and the quarry they have constructed a tramway. There is still another quarry in this neighbourhood deserving notice. It is situated near the apex of the hill at whose base are the Kirkmabreck quarries, and almost in a straight line from these works. It is within the farm of Fell, and hence called Fell Quarry. It is worked by the Scottish Granite Company. Messrs Newall have an extensive polishing establishment. The process of polishing is of recent introduction into Galloway, the first experiments in it having been made by the late Mr Andrew Newall a few years ago. By Messrs D. H. & J. Newall the art has been brought to a state of great perfection, and is extensively carried on by them with a constantly increasing business. From their quarries on Craignair the firm are supplied with part of their material for polished work. But large quantities of red granite are obtained from the island of Mull.

The Scotch freestone quarries differ from the granite quarries in that, while the produce of the latter is exported in large quantities, that of the former is used almost entirely at home. The Old Town of Edinburgh was built of red or grey sandstone, obtained in the neighbourhood without much labour in quarrying. The newer portions of the city have been built of stone drawn chiefly from the extensive quarries of Craigleith, Ravelstone, Redhall, Humbie, Binnie, and Hailes, which are situated on two extensive tracts of sandstone lying to the west of the city. Craigleith Quarry has been worked out to a great depth. The stone is hard and difficult to excavate and dress. Before railways afforded ready means for conveying stone from a distance, Craigleith was a valuable possession, especially about forty or fifty years ago, when the "liver" rock—as the great unstratified beds in quarries are called—was abundant; but now little stone is being drawn from it, the demand being met more cheaply by softer stone obtained from various quarters. It is still preferred for steps and plats in staircases. Among buildings composed of Craigleith stone may be mentioned the University and the houses in Charlotte Square and Randolph Cliff, Edinburgh. The same quarry supplied some of the material used in the construction of Waterloo Bridge, London, and large numbers of plats and steps for houses in London and elsewhere in England. The Ravelstone Quarry yields stone similar in quality to that of Craigleith. At Redhall the rock which is being worked belongs to the lowest sandstones of the coal formation. The stone is of fine quality, and very durable—its durability depending chiefly on the large proportion of silica that it contains. So abundant is the silica that it is frequently found concentrated in nodules. These nodules give great trouble to the quarrymen, who call them "white whin." The nodules are among the curiosities of geology, and the minds of geologists have been puzzled to account for their appearance, because they are evidently of the same origin as the softer stone in which they are embedded, and contain the same fossils. As specimens of Redhall stone in Edinburgh, St John's and St Paul's Episcopal Churches, and the houses of Heriot Row, may be mentioned. It was in Redhall Quarry that an improvement on the primitive method of breaking out the stone by means of wedges and small charges of gunpowder was made. The quarry belongs to Mr James Gowans, of Rockville, who is the most extensive lessee of quarries in Scotland. About eighteen years ago Mr Gowans introduced the method of boring, at a suitable distance from the "face" of the quarry, a row of holes of great depth, and charging these with gunpowder, which was ignited by a galvanic battery. The holes, which are from three to four inches in diameter, are bored by a machine. Immense quantities of stone are in this way obtained by a single operation, and at less cost than by the old mode. Humbie Quarry, near Kirkliston, was worked for many years, and supplied some good building material; but it is now abandoned, on account of the cost of raising the stone being unfavourable to its introduction into the market in competition with the produce of quarries more advantageously situated. Since this quarry was closed another has been opened further north on the Hopetoun property, which is called New Humbie. The stone from it has been used in street buildings, especially in the Newington district. The stone from Kingscavil Quarry, near Linlithgow, has been long in use for foundations and inside work. The Binnie Quarry, near Uphall, produces stone of equal durability with those already mentioned, but differing in certain respects from all. Binnie stone was extensively used in the buildings erected in Edinburgh and the neighbourhood about the year 1835. The Scott Monument, Donaldson's Hospital, the National Gallery, the Commercial and British Linen Company's Banks, Edinburgh, are built with this stone, and the latest great work in it is the Bank of Scotland. Binnie stone, on account of the bitumen which it contains, has generally a freckled appearance; and wherever this is most distinctly marked, the stone is most durable. In some specimens the bitumen is so abundant that, if a small piece of the stone be placed in the fire, the bitumen will be seen to bubble up. The quarry is situated in a locality abounding in shale, and the bitumen is supposed to have flowed in from the shale beds, and the stone become impregnated with it. Fissures in the quarry are filled with bitumen, some of which was collected by Mr Gowans in 1850, and made into candles, which were shown at the Exhibition of 1851. He was thus on the verge of demonstrating the value of the shale beds of Scotland, which have since become so valuable and important; but the prize fell into other hands. In the buildings of Binnie stone which stand in a comparatively pure atmosphere—such as the National Gallery and Donaldson's Hospital—the stone is of a brown tinge, and has reddish streaks through it. These are caused by the presence of iron. In the case of the Commercial Bank and the British Linen Company's Bank, the stone has been smoked until it presents a uniformly dark surface on which the streaking is not apparent, though the stone appears to be equally durable. The Hailes Quarry produces stone which, though not adapted for the external parts of buildings, is useful from its laminated structure for making plats and steps. The foundations and inner walls of houses are also built of it. For such purposes it is of great value, as it can be worked at a comparatively small cost.

A number of extensive quarries have been opened in Stirlingshire since the advent of railways. These have competed with the local quarries in supplying stone to both Edinburgh and Glasgow. The most important are those of Dunmore, Polmaise, and Plean, in the neighbourhood of Bannockburn, where the coal measure sandstone terminates. The stone obtained from these quarries is durable, but it is not of such a fine quality as that supplied by the quarries in the neighbourhood of Edinburgh. A number of the houses recently built on the south side of Edinburgh have been constructed of stone from Stirlingshire.

In 1867 Mr Gowans sent from the Plean Quarry a large quantity of stone to be used in the erection of a new warehouse in Paternoster Row, London, for Messrs Nelson & Sons, publishers. It is expected that this specimen of work in Scotch stone will lead to further orders; for both its beauty and the ease with which it can be worked will be sure to commend it to the favourable notice of architects and builders. It would be interesting to note the effect of the London atmosphere on the stone of Messrs Nelsons' warehouse, which is composed entirely of silica, as compared with the limestone generally used in London. Had the new Houses of Parliament been built of Scotch stone, there is every reason to believe that the nation would have been spared the regret caused by the premature decay of that costly edifice. The following analyses will convey an idea of the difference in composition of some of the principal Scotch and English building stones :—

CRAIGLEITH STONE.

Silica,	98·3
Carbonate of Lime,	1·1
Iron and alumina,	0·6
	100·00

BOLSOVER STONE USED IN HOUSES OF PARLIAMENT.

Silica,	3·60
Carbonate of lime,	51·10
Carbonate of Magnesia,	40·2
Iron and alumina,	1·8
Water, &c.,	3·3
	100·00

PORTLAND STONE USED IN ST PAUL'S CATHEDRAL.

Silica, ...	1·20
Carbonate of lime, ...	95·16
Carbonate of magnesia, ...	1·20
Iron and alumina, ...	0·50
Water, &c., ...	1·94
	100·00

In Fifeshire there are a number of quarries, the most important being the Grange and Callalo Quarries, near Burntisland, from which the stone employed in building Fettes College, Edinburgh, is being obtained. The Unitarian Chapel in Castle Terrace and some streets on the north side of Edinburgh have been built of the same material, which was also used in the restoration of St Giles' Cathedral. On exposure to the atmosphere the stone assumes a dark colour, and appears to require careful selection on the part of the builders. A variety of freestone obtained from the coal measures in the Wishaw district has recently been introduced by the Edinburgh builders, and is being used in the erection of villas and streets at the south side. Most of the new houses bordering on the West Meadow are composed of it. Cockburn Street is built of a stone also brought from the west, the quarries from which it came being at Bishopbriggs, on the Edinburgh and Glasgow Railway. The houses of Glasgow have been chiefly built of stone obtained in the immediate neighbourhood of the city. Good building stone is abundant all through the west of Scotland from the Clyde to Dumfries, and patches exist at various places in the east and south-east. To the north of the Ochil range the rock for a long distance belongs to the old red sandstone. The Scottish Central Railway passes over that formation during its entire route from Stirling to Perth, and the North-Eastern runs on it from Perth to where the line enters the Grampian range. The stone all through the district referred to is suitable for building purposes. It is durable if used with the strata lying horizontally; but if laid with the strata in a perpendicular position it wears away rapidly. Cases of decay arising from ignorance or neglect of this peculiarity may be seen in the houses of Perth.

About 2000 men are employed in the freestone quarries. They receive from 20s. to 25s. a-week, which is an increase of 25 per cent. on the wages paid twenty years ago. Steam-power is extensively used in the quarries to work cranes, pumps, and inclined planes. A large number of horses are also employed.

Several attempts have been made to dress or prepare the stone by machinery, but one after the other these have failed, except in the dressing of the old red sandstone of Forfarshire, where the planing process has been very successful. The nearest approach to success in local freestone has been made by Messrs Watherstone, builders, Edinburgh. The stones are fixed to movable parts of the machine and their surfaces are dressed by rubbing one against the other.

There are two counties in Scotland which produce paving-stones —Forfar and Caithness. The pavement quarries of Forfarshire are the oldest and most extensive in the kingdom. Of late years the trade has been largely developed by the introduction of planing, cutting, and dressing machines. The planing-machine now in use in almost all pavement quarries was invented by the late Mr James Hunter, of Leysmill, near Arbroath, who was for many years manager of the quarries belonging to the late Mr Lindsay Carnegie, of Spynie and Boysack, and now to his son and successor in the estate. Mr Carnegie, a public-spirited and kindly gentleman, took great interest in the quarries, as well as in the workmen employed about them ; and it was chiefly owing to the encouragement received from him that Mr Hunter was able to complete his useful invention. But, important as the planing-machine has proved to be, it has been eclipsed by the stone-cutting and dressing-machines invented by Mr Hunter's son, Mr George Hunter, now of the Welsh quarries, in conjunction with Messrs Munro & Company, of the Arbroath Foundry. The machines, which are sent to all parts of the country where there are quarries, form a considerable article of manufacture at Arbroath. Had it not been for them, the pavement trade could never have attained its present dimensions. The cutting-machines are of great strength, and consist of a series of chilled iron tools placed in revolving discs, by means of which great slabs of stone are cut through or split up with as much ease as if they were pieces of card-board. The dressing-machines are used for dressing stair-steps, coping, and much of the general mason-work still commonly done by manual labour.

The Forfarshire pavement quarries, which these machines have been largely instrumental in making famous, extend from Leysmill, in the east, which is situated about three miles inland, to near the fine old castle of Glamis, in the west—a distance of about sixteen miles. From north to south, they stretch from Montrewmont Muir to the Sidlaw Hills—about twenty miles. The quarries are thus all situated in the southern or more lowland portion of the county. The pavement produced throughout this district, as well as that worked in the neighbourhood of Arbroath, is known in commerce as "Arbroath pavement." The material is a freestone, and is found in the lower beds of the Devonian or Old Red Sandstone formation. It is solid in composition, and therefore very durable, but is easily worked and dressed. It is extensively used, not only for paving streets, but for all inside work of houses, and for other purposes.

At the eastern point of the geographical line described are the Leysmill Quarries. These have been worked for about half a century, and for the greater part of that time by the late Mr Carnegie, of Boysack, the proprietor. The present lessees are Messrs Straton and Cargill, who employ about sixty men, belonging chiefly to the villages of Friockheim and Leysmill. On the Guthrie estate is Montrewmont Quarry, tenanted by Messrs Baxter & Mann, who have three planing-machines, and employ from fifty to sixty men. Bordering on this, on the estate of Turin, there is a quarry called the Dub, which is tenanted by Mr James Hall, the oldest quarry-master in Forfarshire. Mr Hall has one planing-machine, and employs twenty men. The two quarries just mentioned were opened about ten years ago. The next quarry to the westward is Tilly-whandland. It is situated in the parish of Aberlemno, belongs to Mr P. H. Thoms, and is leased by Mr David Wilkie. Further on is the Myrestone Quarry, on the Pitscandly estate, tenanted by Mr John Gray. Each of these quarries employs one planing-machine and twenty or thirty men; in the latter there is also a ridge-stone machine. In this district a durable but heavy grey slate for roofing was formerly worked; but it is now generally discarded in favour of the thin blue slates. The Carsegownie, Tolbooth, and Balma-shanar Quarries, the latter of which is one of the oldest in Scotland, are situated in the neighbourhood of the town of Forfar. Carse-gownie is worked by Mr David Murray; Tolbooth, by Messrs Thomson & Brown; and Balmashanar, by Mr David Barry. Each employs from twenty to forty men. With the exception of Leys-mill, the pavement from the whole of these quarries has to be carted a distance of from three to four miles to the nearest railway station.

Further south are the Carmylie Quarries, which are the most extensive in the country. They are situated six miles from Arbroath, in the parish of Carmylie, which forms part of the south-eastern breast of the Sidlaw Hills, and consists chiefly of a series of high grounds scarcely approaching to hills, with their intervening valleys running from south-west to east. The quarries belong to the Earl of Dalhousie, and are at present leased by Messrs Duncan, Falconer,

& Co.   They have been worked for several centuries all over the parish.   About seventy years ago the parish was tenanted by an industrious and intelligent race of workmen.   At that time the land was still divided into small farms and pendicles, and each farmer was allowed the liberty of quarrying on his own farm.   The farmers generally worked in the quarries from the time they got the seed into the ground until harvest.   The principal quarrying in those early times was for slates, which were carted in large quantities to Dundee and the neighbouring towns, and also shipped to Leith, to be used in covering houses in that town and in Edinburgh.   It was in the beginning of the present century that the Carmylie Quarries began to be worked for pavement to any considerable extent.   The farmers, who were still also the quarrymasters, quarried the stones from 1 to 1½ inch thick, and carted them in their milk-carts to Dundee and Arbroath, where they were sold at the rate of 2s. 6d. a load.   About sixty years ago the farmer of East Hills of Carmylie, on whose farm the principal quarries were and still are situated, had seven or eight men quarrying.   After they had made several searches for good pavement, the foreman told his men to take up their tools and go home to their master to tell him that the quarries were done !   Soon after that all the quarries were let to one tenant, who shipped to Leith and Glasgow.   The late Mr Fyfe, father of Mr Fyfe of Messrs Mather & Fyfe, pavement merchants, Glasgow, was the first to sell Carmylie pavement in Glasgow.   He was then the only pavement merchant in that city.   At that time the quarries began to be worked on a more extensive scale.   Windmills, with movable wooden frames, were erected for the purpose of pumping the water ; but that arrangement, which was long in use, was subject to the drawback that, if there was no wind to drive the mills, the quarrymen could not work.   About forty years ago there was so much difficulty in getting rid of the water that a drain, running through a large part of the parish, and a portion of it tunnelled through a hard sandstone rock, was constructed, at a cost of L.3000, for the purpose of draining the principal quarry.   The machinery employed at that time consisted of a small single-power crane, which with care might be made to lift a stone a ton in weight.

The present state of things at Carmylie is much different from what it was even twenty or thirty years ago.   There are about 300 men in constant employment at the place.   The machinery consists of eight planing-machines, several cutting-machines, eight saws for jointing pavement, one machine for making steps, coping, and tabling ; two polishing-machines, six steam-engines, and from twelve to fourteen steam and other cranes. The quarries turn out almost any weight or size of stones—some weighing twenty tons and measuring 200 feet superficial. The Carmylie stones are famous for paving, and for steps; cisterns for paper-makers, chemical works, and bleach-fields; columns, balustrades, and other architectural ornaments. For these and similar purposes they are widely exported to the chief cities of the United Kingdom, to many countries in Europe, to the United States, Australia, and the colonies generally. Many towns in Scotland are paved with Carmylie stones. They have been, or are being, used as paving, steps, copes, &c., at a large number of public buildings—such as the new University of Glasgow, the New College, Edinburgh, the Bank of Scotland, Edinburgh, all the principal buildings in Aberdeen, the Perth railway station, &c. The weight of stones sent away from the Carmylie Quarries daily is about 150 tons. A single line of railway extends from the quarries—the rails going to the bottom—to Elliot Junction, near Arbroath, a distance of five miles. The Carmylie Railway, as it is called, passes near the village of Arbirlot, and skirts the beautiful Kelly Den, which is of so much interest to the geologist. The line was constructed by the Earl of Dalhousie, who sold it some years ago to the Caledonian Railway Company. It is used exclusively for the carriage of paving and other stones from the Carmylie Quarries.

The quarries are a great benefit to the parish, as most of the workmen reside there. They are an industrious, sober, and saving class. The greater number of them have a piece of land, and keep a cow or two. Intoxicating drinks are not allowed to be sold in the parish.

South from the Carmylie Quarries are those of the Gwynd, which belong to Mr Pierson, and are leased by Messrs Barrie & Galloway. There are three planing-machines at these quarries, and from sixty to eighty men are employed. The Pitairlie Quarries, in Monikie, belong to Lord Dalhousie, are tenanted by Mr David Barry, and about fifty men are employed. Westward are the Gagie, Wellbank, Kingenny, and Duntrune Quarries, each employing a considerable number of men. Duntrune produces pavement, but yields thick rock in greater abundance. With this exception, all the others mentioned are strictly pavement quarries.

Although not immediately belonging to the industrial part of the history of these flagstones, it may not be uninteresting to note the fossils of Carmylie—especially the "Seraphim" of the Forfarshire quarryman, this being one of the earliest discovered. A figure of it is given in Plate IX. of Mr Hugh Miller's "Old Red Sandstone." At the time that lamented gentleman wrote his fascinating book, nothing was known about these puzzles beyond the quarryman's name, until the genius of Agassiz, on having a lot of the fragments laid before him, pronounced them to be "remains of a huge lobster;" and the correctness of that opinion has since been demonstrated by some almost entire specimens of these crustaceans having been turned up; one of the largest—or rather a good portion of one—discovered in Carmylie Quarry in 1861, showing tail-plate and eight post thoracic segments. This fragment measures in the cast fully 3½ feet in length. The animal, when entire, would no doubt be 6 feet long. It is now in the museum at Arbroath, and is one of the "lions" there. Cephalaspis, with a few plants and masses of "Puddock's spue," also occurred with the Pterygotus (the name now given to the "Seraphim)." When Hugh Miller wrote these were the only fossils known in the Forfarshire rocks. Now, through the industry of the Rev. Mr Mitchell and others, and especially the zealous labours of Mr Powrie of Reswallie, fossils have been found in abundance in various parts of the county, and many interesting new genera and species of fossil fishes and crustaceans have been added to the list.

The pavement quarries of Caithness are of considerable extent and importance, and their produce has a world-wide reputation. The working of the quarries and the dressing of the stones constitute a considerable part of the industry of the county, employing more persons than any other kind of trade, exclusive of agriculture and fishing. Considering the natural disadvantages which it lies under, the county of Caithness has made great progress during the present century. Agriculture has been advanced, the fisheries have engaged much attention, and the manufacture of pavement has added to the growing prosperity of the people. The county is largely indebted for the development of its industries to, among others, the late well-known Sir John Sinclair, of Ulbster, and the late Mr Traill of Ratter. The son of the last-named gentleman—Mr George Traill, the present much-respected Member of Parliament for the county—has worthily continued the work begun by his father, and has especially taken a deep interest in the working of the pavement quarries. Mr Sinclair, of Forss, also deserves to be mentioned for the part he has taken in developing the pavement trade.

The Caithness flagstones belong to the middle formation of the Old Red Sandstone. Great numbers of fossil fish and plants are found intercalated among the flag-beds of commerce, having been buried in the ancient mud of the Old Red waters. These fossils appear on almost every slab of Caithness pavement, but although the fish remains lie by thousands it is seldom that anything like a perfect specimen is got.   However, they are beautiful in their ruin—their blackened, enamelled, and glistening scales and plates standing out in contrast with the sober grey of the matrix.   The flag-beds have suffered much in geological ages from dislocations by "faults" and other causes.

The geological character of flagstones having, of course, a close relation to their commercial and industrial value, it may not be out of place to quote what an eminent geologist states concerning the flagstones of Caithness.   In his "Siluria," Sir Roderick Murchison says :—" The flagstones of Caithness are in many places impregnated with bitumen, chiefly resulting from the quantity of fishes embedded in them.   Their most durable and best qualities as flagstones are derived from an admixture of this bitumen with finely laminated silicious, calcareous, and argillaceous particles, the whole forming a natural cement more impervious to moisture than any stone with which I am acquainted."   Sir Roderick then gives analyses of several specimens of flagstone and the accompanying bituminous shales from Mr Traill's quarries at Castlehill, and considers that they are of "high value, having been prepared by that distinguished chemist, Dr Hoffman."   These analyses are as follow :—

Mineral Analysed.	Silica and Silicates insoluble in HC.	Oxide of Iron and Alumina.	Carbonate of Lime.	Organic Matter.	Water Loss at 100 C.	Salts of Magnesia, the Alkalies, &c.
No. 16, Top Flag,	68·40	10·21	10·93	3·88	0·42	6·16
No. 7, Middle Flag,	69·45	11·50	10·66	5·79	0·40	2·20
Bituminous Shale,	69·96	8·15	7·72	10·73	0·53	2·91
No. 1, Bottom Flag,	61·39	4·87	21·91	3·40	0·20	8·23

Sir Roderick adds, "These results completely sustain the opinion I was led to form on the spot, that the peculiar tenacity and durability of the flag-stones is due to the manner in which silica and alumina are cemented together by certain proportions of calcareous and bituminous (organic) matter."

A portion of bituminous schist got near Barrogill Castle was analysed by Dr Hoffman, and yielded, of fixed matter (mineral), seventy parts, and of volatile matter (organic), thirty parts.   These oil-yielding shales may some day be considered worthy of special attention.   They exist in great quantities.

As a proof of the durability of the Caithness pavement, a circumstance which occurred a few years ago may be mentioned. A fire broke out in a building in Leith Walk, Edinburgh, and the hose for the engines was laid across the road. The ordinary traffic line being thus impeded, the cabs, carts, and waggons, some of the latter heavily laden, took to the footway, and the consequence was that for nearly 200 yards the sandstone flags—from the north of England and other places—were broken under the unusual weight, whereas a portion, several feet in width, of the same road, laid with Caithness flags, bore all the traffic without being injured.

The principal pavement quarries in the county are situated on a line extending from the parish of Olrig, on the shores of the Pentland Firth, to the Parish of Reay, in the west—a distance of ten or twelve miles. Another run of pavement commences at the seaside four miles south from Wick, and extends westward to the parish of Halkirk, in the centre of the county. The first exportation of pavement was made from quarries on the Crown lands of Scrabster, formerly belonging to the bishopric of Caithness, and situated near Thurso, in one of the most cultivated parts of the county. It is believed that, centuries ago, slates quarried at Scrabster were exported to the south. There were no roads in Caithness at that time, nor any carts, and, until a comparatively recent period, the slates were carried on the backs of the workmen to the shore, and put on board ships lying in Scrabster Roads. They were for the most part shipped to Leith, and many of the houses in that town, and in the old part of Edinburgh, are roofed with slates from the Scrabster quarries as well as from those of Carmylie.

The principal flagstone quarries in Caithness are those of Castlehill, which are worked by the proprietor, Mr Traill, M.P., under the management of Mr M'Beath, who has been in charge for nearly forty years. Although these quarries are the most important in the county, they are not the oldest. They were begun by the present proprietor's father, a gentleman who, as already hinted, took a leading part in everything that was likely to prove beneficial to Caithness, and one of whose chief motives was to find employment for the honest and industrious people around him. Mr Traill introduced many improvements on his estate, and, in order the better to carry these out, he built two harbours—one at Castlehill, in Dunnet Bay, and one at a place called Ham. Quarrying was then begun and a number of workpeople were attracted to the district. For their accommodation, and to encourage them to settle there, this public-spirited gentleman feued off at a cheap rate a portion of ground, and, free of charge, gave stones to his workpeople with which to build houses. In that way was built the large and thriving village of Castleton, situated on the shore of the Pentland Firth. The building of this village has secured a race of steady, industrious workmen for the quarries. The village, which attracts the notice of the traveller by its order and cleanliness, possesses several schools, churches, and a public hall. Mr Traill takes much interest in the working population of the place.

The Castlehill Quarries are very favourably situated, and have the advantage over all the other quarries in the county of being close to the seaside and to Castlehill harbour. It was from the Castlehill Quarries that pavement was first exported from Caithness in large quantities, and they continue to take the lead. The quarrying is easily effected, as the stone readily opens up in the natural layers and seams. Immense slabs are detached from the top of the exposed rock, and these are broken up into square pieces of various dimensions. The stones are then placed in trucks, and conveyed along a railway to the cutting and polishing sheds, which occupy the ground between the quarries and the harbour. The sheds are furnished with sawing and polishing machinery, driven by a combination of steam and water power. Usually there are half a dozen vessels at a time in the harbour loading pavement from Mr Traill's quarries, so that a large amount of capital and labour outside the locality is employed in the trade. In the quarries, and other departments of the works, Mr Traill employs from 350 to 400 men. The great establishment on the River Plate, in which Baron Liebig's extract of meat is manufactured, is floored throughout with stones from the Castlehill Quarries, and the cleanliness secured by their use is specially noticed by all visitors to the place.

In the same beds as those of Castlehill there are several quarries belonging to Mr Smith of Olrig. These are worked by Mr John Swanson, who has another quarry on his own property in the Hill of Forss, near Thurso. Mr Swanson has to cart his pavement to Thurso for shipment—a distance of five miles from Olrig, and three miles and a half from Hill of Forss. All the quarrymasters of the county, with the exception of those of Castlehill, Forss, and Barrogill, have their machinery at Thurso, on either side of the river, and this is an important item in the industry of the most northerly town on the Scottish mainland. Near Mr Swanson's are the quarries on the Ulbster estate, belonging to Sir John G. T. Sinclair, and tenanted by Messrs Craig & Son, Mr Innes Dunbar, and Mr Gerrie. Mr Gerrie and Messrs Craig have a quarry each in the Hill of Forss. Westward are the Forss pavement quarries, worked by the proprietor,

Mr James Sinclair of Forss.   Forss quarries are second only to those of Castlehill in the number of men and the extent of machinery employed, and it is worthy of mention that it was at them that machinery for "jointing" stones was first employed.   Further west is the Reay Quarry, situated in the parish of that name, on the border of Sutherlandshire, and on the estate of Sir Robert C. Sinclair.   All the principal pavement quarries are situated in the three parishes of Olrig, Thurso, and Reay, which form the north-western shoulder of the county.

Returning to the second geographical line described, we come to the Holm Head quarry, on the coast south from Wick.   Nearly forty years ago a few men were quarrying there on the edge of the cliff, when a portion of the rock fell away, and carried two of the men with it.   Both were killed, and since then no work had been done at the quarry till about a year ago, when operations were resumed, though not extensively.   Some twelve miles inland are the quarries of Banniskirk, tenanted by Mr L. Dunbar; and Spittal, leased by Messrs Shearer & Son.   Both are distant from Thurso, the place of shipment, about twelve miles.   The most northern quarry in the county is at Barrogill.   It is worked by the proprietor, the Earl of Caithness.   His lordship has persevered in quarrying—though not, it is believed, very profitably as yet—in order to give employment to the labouring people in the district.

The pavement quarries had a marked effect on the wages of the labouring class in Caithness, which have risen from 7s. or 8s. a-week to from 12s. to 18s.   There are no statistics of the quantities of pavement exported annually from Caithness, but it is roughly estimated that from 500,000 to 600,000 superficial yards are shipped every year, the value of which is from L.70,000 to L.80,000.

The date at which slate for covering houses came into use in Scotland is not known, but there is good reason for believing that it was at least three centuries ago.   Though slate suitable for roofing purposes exists in various localities, the earliest used would appear to have been derived from the island of Easdale, which for upwards of two hundred years has been one of the chief sources of supply.   It is stated that among the ruins of a castle in the north of Scotland, which was built in the fifteenth century, fragments of slate similar to that obtained at Easdale were discovered.   There is evidence that the Falconer's Castle at Appin was in 1631 roofed with Easdale slates.   Ardmaddy Castle, a seat of the Earl of Breadalbane, built in

1676, was covered with slates of the same kind. In the latter case they were fastened with wooden pegs, and have withstood the tempests of nearly two centuries without requiring to be replaced. Of greater extent than the Easdale quarries, and of equal fame, though some years younger, are those of Ballachulish. The other slate quarries in Scotland are not of much account.

The island of Easdale is so small and so uninteresting in appearance that persons passing it on the route to Oban would not devote a moment's attention to it were it not pointed out as the seat of the famous slate quarries. It forms one of the group of small islands which skirt the coast of Argyle between Crinan and Oban, and lies close to the south-west point of the island of Seil. There is an eminence about 130 feet in height towards the north end, but the greater part of the island rises little above the high water level. In outline it is irregular, and its extreme measurements are 850 yards by 760 yards. The island is composed chiefly of clay-slate, the other rocks being limestone and trap. Mr John White, who was manager of the quarries for upwards of twenty years, gives the following account of the disposition and character of the slate:—
" The slate-bands appear in two seams, which are much contorted throughout the whole extent of the island. These seams are made up of different beds, ranging in thickness from a few inches to many feet, like our sandstones and limestones, but frequently so closely united that only a practised eye can detect the line of contact; whilst the superinduced phenomenon of cleavage constitutes a distinct dissimilarity between them and the other rocks referred to. It may be noticed, as an indication of the sedimentary character of these slate rocks, that there is a decided difference in the quality of the upper and lower portions of the thicker beds, the former being fine grained and smooth, and the latter coarse and gritty—the feature which we recognise as analogous to that exhibited by other rocks of sedimentary origin. Although the slate-seams are so much contorted as to be found at various angles with the horizon, the cleavage plane invariably maintains an angle of thirty-seven degrees to a vertical line, from which it may be inferred that the property of cleavage was imparted to the rock subsequently to its being disturbed from the horizontal position. Where the rock has been least disturbed from the horizontal position it is more easily quarried, and yields better slates. The slate beds are much affected by what the quarriers call ' cuts.' These are joints which intersect the strata in a nearly vertical direction. Sometimes their presence is advantageous, but when they are too numerous the stone breaks up into useless frag-

ments." The slate-seams extend across the channel, 150 yards in width, which separates Easdale from Seil, and crop up on the shore of the latter island, where they are worked at two points, known locally as the Ellenabeich Quarries.

The first visitors to Easdale found the shores strewn with slates quarried and split by the action of the waves; and from the supply thus provided was taken all that was required for many years. By the time the shores were cleared of the loose stones the value of the slate had come to be more fully realised, and to meet the demand which had arisen quarrying was resorted to. The exposed strata on the sea-shore were first worked, and the process of quarrying, as handed down by tradition, was this: At low water wedges of seasoned oak were driven into the cleavage seams, and when the tide flowed the wedges expanded and detached the superincumbent rock. That was a slow and uncertain process, and in course of time was abandoned for blasting by gunpowder. When the seams of slate had been worked down as far as the tide would allow, the persons engaged in quarrying began to look upon their occupation as in danger of extinction, and some of them actually left the island. The trap dykes which occur in the seams and divide them into compartments had, when reached, evidently led the quarrymen to conclude that they had got to the limits of the slate deposits. Those of the workmen who remained on the island "explored" the ground, and opened a quarry a little distance from the shore. Hydraulic engineering was but little understood by them, and they were again cast into a state of despair by an accumulation of surface water in the quarry. In order to get rid of the water they cut a trench down to the sea on the level of the ebb-tide line, and by means of that ran off the water that accumulated between the ebbs, the rising tide being prevented from getting into the works by a sluice fixed in the channel. The men never dreamed of being able to carry on their work at a lower point than that which their drain enabled them to reach; and it is related that, on the occasion of an unusually low tide, one of the workmen fired a blast which he boastfully exclaimed was "the lowest that would ever be set off in Easdale." How much this prediction has come short of verification will be seen further on. In those early days of the quarries the men were assisted by their wives and families, the slates being carried from place to place in "creels" on the backs of the women.

The mode of working the quarries was revolutionised some time in the course of last century by the introduction of a pump which was obtained from the wreck of a castaway ship. As the pump enabled the workmen to keep down the accumulations of surface-water, the quarries were sunk beneath the sea-level, and thus the beds of slate which had been abandoned for ever, as the quarriers thought, were got out. Finding that the employment of the old pump gave them such a great advantage, the men next obtained a crank-pump fitted with a fly-wheel—a machine which was regarded with admiration by the islanders. The slate strata in the interior of the island were next broken into, and a number of quarries were opened, for which pumping machinery of an improved kind was introduced. Among the new appliances was one of Newcomen's atmospheric engines, which, when it was set to work, excited much wonder among the workpeople. The engine would appear to have been of very rude construction. The boiler consisted of a square box of cast iron one inch in thickness, and the piston was "packed" with leather. As the engine could not be worked profitably, horse power was substituted in the beginning of this century, and the introduction of horses for that purpose suggested the substitution of carts for wheelbarrows in drawing the slates out of the quarries. About the year 1807 a windmill was erected to work the pump of the principal quarry, and gave great satisfaction. With a moderate breeze it worked very effectively, raising the water through a seven-inch pipe from a depth of fifty feet. The windmill continued in use for twenty years, and the part of the works with which it was connected still bears the name of the Windmill Quarry. In 1826 a powerful steam-engine was erected in such a position as to be available for pumping three quarries. About the same time wharves were erected for loading ships, and these were found to be a great convenience, the practice previously being to anchor the vessels in the sound, and carry the slates to them in boats. Railway inclines were subsequently constructed in all the quarries. These were at first worked by horses, but for a number of years past steam-power has been employed.

For a long period prior to 1841 the quarries had been leased to a company in which the successive representatives of the house of Breadalbane were shareholders. On the expiry of the lease in the year named, the late marquis determined to work the quarries on his own account, and continued to do so up till his death in 1862. Many improvements in the mode of working the quarries and in the condition of the workpeople were effected during the rule of the marquis. When the quarries were worked by a company, the workmen were paid only once a-year, and then only according to the quantity of slates that had been sold, no reckoning being made for the quantity added to stock. The marquis introduced the system, which still prevails, of paying at short intervals without reference to sales. When the present marquis came into possession, he undertook the working of the quarries, but under great disadvantages. Some of the embankments had given way, and a large portion of the workings was flooded. In 1866 the marquis let the quarries to a company of the workmen formed on co-operative principles. Though they obtained the quarries on highly favourable terms, the company did not succeed ; and their affairs were wound up after a year's trial. The quarries were then acquired by a company of slate merchants, who have continued to work them with great vigour. The principal quarry at Ellenabeich is about 450 feet in length, 250 feet in width, and 160 feet in depth. The Windmill Quarry in Easdale is 250 feet in length and breadth, and 120 feet in depth. Two powerful steam-engines are employed for raising the material and keeping the quarries clear of water. About 300 men and boys are employed. The number of slates turned out annually cannot be less than from 7,000,000 to 9,000,000 ; the average from 1842 to 1861 was 7,000,000. Taking the value of the slates at the quarries to be L.2 a thousand, the total produce will, according to the lowest computation, be worth about L.14,000 a-year. The slates are well known in the market, and command a ready sale—it being no uncommon thing to find above a dozen vessels, of an aggregate burthen of over 1500 tons, waiting their turn for loading.

The population of the village of Ellenabeich is about 350, and of Easdale 450, all of whom are dependent upon the quarries. Almost without exception, they are natives of the locality, and, as is usual in the case of communities remote from the great centres of population and industry, they are somewhat primitive in their habits and modes of life, and not quite free from superstitious beliefs. The men are strong and hardy, and there are no more skilful or daring boatmen on the British coast. All the adults can read and write, and the rising generation are well provided with the means of education. Some years ago a Young Men's Association was established and a library formed. Lectures have at various times been delivered on scientific and popular subjects. Gaelic is the language commonly spoken, but most of the people know English to some extent. It is said that so little are the quarry populations given to changes of residence, that the removal of a family from one side of the sound to the other is quite an affecting incident, accompanied by no little shaking of hands and even shedding of tears. Visitors to the quarries will have their attention arrested by slabs of stone set up on end near the houses. These are tombstones which the men have prepared to mark their own graves. In the course of their operations in the quarry the men occasionally meet with stout slabs of hard stone, and these are purchased and preserved as indicated until required.

The Ballachulish Slate Quarries are situated in the north of Argyleshire, on the shores of Loch Leven, about two or three miles from the scene of the tragic massacre of Glencoe. There are two quarries about half a mile apart, and named respectively the East Quarry and the West Quarry, the latter being the smaller of the two, although the first worked in the district. It was opened in or about the year 1697 by the proprietor, Mr Stewart of Ballachulish, from whom Sir John Stewart, the late vice-chancellor, is descended. On slate being discovered, men were procured from Easdale to teach the people of the district how to quarry and dress it. The West Quarry was worked on a small scale by Mr Stewart for many years. When the discovery of another vein of slate, further to the east, was made about the year 1780, the old workings were stopped for a time, and the East Quarry was opened. Subsequently the quarries were let for a number of years, first to Mr Stevenson, and afterwards to Mr Rawse. At the end of Mr Rawse's lease, the proprietor (a descendant of Mr Stewart, who opened the quarries) took them into his own possession, and kept them working until, from pecuniary difficulties, the whole of the Ballachulish estate went into the market, and was purchased, in 1862, by Mr Robert Tennant of Leeds, the present proprietor. Mr Tennant let the quarries to Mr Alexander Pitcairn, who re-opened the old workings at the West Quarry in 1863. After much labour has been expended in clearing away the rubbish left by the previous workers, the West Quarry promises to produce excellent slate. The East Quarry consists of five working levels, four rising one above the other from the ground level, and one being sunk about sixty feet below the surface. The face of the quarry, measuring from the top of the highest level to the ground line, is about 300 feet; and when it is taken into consideration that the ground area already quarried measures from five to six acres, some idea may be formed of the vast amount of solid rock that has been cut out. The rubbish from the quarry has been run out into the sea, and forms two great banks. A third bank runs out from the West Quarry. There are two upper levels and one sunk level in the West Quarry.

The men who work in the quarry are divided into " crews " of

five, six, or seven.   Each crew choose a certain part of the rock face
on one of the levels, and make a kind of bargain with their employer
to work that particular spot for a year, receiving a sum of money
agreed upon for every thousand slates made by them, so that their
wages depend upon the diligence with which they work as well as
upon the quality of the rock they may select.   The slates are divided
into four classes—viz., "Duchess," "Countess," "sizable," and
"undersized."   The first mentioned are the largest, being twenty-four
inches long by twelve broad ; while the "Countess" slates are
twenty by ten inches.   The other two classes are smaller.   Each
crew makes the four kinds, should the quality of the rock allow it,
and receives a different sum per thousand for each sort.   In working, some of the crew stay always in the quarry and blast the rock
face, while the others are out upon the rubbish bank splitting up
the blocks furnished by their comrades.   The Ballachulish quarriers
are said to be the best blasters of rock in Scotland.   Slung up the
face of the perpendicular rock on a kind of chair attached to a rope
fixed at the summit of the level, the worker hammers away at a
sharpened iron rod known as a "jumper."   When he has finished
the " bore," a charge of gunpowder is inserted, and fired with a slow
match.   So well is everything arranged, that in the memory of the
present generation only one man has been killed by a blast, and in
that exceptional case it was entirely the man's own fault.   After
the charge is expended, the workman climbs up the rope, which he
fixes round his leg in a peculiar manner, and placing his feet against
the perpendicular face of the rock, loosens with a crowbar the masses
of stone that have been shattered by the explosion, and sends them
crashing down the steep.   When a large bore has to be made, a
staging is fixed in the face of the rock, on which three men work,
one holding the "jumper" while the other two strike it with heavy
hammers.   After all the loosened rock has been brought down, the
quarry detachment of the crew break the large masses into smaller
blocks.   Those of good quality are then selected and sent in waggons
out to the bank, where they are split and dressed.   The splitters are
very expert at the work.   Two good hands, the one splitting and
the other dressing, provided they are supplied with blocks of a good
quality, can easily make from 1800 to 2000 slates a-day in summer.
The rubbish is sent away in waggons, to be thrown over the edge of
the bank.   At that spot are stationed numbers of boys, from eight
to fourteen years of age, who watch the rubbish projected from the
waggons, and, pouncing upon a block that may have a promising
appearance, split it into slates.   The only kind of slates made by the

boys are the undersized. Iron rails or tramways extend in every direction through the quarry, along which horses draw the waggons laden with slates, rubbish, or blocks. If placed in a straight line, the railways would measure above six miles in length. There are several inclines, with machines attached for letting the waggons down from the upper levels. The largest of these inclines is 530 feet long, and exceedingly steep. There are also three steam-engines used—two at the sinking level in the East Quarry, and the other at the West Quarry.

There is a small harbour between the two rubbish banks at the East Quarry, where the slate vessels are laden. The number of men and boys employed is about 400, and 15,000,000 slates are turned out annually.

The workmen are all natives of Ballachulish or of the hamlet of Glencoe, which lies about a mile inland. The only strangers are a few Easdale families who have settled in the place. One reason why there are so few strangers is, that no adult can learn to split and dress slates so expertly as those who have been trained to the work from boyhood. The boys of the place are put to split slates when about eight years of age, or as soon as they can handle the chisel, slate-knife, and hammer; and as there is no opportunity of learning other trades, they stick to the quarry all their lives. The workmen are in general strong and healthy. The population of Ballachulish is nearly 2000. The people intermarry among themselves, very few men having wives from places farther away than Appin. There is no inn or publichouse in the village, and, except on the day or two succeeding the pay (which is monthly), when a little drinking occurs, no intoxicated person is to be seen. Taking into consideration the extent of the village, and the social condition of the population, the drinking is nothing to what it is in Lowland mining villages of the same size, although once or twice the people have been maligned in the public prints by tourists who happened to pass through the village on a pay day. Thieving is a thing quite unknown in the locality. A peculiar trait that the people possess, in common with other Highlanders, is love of their birth-place. Most of those who leave the place find their way back again after a time, even although they may have been better off elsewhere. A great number of them are still very superstitious, firmly believing in ghosts, in blue lights appearing on water where persons are to be drowned, and in omens of various kinds seen before a death. Persons are to be met with who profess to possess the interesting faculty of "second-sight." A very old custom still lingers in the village of holding a wake over the body of a friend or relative deceased. Another strange fashion is that of firing off a small cannon kept for the purpose when an agreement of marriage has been entered into between a youthful couple. Education is deficient, on account of the boys being sent early to work in the quarry. The women are generally better educated than the men, but with few exceptions all can read the mother-tongue, and make a pretty fair attempt at English. There are four churches in the place, viz., the Scotch Episcopal Church (which is the strongest body), the Established and Free Churches, and a Roman Catholic Chapel, but there are very few members of the latter persuasion. Schools are attached to the first three mentioned denominations, but these are poorly attended. A Mechanics' Institute, of which most of the men are members, has been erected by Mr Tennant at his own expense.

There are also slate quarries at Dunkeld, Luss, Aberfoyle, and Craiglea. Those of Craiglea are the most important. They are situated on the Logiealmond estate (the property of the Earl of Mansfield), and are about fourteen miles north-west from Perth. Craiglea is a hill 1500 feet in height, and the quarries are near the top of it. For local purposes slate was taken from Craiglea many years ago, and for upwards of a century the quarries have been worked to supply a more extended demand. The quarries had, up till three years ago, been let on lease; but the proprietor then took them into his own care, and under the skilful management of Mr John White, who for upwards of twenty years was manager of the Easdale Quarries, they are being opened up on an extensive scale. The slate vein is of excellent quality, and has this peculiarity, that, while one portion of it supplies slates of a dark-blue colour, those obtained from the other portion are of a sea-green hue; but otherwise there is no appreciable difference between them.

BREWING.

THE art of preparing beverages from fermented grain is of great antiquity. It was, acording to Herodotus, practised by the Egyptians; Pliny the elder states that it was known to the western nations; and Tacitus mentions that a fermented liquor extracted from grain was the common beverage of the ancient Germans. One of the pleasures promised to Scandinavian heroes was that in their Valhalla, or Palace of the Gods, they should drink ale out of carved horns. A favourite beverage of the Anglo-Saxons in the fifth century was a kind of ale made from grain; and among the civic officers of the time were "ale-conners," whose business it was to taste the liquor offered for sale, and fix its price. In the parish of Minnigaff, Kirkcudbrightshire, are remains of some kilns which tradition alleges to have been constructed by the Picts for the purpose of brewing ale from heather. The earliest mention of "ale-houses" in England occurs in 1014, and about that period the price of ale was fixed by law.

In Scotland the "broustaris," or brewers, were taken cognisance of in the *Leges Burgorum*, a code of burgh laws sanctioned by the Legislature in the twelfth century. Under those laws a licence-duty at the rate of 4d. a-year was imposed on all persons engaged in brewing. Another clause, entitled "Of the manner of ale brewing be assise," is as follows :—"What woman that wil brew ale to sell sall brew al the yhere thruch eftir the custume of the toune. And gif scho dois nocht scho sall be suspendyt of hir office be the space of a yhere and a day. And scho sall mak gud ale approbabill as

2 E

the tyme askis. And gif scho makis ivil ale and dois agane the custume of the toune and be convykkyt of it, scho sall gif til hir mercyment viii. s., or than thole the lauch of the toune—that is to say, be put on the kukstule, and the ale sall be geyffin to the pure folk the twa part, and the thryd part send to the brethyr of the hospitale. And rycht sic dome sal be done of meid as of ale. And ilk browstare sal put hir alewande ututh hir house at hir wyndow or abune hir dur that it may be seabill communly til al men, the whilk gif scho dois nocht scho sal pay for hir defalt iiij d." This shows that the brewing business was originally in the hands of women, and there is evidence that it continued to be so for many years subsequent to the passing of the laws referred to.

In the fourteenth century a list was drawn up of matters to be inquired into by the Chamberlain of each burgh in his term of office. The following "items" will show that brewers were to be particularly looked after :—"Also, gif the Bailies have executed judgment upon baksters, browster men and women, after they be amerced. Also, gif browster-wives sel aill be quart and be just measures. Also, gif browster-wives brewe and selle aill conform to the price set upon it by the taisters. And gif they selle before the aill has been prised by the taisters. Also, gif browster-wives sell their aill by potsful, and not by sealed measure. Also, how many of the browster-wives were amerced in the year. Also, gif any man keip hand mylnes, other than are burges, and brewes and maks malt, composition not made, and wha manteins them."

In the year 1124 the price of a Scotch gallon of ale was equal to 1s. 6d. of modern money. In 1562 the price of a pint of ale is stated in the Council Register to have been 9d. The records of the Scotch Privy Council show that, in 1666, the price of ale was fixed as follows, in sterling money :—"When rough bere is 10s. per boll, Linlithgow measure, then ale shall be sold, per Scotch pint, at 1d.; with the addition of one-sixth of a penny as excise in country parishes, and one-sixth more in the city of Edinburgh. When bere is at 13s. 4d., the pint of ale shall be $1\frac{2}{3}$d.; when at 16s. 8d., the pint of ale shall be 2d." Ale was in those days an article of consumption in every household and at every meal. The upper classes drank wine, which they obtained at the following rates :—Bordeaux wine, if imported by the east sea, 3s. $11\frac{1}{2}$d. per Scotch pint (equal to about $1\frac{1}{2}$ imperial quarts); ditto, by the west sea, 2s. 6d. Rochelle wine, if imported by the east sea, 2s. 6d.; ditto, by the west sea, 1s. $10\frac{1}{2}$d. An Englishman who visited Edinburgh in 1598 wrote :—"The Scots drink pure wines, not with sugar as the

English; yet at feasts they put comfits in the wine, after the French manner, but they had not our vintners' fraud to mix their wines. I did never see nor hear that they have any public inns with signs hanging out; but the better sort of citizens brew ale, their usual drink (which will distemper a stranger's body)." The usual allowance of ale at table was a chopin (equal to about an imperial quart) to each person.

For many years the inhabitants of the City Parish of Edinburgh paid a tax of 2d. on every Scotch pint of ale they consumed, and that was continued after the imposition of a tax by Government. In the year 1690 this local tax produced L.4000. In 1723 the tax was extended to the parishes of St Cuthbert's, Canongate, and South and North Leith, and with the money which it was expected would flow into the municipal treasury from this source many good things were to be done. The stipends of the city ministers and the salaries of professors were to be augmented, an increased supply of water was to be brought in, and public buildings were to be erected. The imposition was ill-timed, for the people were beginning to cultivate a taste for a more potent preparation from malt, and to find out the virtues of tea. The tax produced L.7939 in the first year of its operation over the extended area; but never again did it reach that figure. In 1776 it had dwindled down to L.2197; and moralists were loud in their wails over the fact that "the use of that destructive spirit (whisky) was increasing among the common people of all ages and sexes with a rapidity which threatened the most important effects upon society." After a time the tax was abolished, but the domestic use of ale has never again been so common as it was a hundred and fifty years ago.

In 1643 a duty was imposed on the ale produced at public breweries in England. This tax was subsequently levied in Scotland, and continued till 1830, when it was repealed. In England malt was subjected to a tax of 6d. a bushel in 1695, and the tax was extended to Scotland in 1725. The Scotch people submitted very unwillingly to the imposition, and strong manifestations were made against it in both Edinburgh and Glasgow. In the latter city riots occurred which resulted in the death of nine and the wounding of seventeen persons. Mr Campbell of Shawfield, the member for the Glasgow district of burghs, had rendered himself obnoxious to a large body of the citizens, by voting in Parliament for the extension of the tax to Scotland, and on the 23d June 1725, the day on which the tax came into operation, a mob assembled, obstructed the excisemen, and assumed such a threatening attitude, that on the evening of the next day, two companies of soldiers, under the command of Captain Bushel, entered the city. The appearance of the military did not overawe the mob, nor deter them from making an attack on Mr Campbell's house, then one of the finest in Glasgow. While the magistrates were spending the evening in a tavern, and the soldiers were at their barracks, a mob marched to Shawfield House, the furniture and fittings of which they completely demolished. Mr Campbell and his family had removed to their country residence a few days previously. The captain of the soldiers having learned what had been done, sent to the Provost to ask for instructions; but the proffered services of the military were declined. Emboldened by the success they had achieved so far in carrying out their designs, the mob next day set themselves to molest the soldiers. After his men had withstood several volleys of stones, Captain Bushel gave the order to fire on the mob, and two persons were killed and several wounded. Finding themselves at a disadvantage against the muskets of the soldiers, the mob broke into the town-house magazine, and carried off the arms. At the request of the Provost, Captain Bushel removed his men towards Dumbarton, but they were overtaken on the way, and a sharp encounter took place. The military fired on the people, and several were killed and wounded. On information of the riots reaching headquarters, General Wade, with a large body of troops, took possession of the city. The Lord Advocate accompanied the General, and made an investigation into the circumstances of the disturbances, the result of which was that nineteen persons were apprehended, bound with ropes, and delivered over to Captain Bushel, who conveyed them to Edinburgh, and lodged them in the Castle. At the same time the whole of the magistrates were apprehended, and taken to Edinburgh. The charge against them was that they had favoured the rioters, and winked at the destruction of Shawfield House; but that charge was not substantiated; and after being a day in custody they were released on bail, and subsequently absolved. The nineteen inferior persons were punished in various ways—two were banished for life, some were sentenced to long terms of imprisonment, and others were whipped through the streets of Glasgow. By order of Parliament the citizens had to pay L.9000 to Mr Campbell as indemnity for his loss.

The riots left many bitter recollections in Glasgow, and did not tend to allay the popular feeling against the malt-tax. In the following year the duty was reduced to 3d. a bushel. Subsequently the rate underwent many fluctuations. In England, in 1760, it was raised to 9¼d. a bushel; in 1780 it was raised to 1s. 4¼d. a bushel, and to 8d. a bushel in Scotland. In 1785 the duty was imposed in Ireland at 7d. a bushel; and raised in 1795 to 1s. 3d. In 1802 it was respectively 2s. 5d. in England, 1s. 8¾d. in Scotland, and 1s. 9½d. in Ireland. In 1804, which was a year of war tax, it was raised to 4s. 5¾d. in England, to 3s. 9½d. in Scotland, and to 2s. 3½d. in Ireland. In 1813 it was raised to 3s. 3¾d. in Ireland; and in 1815 it was further raised to 4s. 5d. In 1816 the duty was reduced to what it had been prior to 1804, namely, 2s. 5d. in England, and 1s. 8¾d. in Scotland, but it was reduced to 1s. 4d. in Ireland. In 1819 it was raised to 3s. 7¼d. in England and Scotland, and 3s. 6¾d. in Ireland. In 1822 the duty was fixed at 2s. 7d. uniformly, at which rate it stands at the present time, with the exception of 5 per cent., which was added in 1840 to the Excise duties generally, making the actual impost 2s. 8½d. During the Crimean war the duty was raised to 4s., and after the war it reverted to 2s. 8½d.

Like other branches of trade which had long been conducted on a small scale in the ordinary dwellings of the people, brewing was about two centuries ago developed into a wholesale manufacture, and carried on in buildings specially fitted up. There are no statistics to show what the extent of the trade was in those early days; but for many years the production was limited to home requirements. In the beginning of last century ale and beer were exported from Leith to several continental countries. Since that time the export trade has gone on extending, and a marked increase has taken place within the past eight or ten years. The brewing trade is becoming concentrated into fewer hands, and operations are in some cases conducted on a gigantic scale. In the year 1835 there were 640 persons licensed to brew beer in Scotland. By 1863 these were reduced to 225, and in 1866 the number was 217, of whom 98 were brewers, and 119 victuallers who brewed their own beer. In 1836 the Scotch brewers consumed 1,137,176 bushels of malt; in 1863, 1,780,919 bushels; and in 1866, 2,499,019 bushels. The exports of ale and beer in 1863 amounted to 47,415 barrels of 36 gallons each, the declared value of which was L.172,140; in 1866 the quantity sent out was 61,723 barrels, valued at L.230,109. In order to show the wide connection which the brewers have established, the places to which the last mentioned quantity of beer was sent may be stated:—
1370 barrels went to Hamburg, 1250 to Mauritius, 13,975 to the continental territories of British India, 1564 to Singapore, 4337 to Victoria, 455 to New South Wales, 557 to Queensland, 1420 to New

Zealand, 1904 to British North America, 8797 to the British West Indies, 5161 to Foreign West Indies, 3346 to the United States, 956 to Chili, 2715 to Brazil, 3636 to Uraguay, and 5965 to the Argentine Republic.   In 1867 there were 66,909 barrels exported.

The Edinburgh brewers have long been famous for the superior quality of their ales and beers, and their trade forms one of the most important branches of manufacturing industry in the city.   The names of Younger, Jeffrey, Drybrough, Campbell, Usher, and others, are familiar wherever Scotch ale is consumed, and that signifies, as shown above, that they are known in every quarter of the world. A description of the malting and brewing establishments of Messrs J. Jeffrey & Co., of the Heriot Brewery, will convey some idea of the mode in which an extensive business of this kind is carried on. The malting premises, bottling-house, and ale stores of this firm are at Roseburn, at the extreme west end of the city, while their brewery is in the Grassmarket.   This separation is a considerable inconvenience; but as the brewery, by repeated extensions, occupied every inch of available ground, it became imperative, when further extension was required, to sever the connection between the malting and brewing departments.   Accordingly, a year or two ago, the firm acquired a site at Roseburn, adjoining the Caledonian Railway, and erected thereon malting premises and stores of great extent, and fitted up in the most complete manner.   The malt-barn is a substantially constructed building of five floors, and measures 320 feet in length by 90 feet in breadth.   On the upper floor the barley is stored in bulk, being raised to that part of the building by means of belt-and-bucket gearing, communicating with horizontal tubes fixed overhead, through which Archimedean screws draw the grain along to any point desired. From the store the grain is transferred, as required, to the floors beneath, by means of tubes or shoots.

The process of malting embraces four operations—namely, steeping, couching, flooring, and kiln-drying—the object of all being to force the barley to germinate, and then to check the germination at a certain point.   Across the end of each of the malting floors is a steep capable of containing 84 quarters of barley.   The grain is run into the steep from the store-loft, and when the steep is partly filled, water is allowed to flow in.   After the grain has been steeped for about sixty hours, the superfluous water is run off, and the barley is thrown out of the steep.   At this stage it is measured by the Excise officers, and charged with malt duty.   It is then "couched," that is, allowed to lie in a heap on the floor for twenty-six hours or so, during which time its temperature rises about ten degrees, and it gives off some of the superfluous water. This " sweating," as it is termed, is the result of the partial germination of the barley. On examining the grain at this stage, it is seen that rootlets have begun to appear, and traces of a stem may be detected beneath the husk. Now is the time for " flooring." The barley is spread in an even layer on the floor, to a depth of six or eight inches, and as it dries it is frequently turned. This operation extends over several days, at the end of which the barley is placed in a kiln and dried thoroughly. The action of the kiln in drying is not confined to expelling the moisture from the germinated grain, but serves to convert into sugar a portion of the starch which remained unchanged. Malt is generally distinguished by its colour—as pale, amber, brown, or black malt—arising from the different degrees of heat and the management in drying. The pale and amber coloured varieties are used for brewing the lighter kinds of beer; a darker variety is used for sweet ale; and the darkest for porter.

A remarkable change takes place in the grain during its conversion into malt, as will appear from the following analysis:—

	Barley.	Malt.
Hordein (a form of starch), . .	55	12
Starch, ...	32	56
Sugar, ...	5	15
Gluten, ...	3	1
Gum, ...	4	15
Resin, ...	1	1
	100	100

These figures show that the amount of the convertible starch and sugar has been nearly doubled at the expense of the hordein, a portion of which has also passed into the condition of mucilage, or a soluble gum, while the gluten is reduced to one-third of its original quantity. In converting barley into malt a loss of material occurs. Thus, 100 ℔. of barley yield only 80 ℔. of malt; but, on the other hand, there is an increase in bulk, 100 measures of barley yielding 101 to 109 measures of malt. This change in weight and bulk may be tested by casting some grains of barley and malt into water, when it will be seen that, while the barley sinks at once, the malt keeps afloat.

In order to see how beer is made, the malt must be followed to the brewery. The Heriot Brewery has been in operation for a century, and for upwards of thirty years it has been in the possession of Messrs Jeffrey. Like most works which have been gradually extended from small beginnings on limited sites, the brewery is not arranged according to modern ideas of such establishments; but that drawback apart, the place is complete in all its appointments, and the more recent additions have been made according to the most advanced views of the business. The malt is raised to a large store-room on an upper floor, and thence it is withdrawn to supply the mill. The latter consists of a pair of steel cylinders which bruise the malt—bruising being preferred to grinding, which would make the malt become pasty when mixed with water. From the mill the malt descends to the mashing-room. A new mode of mashing recently introduced is here at work. Formerly the bruised malt was placed in the mash-tun with a certain quantity of hot water, and there stirred about by revolving rakes till all the saccharine matter was dissolved. By the new method the malt escapes from a hopper into a horizontal cylinder having a series of revolving arms inside. At the same time the proper supply of water is allowed to flow in, and, as the malt and water pass through the cylinder, they are so completely mixed that they require no further mechanical treatment for the production of "wort," as the extract of malt is called. The mash-tuns are fitted with false bottoms, through which the liquid percolates, leaving the malt behind. The wort is drawn off into large vats called "underbacks," situated beneath the mash-tuns, and fresh water, at a higher temperature than that first used, is run upon the malt. This is styled the "second mash," and it is effectual in extracting a further quantity of sugar from the grain. The produce of those two mashes is mixed, but that of a third mash, which is sometimes made, is kept apart and used either in brewing small beer or in treating the malt in a first mash. The residue of the malt, under the name of "draff," is used as food for cattle.

The wort having been reduced to the proper strength, is pumped from the "underbacks" to the boiling-house, which is occupied by two copper boilers, each capable of containing 5500 gallons. At this stage the hops are added according to the kind of beer that is being made. The proportion of hops varies from 4 to 14 ℔. to the quarter of malt. The boiling is continued until the aromatic and bitter principles of the hops have been extracted, and the liquid has been concentrated to the required degree. A tap in the bottom of the boiler is then opened, and the liquid is run off into the "hop-back," a large iron cistern with a perforated bottom. As the wort percolates through the bottom of the cistern it runs into the "coolers," shallow troughs of iron covered with a roof but open at the sides. There the liquid cools rapidly, but not so rapidly as desired sometimes, and means are taken by fans and other contrivances to send currents of air over the surface. Systems of tubes called "refrigerators" are also used. The tubes are arranged in the form of a vertical screen, and as a current of cold water flows through them, the wort is poured over them from above, and allowed to trickle from one to the other. If the cooling be not effected rapidly, the sugar in the wort becomes partially converted into acetic acid, and the quality of the beer is thereby deteriorated. When the liquid has been cooled down to about 60° it is ready for the next process, which is fermentation. The apartments in which the process is conducted are called tun-rooms, and each contains a dozen large tuns or vats ranged along the sides. The tuns are capable of holding 2000 gallons each. When the tuns are filled yeast is added to the wort, in order to start the fermentation. In a short time carbonic acid gas is evolved, and the liquid becomes covered with froth. The gas is so abundant that it becomes dangerous to breathe over the tuns. Even after the vats have been emptied the gas hangs about, and workmen entering them without first ascertaining whether the fatal gas had disappeared have fallen victims to their negligence. Great skill is required in determining the temperature to which the wort should be reduced before adding the yeast. In summer it is usual to cool it to some twenty degrees below the temperature of the tun-room, while in winter it is worked at several degrees above the temperature of the room. For the proper modification of the temperature the tuns are fitted with tubes inside through which warm or cold water may be made to flow. The pale amber colour and mild balsamic flavour which characterise Scotch beer are owing in some degree to the low temperature at which it is fermented. The process of fermentation is completed in from three to eight days, and then the yeast is skimmed off and the beer "cleared" by being subjected to a filtering and settling process, which removes all traces of fermentation. That completes the manufacturing operations, and the beer is run into casks, and either sent out to order or stored. The stock of porter is kept at the brewery in great vats upwards of twenty feet in depth, but the ale is stored at Roseburn. All varieties of beer, ale, and porter are made by processes similar to those above described. The liquor may differ in strength according to the quantity of water used, or in colour from the malt being more or less charred in drying.

Messrs Jeffrey's ale store at Roseburn consists of two floors, each open throughout, and measuring 600 feet in length by 120 feet in width. A portion of the ground floor at one end is devoted to the bottling of ales for export, which forms a considerable item in the business of the firm. The remainder of the floor is piled with five or six tiers of large store casks, in which the beer is placed to mellow until required to be bottled or otherwise disposed of. Cranes, tramways, &c., are provided for moving the casks; and in this, as in the other parts of the establishment, hand labour is reduced to a minimum. On the floor above the empty casks are kept. The bottling operations are conducted with great expedition. The bottles arrive in crates, on being taken from which they are thoroughly rinsed, placed on hand-trucks, and brought forward to the bottlers. A boy sits in front of each bottling-machine, which consists of a series of six taps arranged on the syphon principle. Beginning at the left-hand side of the machine, he places a bottle on each tap, and by the time he gets the sixth bottle attached the first is full. Removing the full bottle he replaces it by an empty one, and so on. The rapidity of his movements may be judged from the fact, that in the course of a day of ten hours he fills 12,000 bottles, which is equal to an average rate of twenty a-minute. For the service of each bottling-machine there is a corking-machine, worked by a stout lad assisted by a boy; and a staff of " wirers" and " foilers." The corking goes on at the same rate as the filling; but as the wiring is a slower process, two nimble-fingered boys are required to wire for each machine. The wiring is indispensable in the case of beer intended for export. In some cases the cork and neck of the bottle are covered with tinfoil, and in others by metallic capsules, which are attached with great expedition. When all these operations are completed, the bottles pass to the labeller, who, holding a bunch of gummed labels in one hand, applies them with the other to the bottles, which are always moist enough to make the paper adhere firmly. Thus got up, the bottles have rather a gaudy appearance. As they are subjected to the successive operations described, the bottles are passing from one side of the bottling-house to the other ; and as they leave the hands of the labellers, they reach the packers. One set of men in this department twist a layer of straw round each bottle, and another set pack the bottles into dryware barrels which contain two or four dozen each. When the barrels are filled, they are taken charge of by coopers, who put in the heads and make all secure. The barrels are then rolled across the yard to a steam-hoist, by which they are elevated to a platform on the railway siding belonging to the works, and thence taken to the port of shipment.

Adjoining the stores is the cooperage, which is on a scale commensurate with the other departments of the establishment. The beer-casks are made of seasoned oak, which is cut up and shaped by steam-machines. The old system of firing has been abolished, and after the staves are "set up,"—that is, arranged in a circular form within a strong iron hoop,—they are placed under a hollow iron cone, and subjected to the action of steam, which speedily makes them pliable; and while in that condition, they are "trussed" into form. Making large casks is heavy work. The dryware barrels are of slighter construction and are hooped with wood. Of these an immense number are turned out weekly.

Notwithstanding the numerous mechanical appliances which exist in the various departments of Messrs Jeffrey's establishments, they require the services of 250 men.

DISTILLING.

DISTILLING has been described as " the art of evoking the fiery demon of drunkenness from his attempered state in wine and beer." Though the Arabians from the remotest ages extracted the aromatic essences of plants by distillation, no mention of the production of an intoxicating spirit by the same mode occurs until the eleventh century. It is first alluded to by an Arabian physician; but the discovery is believed by some authors to have been made in one of the northern countries of Europe. Another physician, who wrote in the thirteenth century, refers explicitly to an intoxicating spirit obtained by the distillation of wine, and he describes it as a recent discovery. So delighted was the physician by the effect of the spirit, that he pronounced it to be the *panacea*, or cure for all evils and disorders, so long sought after in vain. It need hardly be said that many persons still hold a similar belief in the virtues of alcohol. Raymond Lully, the famous chemist of Majorca, who was a disciple of the physician last referred to, claims for alcohol an important mission. He describes " the admirable essence" to be " an emanation of the divinity—an element newly revealed to man, but hid from antiquity, because the human race were then too young to need this beverage, destined to revive the energies of modern decrepitude." He further imagined that the discovery of the *aqua vitæ*, as it was called, indicated the approaching consummation of all things—the end of the world. The process of distillation was thus described by Lully :— " Limpid and well-flavoured red or white wine is to be digested during twenty days in a close vessel by the heat of fermenting horse-dung, and to be then distilled in a sand-bath with a very gentle fire. The true water of life will come over in precious drops, which, being rectified by three or four successive distillations, will afford the wonderful quintessence of wine."

From its birthplace, the art of producing ardent spirits by distillation slowly extended over Europe. No reliable information exists as to the date of its introduction into Britain. It is certain, however, that spirits were imported as early as 1430. The French applied themselves to the distillation of brandy from wine, and were so successful that their country came to be spoken of as the great still-house of Europe, England being one of their best customers. Meanwhile, the people of England had made some progress in agriculture; and grain having in consequence become plentiful, they began to distil spirits from it. The home-made article appeared to suit their palates better than the French product; and the manufacture of it was gradually developed into an extensive branch of industry, which was visited with legislative patronage in the reign of Charles II., when a duty of twopence was imposed on every gallon of spirits.

The people of Scotland were not far behind their neighbours in turning attention to distilling. They preferred spirits to the wines which they obtained from the Continent and to the beer which they brewed at home. No record exists of the introduction or progress of the trade prior to 1708, when 50,844 gallons of spirits were produced. A duty had been levied on the article before that time, and acted as a partial check on its production. Though the trade did not increase so rapidly as it probably would have done if there had been no tax, yet the progress made by distillers was astonishing. In 1756 they made 433,811 gallons of spirits; but an increase in the rate of duty in that year had the effect of causing a considerable falling off in the quantity produced. About the year 1776 a demand for Scotch spirits sprang up in England, and large quantities were sent thither. An import duty of 2s. 6d. a gallon was charged in England; and an extensive system of smuggling also sprang up. It is stated that in 1787 upwards of 300,000 gallons crossed the Border without the knowledge of the Excise. The mode of charging duty on the spirits made by the distillers gave place, in 1786, to a license duty according to the capacity of the stills. The distillers soon found that, by altering the form of the stills, they could increase the rate of production immensely. Government, becoming aware of the ingenious device of the Scotch distillers, raised the amount of license step by step, until, before the end of last century, it amounted to L.64, 16s. 4d. per gallon of still contents in the Lowlands, and to L.3 per gallon in the Highlands. Passing over many changes that have taken place in the interval, it may be sufficient to state here that the license duty at present payable by distillers is L.10, 10s., with 10s. of spirit-duty for every gallon of whisky sent out for home consumption.

Eighty or ninety years ago the illicit manufacture of whisky was common throughout the Highlands. At first the " sma' stills" were set to work to produce a supply of whisky for the use of the owners and their friends; but as the restrictions on licensed distillers were increased, the proprietors of the unlicensed stills were encouraged to ·extend their operations, and to enter into competition with the legal manufacturers. Then began that system of smuggling which made a certain class of Highlanders so notorious, and gave so much trouble to the Excise department. The wild glens of the north afforded secure retreats for the working of the stills; and many ingenious modes of conveying the produce to market were devised. The tendency was to demoralise the smugglers, and cast them back towards barbarism. They became reckless and daring to an extraordinary degree, and the stories of smuggling adventures record the performance of acts which, had they been rendered in a legitimate service, would have conferred undying honour on the actors. A man who could " jink the gauger" was a hero in the little circle in which he moved, and the people of the rural districts generally hailed with delight the performance of any deed which set the Excise laws at defiance. Even persons in authority winked at the breach of those laws. The great strongholds of the smugglers in the north were Glenlivet, Strathden, and the Glen of New Mill. The proprietor of the only distillery now in Glenlivet recollects seeing 200 illicit stills at work in Glenlivet alone. Owing to the quality of the water and other causes, the whisky made in the Glen became famous—indeed, smuggled whisky generally was preferred by consumers, on account of its mildness and fine flavour. The solitary distillery in the Glen has an interesting history, and the spirits made at it retain the old renown. Mr George Smith, the proprietor of the distillery, was the pioneer of licensed distilling in the Highlands, and he recently supplied to a correspondent of the "London Scotsman" the following account of the origin of his establishment, which is worth reproducing:—

" About this time (1820), the Government, giving its mind to internal reforms, began to awaken to the fact that it might be possible to realise a considerable revenue from the whisky duty north of the

Grampians. No doubt they were helped to this conviction by the grumbling of the south country distillers, whose profits were destroyed by the quantity of kegs which used to come streaming down the mountain passes. But through long impunity the Highlands had become demoralised, and the authorities thought it would be 'safer to use policy than force. The question was frequently debated in both Houses of Parliament, and strong representations made to the north country proprietors to use their influence in the cause of law and order. Pressure of this sort was brought to bear very strongly upon Alexander, Duke of Gordon, who at length was stirred up to make a reply. The Highlanders, he said, were born distillers; whisky was their beverage from time immemorial, and they would have it, and would sell it too, when tempted by so large a duty; but, said Duke Alexander, if the Legislature would pass an Act, affording an opening for the manufacture of whisky as good as the smuggled product, at a reasonable duty easily payable, he and his brother proprietors of the Highlands would use their best endeavours to put down smuggling and to encourage legal distillation. As the outcome of this pledge, a bill was passed in 1823, to include Scotland, sanctioning legal distillation at a duty of 2s. 3d. per wine gallon proof spirit, with L.10 license for any seized still above forty gallons; none under that seize being allowed.

"This would seem a heavy blow to smuggling; and for a year or two before the farce of an attempt had been made to inflict a L.20 penalty where any quantity of smuggled whisky was found manufactured or in process of manufacture. But there were no means of enforcing such a penalty, for the smugglers laughed at attempts of seizure; and when the new Act was heard of, both in Glenlivet and in the Highlands of Aberdeenshire, they ridiculed the idea that any one would be found daring enough to commence legal distillation in their midst. The proprietors were very anxious to fulfil their pledges to Government, and did everything they could to encourage the commencement of legal distillation; but the desperate character of the smugglers and the violence of their threats deterred any one for some time. At length, in 1824, I, George Smith, who was then a powerful robust young fellow, and not given to be easily 'fleggit,' determined to chance it. I was already a tenant of the Duke, and received every encouragement in my undertaking from his Grace himself, and his factor, Mr Skinner. The lookout was an ugly one, though. I was warned before I began by my civil neighbours that they meant to burn the new distillery to the ground, and me in the heart of it. The laird of Aberlour presented me with a pair of hair-trigger pistols, worth ten guineas, and they were never out of my belt for years. I got together two or three stout fellows for servants, armed them with pistols, and let it be known everywhere that I would fight for my place till the last shot. I had a pretty good character as a man of my word, and through watching, by turns, every night for years, we contrived to save the distillery from the fate so freely predicted for it. But I often, both at kirk and market, had rough times of it among the glen people; and if it had not been for the laird of Aberlour's pistols, I don't think I should have been telling you this story now. In 1825 and '26 three more small legal distilleries were commenced in the Glen; but the smugglers succeeded very soon in frightening away their occupants, none of whom ventured to hang on a single year in the face of the threats uttered so freely against them. Threats were not the only weapons used. In 1825 a distillery which had just been started at the head of Aberdeenshire, near the Banks o' Dee, was burned to the ground with all its out-buildings and appliances, and the distiller had a very narrow escape from being roasted in his own kiln. The country was in a desperately lawless state at this time. The riding officers of the revenue were the mere sport of the smugglers, and nothing was more common than for them to be shown a still at work, and then coolly defied to make a seizure."

Though smuggling was most extensively carried on in the Highland regions, the trade was not limited to them, and unlicensed stills have been discovered in situations which might be considered beyond the breath of suspicion. In Arnot's " History of Edinburgh," it is stated that, while in 1777 there were only 8 licensed stills in Edinburgh, the unlicensed numbered 400. In July 1815, a "private" distillery of considerable extent, which had been in operation for eighteen months, was discovered under an arch of the South Bridge in Edinburgh. The bridge consists of a large number of arches, only one of which is open, the others being blocked up by the houses which line the bridge on both sides. It was in one of the arches adjoining the open one that the distillery had been set up. All the arrangements for conducting the business without the knowledge of the Excise officers were of the most complete kind. The only entrance to the place was by a doorway situated behind the grate of a bedroom in a house on one of the lower flats adjoining the bridge. Between this doorway and the distillery communication was established by means of a ladder and trap-door. A supply of water was obtained from a branch attached to one of the mains of the Water Company which passed overhead, and the smoke and waste were got rid of by making an opening in the chimney of one of the adjoining houses, and establishing a communication with the soil-pipes. The spirits were sent out in a tin case capable of containing two or three gallons. The case was placed in a bag, and taken to the customers by a woman in the service. When the place was entered by the officers of the law, a large quantity of material and all the appliances employed in the manufacture of whisky were found. A number of years ago a secret distillery was discovered in the cellars under the Free Tron Church, in the High Street of Edinburgh ; and soon afterwards, an extensive concern of the same kind was revealed at Marionville, between Edinburgh and Portobello. More recently, a still was seized at work almost within a stone-throw of the Excise office in Aberdeen, and subsequently there was a " seizure " in a close in the centre of Leith. Not long since, a still was discovered in England beneath the pulpit of a church; and only the other day a thriving business in Nottingham was interrupted in consequence of the absence of the all-important license.

In the year 1799 there were 87 licensed distillers in Scotland, who paid duty on spirits retained for home consumption to the amount of L.1,620,388. That was the first year of the change in the mode of levying duty. Previously so much was paid according to the capacity of the still, but now a 4s. 10¼d. duty was laid on every gallon of spirits made for home consumption. The change was not approved of by the distillers, about a third of whom gave up business in the following year, and the duty decreased to L.775,750. The lowering of the duty to 3s. 10½d. in 1802 revived the trade, and the returns for 1803 showed 88 distillers paying L.2,022,409. In 1804 progress was checked by another advance in the duty, and the number of distillers dwindled down until, in 1813, there were only 24. The duty reached 9s. 4½d. a-gallon in 1815, but the produce was considerably under a million pounds. It is probable, however, that the quantity of whisky actually made in the country was greater than at any previous time, the high duty tending, as already stated, to foster illicit distillation and smuggling. The lowering of the duty to 2s. 4¾d. in 1823 had the effect of giving an impetus to the trade. The number of licensed distillers greatly increased, and the revenue rose steadily. There were 243 distillers in 1833, who paid duty to the amount of L.5,988,556, the rate then being 3s. 4d. a-gallon.

The following table shows the rates of duty in England, Scotland, and Ireland respectively at various periods, and the quantity of spirits charged with duty for consumption in Scotland :—

2 F

Years	England.			Scotland.						Ireland.			Paid Duty in Scotland for Consumption.
	Duty per Gallon.			Lowland.			Highland.			Duty per Gallon.			
				Duty per Gallon of Still Contents.									
	£	s.	d.	£	s.	d.	£	s.	d.	£	s.	d.	Gallons.
1791	0	3	4¾	3	12	0	1	4	0	0	1	1¼	
1794	0	3	10¾	10	16	0½	1	16	0	0	1	1¼	
1797	0	4	10¼	64	16	4	3	0	0	0	1	5¼	
1800	0	5	4¼	64	16	4	7	16	0¼	0	2	4¼	1,277,596
				Per Gallon Spirits made.									
1802	0	5	4¼	0	3	10½	0	3	4¼	0	2	10¼	1,158,558
1804	0	8	0½	0	5	9¾	0	5	0½	0	4	1	1,189,757
1807	0	8	0½	0	5	8¾	0	4	11¼	0	4	1	2,653,478
1811	0	10	2¾	0	8	0¼	0	6	7½	0	2	6½	1,951,092
				£	s.	d.							
1815	0	10	2¾	0	9	4½				0	6	1½	1,591,148
1817	0	10	2¾	0	6	2				0	5	7¼	1,906,950
1823	0	11	8¼	0	2	4¾				0	2	4¾	2,303,286
1825	0	11	8¼	0	2	4¾				0	2	4¾	5,981,549
1826	0	7	0	0	2	10				0	2	10	3,988,788
1830	0	7	6	0	3	4				0	3	4	6,007,631
1840	0	7	10	0	3	8				0	2	8	6,180,138
1852	0	7	10	0	3	8				0	2	8	7,172,015
1853	0	7	10	0	4	8				0	3	4	6,584,648
1854	0	7	10	0	6	0				0	4	0	6,553,239
1856	0	8	0	0	8	0				0	6	2	7,175,939
1860	0	10	0	0	10	0				0	10	0	4,729,705
1864	0	10	0	0	10	0				0	10	0	5,014,121
1867	0	10	0	0	10	0				0	10	0	4,983,009

The following table shows the quantity of whisky made in Scotland in the years given, and also the number of licensed distillers :—

Years.	Number of Distillers.	Gallons.
1824	120	5,908,373
1830	249	9,815,257
1840	205	9,032,353
1850	161	11,638,429
1855	137	11,283,636
1861	122	11,879,436
1862	119	13,113,384
1863	...	13,228,547
1864	118	14,869,564
1865	116	13,445,752
1866	112	11,999,119
1867	111	10,813,996

The two most extensive distilleries in Scotland are the Port-Dundas Distillery in Glasgow, which belongs to Messrs M. M'Farlane and Co., and the Caledonian Distillery, in Edinburgh, owned by Messrs Menzies, Bernard, & Co. The latter has been chosen to illustrate the processes of distillation. The establishment covers five acres of ground at the west end of the city, in a situation most convenient for carrying on a large trade. It is one of the most recently erected distilleries in the country, having been built in 1855 ; and every part of it has been constructed according to the latest improvements in the trade. All the principal buildings are five storeys in height, and are so arranged that the labour of carrying the materials through the various stages of manufacture is reduced to the smallest amount. A branch line from the Caledonian and another from the North British Railway converge in the centre of the works, and afford ready and convenient accommodation for bringing in the raw materials and sending out the products. The extent of this traffic may be judged from the facts that 2000 qrs. of grain and 200 tons of coal are used every week, while the quantity of spirits sent out in the same time is 40,000 gallons, the duty on which is L.20,000, or at the rate of L.1,040,000 a-year. The machinery is propelled by five steam-engines of from 5 to 150 horse power, for the service of which, and supplying the steam used in distilling, there are nine large steam-boilers.

The Caledonian is one of the eight distilleries in Scotland which produce "grain whisky," the others making malt whisky only. The kinds of grain used are maize, rye, buckwheat, oats, and barley. The latter is converted into malt, which is used in certain proportions with the other grains in a raw state. Most of the grain is imported into Granton, whence it is brought, by rail, in special trucks to the distillery. The trucks hold fifty quarters each, and when they arrive they are taken one after the other to a part of the line adjoining the stores and malting premises. Beneath the spot to which they are brought is a pit, forming the terminus of an Archimedean screw and tunnel. The opening of a valve in the bottom of the truck allows the grain to run into the pit, from which the screw draws it along to a hopper inside the building. From the hopper it is carried by belt and bucket apparatus to the store on the upper floor, where other sets of screws carry it along and deposit it in an even layer on the floor—so that no manual labour is required in this part of the work. The contents of two trucks, or 100 quarters, are thus disposed of in an hour. The stores have accommodation for 10,000 quarters. As most of the grain used has to be dried as well as the malt, the kilns are very extensive, and are capable of drying about 400 quarters a-day. One of the kilns is heated by the waste steam from one of the engines, and the others are fired with coke. The malt-barns are capable of producing about 600 quarters a-week. From the kilns the malt and grain are transferred to the mill, where the former is bruised between rollers, and the latter ground into meal by means of common mill-stones.

The next process is the mashing, which works a wonderful change on the constituents of the grain, converting a large proportion of them into starch-sugar, which is the prime source of alcohol. In the mashing department are four wort and water tanks, capable of containing 30,000 gallons each, and five mash-tons of from 10,000 to 30,000 gallons. Water, at a temperature of about 160°, is run into the tons, and then the "grist" is added, the compound being mixed thoroughly by a set of revolving rakes. Many nice points have to be considered in this as in the other processes of spirit manufacture. The proportion of different kinds of grain used in the mash determines the temperature to which the water should be raised, and also the duration of the process. Malt is more easily mashed than a mixture with raw grain; but the latter produces a greater proportion of spirit. The saccharine matter extracted in mashing is held in solution by the water, and so drawn off. When the mashing is completed, the liquid is run into large tons called "underbacks," leaving the spent grains, or draff, in the mash-ton. At this stage the liquid is known as "wort." The wort contains the elements of alcohol; but another process—that of fermentation—is necessary for the entire conversion. Before being subjected to fermentation, the wort must, as speedily as possible, be cooled to a certain point. For this purpose it is pumped from the underbacks to a large cistern at the top of the building, from which it flows into refrigerators. Cooling used to be effected by running the liquid into shallow iron troughs, and causing currents of cold air to sweep over it. In order to cool the produce at this great establishment, several acres of such coolers would have been required; but the invention of the refrigerator has made it possible to do all the cooling in very little space, and in much less time than by the old method. The refrigerator consists of a range of tall copper cylinders, inside of which a number of tubes are so arranged that while the wort flows through them, a current of cold water passes over the outside, so that, as the liquid traverses the cylinders it is cooled to the requisite degree. From the refrigerator the wort passes to the fermenting-room—a great apartment occupied by twelve tuns of 50,000 gallons each. Distillers have to work under regulations designed for the convenience of the Excise department, and certain things have to be done on certain days. Thus the distilling process is carried on during Monday and Tuesday, while the mashing and brewing occupy the remainder of the week. The distillers consider this compulsory idleness of their fermenting plant while the distilling is in progress to be a great hardship, and a movement has been made on several occasions to obtain power to carry on the manufacture of beer as well as of whisky, but the privilege has hitherto been refused. When the tuns are filled with wort, yeast is added, and the process of fermentation goes on, under survey of Excise officers, of whom no fewer than eleven are stationed on the Caledonian Distillery. By fermentation and the decomposition of the sugar alcohol is developed in the wort. When the fermentation is completed, the liquid, which then receives the name of " wash," is measured, and its strength ascertained by the Excise officers.

All that now remains to be done is to extract the alcohol from the wash. This is done by distillation. The stills first used were exceedingly simple in their form and mode of action, but were suitable only for dealing with small quantities of liquor. The common still was brought to the highest degree of perfection in Scotland during the time that the method of charging duty according to the capacity of the stills prevailed. Charging in that way led the distillers to devise stills which would produce a greater quantity of spirit in a given time than those of the old form were equal to. They made their stills broad and shallow, so that a larger surface was exposed to the fire. But an entire change in the form of the apparatus was subsequently effected. In 1832 Mr Æneas Coffey, Inspector-General of Excise in Ireland, patented a still which, as finally improved, is the best apparatus in use for the production on a large scale of a rectified spirit of high strength, with the greatest economy of time and fuel. In most of the extensive distilleries Coffey's still is employed. The Caledonian Distillery contains the largest still in Scotland. Before describing its mode of action, it may be well to state generally the scientific principle involved in the process of distillation. Alcohol is more volatile than water. Pure alcohol boils at 173°, water at 212°, while mixtures of the two liquids boil at an intermediate temperature. Therefore, when the wash is raised to the boiling point of alcohol, the latter is converted into vapour, and passes off into a condenser. Some water and a portion of fusel oil are carried along with the alcoholic vapour in distillation after the old method, and have to be separated by the "low wines," as the produce of the first operation is called, being passed through the still again and again until the desired degree of purity is attained. With Coffey's apparatus the distillation is completed at one operation. The cold wash flows in at one part of the still, and at another a strong sparkling spirit gushes forth at the rate of 1000 gallons an hour. Coffey's still consists of two columns—one called the analyzer and the other the rectifier—each forty feet in height; and the wash, in passing through them, loses its alcohol by evaporation, which is due to the action of steam on a system of copper pipes and perforated trays. The apparatus is so constructed that it is impossible for spirit of less than a certain degree of strength and purity to pass out of it. The product of this still is a "neutral spirit," which, being deprived of all essential oils, is, when matured by age, considered by some consumers the most wholesome. A great quantity of it is sent to London, where it is converted into brandy and gin by the addition of flavouring ingredients. In the still-room is a large rectifier, by means of which some of the spirit from Coffey's still is brought to a higher degree of purity.

In order to meet a growing demand for the variety of whisky known as "Irish," the proprietors of the Caledonian distillery, about two years ago, fitted up two large stills of the old pattern, with which they manufacture whisky similar to that made in Dublin. In connection with this branch of the business, stores capable of accommodating about 5000 puncheons have recently been constructed, in which the various kinds of whisky are allowed to lie for some time before being sent out.

In proportion to the amount of money turned over, fewer workpeople are engaged in distilling and brewing than perhaps in any other branch of manufacture, as is proved by the fact that in this great establishment only 150 men are employed. The plant and stock, however, represent immense capital.

SUGAR-REFINING.

SUGAR is so extensively used among all classes of society, that it is
regarded as almost a necessary. In point of commercial importance
it ranks very high, and the duty drawn from it forms a considerable item in the national revenue. Yet, up till what must be
considered a recent period, in speaking of what is a common article
of food, sugar was mentioned only as a medicine, and subsequently
as a luxury, to which the wealthy alone could aspire. The first
reference to sugar occurs in the works of Theophrastus, who lived
320 years before the Christian era, where it is called "a sort of
honey extracted from canes or reeds." Subsequent writers among
the ancients refer to sugar as an object of curiosity, and Pliny expressly states that it was used in medicine only. The Chinese were
acquainted from a remote period with the process of making sugar-
candy, and the Greeks and Romans obtained the little that came
under their notice from China. It would appear that Europe is
indebted to the Saracens not only for the first considerable supplies
of sugar, but for the earliest example of its manufacture. After
conquering the islands of Rhodes, Cyprus, Sicily, and Crete in
the ninth century, the Saracens introduced into them the sugar-
cane; and from that source Europeans drew their first supplies.
The same people began the cultivation of the sugar-cane in Spain
as soon as they obtained a footing in it. About 1420 the Portuguese took the sugar-cane from Sicily to Madeira and the Canary
Islands; and from these places the cultivation of the cane and the
art of making sugar were extended to the West Indian Islands and
the Brazils. The first British settlement in the West Indies was

Barbadoes, which was taken possession of in 1627. Twenty years afterwards sugar began to be exported from the island, and that was the first sugar produced in British possessions, the supply previously being obtained chiefly from the Portuguese settlements in Brazil. The British planters in Barbadoes, after becoming thoroughly acquainted with the modes of cultivating and extracting the sugar, did a large trade, and thereby accumulated much wealth. In 1676 they employed in their sugar trade a fleet of four hundred vessels. The first mention of sugar being imported into England is found in Marin's "History of the Commerce of Venice," which refers to a shipment at Venice for England in the year 1319 of about forty tons of sugar and four tons of sugar-candy. In those early days honey was the principal ingredient used in sweetening liquors and dishes ; and for many years after sugar had been introduced, it was used only in the houses of the wealthy. Not till the latter part of the seventeenth century, when coffee and tea began to be consumed, did sugar come into general use.

For upwards of two centuries sugar has been subject to an import duty. In 1661, when it was first levied, the duty amounted to 1s. 6d. per cwt. on all sugar brought from British plantations. Eight years afterwards it was raised to 3s. By a series of augmentations the duty rose to 30s. per cwt. in 1806. From 1793 to 1803 the duty on East India sugar was 37 and 38 per cent. *ad valorem*, and for many years afterwards it was 11s. and 8s. per cwt. higher than the duty on West India sugar. The rates of duty were assimilated in 1836 by the reduction of the charge on East India sugar from 32s. to 24s. per cwt. A prohibitory duty of 60s. to 63s. per cwt. was imposed on foreign sugars up till 1844, when it was, under certain circumstances, lowered to 34s.; and in some favoured instances (to encourage free instead of slave labour) the charge was fixed as low as 23s. 6d. Sugar from British possessions was at that time paying from 14s. to 21s., according to quality. By Acts passed in 1846 and 1848, the duties on British and foreign sugar were gradually equalised. The equalisation was completed in 1854, in which year all sugars paid from 10s. to 13s. 4d., according to quality. Those rates were increased during the Russian war, and, by the tariff of 1860, another advance was made to from 12s. 8d. to 18s. 4d. In 1867 the duty was lowered to from 8s. to 12s. per cwt.

The quantity of sugar consumed in Britain in the year 1700 was only 10,000 tons, while in 1868 it was nearly 700,000 tons. The following table, though not quite perfect, shows correctly enough for general readers the quantity consumed in various years, the number

of pounds per head of the population, the price per cwt. including
duty, and the amount of duty paid on all descriptions of sugar :—

Years.	Quantity Consumed. Tons.	Lb. per Head.	Average Rate of Duty per Cwt.		Average Price, including Duty.		Total of Duty Paid.
			s.	*d.*	*s.*	*d.*	£
1700	10,000	3	...		...		...
1754	53,270	12	...		...		...
1801	159,916	22	20	0	79	5	3,066,163
1821	170,612	18	27	0	60	2	4,188,997
1841	202,899	17	25	2	64	1	5,114,390
1851	328,581	26	12	0	37	6	3,979,141
1861	446,865	34	13	10	40	4	6,104,325
1867	625,000	47	10	0	40	0	5,816,445

It is estimated that the annual production of sugar over the world
amounts to 3,000,000 tons. Roughly dividing the nations into
groups, the interesting fact is shown that the Anglo-Saxon races—
Great Britain and her colonies and the United States—are the most
important consumers, as they use 1,142,000 tons of sugar per annum,
or 41·40 ℔ per head. The Latin races come next. France, Italy,
Spain, Belgium, Portugal, and Switzerland use 506,000 tons per
annum, or 12·34 ℔. per head. The Zollverein, Austria, Holland,
the Hanseatic League, and Denmark, consume 262,000 tons per
annum, or 7·30 ℔. per head. Last come the vast but poverty-
stricken districts ruled by Russia, together with the semi-barbarous
Ottoman Empire, and the kingdom of Greece. Russia, Poland,
Turkey, and Greece consume less than half of what is used by the
smallest civilised European consumers, and the deliveries in these
countries amount to only 125,000 tons, or 3·30 ℔. per head.

In Stow's "Survey of London," it is stated that the art of refining
sugar was commenced in that city about the year 1544, by the erection of two refineries. After twenty years of varied fortune, those
refineries turned out to be very remunerative to their owners, and
many other persons embarked in the business. For a considerable
time the London refineries enjoyed a monopoly of the trade, which
was subsequently shared with Liverpool and Bristol. Scotland was
long behind the sister kingdom in engaging in sugar-refining; but
now the Scotch refiners are rivalling those of England, and have,
during the past ten years, made marvellous progress. There are in
Scotland twenty sugar refineries—fourteen at Greenock, three at
Glasgow, two at Leith, and one at Port-Glasgow ; but several of
these, owing to various causes, are not in operation at present.

The first sugar-refinery in Greenock was started in 1765 by a few

West Indian merchants, who formed themselves into a company for that purpose. They fixed upon Greenock, which at that time was the port of Glasgow, as the most suitable place for a sugar-house. A site at the foot of Sugar-House Lane, adjoining the West Burn, and near the West Harbour, was secured; and a sugar-house, which at the time was considered a great undertaking, was erected. This building is still in existence, but it has been repeatedly enlarged and improved. It is now occupied by Messrs A. Currie & Co. The second sugar-house was started in 1787 by a company of Greenock merchants. A large substantial dwelling-house, at the head of Sugar-House Lane, was purchased by them, and converted into a refinery. It had two pans at starting, but the number was afterwards increased. That house was in operation up till within the last five years, when it was sold and converted into a store. A third sugar-house was built in 1802, in Bogle Street, by Messrs Robert Macfie & Sons, the founders of extensive refineries in Liverpool and Leith. It contained two pans at starting, but a third was added eight years afterwards; and it was then considered an extensive concern. This refinery has been in the market for some time, and a part of it was recently let as a store. The fourth sugar-house was built by Messrs James Fairrie & Co., on Cartsburn, in 1809. It was burned many years ago, but another house was built farther up the burn, and is now occupied by a young firm. The fifth sugar-house was built in 1812, by Messrs Wm. Leitch & Co., in the Glebe. This house is not now a refinery, but the name has been adopted by the Glebe Sugar-Refinery Company, now the most extensive refiners in Greenock. It thus appears that in the forty-seven years up till 1812, five sugar-houses had been erected in Greenock, the third, fourth, and fifth having been built within a period of ten years. For the next fourteen years there was a lull, and it was not till 1826 that the sixth house was built, on a site at the High Bridge. Three years later Messrs Tasker built a refinery in Dallingburn Street, which is carried on by Messrs Crawhalls & Co. From 1830 till 1860, refineries were built for or leased by the following firms :—Messrs Pattens; Walkers & Co.; Richardsons & Co.; Anderson, Orr, & Co.; A. Anderson & Co.; Blair, Reid, & Steele; Neill & Dempster; Alexander Scott & Co.; and Ballantyne, Adam, & Rowan. The houses of the two last named firms were built about eleven years ago. During the past ten years only one new house has been built—namely, the Orchards Sugar-Refinery, erected in 1863; but two or three of the old ones have been rebuilt, and two of the houses mentioned are being rebuilt—one of them on a new site. Of the Greenock refineries ten are in operation, two closed temporarily, and two are being rebuilt. The most extensive works are those of the Glebe Sugar-Refining Company, and of Messrs Walker & Company. Each of these, it is said, turns out about 700 tons of sugar a-week. The Glebe Sugar-Refining Company paid, in 1868, to the Customs, as duty on the sugar refined in their establishment, L.350,000. They employ about 300 hands.

At Glasgow there are three refineries. Two are at present working, and one has been stopped, owing to action taken by the creditors of the concern. A sugar-house which was carried on for many years in Port-Glasgow has been sold, and the proprietors have in contemplation a large extension of their refinery in Greenock.

The great improvements made in the machinery used in the refining process have, in the course of ten years, and without any addition to the number of refineries in operation, increased threefold the quantity of sugar refined—the raw sugar consumed by all the refineries on the Clyde in 1858 being 56,769 tons—in 1868 it rose to 171,643 tons. The average quantity of refined sugar turned out by the refineries in Greenock is about 60 tons a-day, when working full time, so that the fourteen refineries are equal to the production of 840 tons a-day. The amount of shipping employed in the sugar trade is very large. Last year there arrived in the Clyde 416 vessels, of about 140,000 tons, or an average of 335 tons. About 400 of these discharged their cargoes at Greenock. With the exception of 16 vessels from Mauritius, 2 from Java, and 30 from Dunkirk and Antwerp, the cargoes were from the West Indies and Brazils, 184 vessels being from Cuba alone.

The following statement of the imports into the Clyde during the past twelve years will show the progress of the trade:—

	Sugar. Tons.	Molasses. Tons.
1857, ...	38,336	9,531
1858, ...	56,769	15,077
1859, ...	69,991	9,720
1860, ...	75,087	4,952
1861, ...	88,694	18,229
1862, ...	106,748	14,802
1863, ...	121,044	5,664
1864, ...	126,061	3,483
1865, ...	136,540	3,708
1866, ...	162,368	2,912
1867, ...	178,013	2,919
1868, ...	171,643	4,880
Decreased consumption of raw sugar in 1868,	6,370	...
Increased consumption of molasses in 1868,	...	1,961

It is evident, from recent articles in trade newspapers, that the London refiners are finding those on the Clyde to be very formidable competitors. Everywhere the Clyde sugars are preferred by consumers, as being superior in quality to, and quite as cheap as, those of London. The spirit of enterprise which has made the banks of the Clyde famous in the annals of commerce and manufactures has been more manifest in nothing than in the manner in which the trade of sugar-refining has been taken up and is now carried on. Every promising invention tending to perfect and expedite the processes is taken advantage of, and the Clyde refiners are in all respects the best in existence. The London refiners would appear to have felt secure in the monopoly which they long enjoyed of supplying sugar to the millions of consumers in the metropolis, and they have in most cases neglected to keep up to the march of improvement. Old and costly processes are retained, and the produce is losing caste in the markets where it comes into competition with Clyde sugar. Up till about three years ago, when a refinery was started in Ireland, nearly all the refined sugar consumed in that division of the kingdom was supplied from Greenock; and it is a remarkable fact that even yet the steamers trading to Dublin, Cork, Londonderry, and Belfast, draw the greater part of their freight from sugar, from 250 to 600 tierces being carried by them weekly.

With reference to the early days of sugar-refining in the east of Scotland we find the following in Arnot's " History of Edinburgh:" —" There is hardly any branch of manufacture which in speculation affords a more undoubted success to the manufacturer, and more general benefit to the country, than the baking or refining of sugars; and we will venture to say that it has been owing alone to the want of capital and conduct in its managers that it has not hitherto been attended with remarkable success. Were this manufacture properly conducted a trade might be established between the West Indies and the east coast of Scotland. Sugars might be afforded to the consumer at an easier rate, the planter and manufacturer might carry on an advantageous species of traffic, and a great sum of money might be saved to the country which is annually remitted to London for baked sugars. There are four sugar-houses on the east coast of Scotland—at Edinburgh, Leith, Dundee, and Aberdeen. These at present are mostly supplied from Glasgow. Now, supposing every house to use 500 hogsheads annually, these, amounting to 2000 hogsheads, with the usual proportions of rum, cotton, coffee, mahogany, &c., would make cargoes for ten or twelve sail of good ships; and these might return with cargoes of linen, negroes' clothing, and the various other articles for which there is a demand in the West Indies. Leith is the most centrical port for carrying on such trade. Vessels can be fitted out there easier than from the Clyde, and greatly lower than from London. Thus a saving would be made on the article of freight; other charges would be likewise more moderate than either in the Clyde or at London; and the sugars, when landed, would be worth from four to five per cent. more to the sugar-houses than if landed either at Greenock or London. This, added to the savings on freight and charges, would amount to a valuable consideration to the West India planter, and should, no doubt, encourage him to make consignments to the port of Leith. A house for baking of sugars was set up in Edinburgh A.D. 1751, and the manufacture is still carried on by the company who instituted it. That of Leith was begun in the year 1757 by a company consisting mostly of bankers in Edinburgh; but in five years their capital was totally lost. For some time the sugar-house remained unoccupied, till some gentlemen from England took a lease of the subject, and revived the manufacture; but as these wanted capital altogether, and were consequently obliged to fall upon ruinous schemes for supporting a fictitious credit, they were speedily involved in destruction. To them succeeded the Messrs Parkers, who kept up the manufacture above five years. The house was then purchased by a set of merchants in Leith, who, as they began with a sufficient capital, as they have employed in the work the best refiners of sugar that could be procured in London, and who, as they pay due attention to the business, promise to conduct it with every prospect of success."

The father of Mr Macfie, M.P. for the Leith Burghs, was a sugar-refiner many years ago in Leith. His establishment, which stood in Elbe Street, South Leith, was destroyed by fire. An extensive sugar-refining business was long carried on in North Leith by Messrs Schultze, and afterwards by Messrs Ferguson; but the house has been shut for eight or ten years. About four years ago a number of merchants in Edinburgh and Leith formed themselves into a company, and built a sugar-refinery in Breadalbane Street, Leith, where, under the designation of the Bonnington Sugar-Refining Company, they carry on an extensive and thriving business. Their refinery is the most recently erected in Scotland, and embraces all the latest improvements in the processes and appliances of the trade. The refinery was partly built in 1865, but owing to an accident it was not in operation until the following year. It is a substantial-looking brick structure, arranged in blocks to suit the requirements of the various departments. The raw sugar is deposited on arrival in a large detached building, where it lies under Customs' bond, and whence it is withdrawn as required for the supply of the works. The great casks in which the sugar is imported weigh when full more than a ton each, but a powerful steam elevator makes light work of transferring them to the upper floor of the refinery. To follow them thither the visitor must ascend a narrow iron stair, which winds round the well in which the elevator works. The steps are black and clammy with the saccharine matter which finds its way most persistently to all parts of such establishments. Treacle seems to ooze from the iron hand-rail and from the brick lining of the stair-case, so that contact with either becomes most undesirable. Up and still up the stair winds, until from its summit a peep over the railing would make one giddy. Stepping through an arched open-ing, fitted with iron doors, the upper floor is reached. There lie the great casks in an atmosphere quite as hot as that which surrounded them when they were filled in the Indies. Men are busy forcing out the "heads" of the casks, and discharging the sugar upon the floor, through grated slits in which it quickly disappears. There are several recognised qualities of sugar, some being so light-coloured and clean that it would seem almost superfluous to subject it to refining; while in other cases the stuff has a dark and uninviting appearance, which would ensure its safety were the casks left open in the play-ground of any school. As the casks are emptied they are placed into a large iron tank, and subjected to the action of steam, which detaches all the sugar and thoroughly cleanses the casks. The refined sugars are sent out in the casks in which the raw material arrives, and, while the sugar is being operated upon in the refinery, the casks are taken to the cooperage, where they receive any necessary repair.

In order to ascertain what has become of the sugar that has dis-appeared so mysteriously through the hatchways, it is necessary to descend to the floor beneath. Here are two large circular iron vessels, called "blow-up cisterns," into which the raw sugar descends from the floor above. Hot water is let into the cistern, and an agitator, worked by steam, is kept going until the soluble matter is completely reduced. The liquid is then drawn off and filtered, in order that all insoluble matter may be removed. The filters con-sist of bags of stout twilled cotton, which are suspended in iron chambers on a lower floor. When the filters are in action steam is admitted to the chambers, which facilitates the process. From this ordeal the liquor comes forth clear and sparkling, but displaying a reddish tinge, the removal of which is the next care of the refiner, and causes him nearly as much trouble and expense as all the other operations put together.

The syrup is deprived of its colour by being passed through powdered charcoal. This operation is carried on in the charcoal-house, which adjoins and is nearly equal in extent to the main building. Animal charcoal, prepared by burning bones, is the substance used, and several hundreds of tons are constantly in use. The filtering-vessels are made of cast iron, and are cylindrical in form, each measuring fifteen feet in depth by nine in diameter. Into each filter are run about twenty tons of charcoal, in grain resembling coarse sand. Liquor from the bag filters is then allowed to flow in until the vessels are filled, when a tap in the bottom is opened, and the syrup runs off in a colourless stream. The charcoal loses its power of decolouring after being in use for two days, and then it has to be revivified. It is first washed by making hot water flow through it, then withdrawn from the filter, and brought to a red heat in peculiarly constructed kilns or furnaces. This treatment completely restores its powers. From the charcoal-house the syrup is pumped into great iron cisterns placed on one of the floors of the main building, and thence it is withdrawn to supply the pans.

Prior to the year 1812 the only mode practised for concentrating the syrup, or bringing it to the crystallising point, was boiling it in open copper vessels. In the year mentioned, Mr Howard patented his vacuum pan, one of the most important inventions in connection with sugar-refining. A high temperature destroys the crystallising property of sugar, and converts it into treacle; and when the old method of concentration was in use, great loss occurred in consequence of the large proportion of treacle which was inevitably produced. Mr Howard, having become acquainted with the well-known fact that liquids boil in vacuum at a much lower temperature than when exposed to the air, applied it to the concentration of sugar with the most complete success, and his pans are now universally employed. By the exercise of ordinary diligence the formation of treacle is almost entirely prevented, and the sugar is brought to much higher condition at one operation than it could formerly be by repeated refining. The Bonnington Company have two pans of the largest size and most perfect construction. Each is capable of boiling twelve tons of sugar at a time, and the cost of the pair was, we believe, L.3000. The body of a pan consists of a spherical copper vessel nine or ten feet in diameter. The bottom is double, leaving a space of an inch or two for the admission of steam. Inside the pan is a large copper pipe of a spiral form, through which a current of steam flows while the pan is in action. From the top of the pan a pipe leads to a large cylindrical vessel whither all the vapour given off by the sugar is drawn by an air pump and condensed by a jet of cold water. An immense quantity of water is required to supply the condensers, and it is drawn from cisterns on the roof of the building. As the supply is limited, the water, after performing its duty in the condensers, is pumped back to the cistern, and the same process is repeated. After the pan is charged with syrup the heat is gradually raised by increasing the flow of steam through the bottom casing and worm. The liquid boils at a temperature of 150°, and care must be taken not to let the heat get beyond what is barely essential for ebullition. The sugar-boiler must exercise constant vigilance, especially when the liquid begins to thicken. The pan is .fitted with instruments showing the temperature and the degree of rarity of the air; and there is a peep-hole through which the sugar inside the pan may be seen. As the boiling proceeds, the attendant withdraws samples by means of a peculiarly constructed test-rod, which extracts some of the contents of the pan without letting in air. By taking a drop of the syrup between his finger and thumb, and drawing it out into a thin film, he is able to note the progress of crystallisation. When a certain degree of consistency has been attained, the greater portion of the contents of the pan is drawn off into open copper vessels called "granulators," and fresh liquor is run into the pan.

Until a recent period, it was customary to complete the process of crystallisation by filling the pulpy stuff from the granulators into conical metallic moulds, which were set with the apex down in an earthenware jar. An orifice in the point of the cone was then opened, through which the uncrystallised syrup escaped. When the sugar was drained sufficiently dry, all that remained to do was to put it into casks and send it out. A much more expeditious mode of drying is now in general use. A Belgian sugar-refiner conceived the idea of fastening a number of the conical moulds on the rim of a large horizontal wheel, and driving off the syrup by centrifugal force as the wheel was spun round at a high rate of speed. An improvement in this method is the centrifugal drying-machine, now extensively employed in manufactories of various kinds. The machine consists of a circular vessel about four feet in diameter and eighteen inches deep. The bottom is formed of iron, and the sides of fine wire-gauze. Enclosing this vessel, at a distance of a few inches from the gauze, is a strong casing of iron. The vessel is attached to a vertical axle and gearing capable of causing it to spin round at the rate of a thousand revolutions a minute. At the Bonnington Company's Refinery there are six such machines on the floor beneath the pans and granulators. The semi-fluid sugar is conveyed from the granulators to the centrifugal machines in a waggon which runs on a railway placed at a convenient elevation. About a hundredweight is put into each machine, and so speedily is the drying accomplished, that three or four men are constantly employed in emptying the machines in succession. Number one is dry by the time number six is filled, and so the work goes on. The emptying is a speedy process—all that is necessary being to stop the machine, open two valves in the bottom, and shovel the sugar through them, when it falls into a truck placed to receive it. The sugar does not have time to cool before these operations are completed, and as it would not do to press the still warm sugar into casks, it is raised to the cooling-floor, where it is spread out in a layer about a foot in depth, and exposed to the free action of the air. When sufficiently cooled it is packed, and either sent to market or stored. The syrup which is driven off by the drying-machine is mixed with raw sugar in the "blow-up cistern," and thus no particle of crystallisable matter is lost.

The establishment turns out 250 tons of refined sugar every week, in the production of which about 100 men are engaged. For driving the machinery six steam-engines of 80 horse power in the aggregate are employed. Taking all the Scotch sugar-refineries together, the number of workmen is over 2000, whose wages may be stated as follows:—Boilers, L.200 to L.300 a-year; pan-men, 30s. to 40s. a-week; filtermen, 17s.; warehousemen, 18s.; upstair's-men (general hands), 16s. to 17s.; charcoal kilnmen, 16s. to 18s.

MANUFACTURE OF CONFECTIONERY.

THE CONFECTIONERY TRADE IN SCOTLAND—DESCRIPTION OF MESSRS KEILLER & SON'S MANUFACTORY AT DUNDEE—HOW MARMALADE AND OTHER CONFECTIONS ARE MADE.

THE luxurious mode of living which has become fashionable with the advance of civilisation has given rise in modern times to many branches of trade which were unknown to the ancients. An interesting example is the manufacture of confectionery, which has assumed dimensions entitling it to mention in any record of industrial progress. Conserves, or fruits preserved in sugar, are mentioned in the works of Shakspeare and other writers of the sixteenth and seventeenth centuries. The Queen in "Cymbeline," addressing Cornelius, says—

> "Hast thou not learned me to preserve? Yes, so
> That our great King doth woo me oft
> For my confections."

Again, in the bedchamber scene in "Taming the Shrew," conserves are among the delicacies offered to Sly in order to delude him into the belief that he is of noble estate. It is evident from these and other passages which might be quoted, that the making of confections was at one time an art practised by ladies of the highest rank, and that among the wealthiest classes conserves and the like were esteemed as the most delicious of luxuries. The consumption of sweetmeats has increased enormously of recent years, and the great advance that has taken place in the quantity of sugar used by the more enlightened nations, as shown in a previous article, is attributable in no small degree to the growing favour which is being extended to the productions of the confectioner, and to the increasing fondness for home-made "sweets." Paris is the chief seat of the manufacture of saccharine delicacies; and to the ingenuity which the French display in devising all manner of dainties is owing many of the choicest varieties of confectionery. Though thus distinguished, however, Paris has by no means a monopoly of the trade. In all the more important towns of Britain sugar goods are produced.

There are several extensive manufactories of confectionery and preserved fruits in Scotland, and some of these have a world-wide reputation. Among the best known firms are the following :—In Edinburgh—Mr Alexander Ferguson, and Messrs R. Shiels & Son. In Glasgow—Messrs J. Buchanan Brothers, Messrs John Gray & Co., and Messrs Robert Wotherspoon & Co. In Dundee—Messrs James Keiller & Son, and Messrs John Low & Son. It is difficult to form an estimate of the extent of the confectionery trade in Scotland, but, judging from the returns of the principal manufacturers, the value of the goods produced cannot be less than L.1,000,000 per annum. The quantity of sugar used is estimated at from 12,000 to 14,000 tons. Upwards of 2000 persons are employed in the trade.

The most extensive confectionery establishment in Britain is that of Messrs James Keiller & Son, Dundee. The firm have a specialty in marmalade—a conserve which they have been chiefly instrumental in bringing into general use. The history of the firm is brief, but it records a brilliant success. About the beginning of the present century, Dundee, which stands in the neighbourhood of a famous fruit-producing district, was pretty extensively engaged in the manufacture of "preserves," and the late Mr James Keiller was among those engaged in the trade. By way of increasing the variety of his productions, Mr Keiller began to make marmalade, and was the first in the country to produce it as an article of commerce. For some years the demand was limited to the town and district ; but in course of time the new conserve worked its way into the more important towns of Scotland, and subsequently crossed the Border into England. Between thirty and forty years ago. one of the principal grocery firms in London gave marmalade a trial, and soon secured a steadily increasing demand for it. A new market was thus opened up ; and from being a subordinate part of Mr Keiller's business, the manufacture of marmalade took precedence. The little factory set up in an old house near the High Street of Dundee became too small for the increasing business, and more commodious premises were acquired. Successive extensions have been made ; and now the establishment, which occupies several blocks of three-storey buildings, is the largest and finest of the kind in the country ;

while Messrs Keiller & Son's marmalade has become familiar and is relished in all quarters of the earth. Though marmalade constitutes the most important part of the goods manufactured by Messrs Keiller, it is not the only article produced, for they do an extensive business in jams, jellies, and general confectionery ; and in giving a description of the manufactory, the processes by which these are prepared will fall to be noticed.

The visitor to Messrs Keiller's manufactory is first conducted to the stores in which the raw materials are kept. There, in boxes, barrels, and bags, are to be seen the fruits, seeds, sugars, gums, &c., used in the trade. Of course, the green fruits are deposited in the store for the briefest possible time ; but of the other articles, a constant stock is kept on hand. Oranges are usually in season from the beginning of December till the end of March, and the year's supply of marmalade must be made in that time. The oranges used are the bitter variety obtained from Seville in Spain. They are imported in chests containing 2 cwt. each. Messrs Keiller consume 3000 chests annually, from which they produce about 1000 tons of marmalade, the greater part of which is sent out in pots containing one pound each. The term marmalade is supposed to have been derived from an Indian fruit resembling the orange, and named the *Ægle Marmelos*, or Indian beal, from which at one period a similar conserve appears to have been made. Marmalade is of the nature of a jam, as it contains the whole substance of the fruit except the internal fibres and the seeds. The first operation in the process of converting the oranges into marmalade is to remove the skins. The fleshy part of the orange is then squeezed to extract the juice, and the "peel" is sent into the cutting-room, where it is softened by being subjected to the action of steam, and is then sliced into "chips." In the early days of the manufacture the peel was cut by hand ; but now the cutting is done by a simple but thoroughly effective machine, invented by Mr Wedderspoon Keiller, of Perth (a nephew of the late Mr James Keiller), the author of several ingenious contrivances for facilitating the manufacture of confectionery. Each machine consists of a spindle, from which a number of blades or cutters radiate. The blades revolve within, and close to the side of a box ; and in an aperture in the side of the box a tube is fixed. The machines are attended by girls, who take up a handful of peel at a time, and, by pressing it into this tube, bring it into contact with the cutting part of the apparatus, which speedily reduces it to very thin chips.

Certain proportions of the chips, the juice of the orange, and refined sugar are mixed together and boiled to produce marmalade. The boiling-house contains a number of open copper pans about three feet in diameter, and two feet deep. The pans are made double, and the boiling is effected by the admission of steam into the space between the outer and inner vessels. A young woman attends to each pan, the contents of which she has to stir constantly. The time of boiling depends on many circumstances; but the attendants know by experience when the proper point has been reached. The boilers are so worked as to be ready in rotation; and when the contents of one are sufficiently boiled the marmalade is emptied into a pan fixed on a small truck, and conveyed to the filling-room. This is a large apartment, with tables arranged longitudinally, on which thousands of pots and jars are piled. Adjoining the filling-room is a sort of scullery in which the pots are washed by a steam-machine. The jars, which contain from 7 to 14 ℔. each, are filled on a set of scales; but, as the pots are made of a uniform size, holding 1 ℔. each, they are not weighed. When the contents have sufficiently cooled, the pots are raised by a steam-elevator to an upper room, where they are covered. About fifty women and girls are employed in this department. A circular piece of tissue paper is first laid on the surface of the marmalade, and then a piece of vegetable parchment is tied over all. Formerly, animal tissue was used for covering the pots; but now vegetable parchment, a much more cleanly and equally effective material, is being employed. In the course of the season, about a million and a-half of pound-pots of marmalade, besides a considerable number of jars containing from seven to fourteen pounds, are turned out. It is important to know that, for the present price of one pound of butter, three pounds of marmalade may be procured; so that in this, as in so many other cases, what was a luxury with one generation, will probably become a necessary with the next.

When the marmalade season closes in the end of March, the manufacture of candied peel, now so much used in cakes and puddings, is commenced, and lasts till the jam fruits begin to appear, which usually happens about the beginning of June. The kinds of peel candied are orange, lemon, and citron. The fruits for this branch of the manufacture are obtained chiefly from Messina. After the peel is stripped from the fruit and subjected to a preliminary treatment, it is steeped in a syrup of pure sugar for several weeks, and when taken out, candied, and dried, it is ready for the market. Messrs Keiller make a large quantity of the various kinds of peel every season. The jams and jellies are made in the ordinary way from fruit grown in various parts of Scotland, England, Ireland, and the Continent. Many tons of these preserves are produced annually.

The most interesting section of the manufactory is that in which the dry confections, such as lozenges, comfits, candies, and gum-goods are made. The sugar used in this branch is chiefly fine loaf; and the first operation is the reduction of the sugar to an almost impalpable powder. That is effected by crushing or grinding it in a mill with vertical stones. The conversion of sugar into the various kinds of sweets is carried on in separate groups of apartments, by distinct sets of workpeople. In the lozenge department the first step is to mix ground sugar, water, and gum in certain proportions, and so to form a paste or dough. The addition of a small quantity of essence imparts the desired flavour. After being well kneaded by a machine, the dough, in quantities of 2 or 3 ℔. at a time, is passed between polished cylinders, and rolled out into sheets of any required thinness. The sheets are received from the rolling-machine on boards, from which they are transferred to the cutting-machines. Before the invention of the cutting-machine all the lozenges were stamped out with hand-cutters; and for particular kinds of goods cutting is still done by hand. Each machine, attended by a boy and two girls, whose duties are exceedingly light, will execute as much work in a given time as a considerable number of expert hand-cutters. The lozenges are spread out on trays, where they are examined, and all imperfections and scraps removed. Large lozenges bearing mottoes are much in fashion; and these are made from dough prepared as above, but printed and cut by hand. The lozenges are baked by being placed in a hot-room or stove, where they are subjected to a slow heat. Comfits, or confects, as they were originally called, are known in the trade as "pan goods," and are produced by surrounding aromatic seeds with a coating of sugar. The seeds most commonly employed are caraway, coriander, and cassia; but almond kernels and slips of cinnamon bark are also extensively used. After the seeds and cinnamon have been carefully cleaned and picked, they are removed to the pan-room, where there are a dozen large copper pans. Some of the pans are inclined at an angle of forty-five degrees, and revolve slowly; while others maintain a horizontal position, but are violently shaken about. The pans have double bottoms for the admission of steam; and, like the other machinery in the establishment, they are set in motion by a steam-engine. When the seeds are put into a pan, they begin to tumble about in consequence of the peculiar motion which the vessel receives. The pan attendant then pours upon them a quantity of syrup made of pure sugar. The sugar is speedily dried by the heat, and the contents of the pan bcome white. As measure after measure of syrup is added, the comfits grow in size, each appearing to receive an equal share of the sugar. By pouring in syrup at intervals, the comfits are made to assume a smooth exterior; while by letting the syrup flow in constantly from a series of minute jets, a rough or " pearled" appearance is given to them. From the din of the pan-room to the quiet but equally warm retreat of the candy and "rock" makers is but a step. Rocks are made by boiling the sugar in a variety of ways, and mixing certain essences with it. The material is worked up into pieces shaped like loaves, and, two or three of these of different colours being pressed together, the workman begins by rolling the mass to a point at one end, and drawing it out into sticks about two yards long. The colours in the attenuated portion bear the same proportion to each other as they did in the mass. A variety of rock with names or mottoes running through it from one end to the other is made; and many persons are puzzled to know how it is manufactured. As the process is not a trade secret, it may be here explained. The workman, in preparing a block of sugar-paste, embeds in it longitudinal pieces of a different colour from the body of the mass, and having a section corresponding with the letters of the motto. The block, as thus built up, may not measure more than a foot in length, but by drawing it out it is converted into many yards of rock, which, on being broken at any point, shows the motto. Fruit drops are of the candy class, and are made in a variety of forms by passing sheets of prepared sugar between moulded rollers. Gum goods consist chiefly of jujubes, plain and candied. The gums used here are the finest Turkey sorts. After being boiled with a certain quantity of sugar, the gum liquor is poured into trays, and deposited in the hot-room, where it is allowed to consolidate for a week. It is then taken out and cut up into square pieces. The circular jujubes, or pastiles, which are usually coated with sugar, are formed by pouring the gum liquid into indentations in a mould of fine starch spread over a tray.

This is a brief outline of the manufacturing processes to be witnessed in Messrs Keiller's establishment. The packing department remains to be noticed. In it the confections, marmalade, &c., are made up, some in bottles, and others in parcels, boxes, and casks. Many hands are employed in packing—first in making up the goods into small parcels, and then in selecting and packing goods ordered.

Taking all the departments together, the number of workpeople is about 300, most of whom are young women, who earn from 8s. to 14s. a-week. About the whole establishment there is an air of cleanliness and order which the visitor cannot but be gratified to witness; while the appearance of the workpeople is a sufficient proof that their occupation is by no means unhealthy.

MANUFACTURE OF PRESERVED PROVISIONS.

THE INVENTION OF A MEAT-PRESERVING PROCESS, AND WHAT LED TO IT—
VARIOUS MODES OF "CURING" ANIMAL AND VEGETABLE SUBSTANCES—THE
PROVISION-PRESERVING TRADE IN SCOTLAND—DESCRIPTION OF A PRESERVA-
TORY.

ONE of the most troublesome characteristics of articles used for food
is their liability to decomposition. Flesh of all kinds is peculiarly
prone to speedy decay; and, though vegetables are on the whole
more durable, yet most of them also soon become unfit for food.
Prior to the beginning of this century, no mode of preserving animal
food on a large scale was known save that of salting. Consequently,
when vessels set out on a voyage that was to last beyond a few days,
their crews had to be provided with stores of salted meat. Now,
though salt preserves flesh from putrefaction, it deteriorates to a
certain degree the quality; and persons who are compelled to use
much salted meat without fresh vegetables are liable to attacks of
scurvy and other affections of the fluids of the body. When
salt is applied to fresh meat, a saline liquid or brine is deposited
in the vessel. This is due to the power which the salt possesses
of extracting from meat a large portion of the water or other
soluble matter which enters into its composition. On analysing
it, the brine is found to contain all the ingredients of a concentrated infusion of flesh; while only fibre is left for the salt to preserve. Salting, consequently, destroys to a great extent the nutritive
properties of the flesh; and when the brine is poured off as waste,
it really carries with it more than half the nourishment contained
in the meat. Until a recent period, it might be said that more
British seamen were destroyed by scurvy, originating in the want of
vegetable food and in the constant use of salted provisions, than by
all other causes combined. When Lord Anson reached Juan Fer-

nandez with the Centurion, the Gloucester, and the Tryal, his united crews of 961 men had been reduced by that terrible disease to 626, of whom only a small number were fit for duty; and during last century a Spanish ship was picked up at sea with all her crew dead from scurvy. Cases of this kind might be multiplied to show the loss of human life entailed by want of means for preserving provisions in a fresh state. The cause of the disease was well enough understood, but a preventive was long sought in vain; and it must have been a dismal thing for a ship's crew to take their departure on a long voyage with the thought that few of them would see the end of it.

To such a height did scurvy attain in the navies and merchant shipping of Britain, France, and other countries, that the attention of the Governments was seriously roused to the importance of doing something to remedy the terrible evil. The French Government took the initiative by offering, in the year 1809, a premium for the invention of a process for preserving meat, so that it would remain fresh for any length of time, and in any climate. In the following year, M. Appert came forward to claim the prize; and, after due investigation, he received L.480 for the invention of a mode of preserving both animal and vegetable matter by subjecting them to a certain degree of heat, and then sealing them up in air-tight vessels. The principle of Appert's system of preserving was known and practised in this country for years before he was made famous as the supposed discoverer of it; but those who were acquainted with the process did not realise its importance, or dream of the application which he made of it.

Before giving an account of the modes of preserving provisions now practised, it may be interesting to note briefly some of the simpler methods of "curing" animal and vegetable substances which were practised before Appert's time, and are still followed in some cases. Heat, moisture, and air are the influences which chiefly contribute to produce decomposition; and when either of these is got rid of, under certain conditions both animal and vegetable substances are rendered proof against putrefaction. When heat is expelled to such a degree that the juices of flesh and vegetables become frozen, decomposition is impossible. Meat, fish, eggs, and vegetables may therefore be preserved for any length of time if they be completely frozen. The remains of animals of a species believed to have been extinct since the advent of the human race have been discovered in a state of entire preservation embedded in the ice mountains of Siberia. It had long been customary to boil and pickle with vinegar the salmon caught in the Scotch rivers before sending them to London and other distant places; and a boiling-house was an essential department of the more remote fishing stations. The late Mr Dempster of Dunnichen, who about the middle of last century did much to encourage native industry, recommended the use of ice instead of fire as a preserving medium; and some experiments were so convincing by their results, that the boiling-houses were abolished and ice-houses erected. Freezing operates as a preservative by sealing up the fluids, and preventing their undergoing chemical action. Extracting the fluids is an equally effective mode of warding off decay. The hot winds which prevail at certain seasons in many parts of the world have the power of evaporating with great rapidity the fluids from inanimate bodies; and the remains of travellers who have succumbed to the scorching blast have been found years afterwards perfectly dried and free from any trace of decomposition. The North American Indians prepare the meat which they carry on their hunting expeditions by removing the fat, cutting the flesh into stripes, and exposing it to the action of the sun and air, so that it becomes quite hard, and may be kept for a considerable time without losing its nutritive properties. This food the Indians call "pemmican." A somewhat similar process has been tried in South America on a large scale, and "jerked" beef has within the past few years been sent to this country in large quantities; but notwithstanding the exceedingly low prices asked, people have not taken to it kindly. Fish, after being salted to a certain degree, and dried by exposure to the sun and air, are rendered safe against decomposition for a considerable time. Many kinds of fruit and vegetables may be preserved by simply hanging them in a dry place. Another method of preserving animal and vegetable substances is by excluding the air from them. Eggs may be protected against decomposition by covering the shells, which consist of porous earthy matter, with any substance that will completely shut out the air. A coat of strong spirit varnish will make an egg keep for years. While some alterations were being made on a church in Italy, three eggs were found embedded in the lime of the wall; and though they had remained in that position for 300 years, they were perfectly fresh, retaining their natural odour and flavour. The eggs owed their preservation to the fact that the mortar in which they were embedded prevented contact with the air. Confectioners in this country preserve eggs for weeks in a solution of lime. Immersion for a moment in boiling water will also preserve eggs. By contact with the heat, a film in the albumen in the egg is coagulated, and so becomes an air-tight envelope within the shell. The principle of excluding air is that now generally employed in preserving provisions on a large scale. The efficacy of acid gases in preserving fish and flesh has been known for many years. The most familiar illustrations of its use are hams, red herrings, and "Finnan haddies." Wood and turf give off pyroligneous acid during combustion, and becoming incorporated with the substance of the fish or flesh, the acid enables it to resist decomposing influences. Sulphurous acid has at various times been brought into notice as a preserving agent; and quite recently Dr Dewar of Kirkcaldy has demonstrated its importance in this respect.

M. Appert claimed to have discovered—"First, That fire has the peculiar property not only of changing the combination of the constituent parts of vegetable and animal productions, but also of retarding, for many years at least, if not of destroying altogether, the natural tendency of those same products to decomposition; secondly, That the application of fire in a manner variously adapted to various substances, after having with the utmost care, and as completely as possible, deprived them of all contact with the air, effects a perfect preservation of those same productions with all their natural qualities." The operations by which fire is made available as a preserving agent are stated to be—first, enclosing in bottles the substance to be preserved; secondly, corking the bottles with the utmost care; thirdly, submitting the enclosed substances for a greater or less length of time to the action of boiling water in a water-bath; and fourthly, withdrawing the bottles from the water-bath at the period prescribed. In practice it was found that in many cases the bottles broke during the boiling process, while in others the corks were found ineffective for excluding the air; so that a large proportion of the substances sought to be preserved was lost. Still the invention was hailed as a great advance towards a much-desired object, and, as such, was promulgated by the French Government. In 1811 an English patent was taken out for Appert's process of preservation. The patent was purchased for L.1000 by Messrs Donkin, Hall, & Gamble, who, in 1812, erected an extensive preservatory at Bermondsey. It is stated that, after a series of experiments made by the patentees for the purpose of testing the accuracy of the process, and ascertaining how far it might be made applicable in a general way for victualling the maritime service occasionally with fresh meat, they found that the system of preservation, so far as it had then been developed, was too defective and uncertain in its results to be made the vehicle of any safe or profitable commercial enterprise. They then made some experiments with vessels of tin, and these were so successful that the art of preserving food was reduced to a certainty. No sooner was the possibility of preserving provisions demonstrated by this firm, than the ships of the navy and of the East India Company were supplied with some of the prepared food ; and soon a happy change became apparent in the health of those on board. Emigrant ships were subsequently ordered to carry certain proportions of preserved meats, and now no vessel sails on a voyage that is to extend beyond a few days without having such stores in the lockers. The meat-preserving trade has assumed large dimensions ; and the method adopted by Messrs Donkin, Hall, & Gamble is that now followed (with certain modifications in the details of the process) in the great preservatories which supply the shipping of all nations.

The meat-preserving trade was introduced into Scotland in 1822, when Messrs John Moir & Son began business in Aberdeen. There are now nine establishments for the manufacture of preserved provisions in the country. Of these five are in Aberdeen; and it may be said that the granite city enjoys a more wide-spread fame for its preserved meats than it does even for the produce of its quarries— indeed, it would be difficult to discover a place to which civilisation has penetrated where the delicacies sent out from the Aberdeen and other Scotch preservatories are not well and favourably known. The pioneer firm in the trade was founded by the late Mr John Moir, father of the present senior partner. The firm began by preserving salmon for exportation, and subsequently added meats, soups, game, fish, and vegetables. The trade made slow progress for a time, as public prejudice had to be overcome, and a market had to be created. Once the thorough efficacy of the mode of preservation adopted was proved, a demand at home and abroad sprang up, and new manufactories were erected. Messrs Moir have well maintained their position in the trade; and their establishment is the largest of the kind in the country. They have an extensive connection with India, China, and Australia, and supply a large proportion of the provisions required by vessels sailing from London, Liverpool, and Glasgow. During the Crimean war they executed several large contracts for the British and French Governments. The quantity of preserved meat produced by the firm is about 2,500,000 ℔. a-year. They employ a large number of workpeople. Messrs D. Hogarth & Co. were the second firm in the trade at Aberdeen. They carry on a considerable business both at Aberdeen and at Deptford. Among the younger establishments may be mentioned that of Mr Alexander Forbes, Canal Terrace, which was erected in 1860. Mr Forbes has already acquired a prominent position in the branch of industry to which he has devoted himself, and is already favourably known in the home and foreign markets. The other preservatories in Aberdeen belong respectively to Mr Morton, and to Messrs Marshall & Co. Besides these there are three firms engaged in salting meat. Altogether, upwards of 500 persons are employed; and the average annual value of the preserved provisions manufactured in Aberdeen is about L.221,000. The business is still increasing, though the only advantage Aberdeen possesses as a seat of the trade is that of having been the first town in Scotland and one of the first in Britain that engaged in it. The other meat-preserving establishments in Scotland are at Leith, Glasgow, Peterhead, and Burntisland. The last is a small experimental concern.

In 1837 Messrs John Gillon & Co. established a provision preservatory in Mitchell Street, Leith. From a small beginning the place was rapidly extended until it became what it still is—one of the largest establishments of the kind in the country. The trade-list of the firm contains about 500 varieties of preserved meats, soups, vegetables, fish, game, &c., and shows the almost unlimited application of the mode of preservation which is employed. To describe the processes to which the various kinds of foods are subjected would be a tedious task, and it will suffice here to note in a general way what may be seen going on by visitors to the establishment. The preservatory occupies a large but irregular range of buildings, some of which are three and others four storeys in height. Almost the whole of the ground floor is devoted to offices and stores for the completed goods. In the stores hundreds of thousands of canisters painted of a uniform red are piled in bins. Each canister bears a label, which, besides indicating the contents, describes the mode of preparing them for the table. Samples are shown of meats which have been in the premises for twenty or thirty years; and some prepared milk brought home from the Crimea is to be seen on the shelves. When the preserving processes are carefully gone through, the only thing that can affect the articles preserved is damage to the canister. If the canisters were indestructible the contents would remain good for any number of years. On ascending to the first floor the first stages of the work of preservation may be witnessed. A large number of oxen are used weekly, and the carcases of these are brought in dressed from the public slaughterhouses. The sides of meat are laid upon benches, and after being divided into pieces of convenient size have the bones taken out.   A selection of the parts is then made.   Certain portions are preserved for roasting, some for boiling, and others for mincing.   The roasting is done in an oven of peculiar construction, capable of turning out a thousand small roasts at a time.   The roasts are put into tins, with the gravy, have lids soldered upon them, and are then subjected to the preserving process, which is conducted in the upper floor.   In the preserving-room is a range of oblong iron tanks with double bottoms. These are partially filled with oil or water, into which the canisters are lowered on a frame or gridiron.   By turning steam into recesses in the bottoms of the tanks, the contents are heated.   After a certain time the canisters are raised out of the tank, and a small puncture is made in the lid of each.   Through this opening steam flies off with great force, carrying with it all the air that was enclosed in the case. At a particular moment, before fresh air has begun to enter the canisters, the apertures are closed by a drop of solder, and the operation is complete.   When the canisters are raised from the bath the ends are bulged outwards by the expansion of the air and moisture within ; but when they cool, after the tapping, the ends assume a slightly concave form, owing to the pressure of the atmosphere.   In all cases this is what takes place, whether the articles preserved be put up in a cooked or in a raw state—the air is driven off by heat, and shut out by closing the aperture made to allow it to escape.   Essence of beef constitutes what may be called the speciality of the firm.   It is prepared from the first gravy or juice which exudes from the meat in the process of cooking.   Professor Christison thus notes its peculiarities :—"The meat-juice contains only $6\frac{1}{2}$ per cent. of solids. As a mere nutrient, therefore, it is much in the same category with beef-tea.   Sixteen ounces of beef-tea, made with the contents of one tin, yield only 114 grains of solid extract.   It contains no fibrin, no albumen, no gelatine.   It does not even gelatinise, on exposure to the air, for days.   It is ozmazome with the salts and sapid and odorous principles of meat, and materially different from all boiled extracts.   I should add that no good beef-tea can be made so cheap as with this preserved meat-juice.   A tin of four ounces makes sixteen of strong beef-tea."   Mutton and chicken juices are prepared by the same method as that of beef.   Minced meats are cut by a steam-machine capable of turning out 4 cwt. an hour.

The soups and vegetables are cooked in a series of huge hemispherical boilers, which are clustered in the centre of one of the upper rooms.   Here, too, the heat is applied in the form of steam, which is admitted into the double casing of the boilers.   Along one side of the room are arranged eight boilers constructed on the principle of Papin's digester. They are used to extract the fat and other nutritive substances from the bones, and the flesh attached to them. These substances have always been regarded as of high value, and Papin invented his digester in 1681 for the purpose of extracting them. The fat from the marrow-bones, under the name of marrow fat, is extensively used for cooking fish, making stews, pastry, &c.

After the bones are taken out of the extractors, they are still of some value, and are sent to the bone mills, from which they come forth as phosphate of lime—a fertilising agent much used by agriculturists. From the hoofs of the oxen and sheep prussiate of potash, or hartshorn, is made, but not at the preservatory. Nothing is lost; but much that would be wasted under the usual domestic modes of treating food is saved, and turned to profitable account. It is by economising the material that the prices are so low, and that persons may obtain, in some cases, the preserved provisions at a cheaper rate than they require to pay for the same articles in an unprepared state. The term "in season," as applied to articles of food, may be said to have been abolished by this meat-preserving process; for there is no delicacy that may not be had in a perfectly fresh state on any day in the year.

It is unnecessary to go further into detail so far as the preparation of the various kinds of food is concerned; but there is one department of the preservatory which merits notice, as illustrating the effect which machinery of the simplest kind exercises in saving manual labour. In order to make the tin canisters by the usual method, more than 100 tinsmiths would be required; but this part of the work is accomplished by machines worked by men and lads, who can learn the business perfectly in the course of a few weeks. Messrs Gillon & Co., in conjunction with an Aberdeen firm, purchased the patent for canister-making machinery from an American inventor. The sides of the canisters are cut to any size by a guillotine apparatus, and bent in another machine. The ends are stamped out, and have their edges raised by dies. Then there are appliances for soldering the parts together. Many thousands of canisters of various sizes are thus made weekly, without the aid of regularly bred tinsmiths.

The ships of some of the Arctic expeditions were supplied with provisions prepared by Messrs Gillon & Co., and nearly all the European Governments have at various times received supplies from them. During the Crimean war their preparations were extensively used in the hospitals, and on board the war-ships of Britain and

France; and the troops of the Abyssinian expedition were provided with a considerable quantity.   It is a remarkable fact that South America and Australia are among the best markets for preserved meats, in both of which countries companies exist for the purpose of preserving beef and mutton for the European markets.   British residents in India depend largely on preserved provisions for their meals.   Scotchmen abroad insist on having haggis at all their national festivals, and many " chieftains of the pudding race " are sent out in a preserved state by Messrs Gillon & Co.

MANUFACTURE OF MINERAL OILS AND PARAFFIN.

THE HISTORY OF THE PARAFFIN MANUFACTURE—THE BATHGATE AND WEST-CALDER PARAFFIN WORKS—DESCRIPTION OF THE MANUFACTURING PRO-CESSES—PRESENT POSITION OF THE SCOTCH MINERAL OIL TRADE.

No branch of industry has risen more rapidly in Scotland than the manufacture of oils and other products from native shales, nor has any been brought to such a high degree of perfection within an equal time. Paraffin, which is so named on account of its want of affinity with most chemical substances, was discovered by Reichenbach in the year 1830, and about the same time by Dr Christison, of Edinburgh, acting without any knowledge of Reichenbach's investigations. After the discovery several patents were taken out for the making of oil from schistus or shale. In 1833 Butler's Specification describes the process for making oil from shale. Hompesch's Specification, dated 1841, describes the process of making oil from schist, clay-slate, and asphalt. Du Buisson's, dated 1845, describes the process of oil-making from the same material. Sellique, previous to 1845, distilled oils from the shales of Autun, in France.

In the year 1847 Professor Lyon Playfair, while visiting a brother-in-law who owned a coal-pit at Alfreton, Derbyshire, had his attention drawn to a thick, dark, oily fluid which trickled from rents in the roof of the colliery, and was struck by the idea that it might be transformed into some useful substance by proper chemical treatment. He imparted his views to Mr James Young, then a well-known chemist in Manchester, and suggested to that gentleman the advisability of subjecting the crude liquor to chemical investigation, with the view of testing the qualities it possessed. Mr Young took the hint, and, in the course of the experiments which he made, found that the crude fluid, on being distilled, yielded a pale yellow oil containing floating particles of lustrous matter, which, on subsequent examination, proved to be crystals of paraffin. Soon afterwards a factory for the distillation of burning and lubricating oils from the crude petroleum of the coal-mine was established at Alfreton by Mr Young. The enterprise was successful, and the new trade was prosecuted with vigour and energy; but the factory had not been in operation for two years when the supply of raw material ceased, and the works, which were expected to develop a new branch of manufacture, were brought to a stand. This untoward event had the effect of directing Mr Young's energies to the solution of a problem to which he had given much thought, and to the practical realisation of which he now looked for continuing the supply of oil so unexpectedly cut short. Observation and reflection had convinced him that the Alfreton petroleum was the product of very simple natural causes, and these he set himself to investigate. Guided by experience, he was led to the conclusion that the oil had its origin in the distillation of coal by subterranean heat. In the course of two years, during which he made many experiments, he was enabled to prove the correctness of the opinion he had formed. He found that, by distilling coal at a low temperature, he obtained a liquid of an oleaginous kind, similar in its virtue and consistency to the natural oil. The primary difficulties of the undertaking having been thus overcome, it became a question of pressing importance to decide whether the discovery could be made available for the purposes of trade. Unless it could be carried out so as to supply sufficient material for the stills of the manufactories, it was apparent that the discovery would be practically valueless. Subsequent experiments proved that the crude oil could be extracted from any coal of a bituminous nature, and that the largest quantity could be obtained from cannel coal. A suitable mineral, therefore, was all that was now required to bring the scheme into operation, and to obtain that the coal-fields were explored and their qualities tested.

Some bituminous coal obtained from Boghead, near Bathgate, in the county of Linlithgow, was tried by Mr Young in 1850, and found to be peculiarly rich in oil. As the supply was abundant, Mr Young, after taking out a patent for "treating bituminous coal to obtain paraffin and oil containing paraffin," was joined by Messrs Meldrum & Binney. They selected a site near the town of Bathgate, and erected thereon an extensive establishment for extracting oil from coal, and converting it into a variety of useful products. The works were put up under the superintendence of Mr Meldrum, and were conducted under his active management.

Such was the beginning of a branch of trade which speedily assumed great importance, and converted the quiet town of Bath-

gate, together with the adjacent villages, into a great centre of industrial activity. The district was chiefly inhabited by hand-loom
weavers, whose miserable earnings—in many cases not exceeding 4s.
a-week—were barely sufficient to prevent starvation ; and when the
new field of labour was opened up, the weavers gladly relinquished
their looms, and sought employment at the paraffin works. A proof
of the marvellous success that attended Mr Young's enterprise, and
the deep hold that it took on the district, is afforded by the fact
that, though the population of the parish and town of Bathgate had
increased only from 2513 to 3341 between the years 1801 and 1851,
the ten succeeding years witnessed an increase to 10,000. The
manufactory was extended until it covered a great space of ground ;
and the value of its products was recognised throughout the world.

Mr Young's patent expired in 1864, but previous to that time
several works were in operation for distilling oil from shale. In the
parish of West Calder a seam of shale lying on the limestone on the
estate of Mr Hare of Calderhall was worked by Mr Gray, and the
shale distilled at the Leavenseat Oil Works. About 1862 the West
Calder Works were erected on the estate of Gavieside, by Messrs
Fell & Co. About the same time Messrs Raeburn erected retorts
at the Grange, on the estate of Charlesfield. In the parish of
Mid-Calder the Oakbank Works were erected about 1863, on the
estate of Mr Hare of Calderhall, by a limited liability company. In
the parish of Uphall there are several extensive fields of shale. The
Broxburn shales on the estate of the Earl of Buchan are leased by
Mr Bell, who is most energetic in developing the mineral resources
of that estate. About 1860 retorts were erected by Dr Steel of
Wishaw at Broxburn, to distil oil from shale supplied by Mr Bell.
Early in 1862 the Broxburn Shale Oil Company (Limited) was
formed ; but after expending a large sum of money, it was wound
up in about two years. The whole plant was sold, and Mr Fernie,
of the Saltney Oil Works, succeeded the company in the occupation
of the ground. After erecting upwards of 200 retorts, he sold his
work to a company called the Glasgow Shale Oil Company (Limited).
This company, as well as Mr Poynter at his works at Broxburn, are
now producing large quantities of oil from shale supplied by Mr
Bell. In addition to these works, Mr Bell has erected a large number of retorts ; and Mr Hutchison has a small refinery at Broxburn.
It will thus be seen that Broxburn is one of the most important seats
of the oil trade. In the same parish of Uphall there are extensive
fields of shale on the estate of Mr M'Lagan, M.P. for the county
of Linlithgow ; and that gentleman erected retorts previous to

the expiry of the coal oil patent in 1864. Shortly after the dissolution, in 1864, of the copartnery of Messrs E. Meldrum & Co. and of Messrs E. W. Binney & Co., the firms under which Young's patent was worked, Mr Meldrum became associated with Mr M'Lagan and Mr Simpson, of Benhar, in the Uphall Mineral Oil Company, which purchased the works on Mr M'Lagan's estate and leased his shale. Through the experience and skill of Mr Meldrum in the manufacture of paraffin oil, to which the prosperity of the Bathgate Chemical Company was so much due, the oil manufactured at the Uphall Works assumed at the first a position in the market second to none in the valuable points of burning quality, purity, and safety. The Uphall Works have recently been greatly extended, and are now the second largest in this country.

Since 1864 several oil-works have been erected, of which may be mentioned the Dundas Shale Oil Company, on the estate of Mr Dundas of Dundas, at Kirkliston; the Westwood Shale Oil Company, on the estate of Captain Steuart; the Hermand Company, belonging to the Messrs Thornton, on the estate of Mr Maitland of Hermand—all of which manufacture the crude oil only.

After the partnership under which the Bathgate Chemical Works had been established was dissolved, Mr Young carried on the concern by himself for a year, during which time he conceived the idea of creating new works in the neighbourhood of West-Calder—a district particularly rich in bituminous shale. Having acquired necessary leases, &c., Mr Young chose a site on the estate of Addiewell, about a mile west from the village of West-Calder, and began the construction of an establishment on a more extensive scale than that at Bathgate. There was no accommodation in West-Calder for a large body of workpeople, and Mr Young's first care was to provide lodging for his workmen by building a range of houses. A bed of clay was discovered on the property, a brickwork was erected, and soon houses to the number of several hundreds were provided, the building of the manufactory being at the same time pushed forward. After considerable progress had been made, Mr Young organised a company to undertake the working of both the Bathgate and Addiewell establishments. The new copartnery, under the designation of Young's Paraffin Light and Mineral Oil Company (Limited), has been in existence for about three years, Mr Young, besides holding stock to a large amount, occupying the place of general manager. A brief description of the two manufactories of the company will convey some idea of their extent and importance.

The Bathgate Chemical Works are situated about a mile from the town. They occupy twenty-five acres of ground, and are connected with the main lines of railway in the vicinity by branch lines, which afford convenient conveyance for the raw material to any desired point, and for sending out the manufactured goods. The various departments are admirably arranged, and the appliances in all are so completely adapted to their purpose that it is difficult, after examining them, to believe that the manufacture of paraffin is really only a thing of yesterday. From a distance, the establishment has the appearance of a village of irregularly built grimy houses. A nearer view reveals a series of broad thoroughfares lined with retorts, stills, boilers, tanks, &c., some under iron roofs, and others exposed to the weather.

In order that he may understand what is going on around him, the visitor must begin at the coal-breaking shed. The coal used is a hard, rusty-black-coloured mineral. It is brought in from the pits in lumps of considerable size, and the first step in the process of manufacture is the breaking of these into small pieces, which may be conveniently shovelled into the retorts. There is a powerful crushing-machine for performing this part of the work. Contiguous to the breaking shed, the retorts, which are 200 in number, are ranged in sets of four each. The retorts are simply vertical cast-iron pipes twelve feet in length and fourteen inches in diameter. Each set of four retorts is built into a furnace, the lower part of the pipes being embedded in brickwork, while several feet of the upper end are left free. The retorts have funnel-shaped tops, and are fitted with air-tight stoppers. The lower extremity projects downwards through the furnace into a pit filled with water, an arrangement which, while it effectively shuts out the air, admits of the exhausted cinder being withdrawn without interruption. As the work goes on, the waste is withdrawn gradually, and fresh coal is added. The portion of the pipe which passes through the furnace is maintained at a dull red heat, and that is the point at which the distillation actually takes place. Under the influence of the fire, the coal is decomposed, the oil being driven off in the form of vapour, which is collected in a large main pipe having a connection with all the retorts. This main pipe conveys the oil vapour to the condensers, which are similar to those used in gas-works. The condensers stand outside, and, as the vapour passes through them, it is reduced to a liquid form. It is always found that a portion of the vapour is incondensable, and that portion is collected in a gas-holder, and is used for lighting the workshops. The liquid formed in the condensers is run off into a reservoir capable of containing 100,000 gallons. At this stage the oil has a black greasy appearance, like natural rock oil or petroleum, and gives off inflammable vapours at the usual atmospheric temperature. The tank is fitted with an air-tight covering of iron for the purpose of restraining this vapour, and lessening the chances of its ignition. Another precaution against loss by fire consists of a large pipe inserted into the roof of the tank, and so arranged that, in the event of the oil taking fire, a strong jet of steam could be at once sent in, and the flames be thus extinguished.

The crude oil obtained by distilling the coal as described is subjected to various other processes, under which it yields four different products—namely, paraffin oil for burning, paraffin oil for lubricating machinery, a light volatile fluid called naphtha, and solid paraffin; but before any separation of these takes place, the oil has to be thoroughly purified. It is first distilled, which is performed in a range of huge cylindrical stills, laid in a hori-zontal position. When the vapour from the stills is condensed, it is collected in tanks, and presents a wonderful improvement in appear-ance. The black sluggish stuff from the stock-tank has been con-verted into a dark-green limpid fluid. The impurities extracted from the crude oil are removed from the still after each charge. They form a large black lustrous substance, resembling coke, and make excellent fuel. Though much improved by the distillation, the oil is not yet sufficiently pure, and requires further treatment. It is run into circular iron tanks; and, after a certain quantity of sul-phuric acid is added, the liquid is violently agitated by a revolving stirrer. The acid has no affinity for the finer oil, but it has for the foreign substances which the oil holds in solution. At the end of four hours' agitation, the oil is seen to have become of a pale green colour; and on the liquid being allowed to settle, the vitriol and the organic impurities, by reason of their greater weight, collect in the bottom of the tank, and, when drawn off, this sediment is used for fuel. The oil is next transferred to a clean set of tanks, in which it is mixed with a strong solution of caustic soda, and again sub-jected to agitation. The soda neutralises any sulphuric acid that may remain in the oil, and rids it of impurities which were not affected by the vitriol. The oil is then distilled a second time, and the treatment with sulphuric acid and soda is repeated. After these operations, the oil presents a clear, pale, yellow colour, and in that condition it contains the elements of the four products mentioned above. To separate these, and make them available, is the next care of the oil-makers. This is accomplished by distilling the oil at various temperatures.

The first product taken off is naphtha, for the separation of which only a gentle heat is required. Naphtha is a valuable liquid, extensively employed in the arts, and as an illuminator. By raising the temperature of the stills after the naphtha vapour has passed off, paraffin oil is obtained. Before being ready for the market, both the naphtha and the oil are distilled separately, in order to make them perfectly pure. The oil is the most valuable and important of all the articles manufactured at Bathgate, and is extensively known for its illuminating qualities. In country districts, where gas is not manufactured, the paraffin oil has almost entirely superseded the other kinds of oil, and is universally admired for the clearness and brilliancy of the light which it affords. One gallon of the paraffin oil is equal in illuminating power to one and a quarter of American petroleum oil; and it can be produced at a price which gives a light cheaper than English coal gas. Another, and not the least important virtue which paraffin oil possesses, is the safety with which it can be used for domestic purposes. Great care is taken to separate the least trace of naphtha from it, so that there is no risk of explosion by accidental ignition.

When all the finer oil has been distilled over, the heat is increased, and a heavier vapour is driven off. This last produces a thick oil, which, when thoroughly cooled, assumes the consistency of grease. It is in reality a mixture of oil and solid paraffin. When the paraffin has been crystallised by cooling the liquid, it is separated from the oil by a process of filtration under pressure. The department in which the separation is accomplished is fitted up with hydraulic presses. The heavy oleaginous liquid is put into strong canvas bags, and these are placed in the hydraulic presses, and squeezed until all the oil is forced through the texture of the bags. The oil extracted in this way is an excellent fluid for lubricating machinery, and is largely used in cotton mills and other establishments in which machinery of a delicate kind is employed. Two valuable peculiarities of this oil are that it does not become rancid, and that it is free from all tendency to spontaneous combustion.

The solid paraffin is now the only substance left. After the oil is extracted, the paraffin is emptied out of the bags. It is then of a dirty-yellow colour, and requires a deal of purification before it assumes that beautiful wax-like appearance by which it is distinguished. The stuff is placed in iron vessels containing heated naphtha, by which it is dissolved. The naphtha acts on the impurities, and after a certain time the liquid is allowed to cool, when it again assumes a degree of consistency. It is then subjected to filtration in canvas bags as before. This operation of dissolving the paraffin in naphtha, cooling it, and filtering it through bags, is repeated until the substance has acquired the requisite whiteness and purity. On being taken from the filters for the last time, the paraffin is removed to a workshop in which it is subjected to the action of steam, which carries off the odour of the naphtha; and the paraffin, in a liquid state, is run into circular iron moulds, in which it solidifies. As thus finally purified and crystallised, the paraffin is a fine white substance, more transparent than wax, and of a beautiful lustrous structure. It lacks both taste and smell, burns with a white flame, without smoke. Candles made of it are in much favour, both on account of the brilliant light they afford and the clearness with which they burn.

The Addiewell Chemical Works occupy seventy acres of ground, fully one-third of which is covered by buildings, tanks, condensers, &c., while a large portion of the remainder is taken up by railways and roads which give access to all parts of the vast establishment. It is difficult to believe that the many buildings and peculiar-looking iron structures which stud the ground are parts of one concern, and that they are not a gathering of a dozen factories used for widely different purposes. The system of iron pipes which passes overhead, beneath the feet, and crops up in all quarters, would be sufficient, one would think, to carry the gas and water supply of a large town. The retort sheds, taken together, are upwards of 200 yards in length, and each contains a double row of retorts. The main pipe which collects the vapour from the retorts and conducts it to the condensers is nearly a yard in diameter. The condensers are on a like gigantic scale, each containing several miles of piping. As the processes at Addiewell are similar to those at Bathgate, it is unnecessary to explain them further. In addition to the four articles specified as being the chief products of the shale, a fifth substance of recent discovery has to be mentioned. It was discovered that the water which accompanies the oil through certain stages of the manufacture contained traces of ammonia, and by experiment it was found that the ammonia might be profitably extracted. There is now a special department for separating the ammonia, which is done by treating the water with sulphuric acid, and so producing sulphate of ammonia. This substance fetches L.16, 10s. a-ton, so that its preservation was well worthy of attention.

An interesting department of the Addiewell Works is that in which candles are made. Most persons are familiar with the beautiful candles made from Young's paraffin. They are to be found in every market of the world, and are manufactured in various parts of England and abroad, the paraffin being supplied from the works of the company in the large round cakes mentioned above. Before the discovery of these candles, Baron Liebig wrote :—" It would certainly be esteemed one of the greatest discoveries of the age, if any one could succeed in condensing coal gas into a white, dry, solid, odourless substance, portable and capable of being placed in a candlestick." The jury of the Great Exhibition of 1851 say :— " This very problem Mr Young appears to have accomplished, by distilling coal at a low temperature;" and they express an opinion that " the brilliant discoveries of Chevreul, but lately threatened by the splendour of the electric light, may be eclipsed by the general adoption of solidified coal-gas candles." Professor Hoffman, reporting for the jury of the Great Exhibition of 1862, describes " Mr Young as the founder of this industry," and speaks of the specimens exhibited " as realising the great problem which the rare sagacity of Liebig pointed out ten years ago." The candles are made by ingeniously devised machines, in which the paraffin, after being poured into the moulds is rapidly solidified by currents of cold water. Many hundredweights of candles are turned out weekly, though the department is the smallest in the establishment.

In the manufacture of paraffin, a great quantity of illuminating gas of a superior quality is produced. The town of Bathgate obtains its supply from the Paraffin Works, and the people of West-Calder have been considering the advisability of lighting their now rapidly increasing village with gas from Addiewell. It is a somewhat astounding fact that one and a quarter million cubic feet of gas are made every day at the Addiewell Works, for which, after subtracting what is necessary for lighting the establishment, no other use can be found than the heating of boilers and the like. It is stated that an offer was made by the company to supply the city of Edinburgh with gas at the rate of 1s. 6d. per thousand cubic feet.

The company's works at Bathgate are being supplied with raw material from the pits adjoining Addiewell, so that there is a constant traffic between the two places. Five locomotives are kept fully employed in the carrying department. Some of the pits lie close to the manufactory, and none of them are beyond a distance of two miles. The company have four hundred miners, and in the other departments upwards of a thousand persons are employed. A great part of the work is done by unskilled labourers, who receive from 16s. to 19s. a-week. There is a large staff of mechanics who execute most of the ironwork required ; and in addition to these

there are joiners, masons, plumbers, and others who find constant
occupation in the great but still growing place. The houses provided
for the workpeople are commodious and comfortable, and they are
let at very moderate rents. A school, under Government inspection,
is also attached to the works.

Taken collectively, upwards of 3000 persons are employed in the
Scotch mineral oil-works, and an immense sum of money has been
invested in the several concerns. The trade recently experienced a
period of depression, arising from the extensive importation of low-
priced oils from America; but now that the public have had an
opportunity of comparing the native with the foreign article, confidence in the former is being restored, and at no time during the
past three or four years has the trade been in a more healthy state
than now. Several of the oil-works which had been stopped for a
time are in operation; and on all sides extensions are in progress.
Mr Young's little factory at Alfreton was the parent not only of the
Scotch mineral oil trade, but also of that of America; for until he
began operations, oil had never been distilled to produce an article
of commerce. The oil trade of Scotland is worth many hundreds of
thousands of pounds annually, and that of America is estimated at
ten millions. It has been rarely that an inventor has lived to see
such a splendid outcome of his ideas, or to be a partaker, as Mr
Young has been, of the wealth created by his discoveries. To any
one who takes an intelligent interest in the manufacturing industries
of the world, there could be few things more enjoyable than to walk
over the great chemical manufactory at Addiewell in company with
Mr Young, and hear him quietly relate, in answer to your queries,
how he devoted himself to reveal some of the mysteries of nature,
and convert to the use of mankind what were apparently the meanest among the contents of her storehouse. The story of Mr Young's
life and labours is as interesting as almost anything to be found in
the whole range of industrial biography.

PRINTING AND PUBLISHING.

THE art of printing began to be practised in England by Caxton in
the year 1474, and the first printing-press in Scotland was set up in
Edinburgh in 1507 by Walter Chepman and Andrew Myllar, two
merchants in the city. They were encouraged to embark in the
new trade by James IV., who, on 15th September 1507, granted
them exclusive privileges for practising the art. The charter set
forth that Messrs Chepman & Myllar, "at His Majesty's request,
for his pleasure, the honour and profit of his realm and lieges, had
taken on them to furnish and bring hame *ane prent*, with all stuff
belonging thareto, and expert men to use the same for imprinting
within the realm of the books of the laws, Acts of Parliament,
chronicles, mass-books, manuals, matin-books, and fortuus, after the
use of the realm, with additions and legends of Scottish saints now
gathered to be eked thereto, and all other books that shall be seen
necessary, and to sell the same for competent prices, by His Majesty's
advice and discretion, their labours and expenses being considered."
Acting under this privilege, Messrs Chepman & Myllar obtained
materials from France, and began to print tracts of a popular kind,
consisting chiefly of short romances, ballads, and other poems, for
the most part of Scotch composition. The more important work of
publishing Acts of Parliament and books of law was postponed by
their royal patron; and for a time the press was engaged upon
such like minor publications. Until 1788 it was supposed that
all the earlier works of the first Scotch printers had become extinct;

but in the year mentioned a number of the tracts were discovered in a sadly mutilated state somewhere in Ayrshire. These were carefully bound up, and are now among the most valued treasures of the Advocates' Library. A facsimile of them was published in 1827, but only a small number of copies were issued, and the book is consequently rare. The collection, which is entitled " The Knightly Tale of Golagrus and Gawane, and other Poems— printed at Edinburgh by W. Chepman and A. Myllar in the year 1508"—consists of ten poems and an essay; and among the authors are Dunbar, Chaucer, and Henryson. There are two curious engravings in the collection. One may have been intended to represent Adam and Eve. The figures, partly clothed in skins, stand on either side of a tree on which hangs a shield bearing the monogram of Chepman. The other engraving is designed to mark the connection of Myllar with the books, and represents a windmill, up the stair of which a man bearing a sack is toiling. It appears that Bishop Elphinstone, of Aberdeen, took much interest in the work of the printers, and, having composed a breviary for use in his cathedral, he got them to print it. Several copies of this work are in existence. It is in two volumes, one of which was printed in 1509, and the other in the following year. The imprint at the end of the second volume is as follows:—" Edinburgensi impresso jussu et impensis honorabilis viri Walteri Chepman ejusdem oppidi Mercatoris quarto die Junii millesimo cccc decimo." The first of the engravings referred to above occupies one side of the last leaf. A few of the other works of note printed at Edinburgh during the sixteenth century may be mentioned. In 1526 a foreigner, who gives his name as Jodocus Badins Ascensius, produced an edition of " Boetii Historia Scotorum." Fifteen years afterwards the Acts of the Scottish Parliament were ordered to be printed, and the work was entrusted to Thomas Davidson, the King's printer. A copy of this book, printed on vellum, is in the Advocate's Library, and is a fine specimen of typography. Davidson also issued a translation of Boetius printed on paper. A succeeding King's printer named Lekprivik printed the Acts of Parliament from the reign of James I. down to his own time. Lekprivik had, in addition to his Edinburgh establishment, printing offices at St Andrews and Stirling, and he printed a considerable number of books. Thomas Bassandyne published a folio Bible in 1576; and Alexander Arbuthnot, King's printer, a copy of " Buchanini Historia," in 1582.

The progress of printing in those early days was slow, as, owing to its tendency to spread facts and diffuse opinions, the art became an object of jealousy to both Church and State. Those connected with the press felt they had a mighty power at their command, and were disposed to exercise it freely. After the Reformed religion had been established in Scotland, the General Assembly assumed the censorship of the press, and no books of a religious kind were allowed to be printed till they had obtained the approbation of the Church Court, and even then the printer had to obtain a license from the magistrates before he could proceed with the work. There is ground for admitting that some restrictions were perhaps necessary, for it is recorded that one printer, in the dedication of a book which he issued, designated the King "the Supreme Head of the Church," and published an edition of the psalms with an obscene song appended. For these offences the only punishment inflicted was an order to call in all the objectionable books, and to remove the dedication from the one and the last leaf from the other.

The seventeenth century was a troublesome period for the Scotch printers. In 1637 Young, printer to Charles I., printed a Book of Common Prayer in a style unequalled anywhere at the time; but this achievement was an unfortunate one for the printer, as the Covenanters compelled him to fly the kingdom. When the Civil War broke out in the same reign, the prosperous career which appeared to be opening for those engaged in the art of printing received a check. Young's partner, Evan Tyler, became printer to Cromwell, and subsequently sold his patent to a company of stationers in London. The company sent down a manager and some workmen, and opened a printing office at Leith, in which, among other things, they reprinted newspapers obtained from London. The concern did not succeed, and the establishment was broken up, the types, &c., being purchased by a number of stationers, who set up distinct offices, in which they printed, in a poor style, treatises on divinity and schoolbooks. A costly attempt was made to revive and improve the art by Archibald Hyslop and William Carron, who brought materials from Holland, and began to work in a very neat style; but they met with only partial success. The printing trade was thrown into confusion by a Glasgow printer, named Anderson, coming to Edinburgh, and, in the name of the other members of the trade, securing the rights of King's printer. Though Anderson assumed the title of royal printer, the privileges of the office were to be shared equally among the Edinburgh printers. A dispute put an end to the latter part of the arrangement, but not to the patent granted to Anderson, whose widow (he being then dead) assumed a monopoly of the printing business. In terms of the patent, "no one in the kingdom durst print any book, from a Bible to a ballad, without licence from Anderson." Mrs Anderson, determined to make the most of the important privilege she possessed, and as competition was out of the question, quantity and not quality was what she exerted herself to produce. In a " History of Printing," written by a printer of that time, it is stated that " nothing came from the royal press but the most illegible and uncorrect Bibles and books that ever were printed in any one place of the world. She (Mrs Anderson) regarded not the honour of the nation, and never minded the duty that lay upon her as the sovereign's servant. Prentices, instead of best workmen, were generally employed in printing the sacred Word of God." Mrs Anderson made a most tyrannical use of her powers, and prosecuted every printer who dared to exercise his trade. Messrs Reid of Edinburgh, Saunders of Glasgow, and Forbes of Aberdeen, were among those who suffered most ; for, in addition to having their printing offices closed, they were subjected to fines and imprisonment. The popular indignation was at length excited by the restrictions to which the printers were subjected, and the result was that Mrs Anderson's privileges were first curtailed, and ultimately annulled. In the year 1700 some pamphlets reflecting on the Government were printed in Edinburgh, and the result was that all the printers in the city were summoned before the Privy Council, and two of them were sent to prison. An engraving offensive to Government was executed about the same time, and the artist and a person who assisted him were tried for high treason. The engraving represented Caledonia in the figure of a woman from whose mouth issued the words, " Take courage, and act as men that hold their liberty as well as their glory dear." Caledonia was supported by the minority in Parliament, numbering eighty-four members ; and in the lower part of the picture, an angel armed with thunderbolts was driving to perdition a large number of men who were understood to represent the majority in Parliament. The Lords found the libel not relevant to infer treason, but relevant to infer an arbitrary punishment.

Though the art of printing was introduced into Scotland three hundred and sixty years ago, it did not assume importance as a branch of industry until about the middle of last century. In Arnot's " History of Edinburgh," an interesting account is given of the rise and progress of printing in Scotland ; and from that source some of the facts stated have been drawn. Referring to the state of the trade at the time he wrote (1779), Arnot says :—" Till within these forty years, the printing of newspapers and of school-books, of the fanatic effusions of Presbyterian clergymen, and the law papers of the Court of Session, joined to the patent bible printing, gave a scanty employment to four printing houses. Such, however, has been the increase of this trade by the reprinting of English books not protected by the statute concerning literary property, by the additional number of authors, and many lesser causes, that there are now no fewer than twenty-seven printing offices in Edinburgh. It must be confessed, however, that printing at Edinburgh is not, in general, so well executed as in London, and that it is far inferior to the workmanship of the Messrs Foulis, of Glasgow, which, indeed, would do honour to the press of any country." Printing had been introduced into Glasgow in 1630 by George Anderson, who was succeeded by Robert Saunders in 1661. The whole printing business of the west of Scotland (except a newspaper published in Glasgow) was carried on by Mr Saunders and his son till about 1730, when the art was improved, and the trade extended by Robert Urie. In 1740, Robert Foulis began printing in Glasgow, and introduced a style of work which excelled in beauty and correctness. In company with a brother, Mr Foulis printed a series of classical works, which were much esteemed for the accuracy and beauty of their typography. An edition of Horace, printed in 1744, is especially famous for its correctness, which was brought about by Mr Foulis sending proof-sheets to the College, and offering a reward for the discovery of errors.

The first newspaper in Scotland was printed—at Leith, it is supposed—on the 5th August 1651. This was the "Mercurius Scoticus, or a true character of affairs in England, Ireland, Scotland, and other forraign parts, collected for publique satisfaction." It was published weekly, and contained eight small pages of print. Apparently the "Mercury" did not pay, for next year it was superseded by a reprint of a London newspaper, entitled, "A Diurnal of some passages and affairs." One year marked the period of the existence of this second publication; and then appeared the "Mercurius Politicus, comprising the sum of intelligence with the affairs and designs now on foot in the three nations of England, Ireland, and Scotland, in defence of the commonwealth, and for information of the people. *Ita vertere seria.* Printed in London, and reprinted in Leith." An edition of this paper began to be printed in Edinburgh in 1655, and that was the first newspaper published in the city. Five years afterwards the paper was declared to be "published by order of Parliament," and was then printed by Christopher Higgins, in Hart's Close, opposite the Tron Church. In 1661, Mr Thomas Sydserf, son of the Bishop of Orkney, began the publication in Edinburgh of the 'Mercurius Caledonius,' comprising the affairs now in agitation in Scotland, with a survey of foreign intelligence." This also was a weekly journal, but it ceased to appear at the end of three months. Notwithstanding the evil fortune that seemed to wait on the projectors of newspapers thus far, the attempt to establish a paper in Scotland was not abandoned. The "Mercurius Caledonius" was succeeded by "The Kingdoms Intelligencer, of the affairs now in agitation in Scotland, England, and Ireland; together with forraign intelligence. To prevent false news. Published by authority." This journal enjoyed a longer existence than all its predecessors united. The "Edinburgh Gazette," an official paper, published by authority, was established in 1699 by James Watson; and a few years later the "Scots Postman," a thrice a-week journal, was started.

In the west of Scotland the first newspaper published was the "Glasgow Courant," which was started in November 1715. It issued from the printing-office in the College. There was no stamp-duty at the time the "Courant" began its career, and the price of the paper was exceedingly low, being to subscribers one penny per copy, and to non-subscribers three-halfpence. This pioneer journal of the west did not survive beyond a few years. Prior to 1813, thirteen distinct newspapers had been set going in Glasgow, but by that year eight had ceased to exist. Among the extinct journals were two bearing the patriotic titles of the "Caledonian" and the "Scotchman." The "Journal" was started in 1729. The "Herald," which has been the most successful of the Glasgow newspapers, was begun in 1783, and up till 1803 bore the title of "Advertiser." In 1815, when the "Herald" was published twice a-week, the circulation was about 1100 copies. The price was 7d.—3d. for the paper and 4d. of duty. It was thought a wonderful thing that the edition of the paper containing an official announcement of the battle of Waterloo attained a sale of 2122 copies. The number of copies of newspapers of all kinds printed in Glasgow in 1815 was 373,718.

The progress of the newspaper press subsequent to the abolition of the stamp, paper, and advertisement duties has been very great. In 1851 there were in existence in the United Kingdom 563 journals, classified as follows:—Liberal, 231; Conservative, 174; neutral and class papers, 158. In 1868 the number of journals had increased to 1324, of which 85 were published daily. The newspapers were distributed thus:—In England and Wales, 1133; in Scotland, 132; in

Ireland, 124; and in the British Isles, 15. Eleven daily papers were published in Scotland. Of these Edinburgh had 3; Glasgow, 4; and Dundee and Greenock, 2 each. Of the 132 newspapers in Scotland, only 5 were in existence at the beginning of this century, and 75 of them have been started since the year 1850. Of the extinct newspapers which were at one time prominent may be mentioned the "Edinburgh Advertiser"—established in 1764 by Mr James Donaldson, who accumulated a large fortune in the printing and publishing business; and when he died, in 1830, left L.200,000 for the endowment of an hospital, which is one of the chief ornaments of Edinburgh. The publication of the paper ceased eight or nine years ago. The "Edinburgh Weekly Journal," which expired about twenty years ago, had a thriving existence for many years. The oldest existing paper is the "Edinburgh Gazette," which was started, as above stated, in 1699. Next comes the "Evening Courant," the publication of which was sanctioned by the Town Council of Edinburgh in 1718. Two newspapers bearing the name "Courant" had previously been issued, but they had but a brief existence. In 1720 the "Caledonian Mercury" (now incorporated with the "Scotsman") was started. The "Aberdeen Journal" dates from 1748; the "Kelso Mail" from 1797; and the "Greenock Advertiser" from 1799. The Scotch newspapers are well dispersed over the country, from Kirkwall to Kirkcudbright, and there are only two or three counties in which a journal of some kind is not published. A few of the old-established county papers have enjoyed for many years a run of uninterrupted prosperity; but some of the less important have had a somewhat chequered career. Of the papers now published, five only were in existence fifty years ago, and the first daily paper was issued so recently as 1847. The "Scotsman" was established in 1817, and became a daily paper in 1855.

About the year 1730 the "Gentleman's Magazine" and the "London Magazine" began to acquire considerable circulation in Scotland, and an Edinburgh publishing firm bethought them that some home-manufactured literature of that sort might be acceptable. In 1739 they gave form to that idea, and issued the first number of the "Scots Magazine," which had a favourable reception, and continued for many years the sole representative of magazine literature in Scotland. A collection of essays and extracts from newspapers was started by Mr Walter Ruddiman of Edinburgh in 1768, under the title of "The Weekly Magazine." This publication was the subject of an action raised to prove that it was a newspaper, and therefore liable to stamp-duty; and as the verdict went against it, the news part was separated from the miscellany, and Mr Ruddiman continued for a number of years to issue them as distinct publications. In Glasgow numerous attempts were made in the end of last century to establish a magazine, but the people were too much engrossed with commercial and manufacturing pursuits to have much taste or leisure for the perusal of literary trifles, and the attempts failed. Several publishers devoted themselves to issuing Bibles and other religious books in parts, at sixpence or a shilling each; and an extensive trade was done, and is still being done, in that way in the metropolis of the west. About fifty years ago L.45,000 worth of books of this kind were disposed of annually by Scotch firms. A new era dawned on the literature of the country in 1802, when the "Edinburgh Review" made its appearance, and struck terror into the whole body of poets, essayists, and bookmakers. The subtle analysis, the profound learning, and the scathing sarcasm of the Reviewers, set them far above all the magazine writers of the time; and they were respected or feared throughout the whole domain of literature. Who they were few persons knew for a time, because, for obvious reasons, their names were kept profoundly secret; and, in order to avoid suspicion, they reached their rendezvous at Mr Constable's office in Craig's Close by various routes. Of their number were Francis Jeffrey, Sidney Smith, Francis Horner, and Henry Brougham. These names, apart from other associations, will live long in the annals of Scottish literature as those of the founders of a most wholesome form of press censorship. The "Review" was the first of the great critical periodicals which form a distinguishing feature of the literature of the nineteenth century, and has had many imitators, but few equals. The next periodical publication of mark produced in Scotland was "Blackwood's Magazine," which was begun in 1817 by William Blackwood, who found able coadjutors in John Gibson Lockhart and John Wilson. The magazine has had a most successful career, and has been the vehicle by which many men of note in literature have risen to fame. Several high-class magazines were established in England on the model of "Blackwood,"—indeed, it may he said to be the parent of most of what is really worth among the monthlies. Something was wanted, however, after the magazines to which we have referred were set on foot, as the high price at which they were sold put them out of the reach of the great body of the people, and a thirst for knowledge had been awakened in the masses by the extension of education and a more general diffusion of books. The want was liberally supplied by the appearance in Scotland in 1832 of "Chambers's Edinburgh Journal," and in London of the "Penny Magazine," the latter published by the Society for the Diffusion of Useful Knowledge. It would be almost impossible to overestimate the effect which these publications, issued at an unprecedentedly low price, and full of pleasantly written papers of a miscellaneous kind, had in elevating the tastes and extending the knowledge of the masses. Messrs W. & R. Chambers have, besides the "Journal," made many contributions to the popular literature of the country, their last great work being one of the fullest, most accurate, and most handy of encyclopædias yet produced. The periodical publications of the magazine order printed in Scotland are nearly forty in number, and a large portion of these are produced in Edinburgh. This would appear to indicate that the city is a sort of gold-mine for *litterateurs;* and, for the benefit of persons at a distance, it may be well to state that at least three-fourths of the magazines are trivial productions of the ecclesiastical type, or of the kind supplied to Sunday-School children in order to awaken their sympathies for the heathen in foreign lands. Much of the contents are contributions of love ; and, if we make one or two exceptions, such as "Blackwood's Magazine" and the "North British Review," the amount paid for the literary work of the periodicals, other than newspapers, published in Scotland is exceedingly small.

Printing and its allied trades constitute the staple industry of Edinburgh ; and in Glasgow and elsewhere throughout Scotland they afford employment to a large number of persons. Though Scotch printers enjoyed a degree of prosperity towards the close of last century, it was not until Mr Ballantyne of Kelso published Scott's "Minstrelsy of the Scottish Border" that any decided attempt was made to improve the style of book printing. The work referred to was printed from a beautiful new type in the most careful manner, and when it appeared those connected with the metropolitan press could not gainsay the fact that a provincial publisher had produced work infinitely superior to anything they had achieved. There was a general shaking of the dry bones, and a search after better things was begun. Improvements in the form of type and in the mechanism of the press were introduced, and the work now executed in Edinburgh will bear comparison with that of any country in the world. Mr Archibald Constable, the first publisher of the "Edinburgh Review," and of the poems and novels of Sir Walter Scott, did much to improve the trade.

The greatest work issued from the Scotch press is the "Encyclopædia Britannica," the eighth edition of which was published a few years ago. The "Encyclopædia" was first published in 1771, by Mr William Smellie, a printer and man of letters. Messrs A. & C. Black hold the copyright, and have issued the later editions. Another great work was the "Edinburgh Encyclopædia," edited by the late Sir David Brewster.

The following are the more prominent publishing firms in the city, together with the kinds of work in which they are chiefly engaged :—Messrs Bell & Bradfute, law books ; Messrs A. & C. Black, miscellaneous literature; Messrs W. Blackwood & Sons, the same ; Messrs W. & R. Chambers, periodicals and educational treatises ; Messrs T. & T. Clark, law books and translations of eminent theological treatises ; Messrs Edmonston & Douglas, miscellaneous literature and fiction ; Messrs Fullarton & Co., works issued in numbers ; Messrs Gall & Inglis, religious publications ; Messrs Maclachlan & Stewart, medical treatises; Messrs Nelson and Sons, cheap works of a popular and useful kind; Messrs Oliphant and Co., religious publications; Messrs Oliver and Boyd, juvenile and school books; and Messrs W. P. Nimmo, J. Maclaren, Grant and Son, and J. Nichol, miscellaneous literature.

There are a number of extensive printing offices not directly connected with publishing houses, but doing work for them. Of such offices the most extensive are those of Messrs Ballantyne and Co., R. Clark, T. Constable, Murray & Gibb, and Neill & Co. The last-named firm is the oldest in the city, having begun business in 1749. They have printed many important works, chief among which is the latest edition of the "Encyclopædia Britannica." Messrs Neill & Co. have printed the "Acts of the General Assembly of the Church of Scotland" for upwards of a century, and the "Transactions of the Royal Society of Edinburgh" since the foundation of the Society in 1789.

There are engraving and lithographic departments in connection with some of the more extensive publishing houses. Of the engraving and lithographing establishments conducted as separate undertakings, that of Messrs W. & A. K. Johnston occupies the foremost place. Messrs Johnston enjoy a world-wide fame for their geographical works, but in addition to these they produce every species of commercial work, from the designing and engraving of bank-notes to the printing of ordinary circular letters. Messrs Schenck & Macfarlane are well known by their lithographic portraiture works, and Messrs Banks and Co. for their pictorial and fancy-work and commercial engraving. Bookbinding is in like manner conducted both in connection with and distinct from the publishing houses. In this line, Messrs Seton and Mackenzie and Mr G. Macdonald are the principal. Then there are typefounders, die-cutters, &c., without whose assistance the trades above enumerated could not be carried on. Including all the branches of business directly connected with and dependent upon the printing trade, no fewer than 10,000 persons are employed in Scotland, and of these fully one half are in Edinburgh.

Taking printing, publishing, and bookbinding together, the most extensive house in Scotland is that of Messrs T. Nelson & Sons, Hope Park, Edinburgh. The firm began business in the locality occupied by their present establishment nearly a quarter of a century ago, and they have had a most prosperous career. About fifteen years ago they built a range of new offices on a scale surpassing any similar place of business in the city; and since that time they have found it necessary to extend the buildings in various directions. The main part of the premises consists of three conjoined blocks, of neat design, forming as many sides of a square. A portion of the ground in the square is laid out as an ornamental grass plot, and on the remainder a new machine-room was recently erected. There are three floors in the main buildings, and these are appropriated to various branches of the business. In all its appointments the establishment is most complete. Machinery is used wherever it can be made available; and by means of that and a well-organised system of division of labour, the amount of work turned out is wonderful.

The letterpress department embraces a spacious composing-room, a splendidly fitted-up machine-room, a press-room, and a stereotype foundry. As large numbers of most of the works are thrown off, it is usual to print from stereotype plates. The art of stereotyping is one that has tended much to lessen the cost of producing books; and, indeed, it would be almost impossible for a trade such as that of Messrs Nelson to be carried on without it. After the types are finally arranged by the compositors, the pages are removed to the foundry and a cast is taken off them. The types may then be taken down and used in other work. The casts thus made are printed from, and if there is a likelihood of additional copies of the work being required, the plates are preserved, and may be arranged and printed from on the briefest notice. But for this process of casting from the types, printers would either have to keep the types standing or re-set them when further supplies of a work were wanted. Besides what has been stated, certain technical advantages are got by stereotyping. The process, which was invented in the beginning of last century by Mr

William Ged, a goldsmith in Edinburgh, has been brought to great perfection in the place of its birth, and is now universally practised. In the machine and press-room of the letterpress department Messrs Nelson have nineteen machines and seventeen presses at work. Immense quantities of children's books are produced, and a number of machines are kept constantly employed upon these. In many cases the pictures are printed in colours, and the neatness and expedition with which that kind of work is executed excite the admiration of all visitors. From the machine-room the sheets are taken to the drying-room, where they are hung up in layers on screens, which, when filled, are run into a hot-air chamber, where the ink is thoroughly dried in the course of six or eight hours.

The bookbinding department occupies several large rooms, and employs fully two-thirds of the whole workpeople in the establishment. It is furnished in the most complete manner with machines for performing a great variety of operations; and yet a large amount of hand-labour is indispensable. When the sheets are brought from the drying-room, they are taken charge of by young women, who fold them up with great expedition. Several machines have been invented with the view of superseding hand-labour in folding, and two varieties are here at work; but they are only suited for the coarser kinds of work. A staff of girls take the folded sheets and arrange them in order for binding. The sheets of each volume are then squeezed in a powerful press, which makes them quite compact. The notches for the binding cords are cut at a machine, and the work is then passed to the sewers. These are young women, who sit at benches, and use their needles with surprising deftness. Before they are ready for the cases the books are passed through several other hands. Meanwhile, the case-makers are busy preparing the cases. In connection with this department is a cloth dyeing and embossing branch, where the beautifully coloured and embossed binding cloths are prepared. The coloured and enamelled papers for the insides of the boards of books are also made on the premises. The case-makers are divided into half-a-dozen sections, each performing a certain part of the work. After the pasteboard and cloth are cut to the required size, one girl spreads some glue on the cloth, a second lays the boards on the proper place, a third tucks in the cloth all round, and a fourth smooths off the work. When the cases have been dried they are taken to the embossers, who put on the ornamental work. The stamping-presses have the dies fixed to a plate of metal kept hot by a series of gas jets. When gold is employed in decorating the cases, and in the titles, gold-leaf is laid on the parts before the work is put into the stamping-presses. All that now remains to be done is to fix the books in the cases, and send them into the warehouse, where they are packed up and despatched to all corners of the world, but chiefly to London and New York, where the firm have branch establishments.

The lithographic department is on a scale commensurate with the other sections of the establishment. It occupies a number of large rooms, in which sixteen machines and presses are constantly employed. The principal productions of the lithographers are maps, book illustrations, coloured cards, and those beautiful little views of places of interest which Messrs Nelson have helped to make popular. Among the artists who execute the preliminary parts of the work for the lithographers, and the engravings for the other departments, are photographers, draughtsmen, steel, copper, and wood engravers, and electrotypers. By a process patented by Messrs Nelson jointly with Mr Ramage, a drawing or print may be converted into an engraving suitable for printing from by the action of light, and the engravings, either for copperplate or letterpress printing, may be multiplied and made larger or smaller at will. The artists are very prolific, and is proved by the fact that, in addition to innumerable plates, the store-room contains no fewer than 50,000 woodcuts and electrotypes, many of which are of high artistic merit. Much that would be interesting might be written about this great establishment, and especially about the artists and their operations; but it must suffice here to say that the work produced is equal to any done in London or on the Continent.

Messrs Nelson employ 440 workpeople, about one half of whom are young women. All the inks and varnishes used are manufactured on the premises.

The newspaper department of the printing and publishing trade has, as already indicated, undergone many changes, and in recent years has been developed to a wonderful extent. The "Scotsman" is the leading journal in Scotland; indeed, it may be truly said that there is no newspaper out of London, and only one or two in it, which has such a widely felt influence, or which is conducted with so much energy and enterprise. It is becoming, then, that it should receive more than a passing notice in a record of this kind.

During the early years of this century the Scotch newspapers were conducted in the most servile and truckling spirit, the chief care of the editors being apparently to avoid giving offence in high quarters. Abuses of the most flagrant kind were rampant, and no journalist was found valiant enough to denounce them. About the year 1816 the late Mr William Ritchie, S.S.C.—a younger brother of Mr John Ritchie, the venerable head of the present firm of Messrs John Ritchie & Co.—at the request of some friends and clients, drew up a statement regarding the mismanagement of the Royal Infirmary; but no newspaper would undertake to publish the document. This and similar incidents suggested to Mr Ritchie and others the great need for some free organ of public opinion in Scotland. The idea of establishing a weekly newspaper of independent principles first occurred to the late Messrs Charles Maclaren and John Robertson. They consulted Mr Ritchie on the matter, and that gentleman entered warmly into their proposals. He suggested that the title of the new journal should be "The Scotsman," drew up the prospectus, and, in the words of one of his partners, " by his exertions and personal influence contributed more than any other individual to establish the paper." Mr Maclaren undertook the editorship, in which he was soon afterwards joined by the late Mr John Ramsay M'Culloch. The following extract from a memoir of Mr Maclaren, which appeared in the " Scotsman" at his death, illustrates the spirit of the time when the first independent newspaper in Scotland was founded:—" Very few persons can now form any adequate idea of the magnitude of the work which in 1817 Charles Maclaren set himself to do, and how much of it he did—for very few persons are now alive who remember what Scotland and Edinburgh were, politically and socially, half a century ago. Corruption and arrogance were the characteristics of the party in power—in power in a sense of which in these days we know nothing; a cowering fear covered all the rest. The people of Scotland were absolutely without voice either in vote or speech. Parliamentary elections, municipal government, the management of public bodies—everything was in the hands of a few hundreds of persons. In Edinburgh, for instance—and the capital was even too favourable an instance—the member of Parliament was elected and the government of the city carried on by thirty-two persons, and almost all these thirty-two took their directions from the Government of the day, or its proconsul. Public meetings were almost unknown, and a free press may be said to have never had an existence."

The first number of the " Scotsman " was published on 25th January 1817. It consisted of eight pages of less than half the size of the present page, and the price was 10d.—6d. for the paper and 4d. of stamp-duty. From the latest news columns of the first number, some idea of the time occupied in the transmission of news in those days may be gleaned. The latest from London was January 22 ; from Paris, January 15 ; and from New York, December 15. The projectors for a long time declined any advertisements of a miscellaneous kind, opening their columns only to announcements of new books and other literary advertisements. The hold which the new journal had obtained on popular favour induced the proprietors to begin, in 1823, to publish it twice a-week, at the price of 7d. In 1831, the broadsheet form was adopted, and in 1837, when the stamp-duty was reduced to 1d., the price of the paper was lowered to 4d. The size of the sheet was enlarged from time to time, until it reached the fullest dimensions that could be conveniently used for a four-page newspaper. When the stamp-duty was abolished, daily newspapers were established in all the great cities of the empire, and the proprietors of the " Scotsman" began a daily issue on the first day of the new order of things in the year 1855, continuing at the same time the bi-weekly publication. The "Daily Scotsman" was at the outset a tiny sheet, but the public took kindly to it, and the first of a succession of enlargements was made in a month or two after starting. To the daily and bi-weekly editions, a weekly publication, composed of selections from the others, was added in 1860. A few years ago the bi-weekly paper was merged into the daily edition, which most of the subscribers had come to prefer. In its various forms the " Scotsman" has enjoyed a most gratifying run of prosperity.

About eight years ago the offices of the " Scotsman" were removed from the High Street, where they had long been situated, to a range of new and specially constructed buildings having a frontage towards Cockburn Street. No expense has been spared to make the establishment complete in all its appointments. The front block contains five floors. On the street floor is the publishing office, where orders for papers are taken in, and the answers to numbered advertisements received and distributed. This department is under the charge of a staff of female clerks. The floor above is occupied by the counting-room and manager's room. The paper contains from five hundred to fifteen hundred advertisements daily; and in receiving and entering these, and performing the other work of the department, about a dozen clerks are engaged. Over the counting-room are the editorial apartments, a fine suite of eight rooms, opening off a large corridor, and all are fitted with speaking-tubes and bells, which enable the occupiers to communicate with any department of the establishment. There is also in each room a copy-shoot or elevator of ingenious construction, which dispenses with the tormenting visits of the printer's imp. The "copy" is dropped into a small case, which, by pulling a cord, is made to ascend to the composing-room. In the editorial and reporting departments, about a dozen persons (exclusive of the London and provincial reporting staff) are employed. One of the rooms is set apart as a telegraph office, the establishment being in direct communication with London by means of a special wire. Ascending to another floor, the composing-room is entered. It is a well-lighted and well-ventilated apartment, 150 feet in length, by 30 feet in breadth. Three rooms for the "readers" are screened off at one end, and at the other there are a lavatory, cloak-room, and smoking-room for the use of the workmen. About ninety persons are employed in the typographical department. Adjoining the composing-room is the stereotype foundry, in which casts of the types are taken for printing from. A library, containing several thousand volumes, is attached to the composing-room, and all persons employed on the premises have free access thereto.

Behind the street and counting-room floors of the front block are the machine-rooms, two spacious apartments, measuring together 80 feet in length, by 40 feet in breadth, and 25 feet in height. In the principal room are two of Hoe's rotary printing-machines, capable of throwing off about 20,000 sheets an hour; while the other room is occupied by seven of Livesey's folding-machines, each capable of disposing of 2000 sheets an hour. As a provision against accidents, there are two sets of engines and boilers, each being of 15 horsepower. There is also a small printing-machine by Brown of Kirkcaldy, which is used for printing the bill of contents. Adjoining the machine-room is the paper wetting-room. Before being printed the paper is slightly moistened with water. The wetting used to be done by hand, and was a tedious and unpleasant job, especially in the winter season; but about eight years ago Mr Scott, chief of the "Scotsman's" machine department, invented a damping-machine, which effects a great saving of labour, and is now employed in many of the principal newspaper offices throughout Great Britain. Over the folding machine-room is the despatching-room, a large hall, the fittings of which are a compound between a post-office and a railway ticket office. Here the supplies to the country agents are made up and sent out, and the demands of local news-vendors are attended to. Several rooms, in addition to those mentioned, are associated with the machine department, and used for various purposes; and on the east side of Anchor Close is an extensive paper and ink store.

This brief description of the "Scotsman" establishment may suffice to convey an idea of the extent and organisation of the place; but in order that the reader may comprehend how a daily paper is produced, it will be necessary to describe the operations carried on in the various departments. Within an hour or two of the time that the last batch of papers leaves the despatching-room, preparations are in progress in other departments for the production of next day's issue. The counting-room is opened at nine o'clock, and from that time till it closes at nine o'clock there is a constant influx of persons leaving advertisements or making inquiries respecting advertisements and other matters. Representatives of the editorial and reporting corps drop in about ten, while the day brigade of the composing staff are at their posts before that time. A good deal of duty is done in the course of the day; but it is at eight o'clock, when a fresh set of officials go into harness, that the hardest work for next day's paper begins; though an hour before that time the telegraph clerk has been at his post, and has already "taken off" a quantity of "special" news. The evening delivery of letters brings scores of epistles from correspondents, and all the late railway trains fetch fresh bundles of news. The force of country correspondents numbers fully two hundred, and embraces men engaged in nearly as many different occupations. Many public meetings and gatherings of a social kind are held in the evening, and these have to be looked after by reporters. Extraordinary efforts are made to undertake important evening meetings held in distant towns; and it is no unusual thing to engage a special train or steamer to bring home the reporters. Telegraphing is also resorted to freely, and it is very rarely that events of interest occurring in any part of Scotland are not to be found recorded in the paper of the day following that on which they occurred. In all the great towns of England correspondents are engaged; and in London there are a staff of reporters and a sub-editor. Even in New York the paper is represented, and special telegrams from that city have appeared on several occasions. The arrangements with the telegraph companies for the supply of foreign news are most complete. With this vast organisation for collecting news at command, the " Scotsman " daily presents not only a complete record of current events in Scotland, but each copy may be said to be an epitome of the world's history for a day.

The work of preparing the chief part of the copy begins, as stated, about eight o'clock in the evening; and from the shoots communicating with the composing-room a constant stream of copy flows to the desk of the foreman printer, who divides it into portions of about twenty lines each. One of these portions constitutes "a copy," or the supply given to one compositor at a time. The compositors usually "set up" about two "takes," or copies, in an hour. As the "matter" is set, proofs are printed, which the readers go over carefully in comparison with the copy, and mark mistakes on the margins. After the types have been altered according to the first proof, a second impression, or "revise," is taken, and again gone over. Towards midnight, the "up-making" begins. The types are arranged into columns, and the columns into pages; and as each page is ready it is removed to the stereotyping foundry, where a metal cast of it is taken. As it is desirable that the pages should be kept open as long as possible, little time is allowed for stereotyping. By a set of peculiar contrivances, the work is accomplished at the rate of one page in twenty minutes. The stereotypers begin by laying the page of type on a metal slab, and spreading a sheet of pulpy paper about one-tenth of an inch thick upon the face of it. The types and paper are then placed under a press on an iron table, heated by a flue, and subjected to pressure for a few minutes. In that way a matrix or mould of the face of the types is obtained. The mould is laid on a slab of iron curved to represent an arc of the circumference of the main cylinders of the printing-machines. A convex piece of metal folds down over the concave plate, leaving a quarter of an inch of space between. Into that space molten type metal is poured, and in a few seconds the mould is opened and the plate withdrawn. The latter is of course equal in thickness to the space between the two parts of the mould, and bears on its convex side an exact copy of the face of the types. Each plate as it is cast is planed round the edges, and has the larger black spaces pared down to prevent "blurrs."

In the case of an eight-page paper like the "Scotsman," each sheet has to go through the Hoe machine twice, four pages being printed at a time. Things are so arranged that the pages first printed are the first, fourth, fifth, and eighth, which occupy what is called the outside of the sheet, and usually contain advertisements and other matter that can be got ready early. Between one and two o'clock the printing of the first side of the sheets is begun; and at half-past four the plates for the second side must be on the machines and all ready to start. Nice calculation is required to have everything finished at the time stated, and during the last hour or so the minutes become more valuable, and are apportioned with great exactness to the work in hand. At five minutes to four o'clock thirty or forty compositors may be engaged with as many pages of a late report, and in twenty minutes after a cast from the types is on its way to the machine-room. A minute or two suffices for fixing the plates. The driving-belt is then turned on, and with a thundering noise the cylinders spin round, and the papers come forth from one machine alone at the rate of 200 a-minute. The folding machines are now got into action, and a scene of bustle and activity prevails. As the papers are folded they are raised by a steam-elevator to the despatching-room, where those which are to go by post are put into wrappers with the addresses ready printed upon them, while those for the country agents are made up in bundles and labelled. Shortly before five o'clock, newsboys and messengers of news-agents crowd to the despatching-room, to take part in the ballot which determines the order in which they are to be supplied. At five o'clock, the copies sent by post are despatched in large bags to the Post Office. The chief despatcher and his twelve assistants then apply themselves with such surprising vigour to the making up and forwarding of the country parcels for the six o'clock trains, that the papers are always cleared off within a few minutes of the time they leave the folding machines. The supply of the town agents—several hundreds in number—is generally completed about six o'clock, and fully 500 parcels for country agents have been counted, checked, made up, and despatched by half-past six, when the publication is generally completed. The whole work, therefore, of printing, folding, and despatching the ordinary daily impression of the " Scotsman," about 30,000 copies, is completed in about two hours ; and so carefully do the arrangements fit into each other, that no parcel ever fails to get off by the proper train. As an instance of the minute arrangements necessary to guard against the possibility of error, it may be mentioned that the labels on which the addresses of the country news-agents are printed are divided into groups differently coloured, each colour representing the branch line of railway by which the parcel is to be carried—a distinction necessary to prevent the occasional despatch of a parcel by a wrong train, and facilitating the work of the porters, who do not require even to read the address to ascertain how the parcel is to be sent.

To the above brief outline of the organisation and working of the " Scotsman " establishment may be added a few statistics. Including all departments, nearly 200 persons are employed on the premises ; and if to these be added paid contributors and others, the number of persons receiving remuneration for their services will be swelled to fully 500, who obtain among them L.17,000 a-year. Of the daily issue of the paper 180,000 copies are printed every week, and of the weekly issue 60,000 copies, which give a circulation of 240,000 a-week, or 12,480,000 a-year. The annual produc-tion would, if spread out, cover seven square miles of ground ; or, if the sheets were placed end to end, they would form a ribbon 10,625 miles long and 4 feet broad. The quantity of paper used considerably exceeds the entire produce of an ordinary paper-mill. Another fact also worthy of mention is, that one copy of the paper contains nearly as much print as a three-volume novel, got up in the usual style, and sold at the fashionable price of 31s. 6d.

FISHERIES.

From the earliest times, fish have constituted an important item in
the food supplies and commerce of the people of Scotland. Within
the last century the fisheries have been developed until they have
become one of the greatest branches of industry in the country,
employing upwards of 100,000 persons, and affording profitable investment for several millions of pounds of capital. As early as the
ninth century, the sea fisheries were sufficiently extensive to merit
special mention by historians; and in each succeeding century references to them became more frequent. The river fisheries also
date from a very early period, and annually contribute a large sum
to the wealth of the country.

The herring fishery is the most valuable branch of the piscatorial
industry of Scotland, and as such merits precedence in this place.
It would appear that the attention of the Legislature was first drawn
to the importance of fostering the herring and other sea fisheries in
the year 1474, when an Act of Parliament was passed, ordaining
that certain Lords Spiritual and Temporal, and the authorities of
burghs, should provide ships, busses, and boats, with nets and other
pertinents, for fishing. In the succeeding reign this statute was confirmed by an Act setting forth "that ships and busses, with all their
pertinents for fishing, be made in each burgh, in number according
to the substance of the burgh, the least of them to be of twenty ton;
and that all idle men be compelled by the Sheriffs in the county,
and by Bailies in burghs, to pass therein for their wages, under the

pain of banishment out of their bounds, and that the Sheriff or officer in burgh negligent shall pay twenty pounds to the King." By an Act of the Parliament of James V., a time for selling fish was appointed, and it was made illegal to send fish out of the kingdom ; but strangers were not prohibited from coming into the country and making purchases. Every person having stocks of fish on hand was bound to sell them for "the service of the lieges." It would appear that more fish were caught than sufficed for the supply of home consumers ; for in 1573 freemen were empowered to buy, salt, and export the fish left after the demands by the lieges had been supplied. The law was re-enacted in 1584, and fishers were prohibited from selling herring to strangers, or those who were not freemen of burghs, and also from exporting. A number of other laws were laid down about that time, among others one determining that the herring and white fish barrel should contain nine gallons, and another that the barrels should be examined and officially marked by " branding." The object which the first legislators who undertook to deal with the fisheries had in view was apparently overlooked by their successors, and instead of inducements being offered to persons to engage in fishing, harassing restrictions were placed upon the trade, which had the effect of preventing its development. Several countries of Europe would have become purchasers for almost any quantity of fish that might be caught by the Scotch fishers; but the supply was denied them, probably because they were prepared to give a better price than home consumers.

The Dutch had long been acquainted with the mine of wealth which existed in the great swarms of fish frequenting the shores of Scotland, and devoted much attention to the working of it. Their rendezvous was Bressay Sound, in Shetland, and there their busses congregated about the beginning of June sometimes to the number of a thousand. The fishing ground is still frequented by a large number of Dutch, Danish, French, and German busses; but they must not approach within a certain distance of the coast. The Dutch mode of fishing was copied by the early Scotch fishermen, and was continued for many generations in what was known as the deep-sea herring fishing.

The author of "The Interest of Scotland Considered" thus describes the manner in which the fishing was carried on before the civil wars which began in the reign of Charles I. distracted attention from the trade. He says:—"The fishing was managed by small busses, from fifteen to thirty tons burden, with close decks,

2 K

and one mast that lowered. Upon this mast one of their nets lay drying in the night-time, while they rode by the other put out in head to catch herrings for bait when they were at the cod-fishing, and lay thus snug in the water, very little exposed to the violence of the winds. In the beginning of March those busses went to the northward and fished cod on the coasts of the Orkneys. The crews salted their fish in the hold, and when the weather was dry they put them ashore and dried them on the beaches in Orkney. They returned in May to the Firth [of Forth] and washed the salt out of their mud-fish and dried them on their own beaches and stages at home, and then sold them partly for home consumpt and partly for export. About the 8th or 10th June, they took in their large nets, salt, and casks, and set out to the fishing of deep-water herring, in the same seas where the Dutch and we now take them. So soon as they had catched as many as their small holds could conveniently stow, besides their fishing equipage and stores, they ran to the coast, and put them ashore, took in a fresh fleet of nets, more salt and casks, and fished on till the end of July. They then returned home, shifted their nets again, and fished across the opening of the Firth so long as the fishing season continued. Here they never failed to fish with success, and gave certain intelligence to the open boats (wherein the same persons were sharers) where to lay their nets for the herring close by the shore in shallow water. When this fishing was over, the same busses, with a fresh fleet of nets each, sailed to the northward round the coasts of Strathnaver to what we call the Lewes fishing, and there fished herring in the deep-water lochs upon the west side of Sutherland, Ross, and Inverness shires till towards Christmas. They then returned home and laid up their busses to be dressed and repaired, the crews, meantime, going to the fishing upon the coast in open boats until the month of March. By this constant practice the men became the most expert fishers in Europe."

A royal company which had been formed for carrying on the fisheries was dissolved by an Act passed in 1690, in consequence of having failed in its mission, and the restrictions which had been imposed upon fishers outside the company were removed. Three years after an " Act anent the loyal curing and packing of herring and salmond fish" was passed. The preamble set forth that the statute was devised out of consideration of "how much the true and loyal curing and packing of herring and salmon fish to be exported forth of this kingdom, contribute to the advancement of trade and the general good of the nation." It was enacted that the barrels in which salmon or herring were to be packed for exportation should be made of well-seasoned knappel or oak timber, of sound quality; that the magistrates of burghs should appoint a qualified person to inspect the barrels, who, if he found them sufficient, was to affix the seal of the burgh to each; and that a man skilled in the curing of fish should be appointed by the magistrates to examine the fish, and to put an official brand upon the barrels which he found to be filled with properly cured fish. Though the penalties attached to breaches of this Act were heavy, the fishing trade began to revive under it. In a manuscript of Sir Robert Sibbald, it is stated that before the Union 600 boats, manned by upwards of 4000 men, might have been seen fishing in the Firth of Clyde alone, and that these afforded 3750 tons of fish for exportation. A district of the Fife coast not above twelve miles in length sent out 168 boats, manned by 1120 men, and exported annually 12,000 barrels of herring. In 1695 the small town of Crail alone exported 2400 barrels. The Parliament of Queen Anne passed, in 1705, an "Act for advancing and establishing the fishing trade in and about this kingdom," which afforded some relief, and encouraged those engaged in the trade.

About the time of the Union the fisheries had a promising appearance; but the enactment of the salt-duties, and the complicated arrangements by which they were managed, had a prejudicial effect. Though the deep-sea fishing had fallen off considerably, the coast fishing of herring in shallow water was in 1733 reported upon as being a trade of "very great importance to the country, and well deserving to be taken care of." It was a good nursery for breeding seamen, and employed many persons on shore—such as carpenters, coopers, twine-spinners, and net-makers. In the Firth of Forth, from 600 to 800 boats were at that time engaged in the herring fishing; and in the Moray Firth the boats were set down at from five to seven hundred. Each boat had a crew of eight or nine men, and carried eight nets. Most of the boats were owned by fishermen who devoted their whole time to one or other branch of the fishing business; and the remainder belonged chiefly to carpenters, who built and hired them out. Usually two experienced fishermen engaged a sufficient number of landsmen to join them in manning a boat for the "drave," as the herring fishing was called. Almost every fisherman owned a net, and what others were required to complete a "fleet" or "drift" were obtained from net-makers and private persons who were willing to share in the profit or loss of the trade. When the fishing was over, the accounts were made up, and the balance that remained, after paying the working expenses, was divided into eight or nine shares or " deals." The proprietor of the boat drew one deal, every man half a deal; and if there happened to be a landsman or two in the boat who had not been at the fishing before, they were entitled to draw a quarter of a deal only. In an average season the quantity of herring caught in the Firths of Forth and Moray was about 80,000 barrels, of which about seven-eighths were exported. The price paid for fish purchased for exportation was 12s. per barrel. At the time referred to the Forth fishermen were loud in their complaints against an impost known as size-duty, which was the money-equivalent of an ancient privilege of the Crown, whereby a certain proportion of the herring taken by every boat had to be set aside " for the service of the King's kitchen." That and other grievances of lesser magnitude were subsequently removed. The fishing season on the coast lasted from the 20th July till the 20th September.

In the Firth of Clyde and along the west coast, upwards of 2000 boats, manned by 14,000 men, were engaged in the herring fishing about the year 1730. The fishing began in Loch Fyne and the other deep lochs on the Argyle side in June, and continued till September. The scene of operations then shifted to the coast of Ayr, and there the fishermen continued their labours till November; by going to the northern lochs and the coasts of the Western Islands, another month's fishing was obtained; so that the west coast fishing may be said to have lasted fully six months in each year. The boats were smaller than those used in the Firths of Forth or Moray. Like their brethren on the east coast, the western fishermen were subject to grievous burdens. They had a size-duty of 16s. 8d. to pay for each boat that engaged in the herring fishing, and that sum had to be paid although the fishing failed. Another exaction was " a night's fishing in the week," taken by all the Highland chiefs and proprietors of land from each boat that landed herring to be cured on their ground. The collectors of this toll usually kept a note of the herring taken on each night of the week, and then seized the produce of the night that yielded the largest quantity. The herring caught on the west coast were chiefly sent to the continental markets, while the lean but firm fish got on the east side were exported to the sugar colonies. Loch Fyne herring were then, as now, held in great esteem, and fetched from six to eight shillings per barrel more than other kinds.

When, under the grievances and restrictions referred to, the herring fishery threatened to become extinct, an effort was made by the State to encourage the public to revive and cultivate the trade. In his speech at the opening of Parliament in 1749, King George II. recommended the adoption of means whereby people might be induced to prosecute the fisheries with increased vigour, to their own profit and the good of the nation. A committee of the House of Commons was appointed to inquire into the matter, and, on their recommendation, L.500,000 was subscribed for carrying on the fisheries under a corporation designated "The Society of Free British Fishery." The Society embraced in its membership the leading men in the country, and the Prince of Wales was elected Governor. The duties on salt used in the fisheries were remitted, and a high tonnage bounty was granted upon every vessel fitted out for the deep-sea fishery. One of the first effects of this order of things was that many vessels were fitted out, not to catch herring, but (as Adam Smith puts it) "to catch the bounty." The bounty amounted to 50s. a ton, and in 1759 the sum of L.159, 7s. 6d. was paid as bounty upon every barrel of herring fit for market that was produced. The Society died of bad management, and in 1786 a new company was established under the patronage of George III. This company did not achieve much, and its history has thus been written—" For a season or, two, busses were fitted out by the Society; but if every herring caught had a ducat in its mouth, the expense of capture would scarcely have been repaid. The bubble ended by the Society for Fishing in the Deep Sea becoming a kind of building society for purchasing ground in situations where curers and fishermen find it convenient to settle; and selling it or letting it in small lots to them at such advance of price as yields something better than fishing profits." This Society is still in existence; but beyond building harbours at several places in connection with their own property, the directors have done little to encourage or extend the. fisheries—indeed, the designation of British Fishery Society is a misnomer.

Finding that little or no good resulted from the operations of the corporate bodies referred to, the appointment of a distinct set of commissioners for superintending all matters connected with the fisheries was authorised by an Act passed in 1808. The commissioners were empowered to appoint officers to be stationed at the different ports where fishing was prosecuted, for the purpose of seeing that the regulations with respect to the gutting, packing, &c., of the herring, and the branding of the barrels, were duly carried into effect. Bounties appeared to be considered indispensable in any attempt to extend the fisheries, and the commissioners began in 1809 by giving a bounty of L.3 a ton for all vessels above sixty tons burden, fitted up for the deep-sea herring fishery. In 1820 a bounty of 20s. a ton was granted on all vessels of from fifteen to sixty tons, fitted out for the shore herring fishery. In addition to the bounties on tonnage, 2s. a barrel was allowed on all herring gutted, packed, and cured in the six years ending 1815, and an additional bounty of 2s. 8d. on exportation. From 1815 till 1826 the bounty on cured herring was 4s. a barrel, and not until the bounty had been fixed at that high rate was there any indication that the commissioners were to prove more successful than the societies which they had superseded. In the year ending 5th April 1811 the total quantity of herring cured was 91,827 barrels; in 1815 it was 160,139 barrels, and thenceforward there was an increase, until the returns for 1826 showed 379,233 barrels cured. A reduction of the bounty was then made by 1s. a-year until, in 1830, it ceased; and since then no premium has been necessary to induce people to engage in fishing. The figures given above relate to England and the Isle of Man as well as to Scotland. Since 1850 no returns relating to England have been obtained. In that year the quantity cured was 770,698 barrels; and next year, when England was excluded, it was 544,009 barrels. The year in which the greatest quantity was cured in Scotland and the Isle of Man was 1862, when 830,904 barrels were produced. The four succeeding years showed a falling off to the extent of a fourth; but in 1867 there was a recovery to 825,589 barrels. The fishing of 1868 failed to a serious extent at nearly all the east-coast stations, and in some cases the public were called upon to afford relief to those whose winter bread was curtailed by the want of success.

The detailed returns for the year 1868 had not been made up at the time of writing; but the following, relating to 1867, will be sufficient to convey an idea of the extent and importance of the Scotch herring fishery :—

Districts.	Herring Cured.	Herring Exported.	Boats.	Fishermen and Boys	Coopers.	Gutters and Packers.	Labourers.	Total Persons Employed.
	Brls.	Brls.	No.	No.	No.	No.	No.	No.
Leith...............	2,467	20,595½	154	626	38	165	74	903
Eyemouth........	47,750	4,114	689	3,364	135	900	236	4,635
Greenock.........	5,135	1,656	70	220	—	—	12	232
Ballantrae,.......	2,222	—	59	141	4	10	25	180
Glasgow, ...	20,706	6,900	300	900	4	10	8	922
Rothesay, ...	28,513	—	450	1,475	25	250	30	1,780
Inverary,.........	36,697	—	617	2,468	35	578	75	3,156
Campbelton,.....	3,720	123	157	480	2	20	—	502
L. Carron & Skye	22,349	194	300	900	—	600	35	1,535
Loch Broom,....	9,047	942½	917	3,668	33	770	570	5,041
Stornoway, ...	101,550¼	65,109½	1225	6,158	221	2,210	124	8,713
Shetland Isles,..	10,008	8,312½	308	1,242	44	513	13	1,812
Orkney Isles,....	12,895	9,565½	234	1,271	47	594	22	1,934
Wick,..............	97,508	81,585¾	970	5,820	293	2,356	237	8,706
Lybster, ...	22,079	17,659½	238	1,266	71	597	27	1,961
Helmsdale,.......	45,302	33,383½	365	1,992	115	1,099	77	3,283
Cromarty,........	23,601	17,412	230	1,200	72	806	73	2,151
Findhorn, ...	41,355½	29,751	284	1,453	99	828	115	2,495
Buckie,...........	23,054	17,551½	166	996	70	506	73	1,645
Banff,..............	24,933	17,823	231	1,155	60	635	110	1,960
Fraserburgh,.....	80,986	65,981½	389	1,945	124	1,061	144	3,274
Peterhead, ...	73,868	51,592	477	2,385	188	1,420	165	4,158
Montrose,........	28,323	14,562½	446	2,345	100	681	76	3,202
Anstruther,......	40,021¼	9,116½	350	1,950	120	993	205	3,268
Total,.........	804,090	473,931¼	9626	45,410	1900	17,582	2526	67,448

It is necessary to explain that the totals given above exceed the actual number of boats and persons employed in the herring fishery. The discrepancy arises from the fact that considerable numbers of boats and men move from one station to another. Thus, the Lewis fishery, which is the earliest, attracts many boats from the east coast, which return in time to attend to the fishery off the parts of the coast to which they belong. In like manner coopers and others shift about, so that a number of them come to be doubly reckoned. A deduction of 10 per cent. from the boats, fishermen, and coopers would probably give something nearer the real numbers. Taking the marketable value of the herring at 25s. a-barrel, and including the quantity of fish sold for immediate consumption, the produce of the fishing of 1867 would be fully L.1,500,000—five-sixths of which sum goes for labour and profit on capital, and one-sixth for salt, wood for barrels, &c. Here, then, is a great contribution to the material wealth of the country, leaving out of view the by no means unimportant fact that the herring fishery affords a cheap and wholesome article of food for the masses.

The chief seat of the herring fishery is on the east coast of Caithness, and Wick is known as the " herring metropolis." The trade was not introduced into that quarter, however, until a date which may be called recent in comparison with the duration of the fisheries on parts of the coast further south. In 1768 an incipient attempt was made, under the inducement of the Government bounty, to fish for herring at Wick, but the effort was attended with little success. Even fourteen years afterwards, the season's catch amounted to only 363 barrels. Between the years 1782 and 1790 the trade came to be better understood, and the prospects grew more cheering. In the latter year upwards of 13,000 barrels were got. Encouraged by this promising state of matters, the British Fishery Society acquired a large space of ground on the south side of the bay of Wick, on which, about the year 1808, they laid out feus, and offered induce-ments for the foundation of a colony of fishers. It was found that this scheme could not be successfully carried out unless the harbour accommodation of the place were extended. The Society received some aid from Government ; and in 1810 built, at a cost of L.10,000, a harbour in the neighbourhood of their property. The trade of the port increased so rapidly that this harbour soon came to be considered too small; and in 1831 a larger space was enclosed, the expense in this case, which was L.40,000, being also defrayed jointly by the Society and the Government. A number of the more convenient creeks along the coast were, about the same time, made safe and commodious by local enterprise. The coast is much exposed, and the operations of the fishermen are frequently interrupted by storms, which burst forth suddenly, and rage with a fury peculiarly violent. With a view to afford against such emergencies a shelter that may be gained in any state of the tide—the old harbours being tidal—a large portion of Wick Bay is being enclosed by a gigantic break-water. The force of the storms in the bay may be judged from the fact that, though the breakwater is of extraordinary dimensions, and stands fully thirty feet above the sea at low water, the waves often dash right over it, and on more than one occasion have threatened to destroy the entire fabric. It is expected that, when the work is completed, the fishing interest will be much benefited, as the certainty of having a safe port to run to will make the fishermen more confi-dent, and enable them to put to sea on many nights when, with no shelter save the old harbours, their boats would be lying aground.

Except during about three months of the year, Wick is a dull enough place, its trade, apart from the fishing, being of very small extent. Indeed, it may be said that the people depend entirely upon the herring harvest; and when that fails, everybody suffers more or less. The depression following on a bad fishing is observable on everything; for no fish means no money. There being no inducement to build new boats, the carpenters go idle; and the stores being filled with unused barrels, the coopers' occupation is gone for a time. Everybody practises retrenchment, and those who manage to make ends meet consider themselves fortunate. On the other hand, when success attends the efforts of the fishermen, a cheerful spirit prevails all through the year. The uncertain nature of the staple industry is a source of constant anxiety to the people, and some additional branches of trade would be most advantageous to the district.

During the fishing season, Wick presents one of the most interesting scenes to be witnessed in the whole range of industry. In the course of the afternoon, the crews of the boats moored in the harbours or anchored in the bay prepare to start for the night's fishing. The nets are got on board, the masts are hoisted, the sails set, and soon the bay becomes shrouded in dark brown canvas. With a breeze from the south-east, the departure of the boats is a splendid sight, for then they have to tack out; and the spectators are favoured by beholding a regatta on a grander scale than any to be witnessed elsewhere. Racing is no part of the fishermen's intentions, but now and again a dozen boats or so draw into a line, and the landsmen may choose a favourite, and become interested in her fortunes until she disappears from view. The movement seaward is simultaneous along the coast, and by the time the last of the fleet get outside the heads of Wick Bay, a dark line of boats extends continuously from Duncansbay Head to the Head of Clyth, a stretch of a dozen miles. Large numbers of persons nightly proceed to the cliffs to see the boats go out, and to watch their dispersion over the fishing ground. Generally those in the boats have no fixed intention as to what spot they shall select for casting out their nets, and taking their draw from Neptune's lottery. If a good haul was previously got at a certain part, those who get it endeavour to return to that part; but in most cases the boats which were successful on the previous night are watched and followed, notwithstanding the fact that it is an exceedingly rare thing for a boat to have two exceptionally successful nights following each other. Having chosen their water, the crew of each boat begin to " shoot " their nets, which, while being " laid " in the boats, were united in a continuous train or drift, by knotting together the " back ropes." Each boat has a train of nets about half-a-mile in length and ten yards in depth. By corks attached at the top and weights at the bottom, the nets are made to float perpendicularly in the water. This wall of netting is suspended from buoys which allow it to sink twenty or thirty feet below the surface. The nets are put into the sea immediately after sunset, and most of the crew then endeavour to snatch "forty winks" of sleep. In the course of an hour or two some of the nets are hauled up and examined to see whether the fish have been "striking." If there should be good signs of fish in the locality, the nets are allowed to lie for some time. The herring are caught by getting fixed in the meshes while trying to pass through. The captain decides the proper time for taking in the nets, and when he gives the word, all hands fall to work. As the nets are got on board, the fish are shaken out of them, and fall into the hold, where, after a gasp or two, they expire. If the night's labour has yielded 20 or 30 barrels of fish, the men think themselves fortunate ; but it is no unusual thing for a boat to bring ashore 80 and even 100 "crans."

The return of the boats in the morning is an event of much more importance and interest to people on shore, and from an early hour anxious inquiries are made respecting the fortunes of the night, while those who have leisure go to make observations from the piers and cliffs. As the boats crowd into the harbours, an opportunity is afforded for judging of the uncertainty of the fisherman's fortunes. A score or two of boats sail swiftly in, with barely as many fish on board as will suffice for the breakfasts of the crews, then at a toilsome pace come one or two boats filled to the thwarts with herring. In one case, the night's labour of six men, and the use and risk of property worth from L.100 to L.200, has produced a return of about 6d.; in the other, of L.60 or L.80. The average catch at Wick in 1868 was 41¾ crans, drawn from returns of individual boats which ranged from one to upwards of 200 crans. When all the boats are in, the harbours are quite crowded; but by mutual arrangement the boats having large quantities of fish to land are allowed to get near the quays. The fish are shovelled into wicker baskets, and then carried to the "station," where they are measured and emptied into the "boxes," or enclosures of wood from twenty to thirty feet square, the sides of which are about thirty inches in height. As soon as a convenient quantity of fish has been deposited in the box, a troop of women, arrayed in canvas and oil-cloth approach, and the "gutting" and "packing" processes begin. The gutters, each armed with a small knife, surround the box, and, taking a herring up in the left hand, operate upon it with the knife held in the right hand. The rapidity of their movements is surprising, a good worker being able to dispose of one thousand fish in an hour. As the fish are gutted, they are dropped into baskets, and handed over to the "packers," who "rouse" them with salt in a large tub, and then arrange them in layers in the barrels. A free use of salt is made, the herrings being first coated with it separately in the rousing process, and the layers in the barrels afterwards thickly overlaid with it. The barrels are temporarily covered and allowed to stand for ten days, during which time the fish settle down considerably. Additional fish are then put in until the barrels are quite full. After being examined and approved by an officer of the Fishery Board, the barrels receive the official brand, which is accepted in the market as a guarantee that the fish are of a certain standard of quality. A large number of coopers and labourers are engaged in preparing and heading up the barrels, and removing them from one place to another.

The above is a brief outline of the operations which may be witnessed at any of the fishing ports. When the fishing is success-ful, the coopers are kept busy during the winter in making barrels, and the carpenters in building boats. The coopers usually make five barrels a-day, for which they receive about 14s. a-week. In the fishing season they are paid extra, as the work they have to do is heavy and protracted. The fishing at Wick employs a large number of men and women from the inland districts of the county, and from the West Highlands, so that the profits of the trade are spread over a considerable area. About three-fourths of the crews of the boats are composed of Highlanders, who receive about L.1 a-week with board and lodging for the season of eight weeks. The herring fishery gives employment to a large amount of shipping. In 1867 the vessels employed in carrying wood and salt to Wick, and herring from that to other ports, measured in the aggregate 18,000 tons, and their crews numbered 1300 men.

Connected with the herring fishery there are many debatable points. The natural history of the fish is very imperfectly under-stood; the system of sales in advance, engagements, bounties, and the like, is open to serious objection; but on none of these ques-tions is it considered advisable to touch in this place. It has been said, with more truth than will be readily admitted by the trade, that our herring fishery is a blunder from beginning to end, and that herring commerce is throughout a mistake. A recent writer in the "Spectator" asks the following questions, which will press for answers some day:—" How is it a merit to capture herring when they are full of roe and milt, while it is a crime to capture salmon at the same period of their lives? What is the difference in the chemistry of these two fishes, that at the time of spawning a gravid salmon should not be esteemed fit for food, whilst only gravid herring can obtain that Government certificate which enables the curer to sell them at the highest price? Further, why is it that Government is asked to certify the proper cure of herring, any more than the proper weaving of cotton or the proper making of cheese?" In criticising the commercial relations of the herring trade, the same writer says :—" Hundreds of thousands of barrels are annually bought and sold long before it is known that a single fish will be taken. Capitalists advance money to curers in order to enable men to build boats and buy nets, and a spirit of gambling generally prevails. Men are engaged to fish, and a bounty is given for their services ten months before they are required. The fishery is throughout a lottery ; a few men succeed, and a large number fail. Some day we shall find out that we have been proceeding on a bad system, and that the herring fishery cannot last in the face of our ignorance of the natural history of the fish, and the blunders that are continually being made in regulating its capture. We do not know at what age that fish becomes reproductive, and on some parts of the coast we have kept up a close-time, whilst we have left other parts open. We have prescribed the kind of nets to be used in this particular fishery. In sober earnest, we have done innumerable things in connection with this and our other fisheries that we ought not to have done, while at the same time we have left undone many of those things that it would have been wise to do."

The " white fishery "—which term embraces the capture of cod, ling, haddock, hake, torsk, and the like—ranks next in importance to the herring fishery. For many years after the people of Scotland set about developing the sea and river fisheries, salmon and herring received almost exclusive attention, and up till the beginning of last century few persons were engaged in the white fishery, which was first practised on a large scale by the people of Orkney and Shetland, who early discovered the excellent fishing grounds on their own coasts and in the vicinity of Iceland. In 1820 Government, realising the importance of fostering this branch of the fisheries, granted bounties which were administered by the Fishery Board ; at the same time an official brand was granted, and the fish were cured under inspection of the officers of the Board. The large vessels received bounty at the rate of L.2, 10s. a-ton from 10th October 1820 till 5th July 1826 ; L.2, 5s. a-ton from thence till

5th July 1827 ; L.2 a-ton till 5th July 1828 ; and L.1, 15s. a-ton till 5th April 1830, when the system of bounty ceased. On fish taken by vessels and boats not on the tonnage bounty, 4s. per cwt. was paid on all cured by drying, and 2s. 6d. per barrel on all cured in pickle. These rates continued unaltered from 1820 till 1830, when the payments were stopped. The statistics of the trade show that the bounties had a most beneficial effect. The statements published by the Board prior to 1826 gave only the quantity of cod, ling, and hake punched or branded, and not the total quantity cured. In subsequent years they gave the quantity cured, "in so far as brought under the cognizance of the officers of the Fishery." In the year ended 5th April 1826 the quantity cured of the kinds of fish specified was 69,136½ cwt. cured dried, and 3634¾ cwts. and 5621 barrels cured in pickle. The figures of 1830, the last year of the bounties, show a great increase, 101,914 cwt. having been cured dried, and 5652½ cwt. and 8836½ barrels cured in pickle. Those engaged in fishing do not seem to have considered their occupation to be sufficiently remunerative without the bounty, as the year ending 5th April 1831 produced only 37,674 cwt. cured dried, and 2950½ barrels cured in pickle. A gradual recovery took place, however; though up till 1850, when the returns from England (which, along with those from Scotland and the Isle of Man, are included in the figures given above) ceased, the quantity of fish caught was considerably less than in the last year of the bounties, being 98,903 cwt. cured dried, and 6588 barrels pickled. In 1852 Scotland and the Isle of Man cured 102,975 cwt. by drying and 7019 barrels by pickling. For the year 1867 the figures were respectively 119,537 cwt. and 10,819 barrels, representing 3,602,117 fish. Though the Isle of Man is associated with Scotland in the returns, its contribution to the above figures was insignificant—the total quantity cured in 1867 being only 10 tons. From 1820 till 1850 the quantity of cured cod, ling, and hake exported from Britain averaged about 2000 tons a-year. In 1867 Scotland alone exported 46,225 cwt., of which 18,849 cwt. went to Ireland, 23,642 cwt. to the Continent, and 3734 to places out of Europe. Taking the value of the dried fish at L.15 a ton, and the pickled at L.1 a barrel, the fish cured in Scotland in 1867 would fetch L.100,474. It is difficult to arrive at even an approximation of the quantity of fish which goes into market in a fresh state ; but if it be taken at three times the quantity cured—which is certain to fall under the mark—the aggregate production would appear to be about 14,000,000 fish, which, at 9d. each, represents a value of L.525,000. This

enumeration does not include haddocks, whitings, soles, skate, floun-
ders, mackerel, and the smaller sea fish used for food in great quantities along the coast. In the various branches of white fishing,
from 4000 to 5000 boats, manned by from 20,000 to 30,000 men, and
carrying L.90,000 worth of lines, are engaged. The fish caught by
all the boats within convenient distance of the great centres of population are disposed of in a fresh state. The places most extensively
engaged in curing white fish are the Orkney and Shetland Islands,
Stornoway, Fraserburgh, Buckie, Lochbroom, Campbelltown, Montrose, and Wick.

The salmon has been an article of food and subject for legislation
with the people of Scotland from the earliest historical periods. The
more remote references to the fish concur in stating it to have been
most abundant in the Scotch rivers ; and it is beyond doubt that, in
the fourteenth century, a considerable quantity of pickled salmon
was exported to France and Flanders. No trustworthy statistics
exist relating to the salmon fishery, so that the extent to which it
was practised a century or two ago can be gleaned only from the
general statements of travellers and others who have put their observations on record.

The author of "The Salmon" thus sets forth the most reliable
evidence on the point :—"Among the oldest statements of what was
to be learned of the extent of the salmon fisheries by travelling in
Scotland, are those given about 200 years ago in the curious book of
the Cromwellian trooper, Captain Francks. Francks (from whose
descriptions, by-the-bye, it is clear that the art of salmon-angling was
practised then almost precisely as it is now) takes occasion at most
of his halting-places to make a short descant on the abundance of the
salmon in Scotland. Thus, of Stirling, he writes—'The Forth relieves
the country with her great plenty of salmon, where the burgomasters,
as in many other parts of Scotland, are compelled to reinforce an
ancient statute, that commands all masters and others not to force or
compel any servant, or an apprentice, to feed upon salmon more than
thrice a-week. . . . The abundance of salmon hereabouts in these
parts is hardly to be credited. And the reader, I fancy, will be of
my persuasion, when he comes to consider that the price of a salmon
formerly exceeded not the value of sixpence sterling.' And a hundred
years later, the English Engineer Officer, Captain Burt, writing from
Inverness, says that the price of salmon there was a penny a-pound,
and that the 'meanest servants who are not on board wages will not
make a meal upon salmon if they can get anything else to eat.' In par-

tial corroboration of these statements about the Ness, it may be mentioned that there is a person still living who held a lease of a fishing in that river, under which he was bound to supply the inhabitants of Inverness, during a considerable portion of the year, with salmon at 2d. a-pound. Indeed, till the present century, almost every traveller that entered Scotland made the 'great plenty of salmon' a subject of remark. Thus Defoe, as soon as he enters the kingdom at Kirkcudbright, writes down—'There is a fine salmon fishing in this river;' and when he reached Aberdeen, he says:—'The rivers Dee and Don afford salmon in the greatest plenty that can be imagined, to that degree that in some of the summer months the servants won't eat them but twice a-week, they are so fat and fulsome; it's almost incredible how they spread; in autumn they engender, and in shallow pools of the river they cast their spawn, and cover it with sand, and then they are so poor and lean that they are only skin and bone; of that spawn, in the spring, comes a fry of tender little fishes, who make directly to the sea, and, growing to their full progress, return to the river where they were spawned.' Defoe wrote this about the same time as Burt wrote; and another traveller, of nearly the same period, describing himself as 'A Gentleman,' begins his book—'The salmon fishery is particularly the delight and the boast of the Scotch, insomuch that for it they too much neglect all the rest.' Speaking of Perth, the same writer says— 'The salmon taken here, and all over the Tay, are extremely good, and the quantity prodigious. They convey them to Edinburgh, and to all the towns where they have no salmon, and barrel-up great quantities for exportation.' Of Aberdeen:—'The quantity of salmon and perches (?) taken in both rivers is a kind of prodigy; the profits are very considerable, the salmon being sent abroad into different parts of the world, particularly into England, France, the Baltic, and several other places.' Of the Ness, he says—'Here is a great salmon fishery;' and he was more interested than gratified by the sight of the 'cruives,' then used by the corporation of the town. These statements—and they might easily be multiplied— are of course good evidences of local plenty, and also of a very considerable export of the fish in a salted state, though it must not be forgotten that at that period, when travellers assigned such great commercial importance to our salmon fisheries, they must be held as speaking in some degree by comparison with other industries, which were then insignificant."

Towards the beginning of the present century various causes tending to lessen the abundance of salmon in the Scotch rivers began to operate; and until the passing a few years ago of some acts bearing on the subject, the work of extermination went steadily on. The chief of the causes referred to were the increase of land drainage, obstructions and pollutions consequent on the rise of population and industry on the banks of rivers, the killing of spawning fish, the brevity or mistiming of the close-season, and over-fishing. As regards the larger rivers, statistics exist which show that the decrease in returns and rental was very serious. The authority above quoted gives a minute analysis of the returns, but makes no attempt to reduce his conclusions to figures. He states generally that, "with the single and partial exception of the Tay, the decline in the Scottish fisheries was, till the legislation of the last three or four years, unusual and alarming, extending over almost every river and district, from the south-western Doon to the north-western Dee; although in one or two cases, such as the Spey and the rivers of Sutherland, where the fisheries are in the hands of one great proprietor, who had resorted to a wise moderation, a great difference for the better was discernible." Under the influence of the recent legislation referred to, the fisheries are improving, though, as some of the prejudicial causes stated above are irremovable, it is not probable that the fish will ever become so abundant as they were before those causes came into operation. The difficulty of ascertaining the number of men employed in the salmon fishery is as great as that of obtaining reliable figures of the produce. The fishers and others connected with the business in one way or other must number several thousands; for the net-fishing of the Tay alone employs 700 men, who receive in wages about L.9000 a-year; and that of the Tweed 350 men, who receive about L.4500. The salmon-fishers are a hardy race; but a serious drawback of their occupation is that it lasts half through the year only, and unfortunately it is available only during the months in which other kinds of out-door labour are abundant, and is suspended during those months when the other kinds of work also fail.

The first vessel sent to the whale fishery from Scotland sailed in 1750, in which year the bounty previously paid to induce British vessels to embark in the trade was increased to 40s. a-ton. This pioneer Scotch ship was of 333 tons burthen, so that she earned L.666 in addition to the value of the fish caught. In 1751 six vessels were despatched, and the number went on increasing till 1756, when sixteen ships sailed for the whale fishing. The aggregate tonnage was 4964, and the bounty L.9315. By the year 1760 the number of ships had decreased to fourteen, and in 1765 there were only eight. The number rose to ten in 1773 ; but from that date a gradual decline took place, until in 1779 only three vessels remained. The bounty was lowered in 1777 to 30s. a ton ; but the falling off in the number of ships was so great that in 1781 the old rate of 40s. a ton was restored. In 1784 the Scotch ships numbered seven, but three years afterwards they had increased to thirty-one. A reduction of the bounty again brought down the numbers, but from 1811 to 1818 there was a steady increase of Scotch ships from twenty-two to fifty-three. The ports to which the ships belonged were Aberdeen, Leith, Dundee, Peterhead, Montrose, Banff, Greenock, Kirkcaldy, and Kirkwall. In the four years ending 1817 the Scotch whaling fleet—consisting on the average of forty-eight and a-half ships annually—captured 1682 whales, which yielded 18,684 tuns of oil. The yearly average for each ship was 8·7 whales, 96·3 tuns oil, and 4·6 tons bone. Other ports were subsequently engaged in the trade, but the ships never exceeded fifty-three in one year. In 1834 forty vessels were despatched to the fishing. Of these eleven belonged to Peterhead, eight to Dundee, six to Aberdeen, five to Kirkcaldy, five to Leith, three to Montrose, and two to Burntisland. They captured 475 whales, which yielded 4515 tuns of oil and 240 tons of bone, the gross value of which was about L.250,000. Peterhead was for many years the principal port engaged in the whale fishery. The first vessel was despatched thence in 1788 ; but though she met with fair success—averaging about five fish each season, up till 1804—a second vessel was not started until that year. In the six following seasons the two vessels captured a total of 225 whales, which yielded 1537 tuns of oil. Seven vessels were sent out in 1814, and one of these, the Resolution, returned with the largest cargo ever brought to Britain by a whaling ship. The cargo consisted of 44 whales, which produced 299 tuns of oil ; value— reckoned at L.32 per tun, the average price at the time—L.9568 ; and when to that sum is added the value of the whalebone and the bounty, the whole would appear to have reached L.11,000. At the present price of oil the cargo would be worth L.3836 additional. In 1821 the Peterhead whaling fleet numbered sixteen ships. The port enjoyed a long run of prosperity, arising from the whale fishery ; and though the trade was decaying in several other places, it remained vigorous and increasing at Peterhead up till 1857, when a fleet of thirty-two vessels was despatched to the seal and whale fishery. The provisioning of this fleet alone cost nearly L.16,500. During the past ten years the number of ships has fallen off very much, owing to the scarcity of fish and the diversion of local enterprise to the herring fishery. In 1868 Peterhead sent twelve vessels to the whale and seal fishery. Though Aberdeen was about the first of the Scotch ports engaged in the trade, no whaling vessel has been despatched from that port for several years. Dundee now occupies the foremost place in the whale and seal fishing business, and sent out in 1868 a fleet of eleven steamers. The trade at best is a hazardous one, and can now be prosecuted with any prospect of success only in steam-vessels. The returns from it do not add much to the industrial revenue of the country, and it provides constant employment for little over a thousand persons. The substitution of coal-gas for oil as an illuminator tended to deprive the whale fishery of much of its importance to the country generally; and though whale oil is of great value in some branches of manufacture, the stoppage of the supply would in all probability be only a temporary inconvenience.

In addition to the fisheries already noticed may be mentioned the lobster fishery, which is carried on to a considerable extent in Orkney, Shetland, and some other places; the oyster fishery, which is prosecuted by the fishermen of Leith and Newhaven; and the pearl fishery, which has at various times drawn attention to some of the northern rivers.

INDEX.